Advanced Analysis of Nontraditional Machining

Hong Hocheng • Hung-Yin Tsai
Editors

Advanced Analysis of Nontraditional Machining

Springer

Editors
Hong Hocheng
Department of Power Mechanical Engineering
National Tsing Hua University
Hsinchu, Taiwan R.O.C.

Hung-Yin Tsai
Department of Power Mechanical Engineering
National Tsing Hua University
Hsinchu, Taiwan R.O.C.

ISBN 978-1-4614-4053-6 ISBN 978-1-4614-4054-3 (eBook)
DOI 10.1007/978-1-4614-4054-3
Springer New York Heidelberg Dordrecht London

Library of Congress Control Number: 2012953517

Printed on acid-free paper

Springer is part of Springer Science+Business Media (www.springer.com)

Preface

Machining is among the most ancient manufacturing technologies human being has employed to advance the civilization. The forms and shapes of a part can be precisely produced. Along with the progress of industry, such as aerospace, electronics, and biomedicine, new functional materials have been developed and applied thanks to their superior mechanical, thermal, chemical, or electrical properties. Nonconventional machining processes have responded to the changes and sprouted in the time frame of the after-World War II booming, including the electrical discharge machining, electrochemical machining, laser machining, ultrasonic machining, and the latest chemical mechanical polishing. These processes have answered many manufacturing challenges of the modern industry using advanced materials. In general, conventional machining refers to the direct contact of tool and workpiece, while mostly nontraditional machining processes do not necessarily use mechanical energy to provide material removal. Therefore, not only hard, strong, or tough workpiece material can be processed but also workpiece that is too flexible to resist cutting forces can be machined by a proper nontraditional machining process. For a condition that the temperature rise or residual stresses are unacceptable or part shape is very complex with external or internal profiles or small holes, nontraditional machining is often a solution. In addition, there are some advantages of nontraditional machining: high accuracy and surface finish, less or even no wear, mostly quiet operation, and long tool life.

The dissemination of this technology is essential for the industry. However, limited number of books can be found in the international community. Besides, these books are of the introductory nature or handbook. Compared to other disciplines in manufacturing sector, such as traditional metal machining, metal forming, casting, welding, automation, etc., the gap of the advanced knowledge of nontraditional machining has yet to be filled. This book partially fills this gap by documenting the latest and frequently cited research results of a few key nontraditional machining processes for the most concerned topics in industrial applications, such as laser machining, electrical discharge machining, electropolishing of die and mold, and wafer processing for integrated circuit manufacturing. For effective utilization of the capabilities of different nontradtional machining processes, there is a need to experimentally or analytically understand different machining processes when a particular shape feature is to be generated on a given work material.

There are eight chapters included in this book. In Chap. 1, the laser machining-induced formation of anisotropic heat affected zones in fiber-reinforced plastics is discussed. In Chap. 2, the authors outline machining characteristics of carbon fiber-reinforced carbon composites and EDM-machined AISI D2 tool steel. A process to erode a hole of hundreds of microns diameter in a metal surface using a moving electrode in electrochemical machining system and the surface roughness of the electropolished internal and external cylindrical surfaces by different electrode designs are discussed in Chap. 3. In Chap. 4, a model of the material removal rate and endpoint monitoring methods are proposed for chemical mechanical planarization. Further, a visualized characterization of the amount and distribution of the fluid film of surry between wafer and pad is analyzed by digital photographs. In Chap. 5, the authors discuss online tool-wear monitoring during ultrasonic machining, the effect of abrasive and drilling parameters on material removal rate, hole clearance, edge quality, tool wear, and surface roughness of composites for application in the manufacturing industry. In Chap. 6, an analytical approach to study the delamination during drilling by water jet piercing is presented. In addition, the feasibility of milling of composite materials and the kerf formation of a ceramic plate cut by an abrasive waterjet are discussed. A laser dragging process capable of ablating a groove pattern, and producing sophisticated 3D features, on a polycarbonate sheet through a shaped mask opening is analyzed in Chap. 7. The same chapter also introduces a sub-micron-structure machining on silicon substrates by a direct writing system using a femtosecond laser. Finally, the modern and emerging technologies related to nano-structure machining by ion and electron beams are developed in Chap. 8.

The authors wish the readers an enjoyable and fruitful reading through the book.

Hsinchu, Taiwan R.O.C Hong Hocheng
Hsinchu, Taiwan R.O.C Hung-Yin Tsai

Contents

Chapter 1
Laser Machining and its Associated Effects

C.T. Pan and H. Hocheng

Abstract Laser machining has a wide range of industrial applications. However, laser energy can cause thermal damage to composite materials during the shaping operation following curing. Such damage leads to poor assembly tolerances and reduces long-term performance. In this study, we investigated the laser machining-induced formation of anisotropic heat-affected zones (HAZs) in fiber-reinforced plastics (FRP). The degree of HAZ is estimated by the isotherm of the matrix char temperature. Analysis revealed that both the laser energy per unit length and the fiber orientation-dependent thermal conductivity are key factors in determining the extent of HAZ. An experimental measurement of anisotropic thermal conductivity for composite materials is developed. Heat conduction is greater along fibers than it is across a fiber section, thus laser scanning direction relative to fiber orientation affects the HAZ geometry. The study also investigated the principal-axis and nonprincipal-axis grooving of unidirectional (UD), [0/90], Mat, and MatUD laminates. An analytical model based on a moving point heat source using the Mirror Image Method and immersed heat source to model principal-axis grooving is adopted to correlate HAZ anisotropy with various process parameters. Finite difference method (FDM) with an isotherm conductivity model and eigenvalue method is applied to simulate the HAZ resulting from nonprincipal-axis grooving.

Keywords Laser • Heat-affected zone • Anisotropic heat conduction • Composite materials

---

C.T. Pan (✉)
Department of Mechanical and Electro-Mechanical Engineering, National Sun Yat-Sen University, No. 70, Lienhai Road, Kaohsiung 80424, Taiwan, ROC
e-mail: panct@mail.nsysu.edu.tw; panct@faculty.nsysu.edu.tw

H. Hocheng
Department of Power Mechanical Engineering, National Tsing Hua University, Hsinchu, Taiwan, ROC

H. Hocheng and H.-Y. Tsai (eds.), *Advanced Analysis of Nontraditional Machining*,
DOI 10.1007/978-1-4614-4054-3_1, © Springer Science+Business Media New York 2013

1.1 Introduction

In general, laser machining produces parts with higher dimensional accuracy and surface quality as well as higher material removal rates than those produced with conventional processes. Materials that can be machined by laser include metals, ceramics, plastics, composites, wood, glass, rubber, and fiber-reinforced plastics (FRP).

The literature includes laser machining, heat-affected zone, thermal conductivity, and moving-source heat transfer, as surveyed in the following.

1.1.1 Laser Machining Applications

Laser drilling can produce holes as small as 0.05 mm in diameter in workpiece. It is used in industry for producing holes in turbine blades, combustion chambers, and aerosol nozzle. Laser cutting is used to produce intricate two-dimensional shapes in workpiece made of materials such as sheet metal and paper with high cutting speeds. Investigations of the potential application of CO_2 laser have been performed on laser drilling and cutting for Kevlar/Epoxy [1].

Laser scribing has been used to create channels in ceramic substrates for cooling and identification labels in finished parts. Evaporative removal of material is achieved by focusing a high power and highly concentrated Gaussian laser beam of continuous and pulsed laser onto the surface of the solid [2]. Material removal mechanism is the fundament for the sophisticated laser machining. Schuocker et al. point out that the erosion takes place at a nearly vertical plane at the momentary end of the cut. That plane is covered by a thin molten layer. The removal of material from that layer is carried out by evaporation and by ejection of molten material [3, 4].

1.1.1.1 One-Dimensional Machining: Drilling

In practical applications of laser drilling, the quality of the drilled hole is the critical factor. A reactive gas jet (O_2) can be used to remove material through oxidation, chemical reactions, etc. [5]. For aerosol nozzles, holes with diameters from 0.15 to 1 mm are required. Conventional molding techniques do not have the flexibility to change hole size in process. Laser drilling combines the flexibility of changeable hole diameter with high production rates [6]. In [7], 0.125 in. diameter holes were drilled in 0.05 in. thick Kevlar/Epoxy for a 1,200 W beam using the trepanning method. Cycle time was approximately 1.5 s per hole. In percussion drilling, a pulsed beam removes material through melting and localized detonation or explosion. In this case, about 90% of the materials are removed through detonation effects [8]. For laser percussion drilling processes on metals, drilling efficiency is strongly dependent on the laser-supported detonation [9].

1.1.1.2 Two-Dimensional Machining: Cutting, Scribing, and Marking

Reactive gas laser cutting was performed with an oxygen jet [10]. During the oxygen-laser cutting process, the kerf zone was heated up to ignition by the laser beam while the oxygen jet burns the material and blows the slag off. These experiments were performed on mild steel and austenitic Cr-Ni steel workpiece using a 5 kW laser operating in a TEM04 mode. In [11], laser cutting was analyzed for metallic workpiece such as stainless steel, tool steel, armor plate, titanium alloy and aluminum with thickness between 10 and 15 mm. Another issue in surface quality is the formation of striations along the kerf walls. In a study on striations [12], two distinct zones on the kerf edge were found. Zone I, the zone of regular striations near the kerf entrance, is thought to be the result of oxygen-assisted laser beam heating. Zone II, the zone of indistinct striations near the kerf bottom, is believed to be caused by a diffusive thermochemical reaction in the absence of direct contact with the laser beam. An increase of Zone I depth was shown to correspond with an increase in beam pulse width. The surface roughness was found to depend on pulse width in the vicinity of the top edge approximately 0.2–0.8 mm from the kerf entrance. Conventional cutting of ceramic materials is usually done with a diamond saw. Although this gives a high surface quality, the cutting speed is low, typically around 20 mm/min [13]. Laser cutting allow cutting speeds up to 1,200 mm/min, a significant increase over conventional machining, without sacrificing surface quality. Laser scribing has been used for high speed marking of identification labels on ceramic parts. Laser scribing has also been used in the electronics industry for producing snap lines in aluminum oxide during fabrication of circuit boards [14].

1.1.1.3 Three-Dimensional Machining

For laser turning, threading, and milling, two laser beams can be used to produce converging grooves on a workpiece. When the two grooves intersect, a volume of material is removed. This technique offers higher energy efficiency, higher material removal rates, and more flexibility than nontraditional machining processes [15]. It has been used in a research effort to implement turning, milling, and threading operations. Chryssolouris presented a concept for performing three-dimensional laser machining on composite materials, using two intersecting laser beams to create groove on composite materials. Laser from Carbon/Teflon and Glass/Teflon materials [16] can produce threaded and turned workpiece. A single beam technique for creating three-dimensional contours was presented using a 450 W CW CO_2 laser to machine overlapping grooves on a ceramic workpiece [17]. Silicon carbide, silicon nitride, alumina, and SiAlON were tested.

1.1.2 Laser-Induced HAZ

A one-dimensional transient thermal model has been developed to predict the thermal response of decomposition of a polymer composite exposed to high temperatures [18]. Tagliaferri [19–21] conducted an experimental study to determine the surface finish characteristics of CFRP and AFRP by CO_2 laser. The heat-affected zone (HAZ) depends strictly on feed rate. The higher speed of laser beam is, the smaller is the volume of damage and the better is the cut finish. An analysis of the groove depth and maximum damage width (W_d) was conducted for the Carbon/Teflon material properties. Close agreement was obtained between model predictions and experimental results [22]. Graphite-reinforced composites are found to be less suitable for laser cutting due to high fiber conductivity and vaporization temperature [20, 23]. The effect of some of the laser beam and machining parameters on the depth of the cut, W_d of the HAZ, and quality of cut on carbon fiber composite materials has been studied [23]. In laser cutting of Graphite/Epoxy, the W_d is larger than groove depth for all experimental conditions. A correlation of peak heat flux data for saturated liquid jets has been provided [24]. Chryssolouris et al. [15] presented a method for reducing thermal damage in laser machining by using water jet in tandem with the laser beam. However, moisture and water content could be undesired for FRP. Cryogenic environment for laser processing can be employed to reduce HAZ. Since the phenomena of thermal damage (including matrix recession, matrix decomposition, and delamination) is thermal interaction between the laser and workpiece, surface quality improvement can be achieved by controlling the heat entering the workpiece through conduction [25].

1.1.3 Optimum Parameters

Experimental results obtained by laser machining of different FRP were reported. A criterion to evaluate laser cutting quality for FRP was proposed and the influence of cutting variables on some of cutting quality parameters has been investigated [19]. Tagliaferri et al. have identified the principal quality criteria as matrix recession, craters, delimitation, slope of the cut surface, and HAZ [20]. There are many parameters associated with optimum laser machining. Olsen [26] presented three groups of major parameters, which affect the cutting process as following.

1.1.3.1 Laser Beam Parameters

There exists an optimum pulse duration for the cut depth of carbon fiber composite materials. Deeper penetration of beam is obtained with compressed air jet, while cleaner edge and less thermal damage are obtained with the argon gas. The ratio of the extent of HAZ to the depth of cut depends on the details of the arrangement of the fibers in the cross-ply laminates [12]. The thermal damage caused during laser cutting of Aramid/Epoxy laminates was investigated. Samples cut with a 500 W CO_2 CW laser using different parameters were examined by both optical and SEM [19]. An analysis of the laser grooving process was conducted for several composite materials. An isotropic analytical model was developed to determine groove depth from process parameters and thermal properties of the material. Experimental measurements of groove depth, groove width, and extent of HAZ were taken for variations in scanning velocity, beam power, and beam passes [22]. The hardness and brittleness of high performance engineering ceramics such as zirconia alumina, silicon carbide and silicon nitride make traditional processing very expensive and time-consuming. Efforts have been made to cut thick commercial ceramics with a continuous wave CO_2 laser. The results obtained so far are very promising [27].

1.1.3.2 Gas Jet Parameters

Na and Yang [27] investigated the influence of the shielding gas pressure on the laser cut quality; a series of experiments was conducted for various cutting conditions. The cutting pressure distribution was measured on the workpiece, and also beneath the workpiece with a prepared kerf for various nozzle pressures and nozzle to workpiece distances. A quantitative analysis of cutting thin stainless steel sheets with CW CO_2 lasers using an oxygen assist gas jet was reported. Results were interpreted theoretically using a point heat source model [28]. An experimental investigation was conducted to clarify the role of an assisting gas during metal drilling for incident laser fluxes. In particular, the effect of change in absorptivity and other thermophysical properties associated with metal oxide formation of laser drilling time was investigated [29].

1.1.3.3 Material Parameters

Composites of Aramid, graphite, and glass cloth-reinforced polyester have been cut by laser. Thermal properties of the fibers and matrix were the principal factors, which affected cutting performance [21]. The optimum performance in laser drilling depended very much upon the proper selection of laser parameters as well as the physical properties of the workpiece material [5]. The above information is summarized in Tables 1.1 and 1.2.

1.1.4 Thermal Conductivity Models

With the increasing use of composite materials, the determination of reliable thermal conductivity data is gaining increasing importance. The thermal conductivity of FRP in the direction parallel to the fibers is well understood in terms of the

Table 1.1 Representative data for laser drilling in literature

Material	Laser type	Laser power	Drilling speed	Hole depth	Hole diameter	References
Copper	Nd:YAG	2 MW/cm^2	N.A.	0.045 mm	0.25 mm	[9]
Nickel	Nd:YAG	10–20 J 1–3.5 ms	1 ms/hole	0.5–1.25 mm	N.A.	[5]
Tantalum	Nd:YAG	10–20 J 1–3.5 ms	1 ms/hole	0.5–1.25 mm	N.A.	[21]
Titanium	Nd:YAG	10–20 J 1–3.5 ms	0.67 ms/hole	0.5–1.25 mm	N.A.	[5]
Kevlar/epoxy	CO_2	185–1,250 W	0.6–1.5 s/hole	0.05 in.	0.125–0.2 in.	[2]

Table 1.2 Representative data for laser cutting in literature

Material	Laser type	Laser power	Cutting speed	Kerf depth	Kerf width	References
Aluminum	Nd:YAG	200 W	0.23 m/min	1.3 mm	0.6 mm	[30]
Copper	Nd:YAG	300–6,000 J/cm^2	N.A.	6.5 mm	N.A.	[9]
Mild steel	CO_2	1 kW	0.8 m/min	10 mm	0.9 mm	[10]
Mild steel	CO_2	2.5 kW	0.85 m/min	10 mm	0.9 mm	[10]
Mild steel	CO_2	4 kW	0.95 m/min	10 mm	0.9 mm	[10]
Mild steel	Nd:YAG	150 W	0.23 m/min	1.5 mm	0.6 mm	[30]
Titanium	CO_2	375 W	100–360 in./min	0.06–0.125 in.	0.03 in.	[6]
Aramid/polyester	CO_2	800 W	0.5 m/min	2 mm	0.6 mm	[21]
Glass/polyester	CO_2	800 W	0.5 m/min	2 mm	N.A.	[21]
Graphite/polyester	CO_2	800 W	0.5 m/min	2 mm	0.5 mm	[21]

conductivity of the fiber and matrix. It is analogous to the electrical resistance, known as Rule of Mixture [31].

$$k_{\text{parallel}} = V_{\text{fiber}} k_{\text{fiber}} + V_{\text{matrix}} k_{\text{matrix}} \tag{1.1}$$

where V_{matrix} and k_{matrix} denote the volume fraction and the thermal conductivity of the matrix, whereas V_{fiber} and k_{fiber} stand for the corresponding properties of the fibers parallel to their longitudinal axis.

However, the thermal conductivity of FRP in the direction transverse to fibers is normally difficult to measure directly due to the small fiber diameter. The transverse conductivity can be two orders of magnitude lower than conductivity parallel to fiber direction [32]. Several prediction models for composite conductivity transverse to fiber direction have been presented [33–35].

$$\frac{k_{\text{transverse}}}{k_{\text{matrix}}} = (1 - \sqrt{V_{\text{fiber}}}) + \frac{\sqrt{V_{\text{fiber}}}}{1 - \sqrt{V_{\text{fiber}}}(1 - (k_{\text{matrix}}/k_{\text{fiber}}))} \tag{1.2}$$

$$k_{\text{transverse}} = k_{\text{matrix}} + \frac{V_{\text{fiber}}}{(1/(k_{\text{matrix}} + k_{\text{fiber}})) + ((1 - V_{\text{fiber}})/2k_{\text{matrix}})} \tag{1.3}$$

$$\frac{k_{\text{transverse}}}{k_{\text{matrix}}} \approx 1 - \frac{2V_{\text{fiber}}}{V' + V_{\text{fiber}} - (0.3058/V')V_{\text{fiber}}^4 - (0.0134/V')V_{\text{fiber}}^8} \tag{1.4}$$

where

$$V' = \frac{((k_{\text{matrix}}/k_{\text{fiber}}) + 1)}{((k_{\text{matrix}}/k_{\text{fiber}}) - 1)} \tag{1.5}$$

A similar approach based on more sophisticated heat transfer calculations has been applied to understand the thermal conductivity in the direction transverse to fiber [31].

$$\frac{1}{k_{\text{transverse}}} = \frac{V_{\text{fiber}}}{k_{\text{fiber}}} + \frac{V_{\text{matrix}}}{k_{\text{matrix}}} \tag{1.6}$$

or

$$k_{\text{transverse}} = \frac{k_{\text{fiber}} k_{\text{matrix}}}{V_{\text{matrix}} k_{\text{fiber}} + V_{\text{fiber}} k_{\text{matrix}}} \tag{1.7}$$

On the other hand, the shape of HAZ developed in anisotropic composite materials can be approximated by isotherms. It has been shown that HAZ was an elliptical shape. The transverse conductivity can be found with the major and minor axis of an ellipse [36].

$$\frac{k_{\text{transverse}}}{k_{\text{parallel}}} = \frac{b}{a} \tag{1.8}$$

where a and b are major and minor axis of that ellipse.

However, since the thermal properties of composite strongly depend on directions relative to fiber axis, there was a conductivity model, based on the view of energy conservation, to calculate conductivities in various directions [37].

$$k_1 \xi_1^2 + k_2 \xi_2^2 + k_3 \xi_3^2 = 1 \tag{1.9}$$

where ξ_1, ξ_2, and ξ_3 are principal axes determined by the fiber axis, k_1, k_2, and k_3 are the principal conductivities.

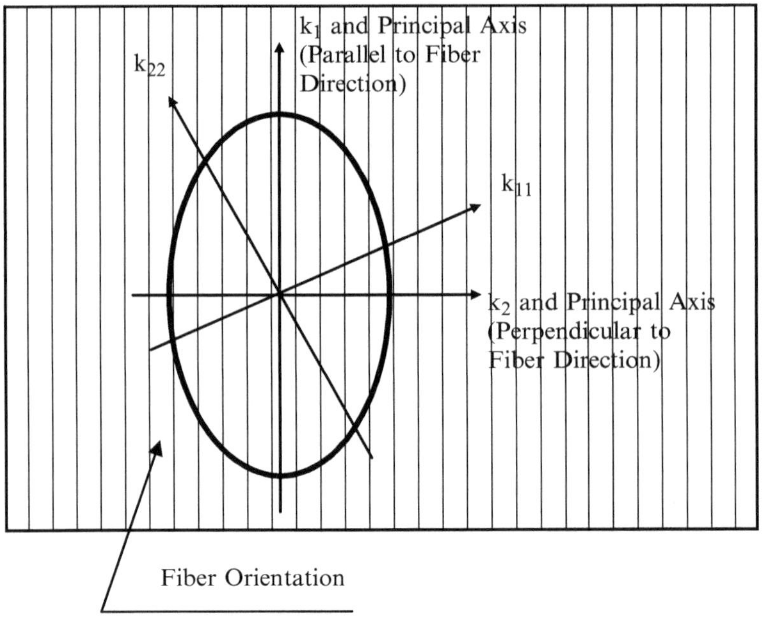

Fig. 1.1 Schematic view of conductivity ellipse in unidirectional composite material

For UD composite materials, (1.9) can be reduced to

$$k_1\xi_1^2 + k_2\xi_2^2 = 1 \tag{1.10}$$

Schematic presentation of a 2D thermal conductivity ellipse can be seen in Fig. 1.1. It can be used to determine k_{11} and k_{22}, the off-principal axis conductivities, for UD and [0/90] composite materials along various directions. However, it requires transverse thermal conductivity of composites in (1.10) before it can work. Therefore, reliable transverse thermal conductivity data are an essential input for such an analysis.

1.1.5 Theory of a Moving Heat Source

There are numerous engineering applications, such as welding, grinding, metal cutting, firing a bullet in a gun barrel, flame or laser hardening of metals, and many others in which the calculation of temperature field in the solid is modeled as a problem of heat conduction involving a moving heat source. Following the pioneering works of Rosenthal [38] on the determination of temperature distribution in a solid during arc welding, many papers appeared on the subject of heat transfer with moving heat sources.

In laser processing, the beam can be viewed as a heat source impinging on the surface of the workpiece. If the beam is stationary such as in the case of laser drilling, then the erosion front moves relative to the laser beam. This results in a temperature distribution, which changes with time. Therefore, laser drilling is a nonsteady process. In laser cutting and scribing, the laser beam is in relative motion with the workpiece. In case of constant scanning velocity, the erosion front and the resulting temperature distribution are constant relative to a coordinate system fixed at the laser beam. A steady temperature field simplified the heat transfer problem to steady-state conduction. Rosenthal [38] transformed the conduction equation into a simple form by assuming a temperature distribution equation. The formation of a deep groove on a moving semi-infinite solid is considered.

1.1.6 Objective

FRP have increasingly widened their use in structural application thanks to its high specific strength and directional properties, making the structure design more versatile. Reinforcements need not be in the form of long fibers. Particles, flakes, whiskers, discontinuous fibers, continuous fibers, and sheets are all practical. Polymer-based composites possess the promising potential for the next-generation materials in industry. In the future, thermoplastic matrices will be more increasingly employed for their improved toughness and processing convenience.

The near-net-shape manufacture through pressing, injection molding of filament winding, is typical for the processing of FRP components. Although in many applications composites are cured to final shape, good accuracy is however very costly in these manufacturing steps. Machining is still required at both the prepreg and product stages.

Conventional machining operations are difficult to perform due to the anisotropy, inhomogeneous composition, hardness, and abrasiveness of composite materials. Excessive tool wear significantly increases the machining time and expense.

Laser is an acronym for "Light Amplification by Stimulated Emission of Radiation." The unique properties of laser light can be quantified by examining the optical properties of monochromaticity, coherence, diffraction, and radiance.

Lasers can be categorized most easily according to the state of their working mediums: gas, liquid, or solid. Furthermore, all laser types operate in one of two temporal modes: continuous wave (CW) and pulsed modes. In CW mode, the laser beam is emitted without interruption. In pulsed mode, the laser beam is emitted periodically.

In principle, laser machining is governed by an energy balance between the irradiation energy from the laser beam and the conduction heat into the workpiece, the energy losses to the environment, and the energy required for phase change in the workpiece. Laser machining can replace for mechanical material removal methods in many industrial applications, particularly in the processing of difficult-to-machine materials such as hardened metals, ceramics, and composites.

Laser machining using high-power density beam offers several advantages over conventional methods, such as zero tool wear and the contact force-induced problem. Laser finds relatively good absorptivity in FRP, but the relevant research of thermal properties of composite materials in laser machining advanced little in the latest decade. Although materials with favorable thermal properties can be successfully cut by laser regardless of their mechanical characteristics, thermal damage associated with laser energy can be produced.

Due to the inhomogeneous and anisotropic nature of the material, the machining of FRP differs in many respects from metal machining. The existing literature is concentrated on machinability and isotropic material removal mechanism in laser machining. Little is done on the anisotropic heat conduction and HAZ in FRP.

The current investigation applies analytical and numerical heat conduction models to the anisotropic material, and performs experiment on the HAZ. Knowledge of the thermal properties of the composite material is of critical importance to predict HAZ. The present study can be seen as an analysis of the thermal-induced problem. Thermal properties and thermal damage of composite materials in laser machining are discussed. The models to determine the HAZ are presented.

1.2 Experimental Preparation

Laser beam (TEM_{oo}) was supplied by a 1,000 and 1,500 W CO_2 laser and directed through an optical assembly to focusing lens. A focused spot of 0.17 and 0.25 mm in diameter was generated. Grooving experiments were conducted using beam power levels from 350 to 1,500 W with continuous beam, scanning velocities from 2.5 to 110 mm/s, and single beam pass. A coaxial N_2 jet was used to protect laser lens from the debris and provides an inert environment for beam/material interaction. The experimental setup is shown in Fig. 1.2, and the modules of the laser system are shown in Fig. 1.3.

In this study, the cryogenic technique was used to produce low temperature environment. A cooling N_2 jet connected to the pressure accumulator and a temperature acquisition unit was established first. The schematic diagram of complete experimental system, as shown in Fig. 1.4, includes a liquid N_2 jet, and the thermocouples attached on the test specimen.

1.2.1 Experimental Materials

To clearly illustrate the anisotropic characteristics of HAZ of FRP, the specimen subject to cutting is designed unidirectional of Carbon/Epoxy and Carbon/PEEK, with volume fractions 50% and 60%, respectively. In this experiment, materials of test specimens were laminated with unidirectionally carbon fiber-reinforced prepreg (UD prepreg laminates). Carbon/Epoxy is 8 mm in thickness (64 laminations)

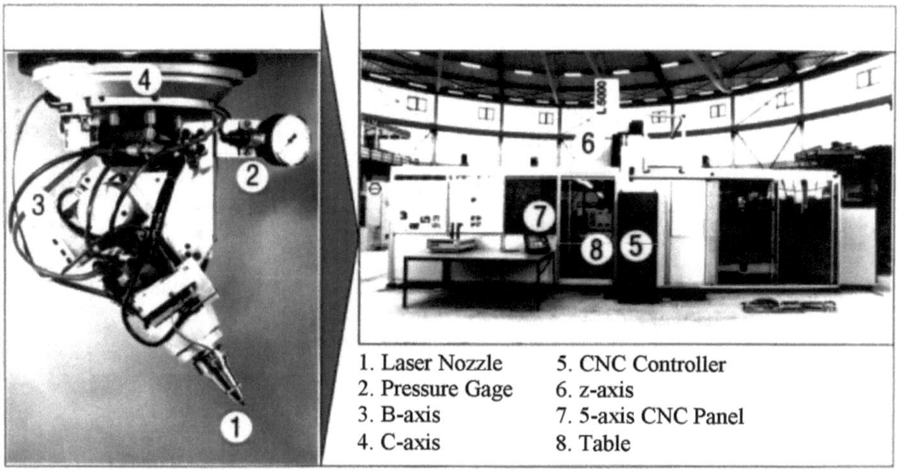

Fig. 1.2 Experimental setup [39]

and 20 mm × 40 mm in area, while Carbon/PEEK is 2.5 mm in thickness (20 laminates) 20 mm × 40 mm in area. The temperature of decomposition for the Carbon/Epoxy determined by thermal gravomatric analysis (TGA) is 360°C (633 K), as shown in Fig. 1.5.

Other ready-made composite materials used in this study were purchased from Strupp and Wenckler in Germany, including UD, [0/90], Mat, and MatUD laminates. The material descriptions are listed in Table 1.3.

Test sample is put in microwave at the temperature between 400 and 800 K until the matrix is evaporated, then the remainder of fiber determines the fiber volume fraction.

$$V_{\text{fiber}} = \frac{\upsilon_{\text{fiber}}}{\upsilon_{\text{composite}}} = \frac{(m_{\text{fiber}}/\rho_{\text{fiber}})}{\upsilon_{\text{composite}}} \tag{1.11}$$

1.2.2 *Experimental Procedures*

Experimental parameters include the laser beam power (P), traverse speed (V), power duty (Q), laser grooving direction relative to fiber axis (θ), thermal conductivity (k), and workpiece temperature (T_o). The experiments are divided into two parts.

1.2.2.1 Non-cryogenic Grooving

Experiments were performed without liquid N_2 (only low-pressure coaxial N_2 gas) to analyze HAZ varying with process parameter, i.e., beam power, power duty, and

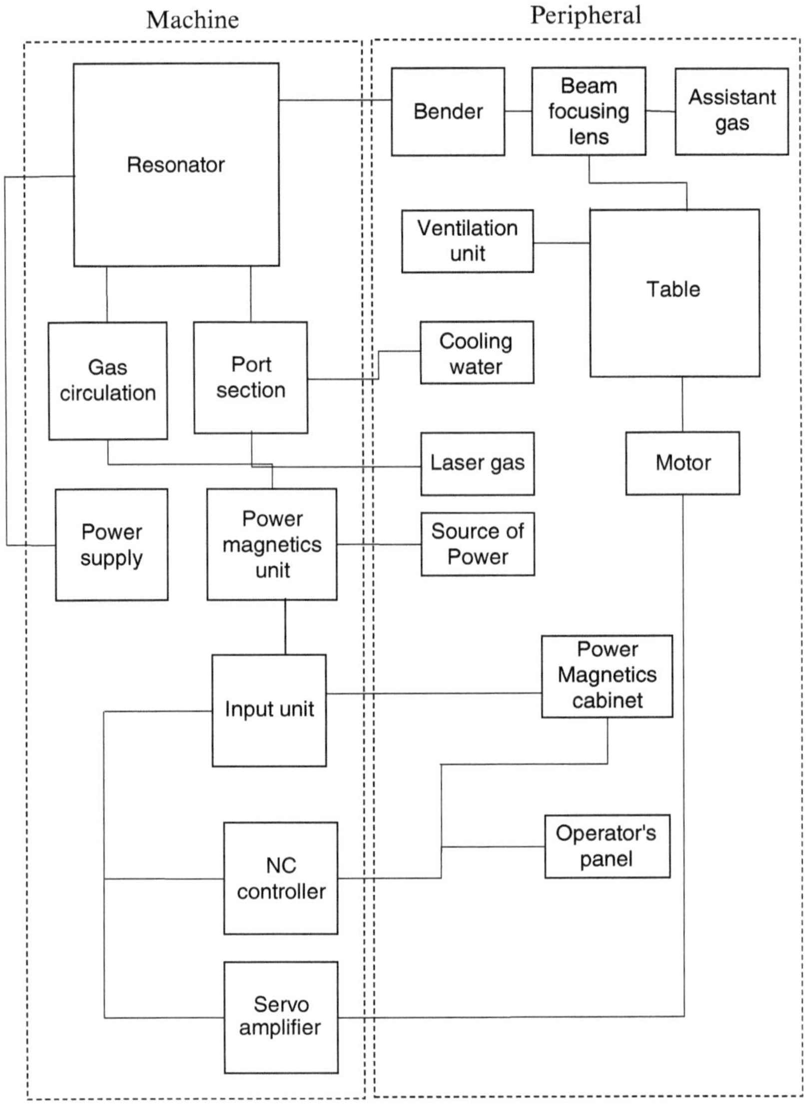

Fig. 1.3 Modules of laser system machinery

transverse velocity as well as grooving parallel and perpendicular to fiber axis, as shown in Fig. 1.6. An energy density parameter (*PQ/V*) was used to incorporate all process conditions into one variable. This parameter correlates the total amount of energy received by a unit grooving length to machining parameters.

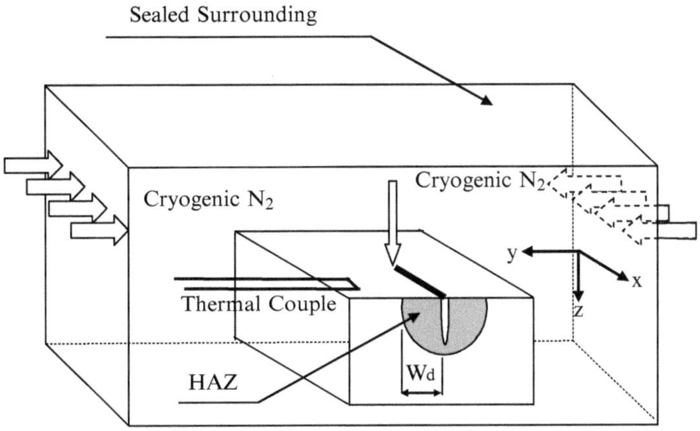

Fig. 1.4 Schematic view of the cryogenic surrounding [40, 41], reprinted with permission

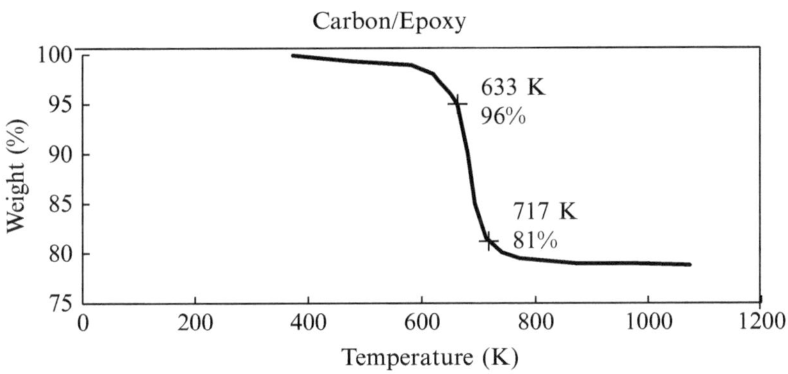

Fig. 1.5 TGA of carbon/epoxy [41, 42], reprinted with permission

Table 1.3 Specification of the investigated composite materials

Composite materials	Volume fraction V_{fiber}/V_{matrix} (%)	Thickness D (mm)	Type
Carbon/PEEK	60/40	2.5	UD
Carbon/epoxy	50/50	8	UD
Carbon/epoxy	53/47	2	UD
Carbon/epoxy	53/47	4	UD
Carbon/epoxy	63/37	2	UD
Carbon/epoxy	63/37	4	UD
GlassMat20/PP	8/92	3.5	Mat
GlassMat30/PP	13/87	3.5	Mat
GlassMat40/PP	19/81	3.5	Mat
GlassMat50/PP	26/74	3.5	Mat
Aramid/PP	18/82	2	[0/90]
Glass/PE	23.3/76.7	3.3	[0/90]
Glass/PP	34/66	2.2	[0/90]
GlassMat20UD20/PP	21.2/78.8	3.7	MatUD

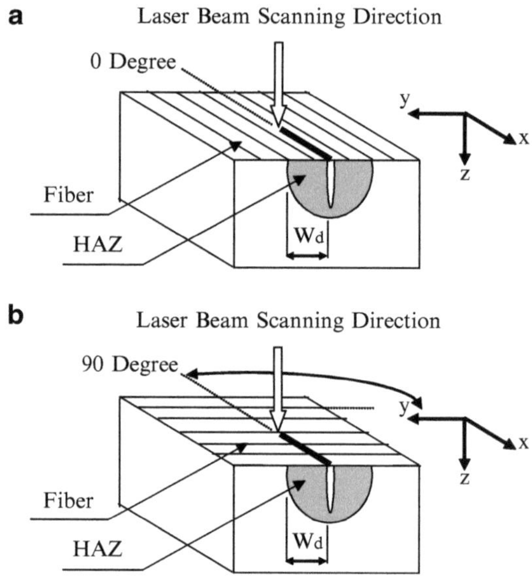

Fig. 1.6 Schematic diagram of laser grooving [43], reprinted with permission. (**a**) 0° relative to fiber axis (i.e., parallel grooving). (**b**) 90° relative to fiber axis (i.e., perpendicular grooving)

1.2.2.2 Cryogenic Grooving

The process parameters were kept the same as above, while a cool N_2 environment was used. After the test specimen had reached an equilibrium temperature, i.e., the preset temperature T_o, 30°C (303 K), −10°C (263 K), −30°C (243 K), −60°C (213 K), and −90°C (183 K), respectively, the laser power system was turned on. By varying the cryogenic temperature, HAZ was examined. Due to its low cost and low temperature, liquid N_2 is chosen as the cooling medium. The setup for the cryogenic surrounding is shown in Fig. 1.4. A simple convergent nozzle is used. The cold N_2 flow arrives at the nozzle from a pressurized accumulator tank. An N_2 gas jet flows coaxial with the laser beam to protect the focusing lens from debris and provide an inert environment for beam/material interaction.

W_d is defined as the maximum distance between the groove front surface and the char temperature isotherm parallel to top surface perpendicular to laser traverse direction. A comparison is made between laser machining with and without the cryogenic surrounding. The W_d of HAZ is examined by optical microscope. The specimen is cross-sectioned by water jet, of which workpiece temperature is about 50°C; hence it causes no concern of changing the HAZ. The experimental results will later be compared with the three-dimensional theoretical prediction of heat conduction in a finite domain as well as numerical computation.

Table 1.4 List of primary variables in dimensional analysis

Primary variables	Dimensions
W_d: Damage width of HAZ	L
P: Peak power	ML^2t^{-3}
Q: Pulse duty	– (%)
V: Traverse speed	Lt^{-1}
k: Thermal conductivity of composite	$MLt^{-3}T^{-1}$
C: Specific heat of composite	$L^2t^{-2}T^{-1}$
α: Thermal diffusivity of composite	L^2t^{-1}
L_v: Fiber latent heat of evaporation	L^2t^{-2}
T_c: Matrix char temperature	T
d: Laser beam diameter	L
λ: Beam pass number	–

1.2.3 Dimensional Analysis

The material removal process in laser grooving involves complex variables; hence experiments of a number of parameters are time-consuming and are liable to incomprehensible results. Dimensional analysis provides a preliminary insight of the processing parameters on reduced expense and time. The dominant parameters involved in the process are first selected, some of which have been analyzed through variance analysis of HAZ [39]. For the extent of HAZ (see Fig. 1.6), the dominant parameters from laser and material are the peak power (P), pulse duty (Q), traverse speed (V), thermal conductivity (k), thermal diffusivity (α), fiber latent heat of evaporation (L_v), material-specific heat (C), incident beam diameter (d), beam pass number (λ), and matrix char evaporation temperature (T_c). Their dimensions are summarized in Table 1.4.

$$W_d = \text{function}(P, Q, V, L_v, T_c, k, C, \rho, \alpha, d) \tag{1.12}$$

and further

$$\frac{W_d L_v^{1/2}}{\alpha} = \text{function}\left(\frac{PQL_v^{1/2}}{VdkT_c}\right) \tag{1.13}$$

The effects of material and processing parameters can be separated as

$$W_d = \text{function}\left(\frac{PQ}{V}\right) \times \text{function(material thermal properties)} \tag{1.14}$$

1.2.4 Preliminary Experimental Results

The dimensional analysis for HAZ summarizes the effect of laser processing parameters as shown in (1.12). It shows that W_d is correlated to PQ/V, called the specific laser energy per unit length, a vital quantity for laser-induced thermal damage. To identify the true functional relationship, experimental results are incorporated. Figure 1.7 illustrates the HAZ from the experiment taken from object projector and SEM. The tolerance of the measurement of the extent of HAZ is 0.1 mm. The char temperature is defined as the value corresponding to matrix

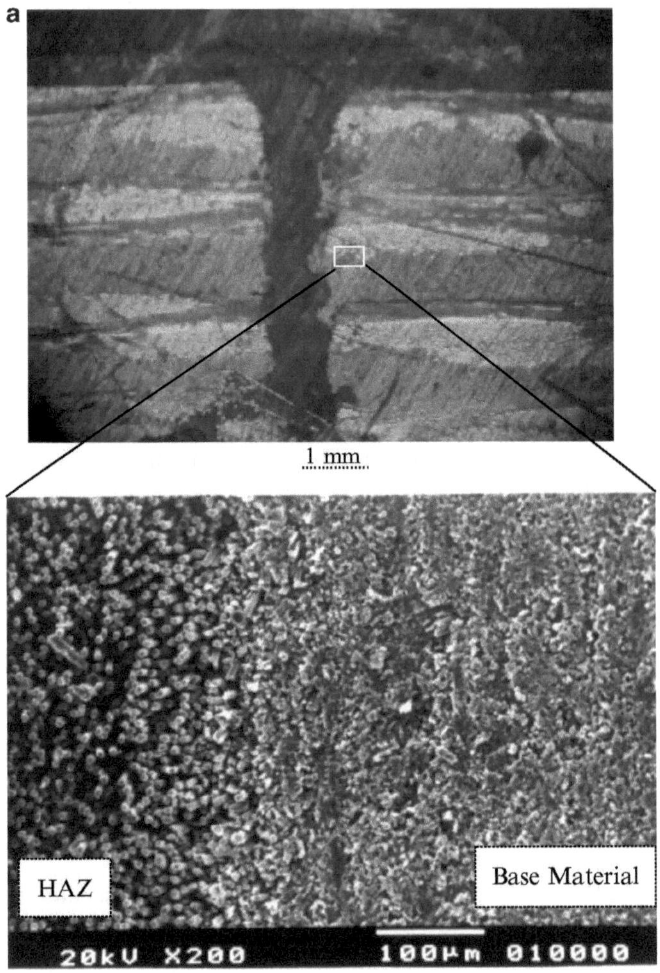

Fig. 1.7 Illustration of HAZ (P = 1,250 W, V = 1,000 mm/min). (**a**) Photograph of aramid/PP, SEM of perpendicular grooving, (**b**) photograph of glass/PP, SEM of parallel grooving

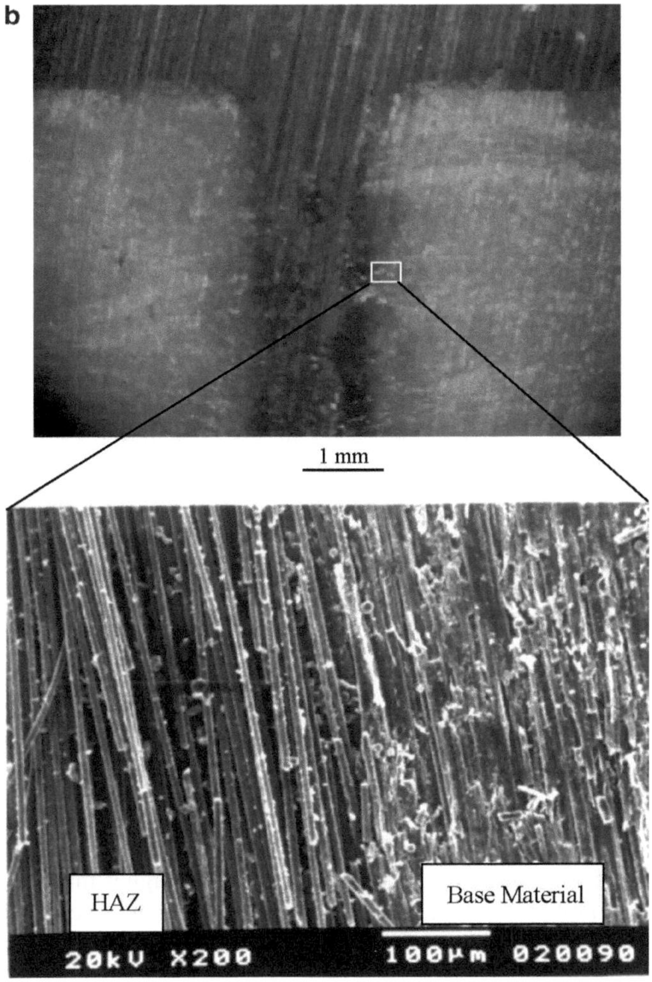

Fig. 1.7 (continued)

weight loss of 4% in Fig. 1.5 [44]. In consideration of the variation of the prediction, the difference between the extent of HAZ induced by matrix char temperature at 4% and 19% matrix weight loss is less than 6%. The matrix material was burned out in the HAZ. The relationship between the laser energy per unit length and W_d of HAZ is experimentally determined as illustrated in Fig. 1.8. The results show that W_d is approximately proportional to the specific laser energy PQ/V. Namely, the increase in the average input power (PQ) or the decrease of the traverse speed (V) of laser beam results in larger damage because the input heat per unit length is decreased.

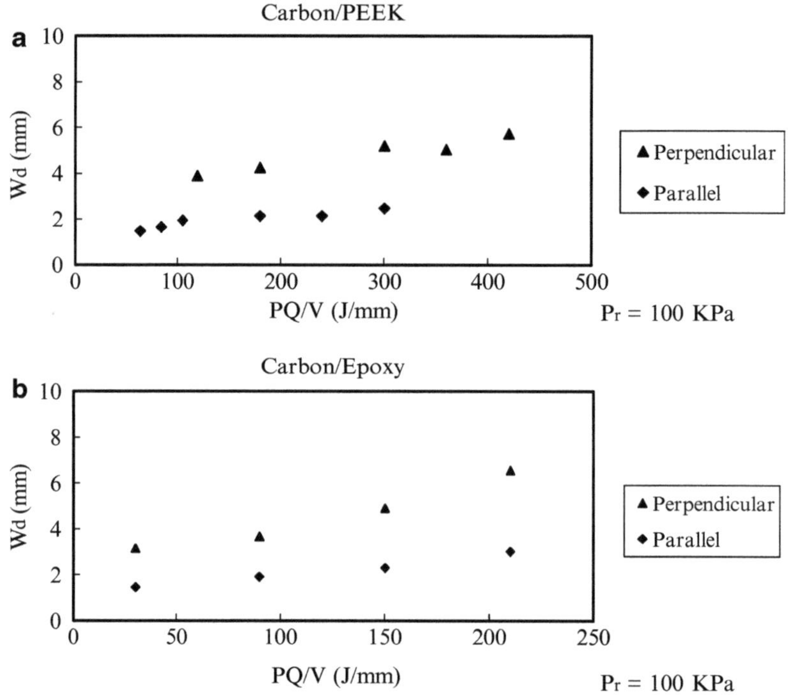

Fig. 1.8 Preliminary experimental results of HAZ varying with specific laser energy. (**a**) Carbon/PEEK, (**b**) carbon/epoxy

The energy for vaporization of fibers is in general higher than that for matrix; hence the laser power required for grooving composites will be dependent on the fiber volume fraction.

1.3 Anisotropic Thermal Conductivity

In the case of principal-axis grooving, an analytical solution for HAZ is developed. It is effective for 0° and 90° grooving of UD, [0/90], Mat, and MatUD laminates.

1.3.1 Experimental Laser Flash Method

To obtain the thermal conductivity, the thermal diffusivity and specific heat of the composite samples were measured by the Laser Flash Method utilizing

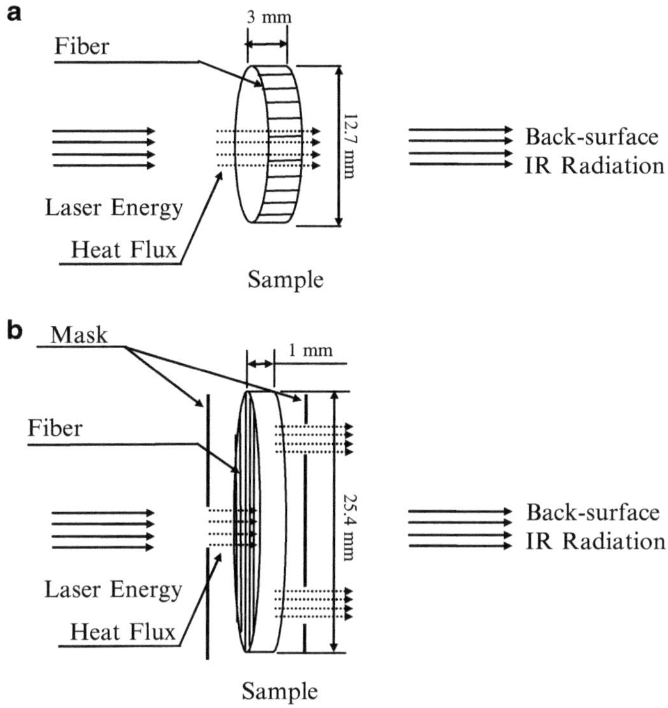

Fig. 1.9 Description of laser flash method for thermal property measurement [45, 46]. (**a**) Through-plane test sample, (**b**) in-plane test sample

a Holometrix Microflash instrument. This instrument and method conform to ASTM E146-92, "Standard Test Method for Thermal Diffusivity of Solids by the Flash Method" [45, 46].

The through-plane and in-plane measurements of the thermal diffusivity of a material are carried out by rapidly heating one side of the sample and measuring the temperature rise curve on the opposite side. The time that it takes for the heat to travel through the sample and to cause the temperature rise on the rear face can be used to measure the diffusivity.

The through-plane sample is a disk with standard diameter of 12.7 mm and thickness of 3 mm, while the in-plane sample is 25.4 mm in diameter and 1 mm in thickness. The Holometrix Thermoflash 22000 Laser Flash System was applied to measure the thermal diffusivity. The sample disk is aligned between a neodymium glass laser (1.06 in wavelength) and an indium antimonite (InSb) IR detector in a tantalum tube furnace. The method is depicted in Fig. 1.9.

The principle of the measurement by Laser Flash Method is described below [45].

1.3.1.1 Thermal Diffusivity

Transient heat transfer problems occur when the temperature distribution changes with time. The fundamental quantity is the thermal diffusivity. It is related to the steady-state thermal conductivity through the equation

$$\alpha = \frac{k}{\rho C} \tag{1.15}$$

Where α is the thermal diffusivity, k is the thermal conductivity, C is the specific heat, and ρ is the density.

The diffusivity is a measure of how quickly a body can change its temperature. It increases with the ability of the body to conduct heat (k) and it decreases with the amount of (ρC).

Thermal properties of materials are determined by the experimental establishment of a boundary value case of heat flow, and by measurement of the temperature or heat flux. The results are then matched to the theoretical solution, as shown in (1.16). The simplest way to measure the thermal conductivity is to set up a steady-state and linear heat flow through the material and to apply Fourier's equation. This approach has led to the development of a number of methods for measuring the thermal conductivity including the Guarded Hot Plate and Linear Rod methods. These methods are time-consuming and can be susceptible to errors arising from the deviation from the assumed boundary or steady-state conditions.

There have been many solutions of this boundary/initial value problem in different forms. The first to apply the solution to the flash diffusivity measurement was Parker et al. [46]. Using the notation of Koski, the derived equation for the rear face temperature as a function of time is,

$$T(D,t) = 2T_{\mathrm{f}} \sum_{n=1}^{\infty} \frac{\gamma_n^2(\gamma_n^2 + L^2)\cos\gamma_n}{\gamma_n^2 + L^2 + 2L} \exp\left(\frac{-\gamma_n^2 \alpha t}{D}\right) \tag{1.16}$$

where is the final sample temperature, t is time, D is the sample thickness, and L is the heat loss factor, is found by solving the transcendental equation

$$\tan\gamma_n = \frac{2\gamma_n L}{\gamma_n^2 - L^2} \tag{1.17}$$

1.3.1.2 Specific Heat

The specific heat of a material is defined as the amount of energy required to raise a unit mass of material by one unit of temperature at constant pressure

Table 1.5 Experimentally measured thermal conductivity by laser flash method at room temperature for unidirectional carbon/epoxy

Material	Density (ρ) kg/m^3	Specific heat (C) J/kg K	Conductivity (k) W/mK	Diffusivity (α) m^2/s
			Parallel to fiber axis	
			$k = 4.50$	$=3.096\text{E} - 6$
Carbon/ epoxy	1,530	950	45° to fiber axis	
			$k = 2.95$	$\alpha = 2.029\text{E} - 6$
			Perpendicular to fiber axis	
			$k = 0.67$	$=4.609\text{E} - 7$

$$C = \frac{E}{m\Delta T} \tag{1.18}$$

where C is specific, m is mass, ΔT is change in temperature, and E is laser energy.

The specific heat can be measured with the Laser Flash Method by the comparison of the temperature rise of the sample to the temperature rise of a reference sample of known specific heat tested under the same conditions.

$$E = \text{absorbed energy} = (mC\Delta T)_{\text{reference}} = (mC\Delta T)_{\text{sample}} \tag{1.19}$$

thus

$$\mathbf{C}_{\text{sample}} = \frac{(mC\Delta T)_{\text{reference}}}{(m\Delta T)_{\text{sample}}} \tag{1.20}$$

1.3.1.3 Conductivity

The sample thermal conductivity can be calculated from (1.15), after the diffusivity and specific heat are measured as described above, and the sample bulk density is calculated from the sample dimensions and mass. The test results are given in Table 1.5.

1.3.2 Model of Resistor

Tsai defined an effective thermal conductivity for a typical unidirectional laminates. The fiber and matrix can be modeled as a thermal resistor. It is widely used to estimate the thermal conductivity of composite material. The advantage of this approach for a given volume fraction is that the heat resistance can be treated as an electrical resistor, as shown in Fig. 1.10a. In the following, thermal conductivities parallel and transverse to fiber axis are analyzed. The thermal conductivity parallel to the fiber direction is obtained by (1.1), i.e., Rule of Mixture, while the transverse thermal conductivity of unidirectional FRP is predicted by (1.7) [31].

In this research, thermal conductivity of the Carbon/Epoxy was measured by the Laser Flash Method. The measured conductivities are listed in Table 1.5. The measured thermal conductivities for UD Carbon/Epoxy can be precisely calculated by (1.1) and (1.7). The comparison between the experimental and the calculated results is shown in Figs. 1.11 and 1.12. The experimental results show satisfactory agreement with the prediction. The results are used in this study to evaluate the conductivity of [0/90] and MatUD laminates.

In the case of [0/90] laminates, the laminae are connected macroscopically. For the upper layer, matrix and fiber are treated as connected in parallel in contrast to that connected in series for the lower layer. The schematic illustration of the resistance theory is shown in Fig. 1.10b.

Consequently, by coupling both cases one obtains:

$$k_x = \frac{k_{\text{series}} + k_{\text{parallel}}}{2} \tag{1.21}$$

$$k_y = \frac{k_{\text{series}} + k_{\text{parallel}}}{2} \tag{1.22}$$

whereas in z direction the resistances are connected in series twice. Hence

$$k_z = k_{\text{series}} \tag{1.23}$$

In the case of GlassMat/PP, semi-isotropic formation was assumed in the study. The following formula [47] is adopted to evaluate the conductivity

$$k_{\text{mat}} = k_{\text{matrix}} \frac{(1 - V_{\text{fiber}})k_{\text{matrix}} + (1 + V_{\text{fiber}})k_{\text{fiber}}}{(1 + V_{\text{fiber}})k_{\text{matrix}} + (1 - V_{\text{fiber}})k_{\text{fiber}}} \tag{1.24}$$

Likewise, in the case of GlassMatUD/PP, it is similar to the case of [0/90]. As far as the y direction is concerned, both parts of Mat and UD material are connected macroscopically. For the upper UD part, matrix and fiber are treated as connected in parallel in contrast to that nearly homogeneous for the lower Mat material (see Fig. 1.10). Modifications are made to calculate the conductivity when grooving 0° relative to fiber axis

$$k_x = \frac{k_{\text{parallel}} + k_{\text{mat}}}{2} \tag{1.25}$$

$$k_y = \frac{k_{\text{series}} + k_{\text{mat}}}{2} \tag{1.26}$$

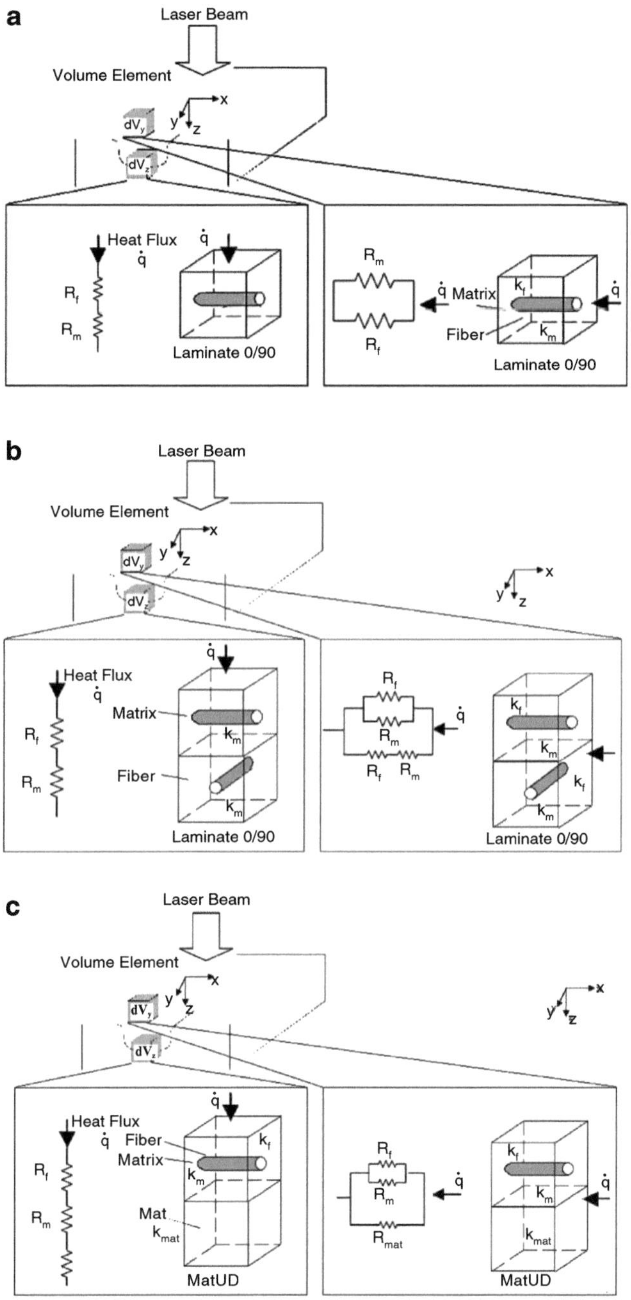

Fig. 1.10 Model of resistor. (**a**) Heat resistance of UD laminates [31]. (**b**) Heat resistance of [0/90] laminates [42], reprinted with permission. (**c**) Heat resistance of MatUD composite material [42], reprinted with permission

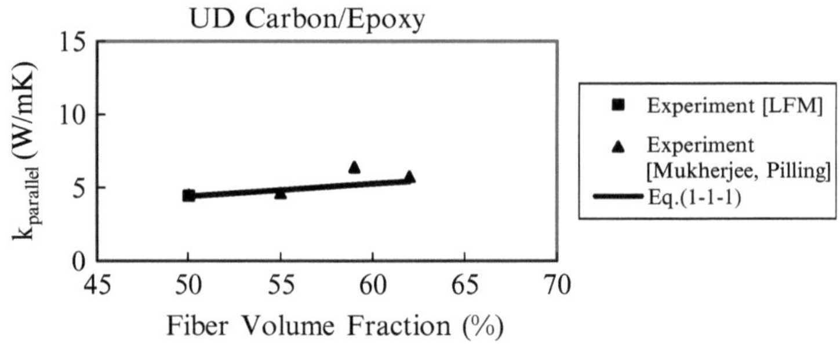

Fig. 1.11 Comparison between calculated and experimental results of thermal conductivity parallel to fiber axis

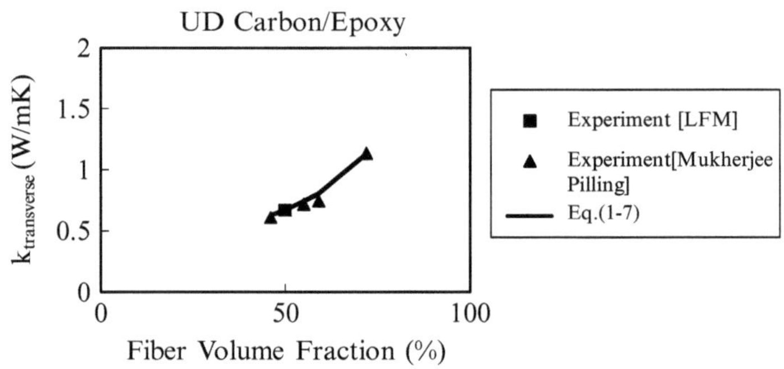

Fig. 1.12 Comparison between calculated and experimental results of thermal conductivity transverse to fiber axis

When grooving 90° relative to the fiber axis:

$$k_x = \frac{k_{\text{series}} + k_{\text{mat}}}{2} \tag{1.27}$$

$$k_y = \frac{k_{\text{parallel}} + k_{\text{mat}}}{2} \tag{1.28}$$

In both cases

$$k_z = 2\frac{k_{\text{series}} k_{\text{mat}}}{k_{\text{series}} + k_{\text{mat}}} \tag{1.29}$$

1.3.3 Prediction of HAZ

1.3.3.1 Preliminary Isotropic Analysis

In this section, a preliminary isotropic thermal conduction model is introduced first. The thermal conductivity transverse to fiber is used for the analysis of HAZ in parallel grooving of a UD composite material, while the thermal conductivity along the fibers is used for perpendicular grooving. It provides a simplified preliminary prediction of HAZ, based on which the effects of anisotropic, finite thickness, and immersed heat source will be incorporated.

The erosion front is simplified and the molten layer is assumed to have a negligible thickness [22]. The absorbed laser beam power is used to conduct into the solid, while radiation and convection can be neglected [16]. The coefficient of laser absorption is assumed to be $\eta = 1$ [48]. Considering a moving point heat source along the x axis, the governing equation is [37]:

$$k\nabla^2 T = \rho c \frac{\partial T}{\partial t} \tag{1.30}$$

The boundary conditions are:

$$\begin{aligned} &T = T_o \quad \text{at } x = \pm\infty,\ y = \pm\infty,\ z = \pm\infty \quad \text{and} \quad \eta\dot{q}Q = -k\nabla T \\ &\quad \text{at } x = Vt, y = 0, \text{ and } z = 0 \end{aligned} \tag{1.31}$$

where η is coefficient absorptivity, and $\dot{q}$ is heat flux.

One can calculate the temperature at (x, y, z, t) from the irradiated heat [37]:

$$T - T_o = \int_0^t \frac{\eta PQ}{8\rho c[\pi(t - t')\alpha]^{3/2}} e^{-(((x - V(t-t'))^2 + y^2 + z^2)/4\alpha(t-t'))}\, dt' \tag{1.32}$$

$$T - T_o = \frac{\eta PQ}{2kR\pi^{3/2}} e^{Vx/2\alpha} \int_\omega^\infty e^{-\omega^2 - (V^2R^2/16\alpha^2\omega^2)}\, d\omega \tag{1.33}$$

where

$$\omega = \frac{R}{2\sqrt{\alpha t}}$$

and

$$R = \sqrt{x^2 + y^2 + z^2} \tag{1.34}$$

This is the solution for temperature field at time t for half of the infinite sphere. Let t approach infinity, the steady-state temperature at (x, y, z) is:

$$T - T_o = \frac{\eta PQ}{2\pi kR} e^{(-V/2\alpha)(R-x)} \tag{1.35}$$

The following assumptions were made:

1. The specimen moves in x direction at constant velocity V.
2. Three-dimensional isotropic conduction in the workpiece occurs normal to the erosion front, and the beam/material interaction results in complete material removal at the vaporization temperature.
3. The constant thermal properties used are the volume fraction average values from the matrix and fiber material.
4. The material properties are independent of temperature and the material surface is relatively smooth.
5. The laser beam has a uniform intensity distribution and is independent of time.
6. The secondary heat source from oxidation reaction is neglected because of inert surrounding, and phase change, heat convection, and radiation are neglected.
7. The evaporated material does not interfere with the laser beam, and multiple reflection of laser radiation within the groove is neglected.
8. Point heat source targets on the surface of workpiece.

The HAZ boundary and W_d can thus be estimated by isotherm of the matrix char temperature $T_c = 360°C$ (633 K), as shown in Fig. 1.5.

$$W_d e^{VW_d/2\alpha} = \frac{\eta PQ}{2\pi k(T_c - T_o)} e^{Vx/2\alpha} \tag{1.36}$$

The differentiation of (1.36) with respect to k is:

$$\frac{\partial W_d}{\partial k} = \frac{\eta PQ e^{Vx/2\alpha}}{2\pi k^2 (T_c - T_o)} \left(e^{VW_d/2\alpha} + \frac{W_d V e^{VW_d/2\alpha}}{2\alpha} \right)^{-1} \left(\frac{V(W_d - x)}{2\alpha} - 1 \right) \tag{1.37}$$

Equation (1.37) shows both positive and negative slopes of W_d varying with k, which will be discussed later.

1.3.3.2 Orthotropic Analysis of Semi-infinite Body

Laser grooving relative to fiber axes is shown in Fig. 1.6. An anisotropic semi-infinite body subject to a moving point heat source is analyzed. The concept of material removal and energy balance is the same as stated in section above. The principal axis heat conduction in the workpiece is governed by [37]:

$$k_1 \frac{\partial^2 T}{\partial x^2} + k_2 \frac{\partial^2 T}{\partial y^2} + k_3 \frac{\partial^2 T}{\partial z^2} = \rho c \frac{\partial T}{\partial t} \tag{1.38}$$

To obtain the temperature field, the additional assumptions are made:

1. The thermal properties measured by laser flash method are independent of temperature.
2. The energy used in the vaporization process is removed before significant heat conduction occurs.
3. The material is assumed continuous since the weight loss at char temperature is only 4%.

Modest [2] expressed the heat balance for the control surface on the erosion front

$$\dot{q}_{\text{laser}} = \dot{q}_{\text{conduction}} + \dot{q}_{\text{convention}} + \dot{q}_{\text{radiation}} + \dot{q}_{\text{phase change}} \tag{1.39}$$

For laser grooving, (1.39) can be reduced to

$$\dot{q}_{\text{laser}} = \dot{q}_{\text{conduction}} + \dot{q}_{\text{phase change}} \tag{1.40a}$$

Equation (1.40a) is rearranged; hence heat flux for conduction is obtained by the following equation

$$\dot{q}_{\text{conduction}} = \dot{q}_{\text{laser}} - \dot{q}_{\text{phase change}} \tag{1.40b}$$

where

$$\dot{q}_{\text{phase change}} \approx \rho L_{\text{v}} V \tag{1.41}$$

The boundary conditions are:

$$\begin{gathered} T = T_{\text{o}} \text{ at } x = \pm\infty,\ y = \pm\infty,\ z = \pm\infty \quad \text{and} \\ -k_1 \frac{\partial T}{\partial x} - k_2 \frac{\partial T}{\partial y} - k_3 \frac{\partial T}{\partial z} = \eta Q \dot{q}_{\text{conduction}} \quad \text{at } x = Vt, y = 0, \text{ and } z = 0 \end{gathered} \tag{1.42}$$

Considering a point heat source moving along the x axis at velocity V subject to the boundary conditions, one can calculate the temperature at (x, y, z, t) resulted from the irradiated heat [37]:

$$T - T_{\text{o}} = \int_0^t \frac{\eta P Q (\rho c)^{1/2}}{8\left[\pi^3 (t - t')^3 k_1 k_2 k_3\right]^{1/2}} e^{-\frac{\rho c \left(\frac{(x - V(t - t'))^2}{k_1} + \frac{y^2}{k_2} + \frac{z^2}{k_3} \right)}{4(t - t')}} \mathrm{d}t' \tag{1.43}$$

In order to solve the anisotropic heat conduction problem, Carslaw and Jaeger made the substitutions of [37]

$$\xi = \sqrt{\frac{\kappa}{k_1}}x, \quad \Psi = \sqrt{\frac{\kappa}{k_2}}y, \quad \zeta = \sqrt{\frac{\kappa}{k_3}}z \quad \text{and} \quad U = \sqrt{\frac{\kappa}{k_1}}V \tag{1.44}$$

where κ is an arbitrary constant, (1.43) is then transformed to

$$T - T_o = \frac{\eta PQ(\rho c\alpha)^{3/2}}{2k\Re(\pi^3 k_1 k_2 k_3)^{1/2}} e^{U\xi/2\alpha} \int_{\omega}^{\infty} e^{-\omega^2-(U^2\Re^2/16\alpha^2\omega^2)}\,\mathrm{d}\omega \tag{1.45}$$

where

$$\omega = \frac{\Re}{2\sqrt{\alpha t}}$$

and

$$\Re = \sqrt{\xi^2 + \Psi^2 + \zeta^2} \tag{1.46}$$

This is the solution for temperature field at time t for half of the infinite sphere. Let t approach infinity, the steady-state temperature at (x, y, z) is:

$$T - T_o = \frac{\eta PQ}{2\pi(k_1k_2k_3)^{1/2}\sqrt{(x^2/k_1) + (y^2/k_2) + (z^2/k_3)}} e^{-(V\rho c(\sqrt{(x^2/k1)+(y^2/k_2)+(z^2/k_3)}-(x/\sqrt{k1}))/2\sqrt{k1})} \tag{1.47}$$

In the current study, the extent of HAZ is then calculated by the isotherm of the matrix char temperature. The modification of the assumed infinite body for a finite specimen is introduced in the next section.

1.3.3.3 Thin-Slab Analysis Using Mirror Image Method

For the workpiece of a thin slab with thickness D, an analytical solution will be sought under the assumption of negligible heat dissipation to outside at the bottom surface with respect to the heat input to the HAZ, i.e., $\partial T/\partial z = 0$ at $z = D$. However, the above analytical model of infinite body with $\partial T/\partial z = 0$ at $z = \infty$ does not provide the desired boundary condition at $z = D$. Hence a modification is needed to satisfy the thin slab boundary condition. Mirror Image Method can be employed for this purpose [49]. Figure 1.13 illustrates the sequence of superposition of mirror

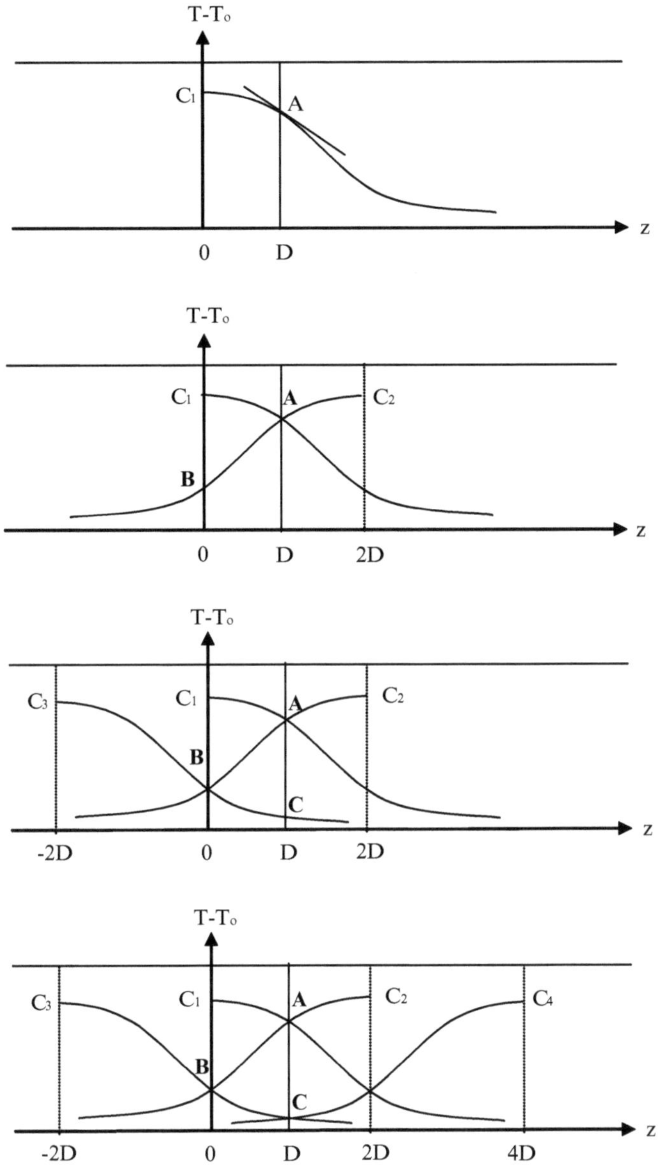

Fig. 1.13 Mirror image method

image to fulfill the required boundary conditions at $z = 0$ and $z = D$ alternatingly (designated in Fig. 1.13 as $A, B, C, \ldots$). The calculated temperature field is:

$$T - T_o = -\Phi_o(x, y, z) + \sum_{n=0}^{\infty} [\Phi_n(x, y, z_n) + \Phi'_n(x, y, z'_n)] \tag{1.48}$$

where

$$z_n = 2nD - z, \quad z'_n = 2nD + z, \quad n = 0, 1, 2, 3, \ldots \tag{1.49}$$

Substitute (1.47) into (1.48), one obtains the temperature

$$T - T_o = \frac{\eta PQ e^{(V\rho cx/2k_1)}}{2\pi(k_1k_2k_3)^{1/2}} \left(\frac{e^{-(V\rho c(\gamma)/2\sqrt{k_1})}}{\gamma} + \sum_{n=1}^{\infty} \left(\frac{e^{-(V\rho c(\gamma 2n)/2\sqrt{k_1})}}{\gamma_{2n}} + \frac{e^{-(V\rho c(\gamma 2n+1)/2\sqrt{k_1})}}{\gamma_{2n+1}} \right) \right) \tag{1.50}$$

where

$$\gamma = \sqrt{\frac{x^2}{k_1} + \frac{y^2}{k_2} + \frac{z^2}{k_3}}, \quad \gamma_{2n} = \sqrt{\frac{x^2}{k_1} + \frac{y^2}{k_2} + \frac{(2nD - z)^2}{k_3}}, \quad \text{and}$$

$$\gamma_{2n+1} = \sqrt{\frac{x^2}{k_1} + \frac{y^2}{k_2} + \frac{(2nD + z)^2}{k_3}} \tag{1.51}$$

1.3.3.4 Analysis Considering Immersed Heat Source

The location of the heat source is often not at $z = 0$ as adopted in previous analysis. The real shape, size, and intensity distribution of the heat source are yet to elaborate involving many unknowns. One approximation is that the laser beam hits on the evaporation depth followed by heat conduction from that location generating material removal and HAZ. For accuracy and simplicity in the following calculation, the heat source is considered moving at evaporation depth (z_o) where material is evaporated by laser energy. Hence (1.47) can be modified as:

$$T - T_o = \frac{\eta PQ}{2\pi(k_1k_2k_3)^{1/2}\sqrt{(x^2/k_1) + (y^2/k_2) + ((z - z_o)^2/k_3)}} e^{-(V\rho c(\sqrt{(x^2/k_1)+(y^2/k_2)+((z-zo)^2/k_3)}-(x/\sqrt{k_1}))/2\sqrt{k_1})} \tag{1.52}$$

and use Mirror Image Method for thin slab

$$T - T_o = -\Phi_o(x, y, z) + \sum_{n=0}^{\infty} [\Phi_n(x, y, z_n) + \Phi'_n(x, y, z'_n)] \tag{1.53}$$

where

$$z_n = 2nD - z, \quad z'_n = 2nD + z, \quad n = 0, 1, 2, 3, \ldots \tag{1.54}$$

Substitute (1.52) into (1.53), one rewrites (1.53) as

$$T - T_o = \frac{\eta P Q e^{(V\rho c x/2k_1)}}{2\pi(k_1 k_2 k_3)^{1/2}} \left(\frac{e^{-(V\rho c(\gamma)/2\sqrt{k_1})}}{\gamma} + \sum_{n=1}^{\infty} \left(\frac{e^{-(V\rho c(\gamma 2n)/2\sqrt{k_1})}}{\gamma 2n} + \frac{e^{-(V\rho c(\gamma 2n+1)/2\sqrt{k_1})}}{\gamma 2n + 1} \right) \right) \tag{1.55}$$

where

$$\gamma = \sqrt{\frac{x^2}{k_1} + \frac{y^2}{k_2} + \frac{(z - z_o)^2}{k_3}}, \quad \gamma 2n = \sqrt{\frac{x^2}{k_1} + \frac{y^2}{k_2} + \frac{(2nD - (z - z_o))^2}{k_3}}, \quad \text{and}$$

$$\gamma 2n + 1 = \sqrt{\frac{x^2}{k_1} + \frac{y^2}{k_2} + \frac{(2nD + (z - z_o))^2}{k_3}} \tag{1.56}$$

Iterations between the groove depth and the extent of HAZ will be needed until a satisfactory convergence is reached.

1.3.3.5 Temperature-Dependent Thermal Conductivity

According to Pilling's study, the conductivities k_1, k_2, and k_3 vary with temperature [50]. The temperature-dependent conductivity, however, has not been taken into consideration in the preceding simulation. It will be considered in the following.

In this section, the nonlinear governing equation considering the temperature-dependent thermal conductivity $k(T)$ is:

$$\frac{\partial}{\partial x}\left(k_1 \frac{\partial T}{\partial x}\right) + \frac{\partial}{\partial y}\left(k_2 \frac{\partial T}{\partial y}\right) + \frac{\partial}{\partial z}\left(k_3 \frac{\partial T}{\partial z}\right) = \rho c \frac{\partial T}{\partial t} \tag{1.57}$$

To derive the temperature field, further assumptions are made:

1. The thermal conductivity (k) is dependent on temperature, while the thermal diffusivity (α) is constant. One notices that the specific heat and density show similar temperature-dependency as conductivity; hence the derived diffusivity varies with temperature much less significantly [50, 51].
2. A proportionality between the anisotropic conductivities exists, or the conductivities vary with temperature parallelly through at different level of magnitude.

In grooving parallel to fiber axis,

$$k_1(T) = \beta \times k_2(T) = \beta \times k_3(T) \tag{1.58}$$

and in the case of grooving perpendicular to fiber axis,

$$k_2(T) = \beta \times k_1(T) = \beta \times k_3(T) \tag{1.59}$$

where β is the proportionality constant, ($\beta = 4.5/0.67 = 6.716$), derived in Table 1.5.

Kirchoff defined a transformed temperature [37]:

$$\Theta = \frac{1}{\bar{k}_o} \int_{T_o}^{T} k(\tau)\, \mathrm{d}\tau \tag{1.60}$$

where $k(\tau)$ is the temperature-dependent thermal conductivity [52], while $\bar{k}_o$ is the thermal conductivity at temperature T_o.

Introducing (1.60) into (1.57), one obtains a concept form of governing equation with the new variable Θ:

$$\bar{k}_1 \frac{\partial^2 \Theta}{\partial x^2} + \bar{k}_2 \frac{\partial^2 \Theta}{\partial y^2} + \bar{k}_3 \frac{\partial^2 \Theta}{\partial z^2} = \rho c \frac{\partial \Theta}{\partial t} \tag{1.61}$$

or

$$\frac{\partial^2 \Theta}{\partial x^2} + \beta \frac{\partial^2 \Theta}{\partial y^2} + \beta \frac{\partial^2 \Theta}{\partial z^2} = \frac{1}{\alpha} \frac{\partial \Theta}{\partial t} \tag{1.62}$$

where

$$\bar{k}_1 = \beta \bar{k}_2 = \beta \bar{k}_3 \tag{1.63}$$

The temperature at (x, y, z, t) produced from the irradiated heat is [37]:

$$\Theta = \int_0^t \frac{\eta PQ(\rho c)^{1/2}}{8\left[\pi^3 (t-t')^3 \bar{k}_1 \bar{k}_2 \bar{k}_3\right]^{1/2}} e^{-\rho c\left(\frac{(x-V(t-t'))^2}{k1}+\frac{y^2}{k2}+\frac{z^2}{k3}\right)/4(t-t')} \cdot \mathrm{d}t' \tag{1.64}$$

In order to solve the anisotropic heat conduction problem, the substitution proposed by Carslaw and Jaeger is used [37]:

$$\bar{\xi} = \sqrt{\frac{\bar{\kappa}}{\bar{k}_1}} x, \quad \bar{\psi} = \sqrt{\frac{\bar{\kappa}}{\bar{k}_2}} y, \quad \bar{\zeta} = \sqrt{\frac{\bar{\kappa}}{\bar{k}_3}} z \quad \text{and} \quad \bar{U} = \sqrt{\frac{\bar{\kappa}}{\bar{k}_1}} V \tag{1.65}$$

where $\bar{\kappa}$ is an arbitrarily constant.

Equation (1.64) is then transformed to:

$$\Theta = \frac{\eta PQ(\rho c\alpha)^{3/2}}{2\bar{\kappa}\bar{\Re}(\pi^3\bar{k}_1\bar{k}_2\bar{k}_3)^{1/2}} e^{\bar{U}\bar{\xi}/2\alpha} \int_{\bar{\omega}}^{\infty} e^{-\omega^2-(\bar{U}^2\bar{\Re}^2/16\alpha^2\omega^2)}\, \mathrm{d}\omega \tag{1.66}$$

where

$$\bar{\omega} = \frac{\bar{\Re}}{2\sqrt{\alpha t}} \quad \text{and} \quad \bar{\Re} = \sqrt{\bar{\xi}^2 + \bar{\psi}^2 + \bar{\zeta}^2} \tag{1.67}$$

This is the solution for temperature field at time t. Let t approach infinity, the steady-state temperature at (x, y, z) is:

$$\Theta = \frac{\eta PQ}{2\pi(\bar{k}_1\bar{k}_2\bar{k}_3)^{1/2}\sqrt{(x^2/\bar{k}_1) + (y^2/\bar{k}_2) + (z^2/\bar{k}_3)}}\, e^{-(V\rho c(\sqrt{(x^2/\bar{k}_1)+(y^2/\bar{k}_2)+(z^2/\bar{k}_3)}-(x/\sqrt{\bar{k}1}))/2\sqrt{\bar{k}1})} \tag{1.68}$$

Use Mirror Image Method for thin slab

$$\Theta = -\Phi_{\mathrm{o}}(x, y, z) + \sum_{n=0}^{\infty} [\Phi_n(x, y, z_n) + \Phi_n(x, y, z_n)] \tag{1.69}$$

where

$$z_n = 2nD - z, \quad z'_n = 2nD + z, \quad n = 0, 1, 2, 3, \ldots \tag{1.70}$$

Substitute (1.68) into (1.69), one obtains the transformed temperature:

$$\Theta = \frac{\eta PQ e^{(V\rho cx/2\bar{k}_1)}}{2\pi(\bar{k}_1\bar{k}_2\bar{k}_3)^{1/2}} \left(\frac{e^{-(V\rho c(\gamma)/2\sqrt{\bar{k}_1})}}{\gamma} + \sum_{n=1}^{\infty} \left(\frac{e^{-(V\rho c(\gamma 2n)/2\sqrt{\bar{k}_1})}}{\gamma 2n} + \frac{e^{-(V\rho c(\gamma 2n+1)/2\sqrt{\bar{k}_1})}}{\gamma 2n+1} \right) \right) \tag{1.71}$$

where

$$\gamma = \sqrt{\frac{x^2}{\bar{k}_1} + \frac{y^2}{\bar{k}_2} + \frac{z^2}{\bar{k}_3}}, \quad \gamma 2n = \sqrt{\frac{x^2}{\bar{k}_1} + \frac{y^2}{\bar{k}_2} + \frac{(2nD - z)^2}{\bar{k}_3}} \quad \text{and}$$

$$\gamma 2n + 1 = \sqrt{\frac{x^2}{\bar{k}_1} + \frac{y^2}{\bar{k}_2} + \frac{(2nD + z)^2}{\bar{k}_3}} \tag{1.72}$$

and the heat source is considered moving at evaporation depth z_o. Hence (1.71) can be modified as:

$$\Theta = \frac{\eta P Q e^{V\rho c x/2\bar{k}_1}}{2\pi(\bar{k}_1\bar{k}_2\bar{k}_3)^{1/2}}\left(\frac{e^{-(V\rho c(\gamma)/2\sqrt{\bar{k}_1})}}{\gamma}+\sum_{n=1}^{\infty}\left(\frac{e^{-(V\rho c(\gamma 2n)/2\sqrt{\bar{k}_1})}}{\gamma 2n}+\frac{e^{-(V\rho c(\gamma 2n+1)/2\sqrt{\bar{k}_1})}}{\gamma 2n+1}\right)\right) \tag{1.73}$$

where

$$\gamma=\sqrt{\frac{x^2}{\bar{k}_1}+\frac{y^2}{\bar{k}_2}+\frac{(z-z_o)^2}{\bar{k}_3}},\quad \gamma 2n=\sqrt{\frac{x^2}{\bar{k}_1}+\frac{y^2}{\bar{k}_2}+\frac{(2nD-(z-z_o))^2}{\bar{k}_3}}\quad \text{and}$$

$$\gamma 2n+1=\sqrt{\frac{x^2}{\bar{k}_1}+\frac{y^2}{\bar{k}_2}+\frac{(2nD+(z-z_o))^2}{\bar{k}_3}} \tag{1.74}$$

The extent of HAZ is then determined by the isotherm of the matrix char temperature. Iterations between the evaporation depth and the extent of HAZ can be carried out until a satisfactory convergence is reached.

1.3.4 HAZ for Orthotropic Machining

1.3.4.1 Isotropic Semi-infinite Body Model

The Overprediction

Equation (1.36) reveals that W_d of theoretical values are larger than that of experimental results when grooving parallel and perpendicular to fiber orientation, as shown in Fig. 1.14. The parameters used in (1.36) are taken from Table 1.5 for the thermal conductivity $k = k_1 = k_2 = k_3 = 0.67$ W/mK (transverse to fibers) for parallel grooving and $k = k_1 = k_2 = k_3 = 4.5$ W/mK (along fibers) for perpendicular grooving. Density (ρ) is 1,530 kg/m^3 and specific heat (C) is 950 J/kg K, as listed in Table 1.5.

The deviation is based on the assumed representative single conductivity for an anisotropic material. Since carbon fiber has a much lower thermal conductivity in the transverse direction, and heat is conducted along the fibers more than the isotropic model expects; hence theoretical values for the W_d of HAZ are overpredicted both in parallel grooving and in perpendicular grooving. These results suggest the need for an anisotropic analysis.

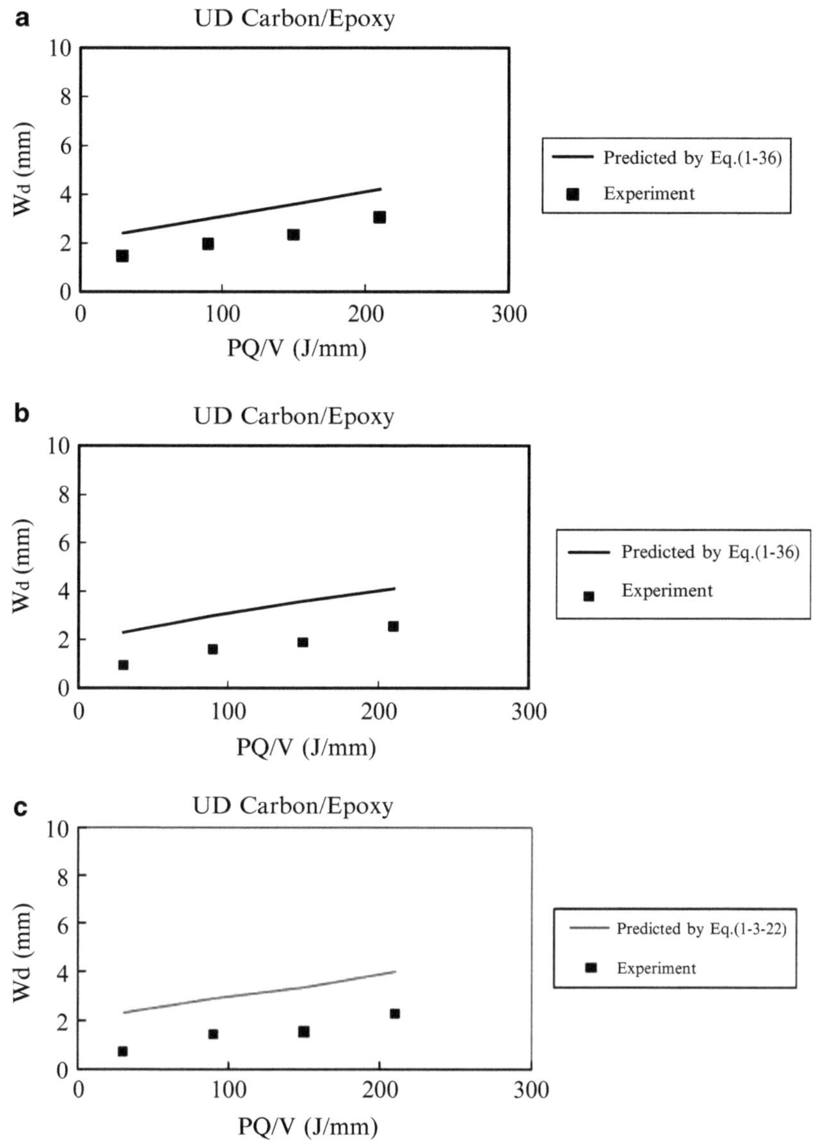

Fig. 1.14 Experimental and simulation of HAZ in laser grooving at different levels of *PQ/V*. (**a**) Parallel grooving ($T_o = 30°C$) [43, 53], reprinted with permission. (**b**) Parallel grooving ($T_o = -30°C$). (**c**) Parallel grooving ($T_o = -90°C$). (**d**) Perpendicular grooving ($T_o = 30°C$) [43, 53], reprinted with permission. (**e**) Perpendicular grooving ($T_o = -30°C$). (**f**) Perpendicular grooving ($T_o = -90°C$)

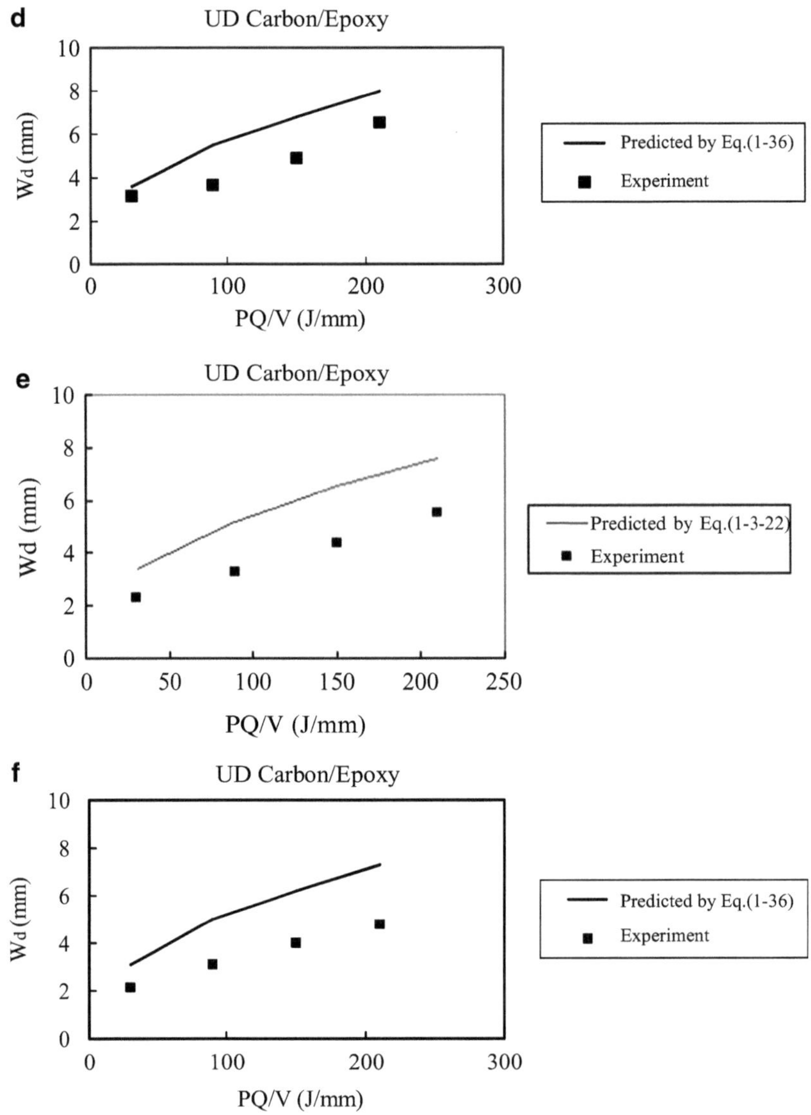

Fig. 1.14 (continued)

The Effects of Thermal Conductivity and Char Temperature

Equation (1.37) shows both positive and negative slopes of W_d varying with k, as shown in Fig. 1.15. The principal conductivities in Table 1.5 all lie on the positive slope; hence grooving perpendicular to fiber axis should produce larger HAZ than parallel grooving. However, from (1.37), one realizes that HAZ is not solely determined by thermal conductivity. The char temperature (T_c) defining the

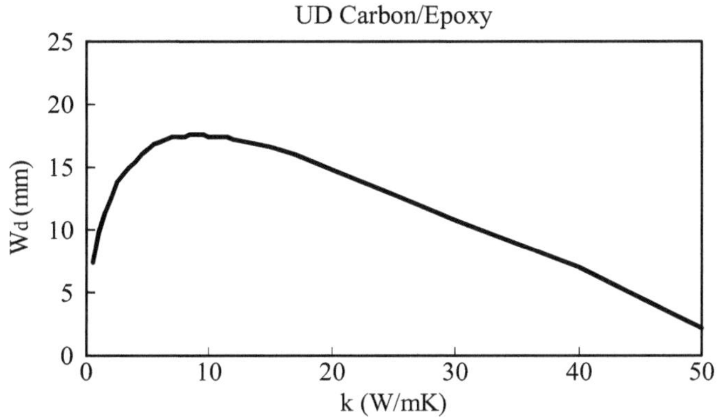

Fig. 1.15 Effects of thermal conductivity on the extent of HAZ ($T_o = 30°C$, $T_c = 360°C$, $P = 525$ W, $Q = 100$ %, $V = 2.5$ mm/s) [43], reprinted with permission

occurrence of HAZ also affects, as discussed in the following. Larger thermal conductivity distributes heat more effectively; the resulted temperature gradient is milder, as shown in Fig. 1.16. In two cases of higher and lower level of the isotherm temperature, the produced W_d of HAZ can increase or decrease with increasing thermal conductivity (k). For the current anisotropic material, different values of k are used in the cases of grooving parallel and perpendicular to fiber axis. Though the current test results show that grooving parallel to fiber will produce less thermal damage to material, care should be taken in considering the effects of both T_c and the conductivity corresponding to various grooving directions. The undesired grooving direction resulting in large HAZ is also a function of T_c.

It shows that grooving perpendicular to fiber orientation is more severely affected than grooving parallel to fiber orientation in the current experiment. Since carbon fiber has a much higher thermal conductivity (nearly two orders of magnitude higher) than the polymer matrix, the thermal conduction is basically along the fiber direction. Consequently, the resulted HAZ is more extensive in that direction.

1.3.4.2 Anisotropic Thin Slab Model

In this study, UD Carbon/Epoxy, GlassMat/PP, and [0/90] laminates, including Aramid/PP, Glass/PE, and Glass/PP, were grooved along the principal axis to evaluate the HAZ.

By the substitution of $T = T_c$, and T_v, as shown in Appendix C, one can solve the extent of HAZ and z_o in setting $x =$ constant by running iterative calculation of the y and z coordinates on the isotherm of $T = T_c$ and T_v. Iterations for the z_o and HAZ are conducted until a satisfactory convergence is reached. Since the specimen is grooved into a certain depth beneath the top surface, the location of the heat source is theoretically at z_o. Hence the heat source is considered moving at groove depth

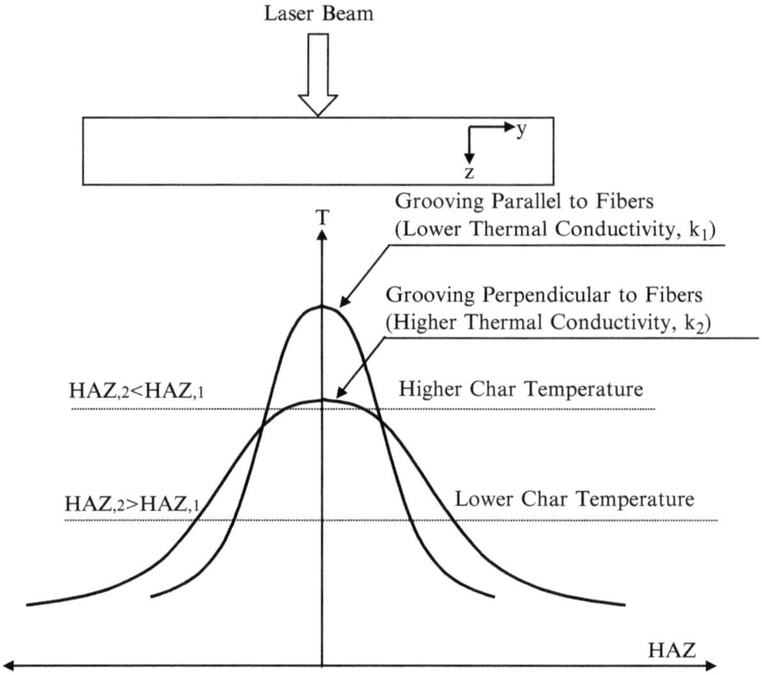

Fig. 1.16 The effects of thermal conductivity and char temperature on the extent of HAZ [43], reprinted with permission

for accuracy and simplicity in the calculation. Figure 1.17b, d shows that the resulted isotherms extend deeper in *z* direction when the effects of immersed heat source are considered.

HAZ seems to remain as long as the workpiece has to be grooved at a certain level of specific laser energy. The existing reference reported that the HAZ is dependent on the laser energy per unit length of grooving (*PQ/V*) [25]. The simulations of HAZ for grooving UD, Mat, and [0/90] laminates are compared with experimental results. Figures 1.18 and 1.19 show that HAZ is indeed approximately proportional to the specific laser energy *PQ/V*.

As far as thermal conductivity is concerned, the thermal conductivity of composite material along the fiber is larger than that transverse to the fiber (refer to Tables 1.5, 1.6, and 1.7). It is worth noticing that Aramid/PP has smallest conductivities of all composite material listed in Table 1.7, while Carbon/Epoxy has largest ones. Thus, grooving of Aramid/PP produces narrower HAZ. Similarly, for a certain composite material system, take Carbon/Epoxy for example, grooving perpendicular to fiber orientation produces larger HAZ than grooving parallel to fiber orientation.

On the other hand, HAZ in grooving of Glass/PE, Glass/PP, and GlassMat/PP is between the two extreme cases of Aramid/PP and Carbon/Epoxy, as shown in Fig. 1.18. The current result shows that Aramid/PP is a more suitable composite material for laser grooving than the others due to its smaller thermal conductivities and the induced HAZ, as discussed above.

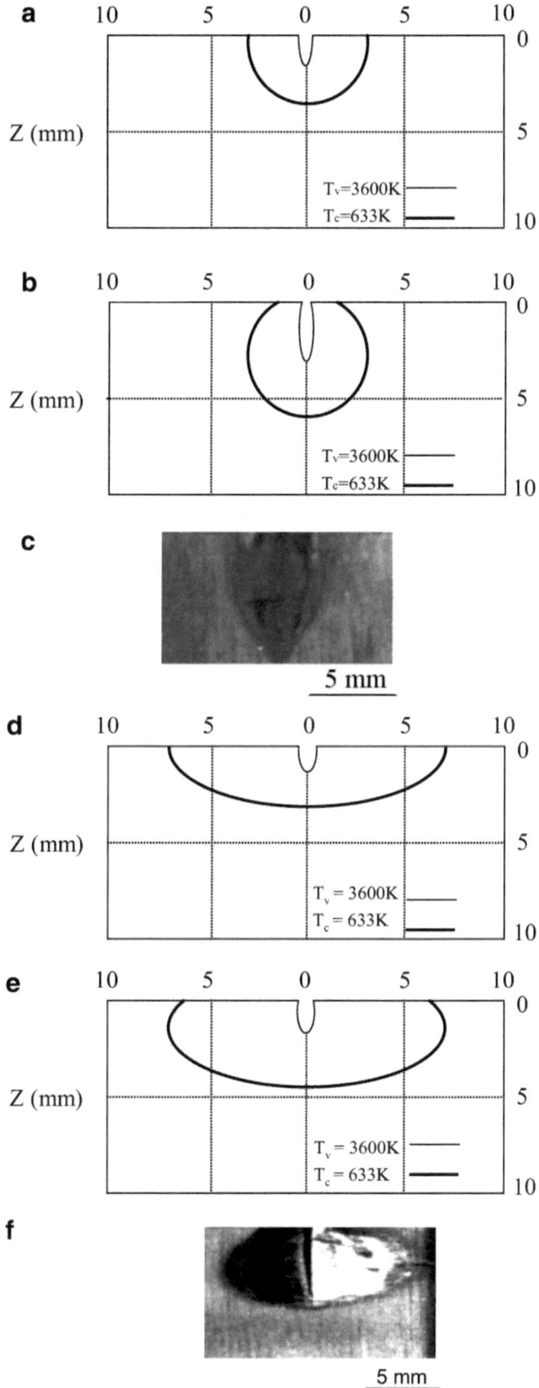

Fig. 1.17 Comparison between simulations of HAZ for carbon/epoxy ($T_o = 30°C$, $PQ/V = 210$ J/mm). (**a**) Simulation of parallel grooving by (1.50). (**b**) Simulation of parallel grooving by (1.55). (**c**) Experimental result of parallel grooving [54], reprinted with permission. (**d**) Simulation of perpendicular grooving by (1.50). (**e**) Simulation of perpendicular grooving by (1.55). (**f**) Experimental result of perpendicular grooving

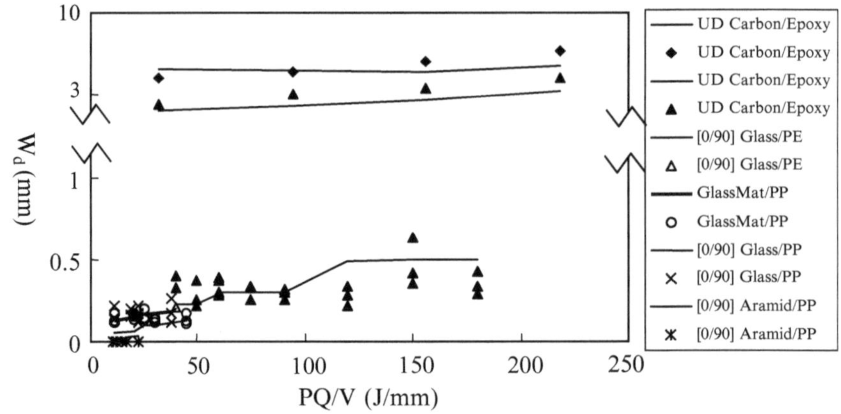

Fig. 1.18 HAZ in laser grooving of various composite materials at $T_o = 30°C$ (simulation by (1.55))

1.3.4.3 Temperature-Dependent Thermal Conductivity

The value of carbon fiber conductivity at temperature exceeding 300°C (573 K) is drastically larger than that at room temperature because of crystallinity [52]. However, the effect of temperature-dependent thermal conductivity was not considered in preceding simulation. The current simulation using constant thermal conductivity shows underpredicted HAZ at higher laser power level and overpredicted HAZ at lower laser power level for both 0° and 90° grooving of Carbon/Epoxy, as shown in Fig. 1.19. Since the laser grooving of Carbon/Epoxy reaches high temperature, the thermal conductivity is significantly increased; hence the actual heat conduction in workpiece is drastically increased at the laser grooving temperature, causing the inaccurate prediction.

Considering the difficulty in measuring the temperature-dependent thermal conductivities of composite materials, a temperature-dependent thermal conductivity based on the existing reference is assumed [52]. Figure 1.19 shows the comparison between the predictions of HAZ by considering the temperature-dependent and the temperature-independent thermal conductivity. It reveals that consideration of the temperature-dependent thermal conductivity on HAZ can improve the prediction, particularly in several cases of low and high levels of laser power.

1.4 FDM Scheme

In this chapter, the numerical finite difference method (FDM) is used to predict the HAZ for this situation. When the cross-derivative terms of thermal conductivity are absent from the heat conduction equation, the analysis of heat transfer can be simplified to the particular case of principal-axis grooving, as discussed in previous section.

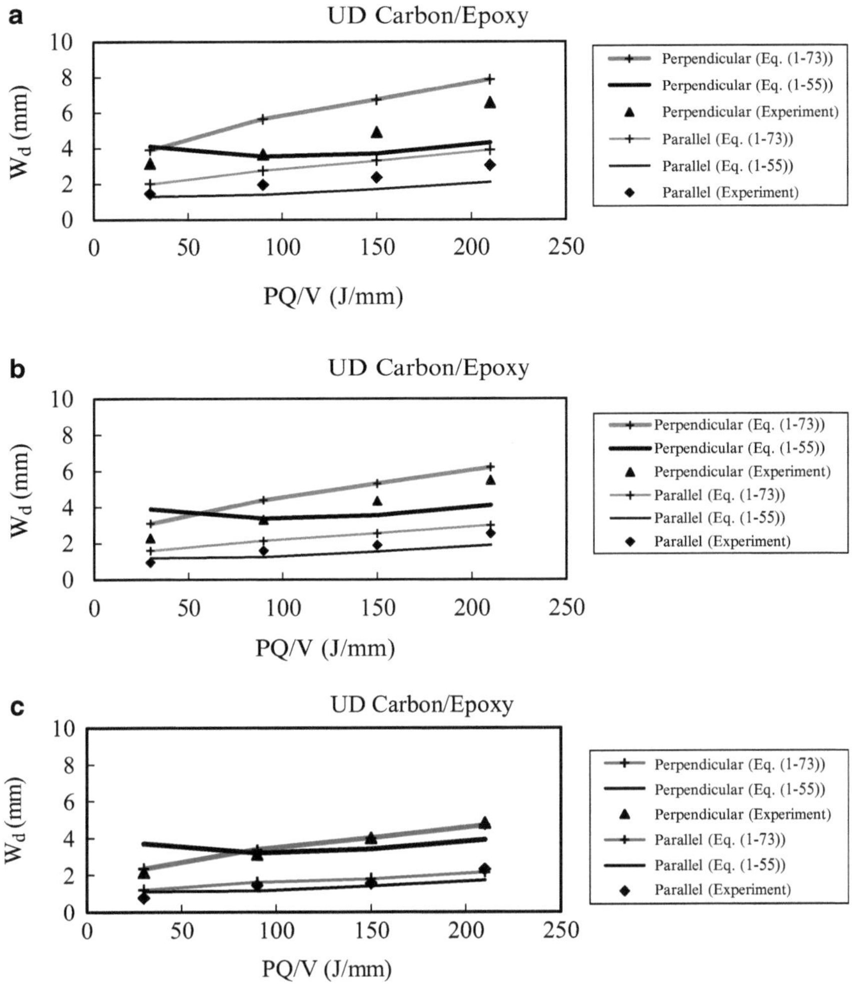

Fig. 1.19 Comparison between experimental and simulation results of HAZ using constant and temperature-dependent thermal conductivity. (**a**) $T_o = 30°C$ [43], reprinted with permission, (**b**) $T_o = -30°C$, (**c**) $T_o = -90°C$

1.4.1 Thermal Conductivity in Nonprincipal Directions

The Laser Flash Method also can be used to measure the thermal conductivity of UD, MatUD, and [0/90] laminates in any direction. However, the experimental setup and preparation for test sample are expensive and time-consuming. In the following section, a general experimental and a theoretical method to determine thermal conductivity is presented.

Table 1.6 Experimentally measured thermal conductivity at room temperature for unidirectional composite material

Materials	Volume fraction V_{fiber}/V_{matrix} (%)	Parallel conductivity $k_{parallel}$ (W/mK)	Transverse conductivity $k_{transverse}$ (W/mK)	References
Carbon/epoxy	46/54	–	0.607	[50]
	55/45	4.62	0.72	[36]
	59/41	6.39	0.747	[50]
	62/38	5.73	–	[50]
	72/28	–	1.133	[50]
Glass/epoxy	55/45	0.685	0.429	[36]

Table 1.7 Calculated thermal conductivity of composite materials [39]

Materials	Volume fraction V_{fiber}/V_{matrix} (%)	Thermal conductivity k_{finer}/k_{matrix} (W/mK)	Composite conductivity k (W/mK)	Method
Aramid/PP	18/82	0.1/0.22	$k_x = 0.189$	Equation (1.21)
			$k_y = 0.189$	Equation (1.22)
			$k_z = 0.181$	Equation (1.23)
Carbon/epoxy (0 grooving)	50/50	8.5/0.35	$k_x = 4.425$	Equation (1.1)
			$k_y = 0.672$	Equation (1.7)
			$k_z = 0.672$	Equation (1.7)
Glass/PE	23.3/76.7	0.8/0.5	$k_x = 0.559$	Equation (1.21)
			$k_y = 0.559$	Equation (1.22)
			$k_z = 0.548$	Equation (1.23)
Glass/PP	34/66	0.8/0.22	$k_x = 0.355$	Equation (1.21)
			$k_y = 0.355$	Equation (1.22)
			$k_z = 0.292$	Equation (1.23)
GlassMat20/PP	8/92	0.8/0.22	$k_x = 0.241$	Equation (1.24)
			$k_y = 0.241$	Equation (1.24)
			$k_z = 0.241$	Equation (1.24)
GlassMat30/PP	13/87	0.8/0.22	$k_x = 0.255$	Equation (1.24)
			$k_y = 0.255$	Equation (1.24)
			$k_z = 0.255$	Equation (1.24)
GlassMat40/PP	19/81	0.8/0.22	$k_x = 0.273$	Equation (1.24)
			$k_y = 0.273$	Equation (1.24)
			$k_z = 0.273$	Equation (1.24)
GlassMat50/PP	26/74	0.8/0.22	$k_x = 0.296$	Equation (1.24)
			$k_y = 0.296$	Equation (1.24)
			$k_z = 0.296$	Equation (1.24)
GlassMat20 UD20/PP (90 grooving)	21.2/78.8	0.8/0.22	$k_x = 0.271$	Equation (1.27)
			$k_y = 0.313$	Equation (1.28)
			$k_z = 0.271$	Equation (1.29)

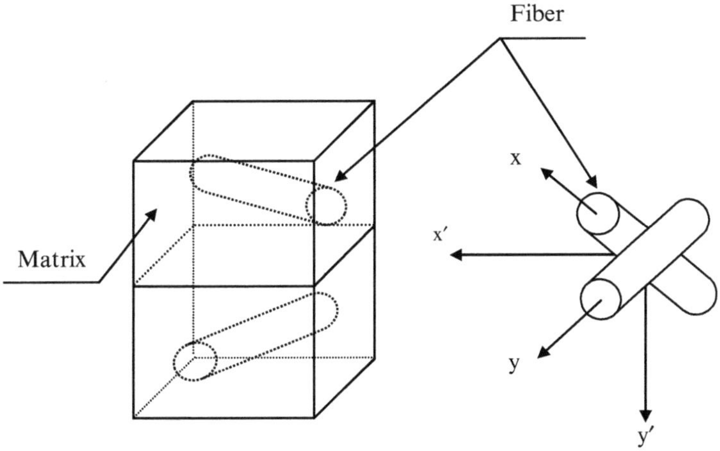

Fig. 1.20 Illustration of the previous axes (x, y) and the new axes (x', y')

1.4.1.1 Generalized Model of Resistor

In the case of nonprincipal axis grooving, thermal conductivity along the grooving direction can be obtained by a coordinate-transformation function based on the resistor model [31], as discussed below.

One considers the heat flux in an anisotropic medium using indicial notation

$$q_i = -k_{ij}\frac{\partial T}{\partial x_j}, \quad i, j = 1,2,3 \tag{1.75}$$

Let C_{ij} denotes the cosine of the angle between the new axis x_i' and the previous axis x_j; the coordinates with respect to the new coordinates, as shown in Fig. 1.20, are given by

$$x_i' = C_{ij}x_j, \quad i, j = 1,2,3 \tag{1.76}$$

Also

$$x_i = C_{ji}x_j', \quad i, j = 1,2,3 \tag{1.77}$$

Similar to (1.76), q_i' in the new coordinates is written in terms of the previous q_m as

$$q_i' = C_{im}q_m, \quad i, m = 1,2,3 \tag{1.78}$$

According to (1.75), q_m is given by

$$q_m = -k_{ml}\frac{\partial T}{\partial x_l}, \quad l,m = 1,2,3 \tag{1.79}$$

where

$$\frac{\partial T}{\partial x_l} = C_{jl}\frac{\partial T}{\partial x'_j}, \quad j,\ l = 1,2,3 \tag{1.80}$$

Equation (1.78), (1.79), and (1.80) give

$$q'_i = C_{im}q_m = -C_{im}k_{ml}\frac{\partial T}{\partial x_l} = -C_{im}k_{ml}C_{jl}\frac{\partial T}{\partial x'_j} \tag{1.81}$$

Compare (1.75) with (1.81), one finds that the new conductivity coefficients k'_{ij} are determined as

$$k'_{ij} = C_{im}C_{jl}k_{ml} \tag{1.82}$$

Equation (1.82) can be transformed to the principal axis [37]

$$k_1\xi_1{}^2 + k_2\xi_2^2 + k_3\xi_3^2 = 1 \tag{1.83}$$

For unidirectional composite materials $k_1 = k_{\text{parallel}}$ and $k_2 = k_3 = k_{\text{transverse}}$, (1.83) directly determines the thermal conductivity in any direction k_θ. One can apply the polar coordinate transformation

$$r_\theta^2 = \frac{1}{k_{\text{parallel}}\cos^2\theta + k_{\text{transverse}}\sin^2\theta} \tag{1.84}$$

where r_θ is radius of polar coordinate system, as shown in Fig. 1.21, hence

$$k_\theta = k_{\text{parallel}}\cos^2\theta + k_{\text{transverse}}\sin^2\theta \tag{1.85}$$

1.4.1.2 Model of Isotherm

Using Laser Flash Method to determine thermal conductivity is time-consuming and costly. In this section, an experimental method to determine the thermal conductivity is introduced.

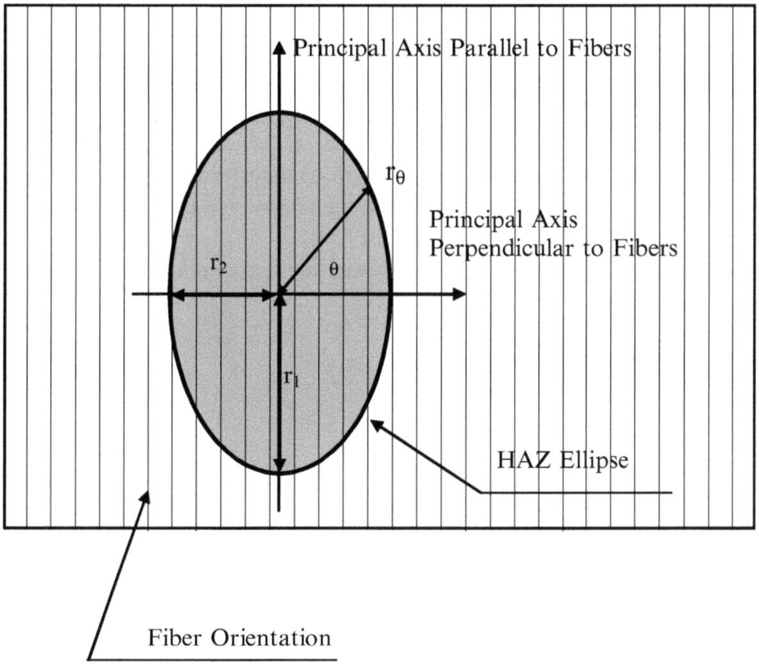

Fig. 1.21 Schematic of the laser-induced HAZ ellipse in UD laminates [41], reprinted with permission

Due to the significant anisotropic thermal conductivity of the composite, the temperature rise and HAZ during laser machining will also be anisotropic, showing an elliptical shape, as illustrated in Fig. 1.21.

On the other hand, the heat conduction occurs as a result of the temperature gradient in workpiece and is dependent on the material thermal diffusivity (α) and the interaction time (t). Both define the thermal penetration depth [24].

$$\delta = \sqrt{\alpha t} \tag{1.86}$$

The ratio between the depths in parallel and transverse directions is

$$\frac{\delta_2}{\delta_1} = \frac{r_2}{r_1} = \frac{\sqrt{\alpha_2 t}}{\sqrt{\alpha_1 t}} = \frac{\sqrt{k_2}}{\sqrt{k_1}} \tag{1.87}$$

One notices that an isotherm is reached at the same time of t. Equation (1.87) can be rewritten as

$$k_2 = \left(\frac{r_2}{r_1}\right)^2 k_1$$

and

$$k_3 = k_2 \tag{1.88}$$

where r_1 and r_2 are the radii of the ellipse, as shown in Fig. 1.21. While k_1 is the longitudinal conductivity (in the fiber direction) determined by Rule of Mixture in terms of the properties of the constituents, the transverse thermal conductivities of composites (k_2 and k_3) are obtained from (1.88).

In general, one can rewrite (1.88) to obtain the heat conductivity in any direction on the UD laminate

$$k_\theta = \frac{k_1}{(r_1/r_\theta)^2} \tag{1.89}$$

where r_1/r_θ is defined as thermal anisotropic factor.

1.4.1.3 Cross-Derivative Coefficient of Thermal Conductivity

When grooving along nonprincipal direction is conducted on an anisotropic composite material, the cross-derivative coefficients of thermal conductivity will be involved in the differential equation of heat conduction. This section presents the model to solve the cross-derivative terms of thermal conductivity.

Equation (1.75) can be written more explicitly in q_x, q_y, and q_z along the x, y, and z directions in the rectangular coordinate system

$$-q_x = k_{11}\frac{\partial T}{\partial x} + k_{12}\frac{\partial T}{\partial y} + k_{13}\frac{\partial T}{\partial z} \tag{1.90}$$

$$-q_y = k_{21}\frac{\partial T}{\partial x} + k_{22}\frac{\partial T}{\partial y} + k_{23}\frac{\partial T}{\partial z} \tag{1.91}$$

$$-q_z = k_{31}\frac{\partial T}{\partial x} + k_{32}\frac{\partial T}{\partial y} + k_{33}\frac{\partial T}{\partial z} \tag{1.92}$$

The differential equation of heat conduction for an anisotropic solid can be derived from (1.90) to (1.92) [37].

$$k_{11}\frac{\partial^2 T}{\partial x^2} + k_{22}\frac{\partial^2 T}{\partial y^2} + k_{33}\frac{\partial^2 T}{\partial z^2} + (k_{12}+k_{21})\frac{\partial^2 T}{\partial x\partial y} + (k_{13}+k_{31})\frac{\partial^2 T}{\partial x\partial z} + (k_{23}+k_{32})\frac{\partial^2 T}{\partial y\partial z} + \dot{q} = \rho C\frac{\partial T}{\partial \tau} \tag{1.93}$$

where the variables are defined as follows:

$k_{ij} = k_{ji}$ by the reciprocity law, Onsager Theorem, and
T: temperature
x,y,z: Cartesian coordinates
$\dot{q}$: energy generated per unit volume
k: thermal conductivity
α: thermal diffusivity
τ: time

The thermal conductivity of an anisotropic solid involves nine components, called the conductivity coefficients, in a second-order symmetric tensor

$$\bar{\bar{k}} = \begin{vmatrix} k_{11} & k_{12} & k_{13} \\ k_{21} & k_{22} & k_{23} \\ k_{31} & k_{32} & k_{33} \end{vmatrix} \tag{1.94}$$

where the principal conductivities k_1, k_2, and k_3 are the eigenvalues of the following equation:

$$\begin{vmatrix} k_{11} - \lambda & k_{12} & k_{13} \\ k_{21} & k_{22} - \lambda & k_{23} \\ k_{31} & k_{32} & k_{33} - \lambda \end{vmatrix} = 0 \tag{1.95}$$

where λ is the principal axis thermal conductivity.

Equation (1.95) is a cubic equation in λ. In the current study, the workpiece is symmetric with respect to z axis, thus, and the principal axis conductivities, such as those shown in Table 1.8, will satisfy

$$\begin{vmatrix} k_{11} - \lambda & k_{12} & 0 \\ k_{21} & k_{22} - \lambda & 0 \\ 0 & 0 & k_{33} - \lambda \end{vmatrix} = 0 \tag{1.96}$$

and through the principal axis transformation, one obtains

$$\bar{\bar{k}}_{\text{principal}} = \begin{vmatrix} k_1 & 0 & 0 \\ 0 & k_2 & 0 \\ 0 & 0 & k_3 \end{vmatrix} \tag{1.97}$$

1.4.1.4 Temperature-Dependent Thermal Conductivity

As discussed in previous section, the effect of temperature-dependent thermal conductivity on HAZ will be considered for nonprincipal-axis grooving of

Table 1.8 Thermal conductivity of UD carbon/epoxy

Grooving direction relative to fiber axis	Conductivity (W/mK)	Method
0°	$k_{11} = 4.50$	Measured by LFM
	$k_{12} = k_{21} = 0$	Equation (1.96)
	$k_{22} = 0.67$	Measured by LFM
30°	$k_{11} = 3.541$	Equation (1.85)
	$k_{11} = 3.243$	Equation (1.89)
	$k_{12} = k_{21} = 1.659$	Equations (1.85) and (1.96)
	$k_{12} = k_{21} = 1.798$	Equations (1.89) and (1.96)
	$k_{22} = 1.628$	Equation (1.85)
	$k_{22} = 1.926$	Equation (1.89)
45°	$k_{11} = 2.585$	Equation (1.85)
	$k_{11} = 2.367$	Equation (1.89)
	$k_{12} = k_{21} = 1.915$	Equations (1.85) and (1.96)
	$k_{12} = k_{21} = 1.697$	Equations (1.89) and (1.96)
	$k_{22} = 2.585$	Equation (1.85)
	$k_{22} = 2.367$	Equation (1.89)
60°	$k_{11} = 1.628$	Equation (1.85)
	$k_{11} = 1.926$	Equation (1.89)
	$k_{12} = k_{21} = 1.659$	Equations (1.85) and (1.96)
	$k_{12} = k_{21} = 1.798$	Equations (1.89) and (1.96)
	$k_{22} = 3.541$	Equation (1.85)
	$k_{22} = 3.243$	Equation (1.89)
90°	$k_{11} = 0.67$	Measured by LFM
	$k_{12} = k_{21} = 0$	Equation (1.96)
	$k_{22} = 4.50$	Measured by LFM

Carbon/Epoxy. The temperature-dependent thermal conductivity $k_{ij}(T)$ along the grooving direction can be obtained by

$$k_{ij}(T) = \beta_{ij} \times k_{\text{transverse}}(T), \quad i,\ j = 1, 2 \tag{1.98}$$

where b_{ij} is proportionality constant from Table 1.8.

1.4.2 Numerical Finite-Difference Approximation

One considers a 3D heat transfer in Fig. 1.22. The boundary conditions on the surfaces are

(I) In x direction
  At $x = 0$

$$-k_{11}\frac{\partial T}{\partial x} - k_{12}\frac{\partial T}{\partial y} + h(T - T_{\infty}) = 0 \tag{1.99}$$

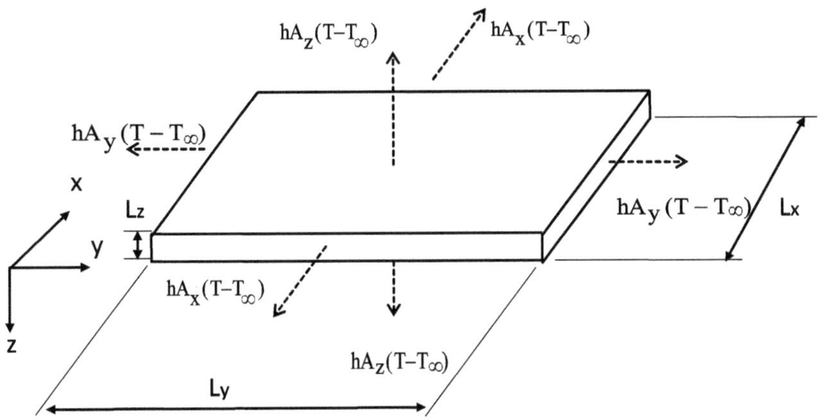

Fig. 1.22 Sketch of the heat flow in three dimensions [40, 41, 43], reprinted with permission

At $x = L_x$

$$k_{11}\frac{\partial T}{\partial x} + k_{12}\frac{\partial T}{\partial y} + h(T - T_\infty) = 0 \tag{1.100}$$

(II) In y direction
At $y = 0$

$$-k_{21}\frac{\partial T}{\partial x} - k_{22}\frac{\partial T}{\partial y} + h(T - T_\infty) = 0 \tag{1.101}$$

At $y = L_y$

$$k_{21}\frac{\partial T}{\partial x} + k_{22}\frac{\partial T}{\partial y} + h(T - T_\infty) = 0 \tag{1.102}$$

(III) In z direction
At $z = 0$

$$-k_{33}\frac{\partial T}{\partial z} + h(T - T_\infty) = 0 \tag{1.103}$$

At $z = L_z$

$$k_{33}\frac{\partial T}{\partial z} + h(T - T_\infty) = 0 \tag{1.104}$$

where h is heat convection coefficient.

The value of h required in FDM is calculated from [39] for the case of vertically impinging jet

$$h = \frac{13\mathrm{Re}^{0.5}\,\mathrm{Pr}\,n^{0.33}k_{\mathrm{gas}}}{B} \tag{1.105a}$$

where B is the jet plate distance,

$$\mathrm{Re} = \frac{V_{\mathrm{gas}}d_{\mathrm{nozzle}}}{\nu}\ \text{(Reynolds number at jet exit)}$$

$$\mathrm{Pr}\,n = \frac{\nu}{\alpha}$$

k_{gas} is conductivity of assisted gas, and ν is kinematic viscosity of the gas.

In order to simplify the mathematics, the Bernoulli equation for a frictionless flow is used to determine the gas velocity at the nozzle exit.

$$V_{\mathrm{gas}} = \sqrt{2\left(\frac{P_{\mathrm{gas}} - P_{\mathrm{ambient}}}{\rho_{\mathrm{gas}}}\right)} \tag{1.105b}$$

where P_{ambient} is the ambient pressure.

The nodal equations for central and mixed partial derivatives are

$$\frac{\partial^2 T}{\partial x^2} = \frac{(T_{m+1,n,l} + T_{m-1,n,l} - 2T_{m,n,l})}{(\Delta x)^2} + O[(\Delta x)^2] \tag{1.106}$$

$$\frac{\partial^2 T}{\partial y^2} = \frac{(T_{m,n+1,l} + T_{m,n-1,l} - 2T_{m,n,l})}{(\Delta y)^2} + O[(\Delta y)^2] \tag{1.107}$$

$$\frac{\partial^2 T}{\partial z^2} = \frac{(T_{m,n,l+1} + T_{m,n,l-1} - 2T_{m,n,l})}{(\Delta z)^2} + O[(\Delta z)^2] \tag{1.108}$$

$$\frac{\partial^2 T}{\partial x \partial y} = \frac{1}{2\Delta x}\left(\frac{T_{m+1,n+1,l} - T_{m+1,n-1,l}}{2\Delta y} - \frac{T_{m-1,n+1,l} - T_{m-1,n-1,l}}{2\Delta y}\right) + O[(\Delta x)^2 + (\Delta y)^2] \tag{1.109}$$

$$\frac{\partial T}{\partial t} = \frac{(T^{p+1}_{m,n,l} - T^{p}_{m,n,l})}{\Delta t} + O[\Delta t] \tag{1.110}$$

where the subscript m, n, and l denote the x, y, and z position, respectively.

One can make a transient energy balance on the node (m, n, l) by setting the sum of the energy conducted and convected into the node equal to the increase in the internal energy of the node. Some representative equations are listed below

(I) For surface of $x = 0$

$$T_{m,n,l}^{p+1} = \frac{\Delta t}{\rho C}\left(\left(\frac{2k_{11}}{(\Delta x)^2} + \frac{2k_{21}}{(\Delta x)(\Delta y)}\right)T_{m+1,n,l}^{p} + \left(\frac{k_{22}}{(\Delta y)^2} + \frac{k_{12}}{(\Delta x)(\Delta y)}\right)T_{m,n+1,l}^{p} + \left(\frac{k_{22}}{(\Delta y)^2} + \frac{k_{12}}{(\Delta x)(\Delta y)}\right)T_{m,n-1,l}^{p} + \left(\frac{k_{33}}{(\Delta z)^2}\right)T_{m,n,l+1}^{p} + \left(\frac{k_{33}}{(\Delta z)^2}\right)T_{m,n,l-1}^{p} + \left(\frac{2h}{\Delta x}\right)T_{\mathrm{o}} + \left(-\frac{2k_{11}}{(\Delta x)^2} - \frac{2k_{12}}{(\Delta x)(\Delta y)} - \frac{2k_{22}}{(\Delta y)^2} - \frac{2k_{21}}{(\Delta x)(\Delta y)} - \frac{2k_{33}}{(\Delta z)^2} - \frac{2h}{\Delta x} + \frac{\rho C}{\Delta t}\right)T_{m,n,l}^{p}\right) \tag{1.111}$$

(II) For surface of $y = L_{\mathrm{y}}$

$$T_{m,n,l}^{p+1} = \frac{\Delta t}{\rho C}\left(\left(\frac{2k_{22}}{(\Delta y)^2} + \frac{2k_{12}}{(\Delta x)(\Delta y)}\right)T_{m,n-1,l}^{p} + \left(\frac{k_{11}}{(\Delta x)^2} + \frac{k_{21}}{(\Delta x)(\Delta y)}\right)T_{m+1,n,l}^{p} + \left(\frac{k_{11}}{(\Delta x)^2} + \frac{k_{21}}{(\Delta x)(\Delta y)}\right)T_{m-1,n,l}^{p} + \left(\frac{k_{33}}{(\Delta z)^2}\right)T_{m,n,l+1}^{p} + \left(\frac{k_{33}}{(\Delta z)^2}\right)T_{m,n,l-1}^{p} + \left(\frac{2h}{\Delta y}\right)T_{\mathrm{o}} + \left(-\frac{2k_{11}}{(\Delta x)^2} - \frac{2k_{12}}{(\Delta x)(\Delta y)} - \frac{2k_{22}}{(\Delta y)^2} - \frac{2k_{21}}{(\Delta x)(\Delta y)} - \frac{2k_{33}}{(\Delta z)^2} - \frac{2h}{\Delta y} + \frac{\rho C}{\Delta t}\right)T_{m,n,l}^{p}\right) \tag{1.112}$$

(III) For surface of $z = 0$

$$T_{m,n,l}^{p+1} = \frac{\Delta t}{\rho C}\left(\left(\frac{k_{11}}{(\Delta x)^2} + \frac{k_{21}}{(\Delta x)(\Delta y)}\right)T_{m+1,n,l}^{p} + \left(\frac{k_{11}}{(\Delta x)^2} + \frac{k_{21}}{(\Delta x)(\Delta y)}\right)T_{m-1,n,l}^{p} + \left(\frac{k_{22}}{(\Delta y)^2} + \frac{k_{12}}{(\Delta x)(\Delta y)}\right)T_{m,n+1,l}^{p} + \left(\frac{k_{22}}{(\Delta y)^2} + \frac{k_{12}}{(\Delta x)(\Delta y)}\right)T_{m,n-1,l}^{p} + \left(\frac{2k_{33}}{(\Delta z)^2}\right)T_{m,n,l+1}^{p} + \left(\frac{2h}{\Delta z}\right)T_{\mathrm{o}} + \left(-\frac{2k_{11}}{(\Delta x)^2} - \frac{2k_{21}}{(\Delta x)(\Delta y)} - \frac{2k_{22}}{(\Delta y)^2} - \frac{2k_{12}}{(\Delta)(\Delta y)} - \frac{2k_{33}}{(\Delta z)^2} - \frac{2h}{\Delta z} + \frac{\rho C}{\Delta t}\right)T_{m,n,l}^{p}\right) \tag{1.113}$$

(IV) For the corner nodal equation at x, y, $z = 0$

$$T_{m,n,l}^{p+1} = \frac{2\Delta t}{\rho C}\left(\left(\frac{k_{11}}{(\Delta x)^2} + \frac{k_{21}}{(\Delta x)(\Delta y)}\right)T_{m+1,n,l}^{p}\right.$$
$$+\left(\frac{k_{22}}{(\Delta y)^2} + \frac{k_{12}}{(\Delta x)(\Delta y)}\right)T_{m,n+1,l}^{p} + \left(\frac{k_{33}}{(\Delta z)^2}\right)T_{m,n,l+1}^{p}$$
$$+\left(\frac{h}{\Delta x} + \frac{h}{\Delta y} + \frac{h}{\Delta z}\right)T_{o} + \left(-\frac{k_{11}}{(\Delta x)^2} - \frac{k_{12}}{(\Delta x)(\Delta y)} - \frac{k_{22}}{(\Delta y)^2}\right.$$
$$\left.\left.-\frac{k_{21}}{(\Delta x)(\Delta y)} - \frac{k_{33}}{(\Delta z)^2} - \frac{h}{\Delta x} - \frac{h}{\Delta y} - \frac{h}{\Delta z} + \frac{\rho C}{2\Delta t}\right)T_{m,n,l}^{p}\right) \tag{1.114}$$

(V) For the edge nodal equation at y, $z = 0$

$$T_{m,n,l}^{p+1} = \frac{2\Delta t}{\rho C}\left(\left(\frac{k_{11}}{2(\Delta x)^2} + \frac{k_{21}}{(\Delta x)(\Delta y)}\right)T_{m+1,n,l}^{p}\right.$$
$$+\left(\frac{k_{11}}{2(\Delta x)^2} + \frac{k_{21}}{(\Delta x)(\Delta y)}\right)T_{m-1,n,l}^{p} + \left(\frac{k_{22}}{(\Delta y)^2} + \frac{k_{12}}{2(\Delta x)(\Delta y)}\right)T_{m,n+1,l}^{p}$$
$$+\left(\frac{k_{33}}{(\Delta z)^2}\right)T_{m,n,l+1}^{p} + \left(\frac{h}{\Delta y} + \frac{h}{\Delta z}\right).$$
$$\left.+\left(-\frac{k_{11}}{(\Delta x)^2} - \frac{k_{12}}{2(\Delta x)(\Delta y)} - \frac{k_{22}}{(\Delta y)^2} - \frac{2k_{21}}{(\Delta x)(\Delta y)} - \frac{k_{33}}{(\Delta z)^2} - \frac{h}{\Delta y} - \frac{h}{\Delta z} + \frac{\rho C}{2\Delta t}\right)T_{m,n,l}^{p}\right). \tag{1.115}$$

(VI) Within the solid body

$$T_{m,n,l}^{p+1} = \frac{\Delta t}{\rho C}\left(k_{11}\left(\frac{T_{m+1,n,l}^{p} + T_{m-1,n,l}^{p} - 2T_{m,n,l}^{p}}{(\Delta x)^2}\right)\right.$$
$$+k_{22}\left(\frac{T_{m1,n+1,l}^{p} + T_{m,n-1,l}^{p} - 2T_{m,n,l}^{p}}{(\Delta y)^2}\right) + k_{33}\left(\frac{T_{m1,n,l+1}^{p} + T_{m,n,l-1}^{p} - 2T_{m,n,l}^{p}}{(\Delta z)^2}\right)$$
$$\left.+k_{12}\frac{(T_{m+1,n+1,l}^{p} - T_{m+1,n-1,l}^{p} - T_{m-1,n+1,l}^{p} + T_{m-1,n-1,l}^{p})}{2\Delta x\Delta y}\right) + T_{m,n,l}^{p} \tag{1.116}$$

(VII) Heat source

$$T_{m,n,l}^{p+1} = \frac{\Delta t}{\rho C}\left(\left(\frac{k_{11}}{(\Delta x)^2}+\frac{k_{21}}{(\Delta x)(\Delta y)}\right)T_{m+1,n,l}^{p}+\left(\frac{k_{11}}{(\Delta x)^2}+\frac{k_{21}}{(\Delta x)(\Delta y)}\right)T_{m-1,n,l}^{p}\right.$$
$$+\left(\frac{k_{22}}{(\Delta y)^2}+\frac{k_{12}}{(\Delta x)(\Delta y)}\right)T_{m,n+1,l}^{p}+\left(\frac{k_{22}}{(\Delta y)^2}+\frac{k_{12}}{(\Delta x)(\Delta y)}\right)T_{m,n-1,l}^{p}$$
$$+\left(\frac{2k_{33}}{(\Delta z)^2}\right)T_{m,n,l+1}^{p}+\left(\frac{2h}{\Delta z}\right)T_{o}+\frac{2P\Delta t}{\rho C\Delta x\Delta y\Delta z}$$
$$\left.+\left(-\frac{2k_{11}}{(\Delta x)^2}-\frac{2k_{21}}{(\Delta x)(\Delta y)}-\frac{2k_{22}}{(\Delta y)^2}-\frac{2k_{12}}{(\Delta x)(\Delta y)}-\frac{2k_{33}}{(\Delta z)^2}-\frac{2h}{\Delta x}+\frac{\rho C}{\Delta t}\right)T_{m,n,l}^{p}\right) \tag{1.117}$$

where Δt is time step, and P is laser power.

The Gauss-Seidel iteration is used to solve the difference equations expressed above through the following procedures:

1. An initial set of values is assumed.
2. The new values of the nodal temperatures are calculated using the above equations.
3. The process is repeated until satisfactory convergence is reached.

1.4.3 Directional HAZ

1.4.3.1 Thermal Conductivity

The experimental measurement of thermal conductivity based on the HAZ isotherm of the Carbon/Epoxy composites is compared with other experimental results [36, 46] including the LFM method, as shown in Fig. 1.23. The results show good agreement with other experimental results. Thus, this approach provides a new experimental technique for a quick determination of the thermal conductivity in any direction of composites, once the longitudinal thermal conductivity based on the Rule of Mixture is obtained.

In this study, extensive experiments were conducted on UD Carbon/Epoxy to determine the conductivity. The comparison between (1.85) and (1.89) is illustrated in Fig. 1.24. It shows the measured thermal conductivity as a function of the orientation relative to fiber axis (θ).

The average values of the conductivity in Fig. 1.24a–d for (1.85) and (1.89), respectively, are used in the simulation of HAZ later, as shown in Fig. 1.25. One will find that the difference in (1.85) and (1.89) causes little concern in the simulation of HAZ. k_{12} and k_{21} of the UD and [0/90] laminates can be calculated through (1.96), as shown in Table 1.8.

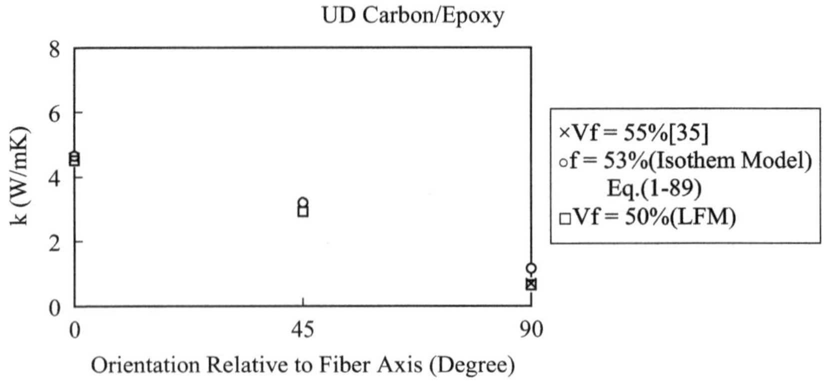

Fig. 1.23 Comparison between experimental results and isotherm model [41], reprinted with permission

1.4.3.2 Extent of HAZ

In current study, experiment of nonprincipal-axis grooving was conducted on UD Carbon/Epoxy. A schematic diagram of anisotropic grooving for UD Carbon/Epoxy relative to fiber axis is shown in Fig. 1.25. Grooving of 30° and 60° relative to fiber axis for UD composite material cannot be analytically solved; instead, the FDM is used to analyze the anisotropic temperature field. Both thermal conductivities obtained from (1.85) and (1.89) are used in FDM simulation for comparison. The other material parameters used in FDM schemes are listed in Table 1.8. Considering the immersed heat source, one can solve the extent of HAZ by Gauss-Seidel iteration until satisfactory convergence is reached.

To verify the validity of FDM schemes developed in this study, the comparison between analytical solution, i.e. (1.55), and FDM derivatives is made. The FDM method shows an excellent agreement with analytical solution, as shown in Fig. 1.26.

The extent of HAZ through numerical computation and experiment is shown in Figs. 1.27 and 1.28. Both simulations using two methods for thermal conductivity predict the HAZ with little difference. It reveals that the two models of thermal conductivity, (1.85) and (1.89), are equally acceptable in prediction of HAZ. The prediction in nonprincipal-direction grooving at both 30° and 60° relative to fiber axis shows agreement with the experimental results. It also reveals that the HAZ in grooving 30° relative to the fiber axis is narrower than that of 60°. The explanation is that carbon fiber has a much higher thermal conductivity in the longitudinal direction, and the heat is mainly conducted along the fiber axis than in transverse direction; thus machining in direction closer to transverse direction produces larger HAZ, as shown in Fig. 1.29.

The temperature-dependent thermal conductivity is further considered for nonprincipal-axis grooving of Carbon/Epoxy. By substituting the nonlinear thermal

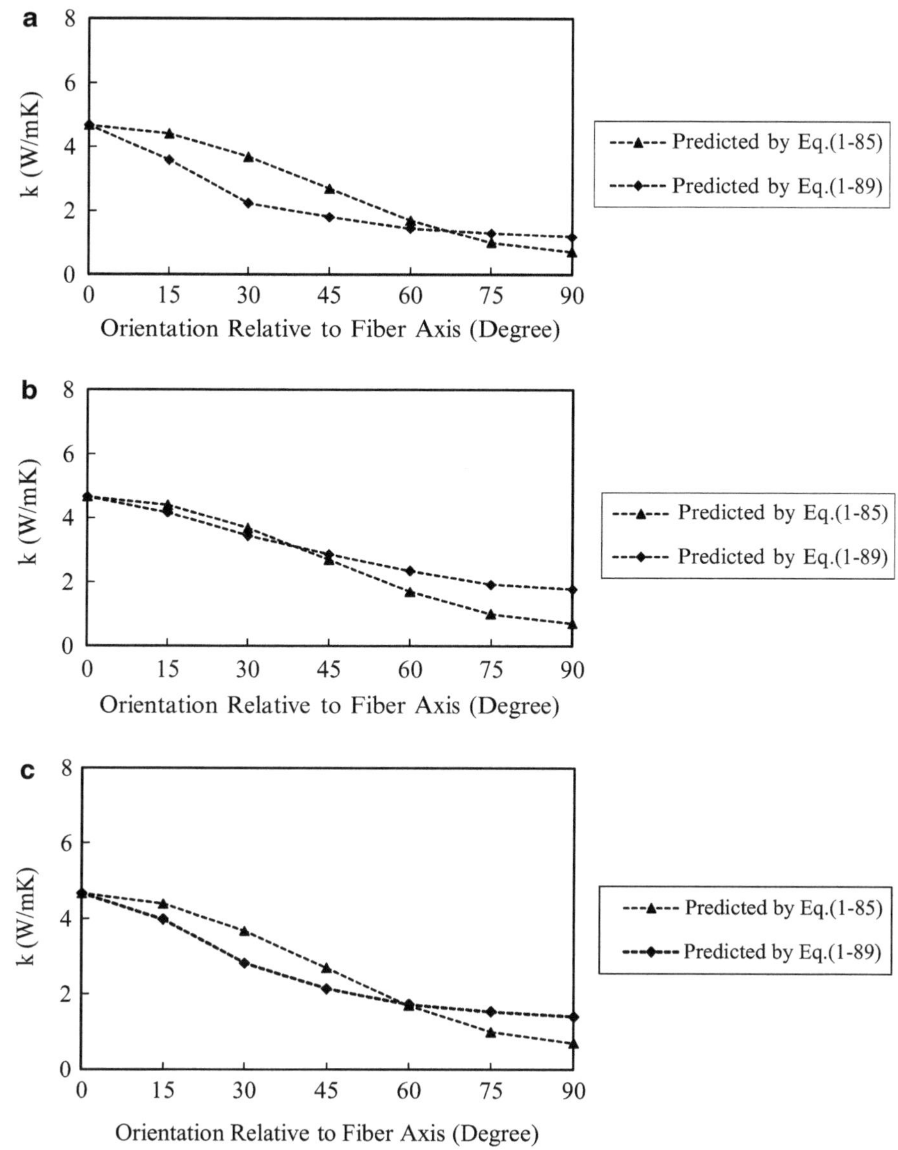

Fig. 1.24 Comparison between thermal conductivity models [41], reprinted with permission. (**a**) $UD_{carbon/epoxy}$, $V_{fiber} = 53$ %, $D = 2$ mm, $P = 100$ W, $\tau = 8$ s. (**b**) $UD_{carbon/epoxy}$, $V_{fiber} = 53$ %, $D = 2$ mm, $P = 200$ W, $\tau = 8$ s. (**c**) $UD_{carbon/epoxy}$, $V_{fiber} = 53$ %, $D = 4$ mm, $P = 100$ W, $\tau = 8$ s. (**d**) $UD_{carbon/epoxy}$, $V_{fiber} = 53$ %, $D = 4$ mm, $P = 200$ W, $\tau = 8$ s. (**e**) $UD_{carbon/epoxy}$, $V_{fiber} = 63$ %, $D = 2$ mm, $P = 100$ W, $\tau = 8$ s. (**f**) $UD_{carbon/epoxy}$, $V_{fiber} = 63$ %, $D = 2$ mm, $P = 200$ W, $\tau = 8$ s. (**g**) $UD_{carbon/epoxy}$, $V_{fiber} = 63$ %, $D = 4$ mm, $P = 100$ W, $\tau = 8$ s. (**h**) $UD_{carbon/epoxy}$, $V_{fiber} = 63$ %, $D = 4$ mm, $P = 200$ W, $\tau = 8$ s

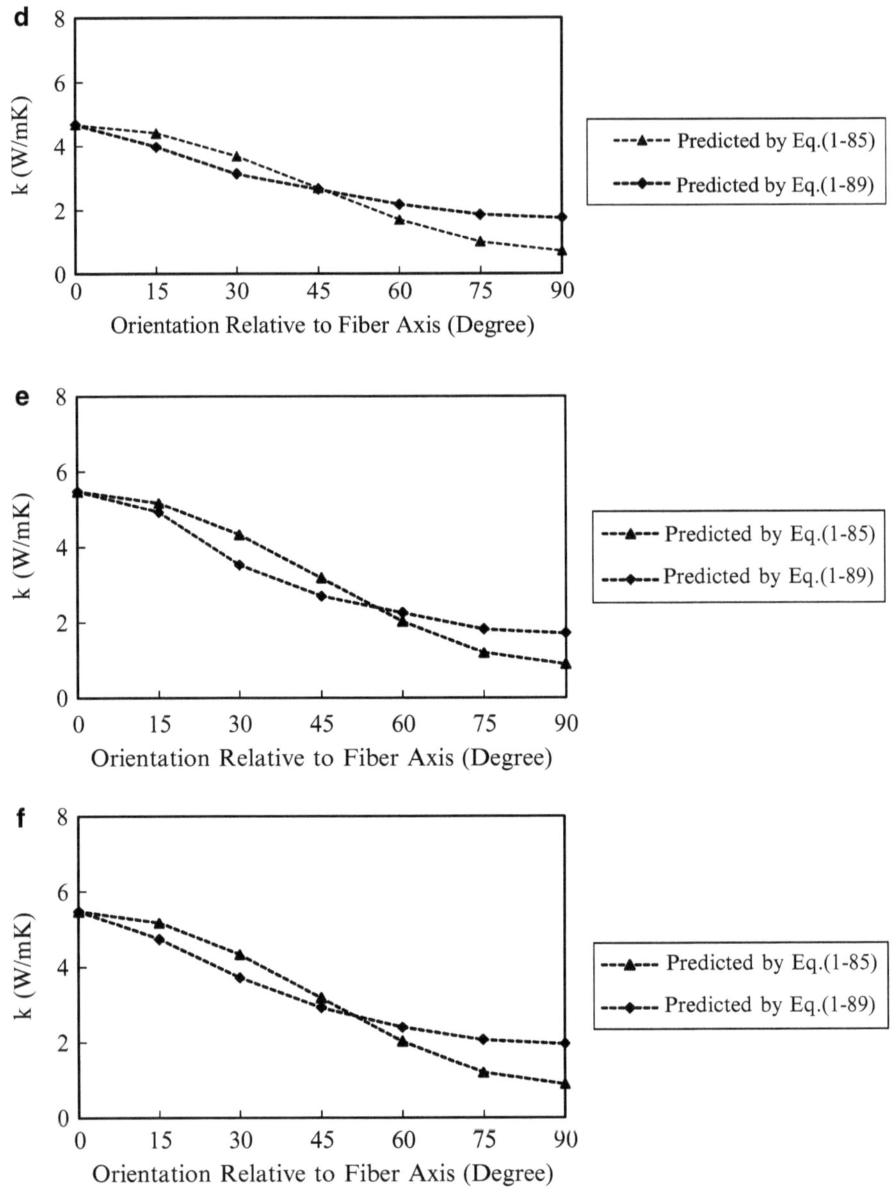

Fig. 1.24 (continued)

conductivity into FDM schemes, one can predict the HAZ. The comparison between the results of experiment and simulation with constant and temperature-dependent thermal conductivity is shown in Figs. 1.27 and 1.28. It shows that the prediction is improved by considering the temperature-dependent thermal conductivity.

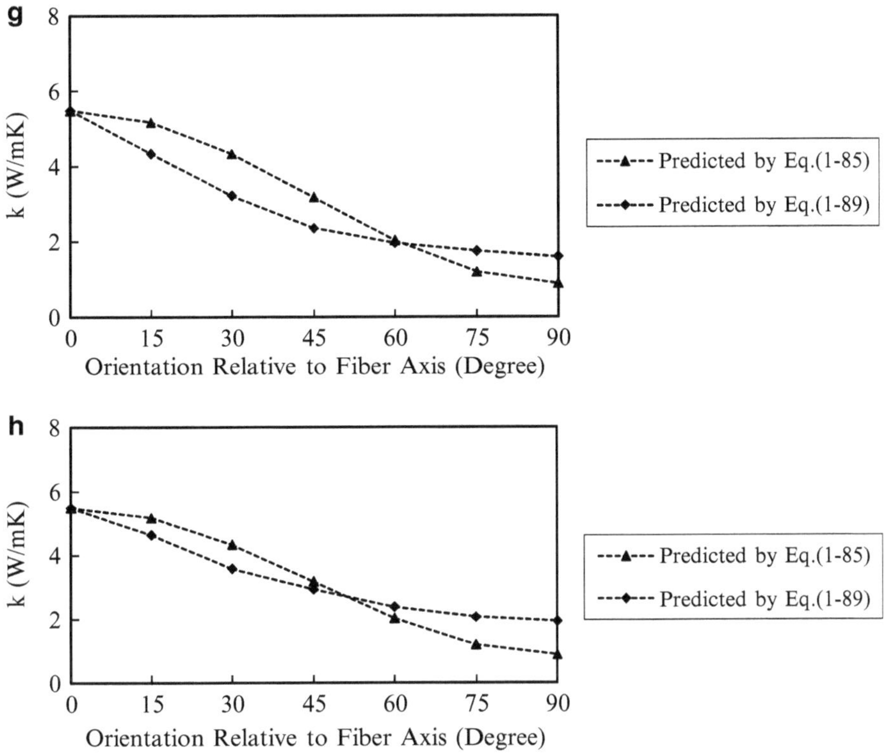

Fig. 1.24 (continued)

1.5 Conclusions

This chapter deals with an analytical evaluation of the HAZ formed during laser grooving along the principal material-directions with respect to workpiece thickness, immersed heat source, and phase changes. The significant anisotropy in composite material thermal conductivity plays an important role in HAZ during laser machining. Thermal conductivity is determined using the Laser Flash Method and Tsai's Resistor Model. UD Carbon/Epoxy, GlassMat/PP, and [0/90] laminates including Aramid/PP, Glass/PE, and Glass/PP were grooved along specific axes to evaluate the HAZ. Numerous experiments have been conducted on UD, [0/90], Mat, and MatUD composite materials to verify the HAZ models developed in this chapter. The effects of temperature-dependent thermal conductivity of anisotropic composites on HAZ are also discussed.

Modeling thermal conductivity along the fiber axis for UD and [0/90] composite materials is discussed. We determined the HAZ of UD and [0/90] composite material machining by the FDM and the associated Eigenvalue Method.

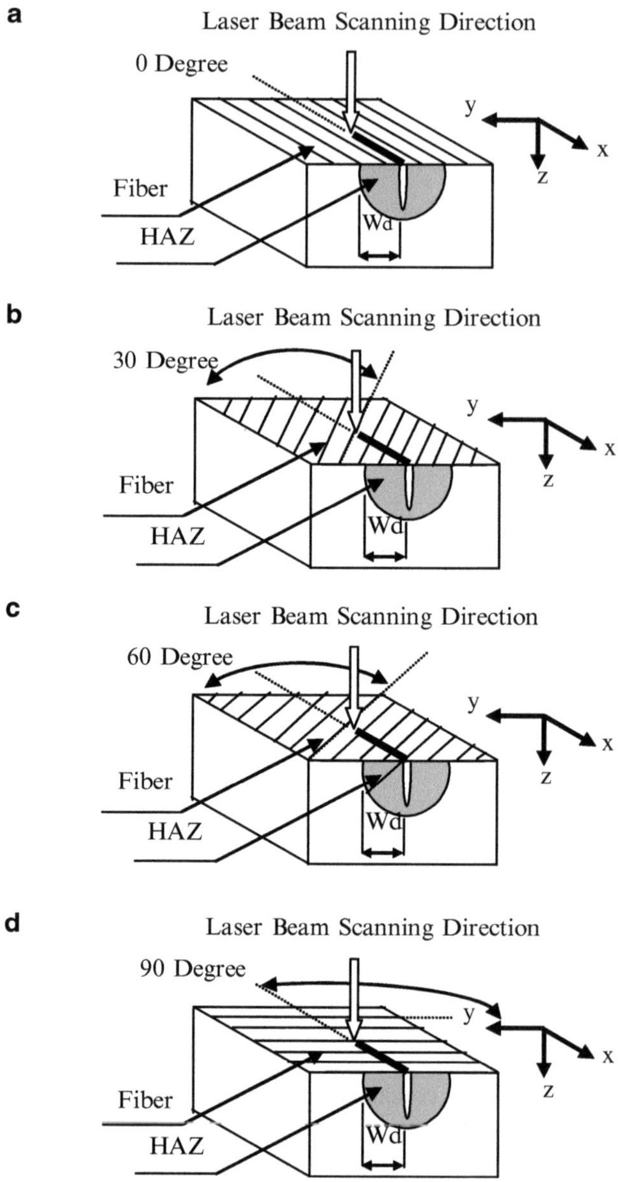

Fig. 1.25 Schematic diagram of laser grooving of UD laminates [43], reprinted with permission. (**a**) 0° relative to fiber axis (i.e., parallel grooving). (**b**) 30° relative to fiber axis. (**c**) 60° relative to fiber axis. (**d**) 90° relative to fiber axis (i.e., perpendicular grooving)

Advantages of the innovative "cryogenic surrounding" method are presented, and experimental studies of UD composite materials are illustrated accordingly. The main points are as follows:

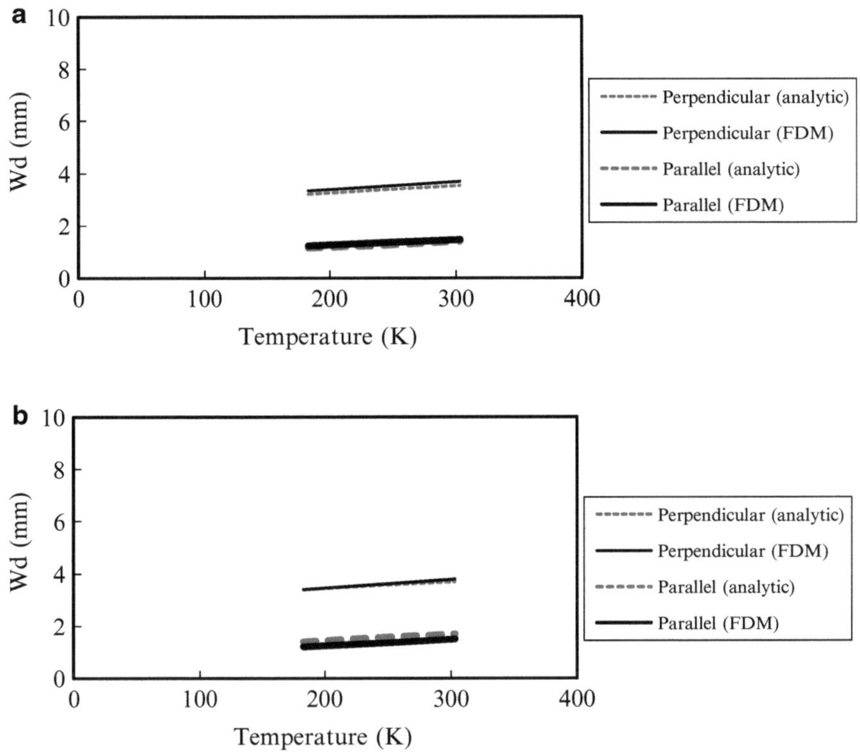

Fig. 1.26 Comparison between analytical and FDM solutions. (**a**) $PQ/V = 90$ (J/mm). (**b**) $PQ/V = 150$ (J/mm)

1. The Laser Flash Method successfully measured the thermal properties of Carbon/Epoxy composite parallel, perpendicular, and 45° relative to the fiber orientation.
2. An experimental method to determine anisotropic conductivity in nonprincipal directions was used together with the Eigenvalue Method to determine the cross-derivative thermal conductivity for any desired direction.
3. A modified resistor model for the thermal conductivity of [0/90] and MatUD laminates was developed.
4. A 3D anisotropic analysis of HAZ including an analytical solution model for the prediction of principal-axis grooving using the FDM for nonprincipal-axis grooving was derived, based on a moving heat source for laser grooving of fiber-reinforced composite materials.
5. A mirror image method assuming an immersed heat source was employed to analyze the HAZ for a workpiece with finite thickness. The temperature-dependent effects of heat conductivity on HAZ were investigated using the Kirchoff transformation.

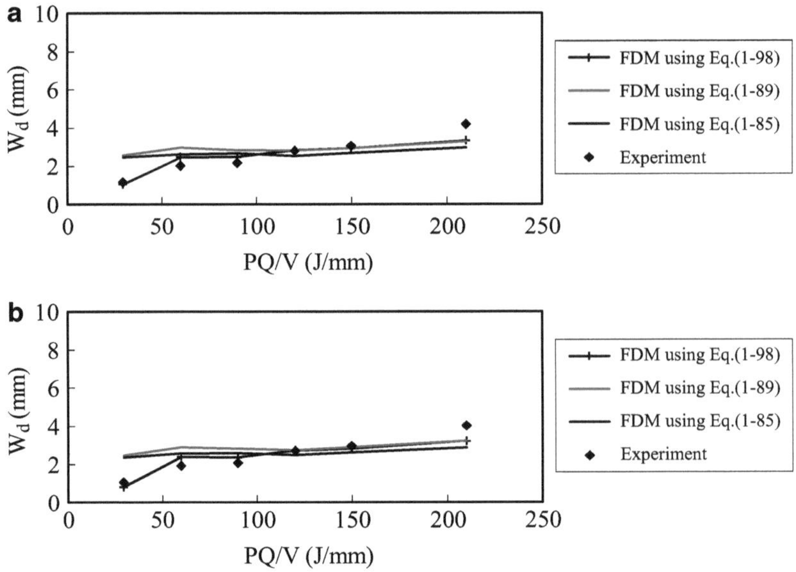

Fig. 1.27 HAZ for UD carbon/epoxy in 30° grooving. (**a**) $T_o = -60°C$. (**b**) $T_o = -90°C$

6. The results of both simulation and experiment reveal that grooving parallel to fiber orientation produces smaller HAZ zones than grooving perpendicular to fiber orientation does, while the HAZ formed during nonprincipal-axis grooving lies between that of parallel and perpendicular orientations under the same working conditions. HAZ dimensions increase with specific laser energy PQ/V, and with ambient temperature T_o.
7. Grooving perpendicular to fiber orientation has a greater impact on HAZ than grooving parallel to fiber orientation has in the current experiment. Carbon fiber has much greater thermal conductivity (by nearly two orders of magnitude) than a polymer matrix has, and thermal conduction propagates principally along the fiber direction; thus, HAZ is more extensive along that path. The most disadvantageous grooving direction for a given material system is determined by both the effective thermal diffusivity and the material's specific char temperature.
8. The application of liquid nitrogen can reduce thermal damage induced by laser grooving, by lowering the ambient temperature and providing an inert atmosphere.

This innovative machining technique finds its theoretical base in derived heat conduction analysis.

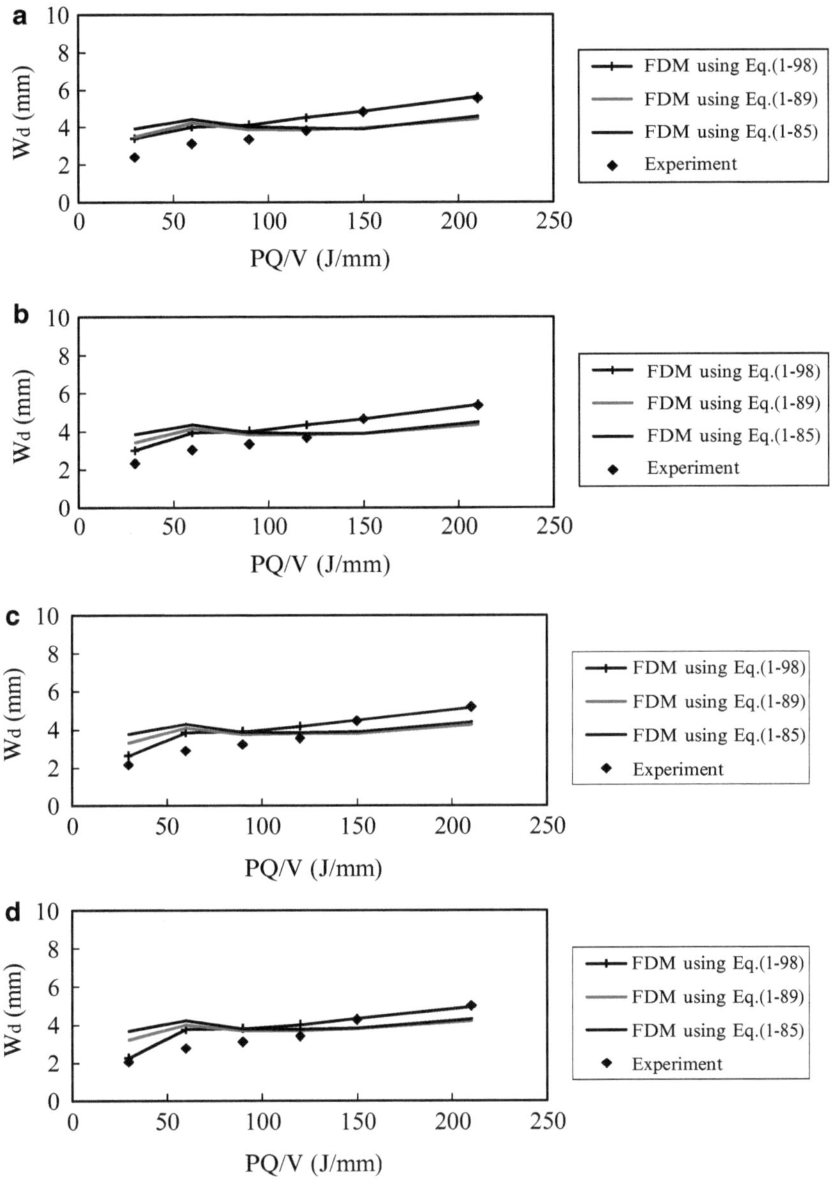

Fig. 1.28 HAZ for UD carbon/epoxy in 60° grooving [43], reprinted with permission. (**a**) $T_o = 30°C$, (**b**) $T_o = 10°C$, (**c**) $T_o = -10°C$, (**d**) $T_o = -30°C$, (**e**) $T_o = -60°C$, (**f**) $T_o = -90°C$

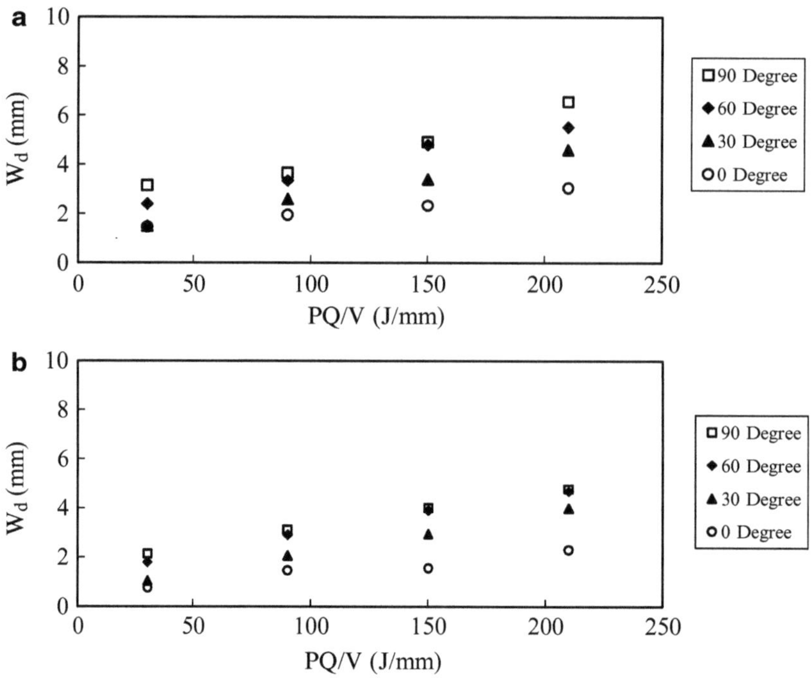

Fig. 1.29 Experimental results of HAZ in 0, 30, 60, and 90° grooving of UD carbon/epoxy [43], reprinted with permission. (**a**) $T_o = 30°C$, (**b**) $T_o = -90°C$

References

1. Dell'Erba M, Galantucci LM, Miglietta S (1992) An experimental study on laser drilling and cutting of composite materials for the aerospace industry using excimer and CO_2 sources. Compos Manuf 3(1):14–19
2. Modest MF, Abakians H (1986) Heat conduction in a moving semi-infinite solid subjected to pulsed laser irradiation. ASME J Heat Transfer 108:597–601
3. Schuocker D (1986) Dynamic phenomena in laser cutting and cut quality. J Appl Phys 40:9–14
4. Schuocker D, Abel W (1983) Material removal mechanism of laser cutting. In: Proceedings of the SPIE, pp 88–95
5. Yilbas BS (1987) Study of affecting parameters in laser hole drilling of sheet metals. Trans ASME 109:282–285
6. Anon., Coherent Inc. (1988) Lasers-operation, equipment, application, and design. McGraw-Hill, New York, pp 19–30
7. Mello MD (1986) Laser cutting of non-metallic composites. In: Proceedings of the SPIE-laser processing: fundamentals, applications, and systems engineering, pp 288–290
8. Luxon JT, Parker DE (1985) Industrial lasers and their applications. Prentice-Hall, Engelwood Cliffs, NY, pp 200–242
9. Sturmer E, Von Allmen M (1978) Influence of laser-supported detonation waves on metal drilling with pulsed CO_2 lasers. J Appl Phys 49(11):5648–5654
10. Sepold G, Rothe R (1983) Laser beam cutting of thick steel. ICALEO, SPIE, pp 156–159

11. Eberhardt G (1983) Survey of high power CO_2 industrial laser applications and latest laser developments. In: Kimmitt MF (ed) Proceedings of first international conference on lasers in manufacturing, IFS Publication, Bedford, pp 13–19
12. Lee CS, Goel A, Osada H (1985) Parametric studies of pulsed-laser cutting of thin metal plate. J Appl Phys 58(3):1339–1343
13. Hamann C, Rosen H (1987) Laser machining of ceramic and silicon. In: SPIE-high power laser and their industrial applications, vol 801, pp 130–137
14. Siekman JG (1979) Analysis of laser drilling and cutting results in Al_2O_3 and ferrites. AIP Conf Proc 50:225–231
15. Chryssolouris G, Bredt J, Kordas S, Wilson E (1988) Theoretical aspect of a laser machine tool. ASME J Eng Industry 110:65–70
16. Chryssolouris G, Sheng P, Choi WC (1990) Three dimensional laser machining of composite materials. ASME J Eng Mater Technol 112:387–392
17. Copley S (1983) Shaping materials with lasers. Laser Mater Process 3:297–336
18. Henderson JB, Wiecek TE (1987) A mathematical model to predict the thermal response of decomposite expanding polymer composite. J Compos Mater 21:373–393
19. Tagliaferri V, Diilio A (1989) Thermal damage in laser cutting of (0/90)2s aramid/epoxy laminates. Composites 20(2):115–119
20. Tagliaferri V, Visconti CI, Diilio A (1987) Machining of fiber reinforced material with laser beam: cut quality evaluation. In: Proceedings of the sixth international conference on composite materials, pp 1.190–1.198
21. Tagliaferri V, Diilio A, Visconti CI (1985) Laser cutting of fiber-reinforced polyester. Composites 16(4):317–325
22. Chryssolouris G, Sheng P, Anastasia N (1993) Laser grooving of composite materials with the aid of a water jet. ASME J Eng Industry 115:62–72
23. Lau WS, Lee WB (1990) Pulsed Nd: laser cutting of carbon fiber composite materials. Ann CIRP 39:179–182
24. Lienhard JH (1981) A heat transfer handbook. Prentice-Hall, Englewood Cliffs, NY
25. Hocheng H, Pan CT (1993) HAZ in laser cutting of carbon fiber-reinforced PEEK. In: Proceedings of ASME winter annual meeting, PED, vol 66, pp 153–159
26. Olsen O (1989) Cutting front formation in laser cutting. Ann CIRP 38:215–218
27. Na SJ, Yang YS (1989) Effect of shielding gas pressure in laser cutting of sheet metals. Trans ASME 111:314–318
28. Duley WW, Gonsalves JN (1974) CO_2 laser cutting of thin metal sheets with gas jet assist. Opt Laser Technol 6(2):78–81
29. Patel RS, Brewster MQ (1991) Gas-assisted laser-metal drilling: theoretical model. J Thermophys 5(1):32–39
30. Lee SL (1989) Weighting function scheme and its application on multidimensional conservation equations. Int J Heat Mass Transfer 32(11):2065–2073
31. Springer GS, Tsai SW (1967) Thermal conductivities of unidirectional materials. J Compos Mater 1:166–173
32. Rolfes R, Hammerschmidet U (1995) Transverse thermal conductivity of CFRP laminates: a numerical and experimental validation of approximation formulae. Compos Sci Technol 54(1):45–54
33. Chawla KK (1987) Composite materials science and engineering. Springer-Verlag, New York
34. Hashin Z (1983) Analysis of composite material—a survey. J Appl Mech 50:481–505
35. Rayleigh L (1982) On the influence of obstacles arranged in rectangular order upon the properties of a medium. Phil Mag 34:481–489
36. Mukherjee K, Khan PAA (1990) Laser machining of graphite, kevlar and glass-reinforced composites. In: Proceedings of the American society for composite fifth technical conference, East Lansing, Michigan, pp 91–104
37. Carslaw HS, Jaeger JC (1959) Conduction of heat in solids. Clarendon, London

38. Rosenthal D (1946) The theory of moving sources of heat and its applications to metal treatments. Trans ASME 68:849–866
39. Liebelt S (1992) Modeling and simulation of laser grooving and cutting for isotropic materials and fiber reinforced plastics. Diplomarbeit IWF, Berlin
40. Hocheng H, Pan CT (1999) The effects of cryogenic surroundings on thermal induced damage in laser grooving of fiber-reinforced plastic. J Mach Sci Technol 3(1):77–90
41. Pan CT, Hocheng H (2001) Evaluation of anisotropic thermal conductivity for unidirectional FRP in laser machining. Compos Part A 32:1657–1667
42. Uhlmann E, Spur G, Hocheng H, Liebelt S, Pan CT (1999) The extent of laser-induced thermal damage of fiber Ud and cross-ply composite laminates. Int J Mach Tool Manuf 39(4):639–650
43. Pan CT, Hocheng H (1998) Prediction of laser-induced thermal damage of fiber mat and fiber matUD-reinforced polymers. J Mater Eng Perform 7(6):751–756
44. Jie BL, Yu BL (1990) Practical thermal analysis. Textile Industry, Taiwan
45. Cowan RD (1969) Pulse method of measuring thermal diffusivity at high temperature. J Appl Phys 34:927–929
46. Parker WJ, Jenkins RJ, Butler CP, Abbott GL (1961) A flash method of determining thermal diffusivity heat capacity, and thermal conductivity. J Appl Phys 32:1679–1684
47. Heber A (1994) Modell zur rheologischen auslegung faserverstärkter thermoplastischer preßbauteile, Dissertation RWTH, Aachen, Germany
48. Carprino G, Tagliaferri V (1988) Maximum cutting speed in laser cutting of fiber reinforced plastics. Int J Mach Tools Manuf 28(4):389–398
49. Tyn Myint U (1973) Partial differential equation of mathematical physics. Elsevier North Holland, New York
50. Pilling MW, Yates B, Black MA, Tattersall P (1979) The thermal conductivity of carbon fiber-reinforced composites. J Mater Sci 14:1326–1338
51. Spur G, Liebelt St. (1997) Modeling of laser cutting composite and comparison with experiment. In: Fourth international conference of composites engineering (ICCE/4), Big Island, pp 599–600
52. Powell RW (1951) Thermal conductivities of metallic conductors and their estimation. In: Proceedings of general discussion heat transfer, ASME-IME, London, pp 290–295
53. Pan CT, Hocheng H (1998) Prediction of extent of heat affected zone in laser grooving of unidirectional fiber-reinforced plastics. J Eng Mater Technol ASME 120:321–327
54. Pan CT, Hocheng H (1996) The anisotropic heat affected zone in laser grooving of fiber reinforced composite material. J Mater Process Technol 62:54–60

Chapter 2
Electrical Discharge Machining

Y.H. Guu and H. Hocheng

Abstract Electrical discharge machining (EDM) is considered suitable for machining materials that are extremely hard or strong, and are wear or temperature resistant. In Chap. 2, we outline EDM characteristics of carbon fiber-reinforced carbon composites and AISI D2 tool steel. This article is organized as follows: first, the effects of EDM processing variables on delamination, the recast layer, surface roughness, and material removal rate (MRR) of carbon fiber-reinforced carbon composites are presented. Rotation of the workpiece allows fresh dielectric material to enter for effective spark discharge, and to provide better machining performance. Therefore, the second section of this chapter is devoted to a study of the effects of rotary EDM parameters on MRR and surface roughness of AISI D2 tool steel. The EDM process produces a damaged layer with different mechanical behaviors from those of the base metal. An understanding of the strength of an EDM sample is required. The third section describes an investigation of EDM AISI D2 tool steel surface characteristics and machining damage. EDM damage was studied using a new damage variable. An understanding of the surface texture of EDM specimens on the nanoscale is required. In the fourth section, surface morphology, surface roughness, and micro-cracks of AISI D2 tool steel machined by the EDM process were analyzed by atomic force microscopy (AFM). Experimental results show that the thickness of the recast layer, micro-crack depth, surface roughness, and residual tensile stress increase with the increase in power input. The EDM process effectively produces excellent surface characteristics in specimens, under low discharge energy conditions.

Y.H. Guu (✉)
Department of Mechanical Engineering, National United University, Miaoli, Taiwan, ROC
e-mail: yhorng@nuu.edu.tw

H. Hocheng
Department of Power Mechanical Engineering, National Tsing Hua University, Hsinchu, Taiwan, ROC
e-mail: hocheng@pme.nthu.edu.tw

H. Hocheng and H.-Y. Tsai (eds.), *Advanced Analysis of Nontraditional Machining*,
DOI 10.1007/978-1-4614-4054-3_2,

Keywords Atomic force microscopy • Carbon fiber-reinforced carbon composites • Damage • Delamination • Electrical discharge machining • Material removal rate • Micro-crack • Recast layer • Residual stress • Rotary EDM • Strain energy density • Surface morphology • Surface roughness

2.1 Introduction

The electrical discharge machining (EDM) process is used widely in machining hard metals and alloys in aerospace and die industries [1, 2]. Its main applications are in pressure casting dies, forging dies, powder metallurgy, and injection mold. This method is commonly used for profile truing of metal bond diamond wheel, micro nozzle fabrication, drilling of composites, and manufacturing of moulds and dies in hardened steels [3, 4]. As the electrical breakdown of the dielectric occurs, an electrical arc is generated between the cathode and anode. During the pulse-on duration, the discharge energy produces a very high temperature at the point of the spark on the surface of the workpiece, and removes the material by melting and vaporization. The top surface of the workpiece resolidifies and cools subsequently at a very high rate. The surface produced by EDM is largely influenced by the discharge energy. This process produces the slightly dimpled surface, namely, the increased surface roughness, which will facilitate crack initiation on the surface. After the electrical-discharge treatment, high tensile residual stresses often induce damage such as micro-cracks or pinholes in the surface layer which reduce the strength. The fracture strength of the material after EDM varies significantly depending on the pulsed current applied [5]. During EDM, the concentration of debris in the gap increases drastically, impeding the further removal of the materials from the gap. Fresh dielectric fluid has to enter for continuous electrical discharge. The basic parameters of conventional EDM are the pulsed current, pulse-on duration, and pulse-off duration. The experimental setup of conventional EDM is shown in Fig. 2.1 [6]. An electrical discharge machine (Charme CD-50M) with servo-control was used to conduct the experiments. During EDM, the tool and the workpiece are separated by a small gap, and submerged in dielectric fluid. The tool electrode geometry is reproduced in the workpiece after the process. A high voltage is applied across the thin gap in a dielectric fluid. As the electrical breakdown of the dielectric occurs, an electrical arc is generated between the cathode and anode. During the period of on-time, the discharge energy produces very high temperature at the point of the spark on the surface of the workpiece, and removes the material by melting and vaporization. The top surface of workpiece resolidifies and cools subsequently at very high rate.

In rotary EDM, an additional parameter, i.e., the rotation of the electrodes, is considered. Researchers used the techniques of applying rotation or orbital motion to the electrode to ensure adequate flushing of the gap and to obtain better machining performance [7, 8]. The rotating motion of the workpiece can rapidly

Fig. 2.1 Schematic diagram of the EDM process (*1* servo-control; *2* electrode; *3* specimen; *4* dielectric fluid; *5* pulsed generator; *6* oscilloscope) [6], reprinted with permission

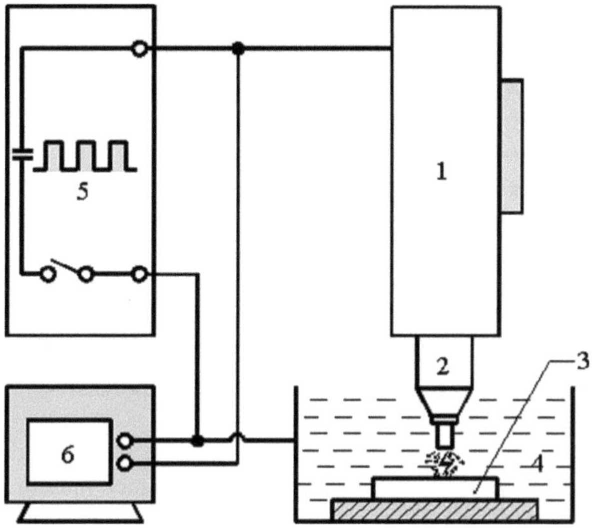

Fig. 2.2 Schematic drawing of rotary EDM (*1* servo-control; *2* electrode; *3* workpiece; *4* dielectric fluid; *5* pulsed generator; *6* oscilloscope; *7* DC motor) [9], reprinted with permission

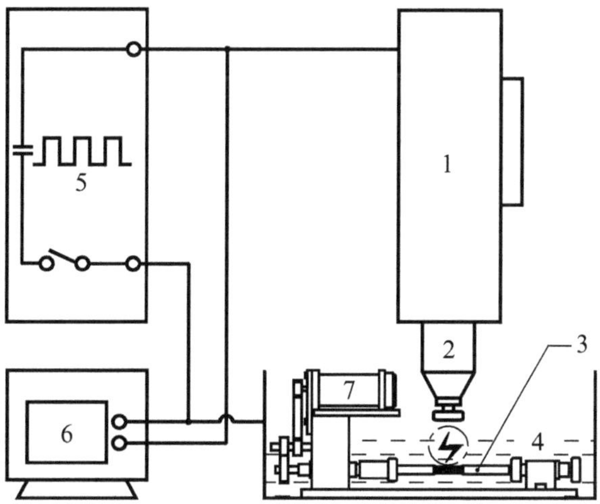

remove debris from the gap, allowing fresh dielectric to enter for effective spark discharge. Figure 2.2 [9] schematically displays the experimental setup of the rotary EDM processes, in which a rotation mechanism was developed and installed on the machining table. A DC motor rotates the workpiece at the desired speed. EDM was applied to the central part of the specimen. The tool electrode geometry was reproduced in the workpiece after the process. This process will benefit certain axial-symmetrical parts used in molds and dies made of difficult-to-machine materials when the parts are electrical discharge machined.

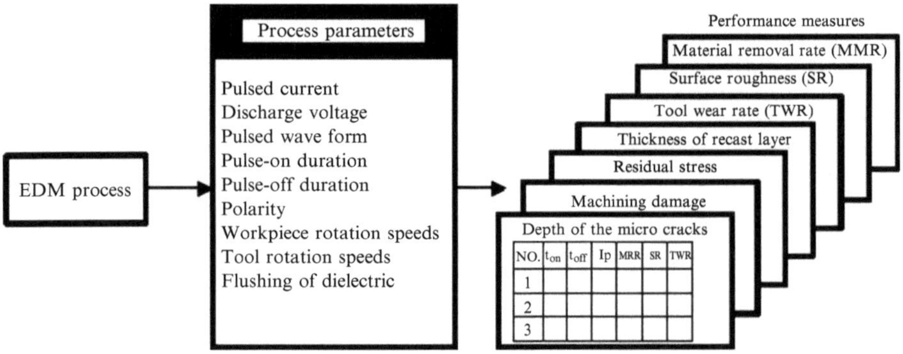

Fig. 2.3 Process parameters and performance measures

The process parameters and performance measures are shown in Fig. 2.3. The primary parameters of this study are pulsed current, pulsed voltage, pulse-on duration, pulse-off duration, polarity, and workpiece rotation speeds. Major performance measures in EDM are material removal rate (MRR), tool wear rate (TWR), surface roughness (SR), thickness of recast layer, residual stress, machining damage, and depth of the micro-cracks. These parameters play a critical role in optimizing performance measures.

Two types of materials are considered in this chapter. The first kind is the carbon fiber-reinforced carbon composites with superior strength-to-weight and stiffness-to-weight ratios and high service temperatures and the second kind is the AISI D2 tool steel which is widely used in industry for the inherent high strength and wear resistance from its high chromium content. The reason to study the first topic is that carbon fiber-reinforced carbon composites machining by conventional techniques would occur chipping, cracking, delamination in the specimen, and create high wear on the cutting tools [10–12]. Although a few studies [13, 14] have examined this research field, clear correlation among processing variables of EDM and the machinability of carbon fiber-reinforced carbon composites remain uncertain. Thus, the characteristics of EDM-machined surfaces and the effects of processing variables on delamination, recast layer, surface roughness, MRR, and mechanisms for material removed are proposed herein.

Thc rotary EDM with a disk-like electrode has a higher MRR due to the improved dielectric circulation in the spark gap [3]. The authors believe that the rotating motion of the workpiece can rapidly remove debris from the gap, allowing fresh dielectric to enter for effective spark discharge. Hence, the second topic of this chapter is to discuss the effect of workpiece rotation on machinability during EDM. The MRR and surface roughness are compared with those from conventional EDM. To prepare the EDM specimen, the workpiece of AISI D2 tool steel is mounted on a rotating table driven at a desired speed (Fig. 2.2).

The fracture strength of the material after EDM varies significantly depending on the pulsed current applied [1, 5]. Ghanem et al. [15] showed that the heat-affected zone generated by EDM produces a harmful influence decreasing the life of the virgin material. Lin and colleagues [16, 17] confirmed that the rupture strength of EDM specimen decreases with the increase of the discharge input energy. Zeid [18], Guu, and Hocheng [19] studied the fatigue strengths of EDM tool steel. They proved that the fatigue strength of mechanical components is dependent on the properties of the surface and near surface regions. Small surface damages are known to reduce the fatigue strength. The EDM process produces a damage layer that has a different mechanical behavior from that of the base metal. Fracture mechanics deals mainly with the load carrying capacity of a body containing macroscopic cracks; it does not predict the damage when microdefects appear in the specimen [20, 21]. A variety of damage variables have been proposed by Lemaitre and colleagues [22, 23]. According to their report, the damage variable can be defined in terms of the variation of elastic modulus and of the ultrasonic waves propagation or expressed by the variation of density and of the micro-hardness. Several damage variables and damage measurement methods have been proposed. None of them is perfect. An understanding of the factors leading to the degradation of strength is required. The purpose of the third topic of this chapter was to study the effects of machining conditions on surface characteristics and to analyze the machining damage in the EDM specimen by means of a new damage variable.

Surface morphology plays an important part in understanding the nature of machined surfaces. Before the 1990s, the surface textures of specimen were evaluated mainly with a contact stylus profiler. This instrument has various limitations including a large stylus radius, a large contact force and low magnification in the plane, which will result in surface damage of the sample and may misrepresent the real surface topography owing to the finite dimension of the stylus tip [24]. On the ultramicroscopic scale of surface measurements, atomic force microscopy (AFM) has been developed to obtain a three-dimensional image of a machined surface on a molecular scale. AFM overcomes almost all the above drawbacks of the stylus profiler. In recent years, AFM has been widely used for semiconductors to obtain the surface image in the nanometer scale [25–27]. During machining, the discharge energy produces very high temperatures at the point of the spark on the surface of the workpiece removing the material by melting and vaporization. The top surface of the workpiece resolidifies and subsequently cools extremely quickly. This process causes a ridged surface and induces damage such as micro-cracks in the surface layer [28]. Lee and Tai [29] used the Digitizing Area-Line Meter to measure the total length of the cracks in the scanning electron microscopy (SEM) photograph and defined a surface crack density to evaluate the severity of cracking. Ramasawmy and Blunt [30, 31] studied the 3D surface topography of the EDM specimen using 3D Talysurf with a 60° diamond conical stylus of 2 μm tip radius. Previous investigations [32, 33] looked into the surface morphology and the crack depth of the cross-section profile in the EDM specimen using SEM. An understanding of the surface texture of the EDM specimens in the nanoscopic sense is required. The fourth topic of this chapter, the three-dimensional images of AISI D2 tool steel machined by the EDM process were analyzed by means of the AFM technique.

2.2 Effect of Electrical Discharge Machining on the Characteristics of Carbon Fiber-Reinforced Carbon Composites

2.2.1 Experimental Setup

2.2.1.1 Preparation of Composites

Plain weave carbon fabrics were impregnated with phenolic resin using a wet dipping technique. The fabrics were then placed in a circulation air oven at 70°C for 2 h to remove excess solvent. The raw materials used in this study are summarized in Table 2.1 [34]. The prepreg hand lay-up method was used to laminate eight layers of impregnated fabrics. The stacked fabrics were then placed in a picture frame mold and compression-molded at 150°C under a pressure of 13.5 MPa for half an hour. Post-cured process was performed in an air circulation oven at 175°C for 36 h. The post-cured carbon/phenolic composites were cut into 50 × 10 × 1 mm specimens using a diamond saw under water cooling. The carbonization process was accomplished at 2,200°C in a high temperature tube furnace for 1 h under an argon atmosphere at a constant gas flow rate of 6×10^{-3} $m^3 h^{-1}$, where the furnace was heated at a rate of 30°C h^{-1}. The density of the fabricated carbon fiber-reinforced carbon composites was 1,470 kg m^{-3}.

2.2.1.2 Parameters in EDM

An electrical discharge sinker machine equipped with a servo-controlled head was employed to perform the machining experiments. Meanwhile, graphite was used as the tool electrode for cutting purposes. The electrode was cylindrical with a diameter of 1 mm. The tool electrode was positively polarized and the workpiece served as negative polarity during the EDM process. Kerosene was used as the dielectric fluid. The working voltage between the composite and the electrode was 35 V. The pulsed current (I_p) varied from 1 to 5 A, with a duty factor of 90%.

Table 2.1 Resin and carbon fabrics materials [34], reprinted with permission

Specification	Code	Supplier
Resin	PF-650	Chung-Chun Plastics Co. Ltd., Taiwan
Resole type phenolic resin		
pH = 8.0–8.5		
Solid content: 62%		
Viscosity at 25°C: 0.10–0.25 Pa s		
Carbon fabrics	TR-30K	Mitsubishi Rayon Co. Ltd., Japan
Plain weave cloth, 2,000 kg m^{-2}		
Fiber diameter: 7.0 μm		

Fig. 2.4 Scanning electron micrograph of the EDM surface (pulsed current 3 A, pulse-on duration 20 μs) [34], reprinted with permission

The pulse-on duration (τ_{on}) from 20 to 220 μs was employed. After each experiment, the change in volume of the workpiece, and the machining time were recorded. The value of the MRR was evaluated for each condition by dividing the measured amount of material removal by the machining time.

2.2.1.3 Surface Examination

The structure of the EDM area was examined with SEM to determine the nature of damage. The thickness of the recast layer was obtained by calculating the average of 25 readings measured at various locations on the cross section. A profilometer was used to measure the surface roughness of the specimens.

2.2.2 Experimental Findings

2.2.2.1 Removal Mechanisms

Figure 2.4 [34] shows the machining surface from the EDM process. This scanning electron micrograph reveals that the randomly distributed white layers and solid particles are resolidified and remain attached to the eroded surface. The micrograph displays that the particles formed by electrical discharge are solid spheres, and range from 3 to 30 μm in diameter. The solid behavior attributed to the discharge channel temperature is higher than that of molten particles ejected from the crater and nucleation generally starts internally. The perfectly spherical particles were apparently resolidified from the gaseous state, whereas irregularly recast layers were solidified from the liquid state. Previous studies of the EDM surface morphology and debris have demonstrated that composite material was removed in both the liquid and gaseous states [35, 36].

2.2.2.2 Hole Quality

Figure 2.5a, b [34] summarizes the effects of pulsed current and pulse-on duration on the hole edge damage produced by 1 mm diameter graphite electrode, and

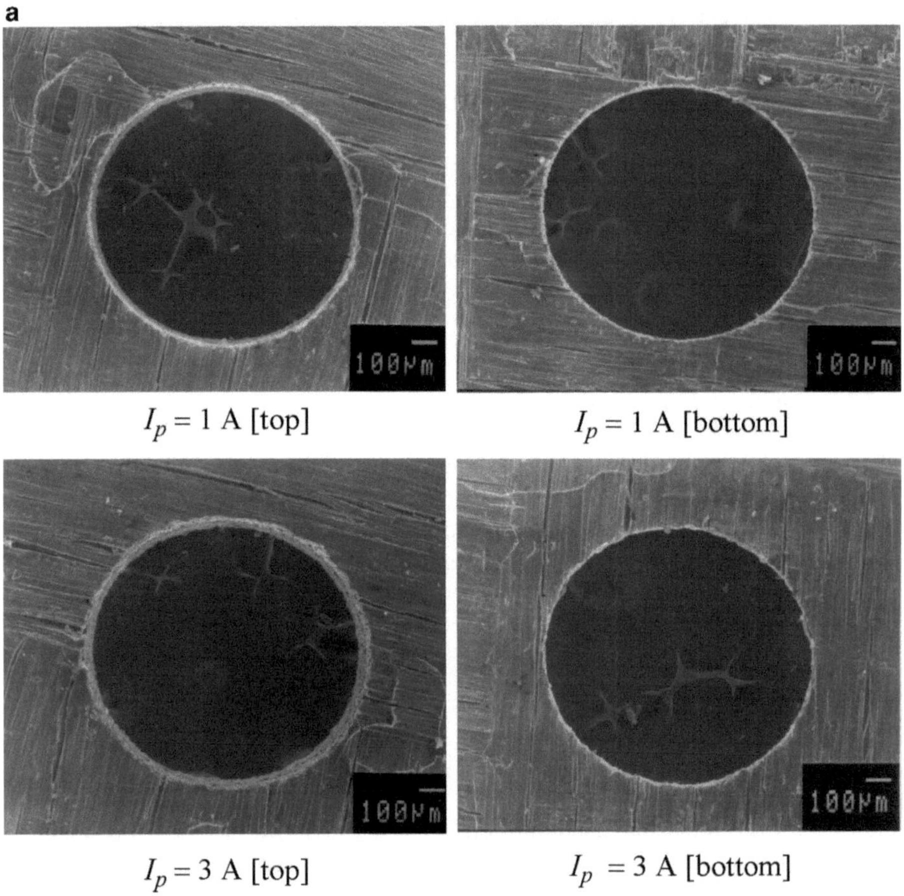

Fig. 2.5 (**a**) SEM photomicrographs of the top/bottom surfaces ((τ_{on}) = 20 μs, electrode diameter 1 mm) [34], reprinted with permission. (**b**) SEM photomicrographs of the top/bottom surfaces (I_p = 1 A, electrode diameter 1 mm) [34], reprinted with permission

b

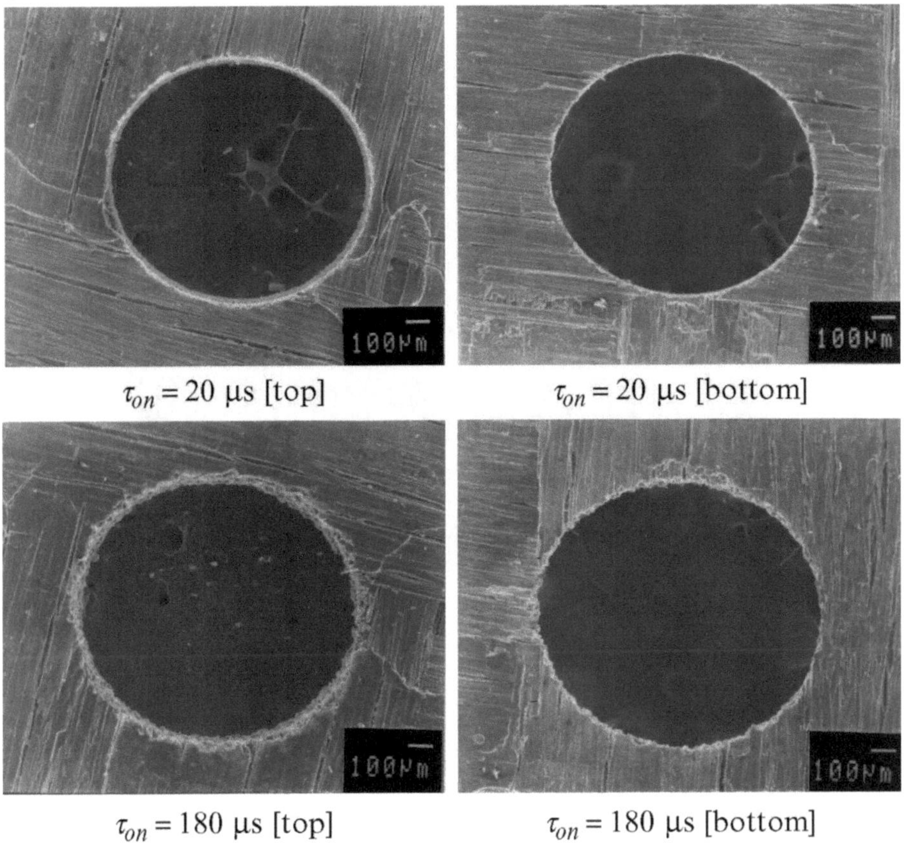

Fig. 2.5 (continued)

indicate the SEM photomicrographs of both the top and bottom surfaces for holes produced in composite material. Clearly, unlike with conventional machining and water jet drilling, the edges of the hole are completely free from burrs. During processing, working conditions can influence the incidence of tear-like or delamination damages. When applying the smaller pulsed current 1 A and the pulse-on duration 20 μs, the hole quality was good and no machining damage was observed in composite laminate on either the top or bottom surfaces. However, as the current and pulse-on duration increased, the damage could reach 100 μm. Such damage extends far beyond the thermally influenced area and into the base material, causing stress concentration along the holes, and eventually decreasing the service strength. These results also demonstrate that the machining quality on the bottom surface is superior to that on the top surface.

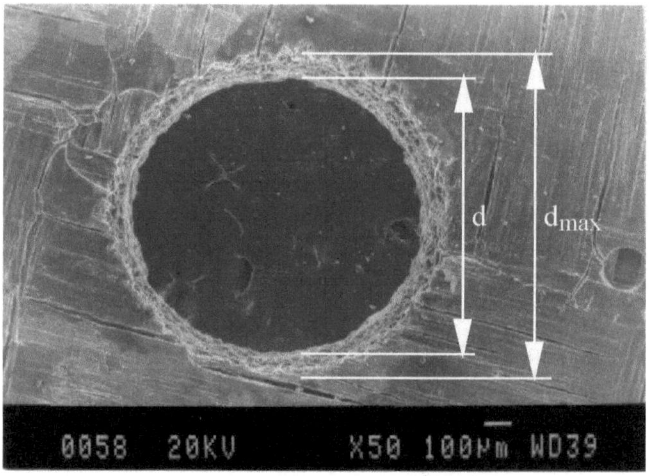

Delamination factor $= \frac{d_{max}}{d}$

Fig. 2.6 Estimation of the delamination factor ($I_p = 3$ A, $\tau_{on} = 100$ μs, $d = 1$ mm) [34], reprinted with permission

2.2.2.3 Delamination Factor

During the EDM process, the extremely high temperature produced by electrical discharge sparks rapidly vaporizes the dielectric fluid and creates pressure impulses around the tool electrode. The impact pressure and high thermal stresses generated by the discharge produces the delamination in the machining surface. The extent of delamination can be determined according to the maximum hole diameter in the damage zone (Fig. 2.6 [34]). Herein, the delamination factor is proposed to facilitate analysis of the delamination degree and comparison with different machining conditions. The delamination factor is defined as

$$D_f = d_{max}/d \tag{2.1}$$

where D_f denotes the delamination factor, d_{max} represents the maximum diameter of the damage zone, and d is the hole diameter. Figure 2.7 [34] shows the correlation between the delamination factor and machining parameters for the drilling of the composites. The delamination factor clearly increases with the energy supply. This increase occurs because the impulse force increases with increasing discharge energy, thus increasing the delamination degree. The variation in the maximum diameter of the damage zone under coarse conditions was higher than that under fine cutting conditions. Therefore, properly choosing operating parameters can eliminate the damage to the workpiece. Regression analysis of the experimental data produces the empirical formula

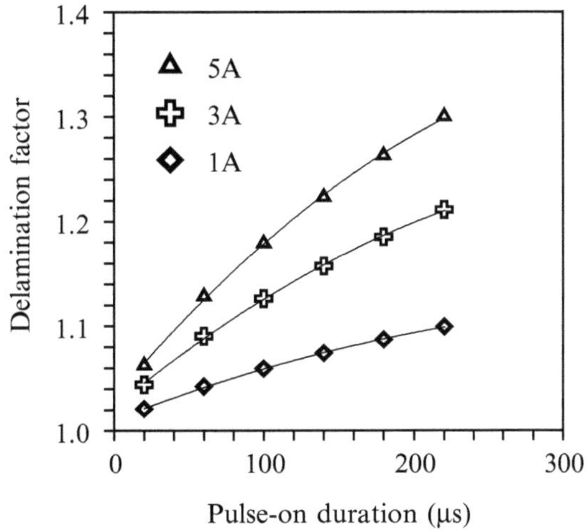

Fig. 2.7 The correlation between the delamination factor and machining conditions [34], reprinted with permission

$$D_{\mathrm{f}} = 0.815 I_{\mathrm{p}}^{0.0681} \tau_{\mathrm{on}}^{0.0579} \tag{2.2}$$

where I_{p} denotes the pulsed current and τ_{on} represents the pulse-on duration. Equation (2.2) gives average error of 6.3% with respect to the experimental results.

2.2.2.4 Recast Layer

The surface of the workpiece closest to the electrode in EDM is subjected to a temperature rise of up to 10,000°C [37]. The upper material in the extremely high temperature region will vaporize, while the lower material will melt. The present investigation attempted to evaluate the effect of process parameters on the recast layer of the composite material by examining a cross section of the specimen. Figure 2.8 [34] depicts SEM observation of the composites. Obviously, two regions are present in the composites: one is the original material and the other is the recast layer (which has a white color). A smaller thermal gradient occurs at the lower pulsed current, thus forming a thinner recast layer. The recast layers appear thicker as pulsed current increases, since at a higher pulsed current, a steeper thermal gradient builds up in the composites, possibly causing a thermal effect beneath the melting zone. This phenomenon leads to a greater residual melted layer that is not flushed out by the dielectric fluid, and that then resolidifies and remains attached to the machined surface. Figure 2.9 [34] shows the thickness of the recast layer on the machined surface under various machined conditions. The figure shows that the pulsed current has a greater effect on the surface roughness than the pulse-on duration. The correlation between the thickness of the recast layer (d_{t}) and machining conditions using regression analysis yields

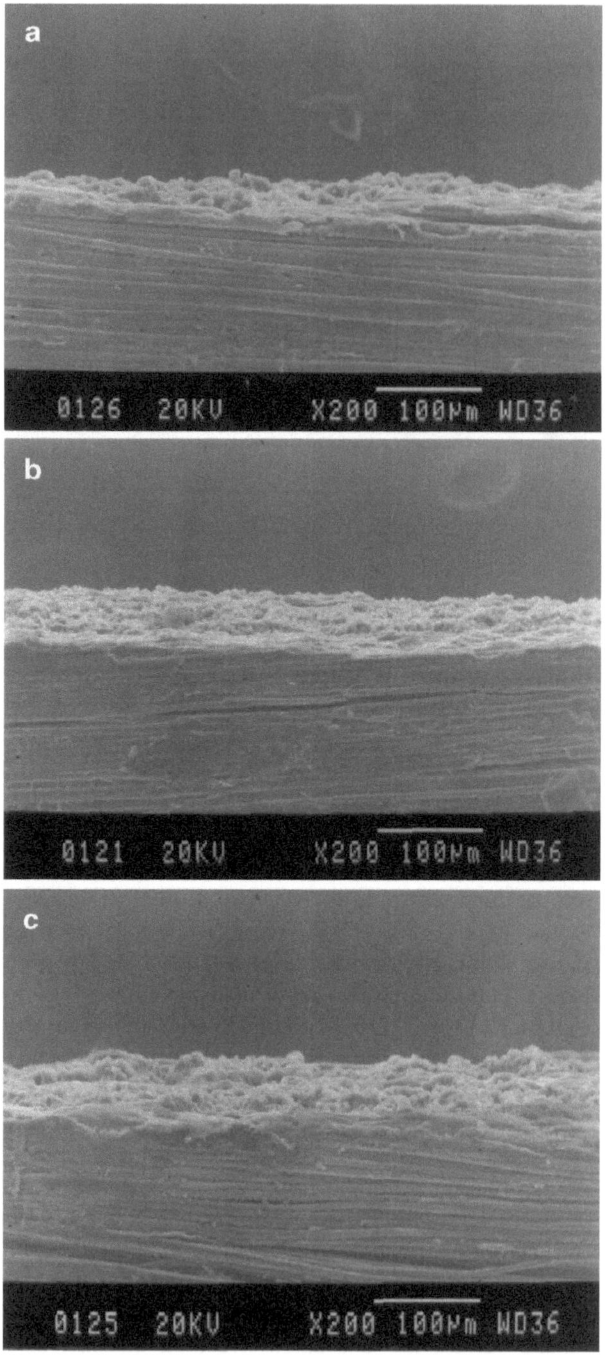

Fig. 2.8 SEM photomicrographs of the recast layer of the composites after EDM at (**a**) 1 A, (**b**) 3 A, and (**c**) 5 A. The pulse-on duration is 100 μs [34], reprinted with permission

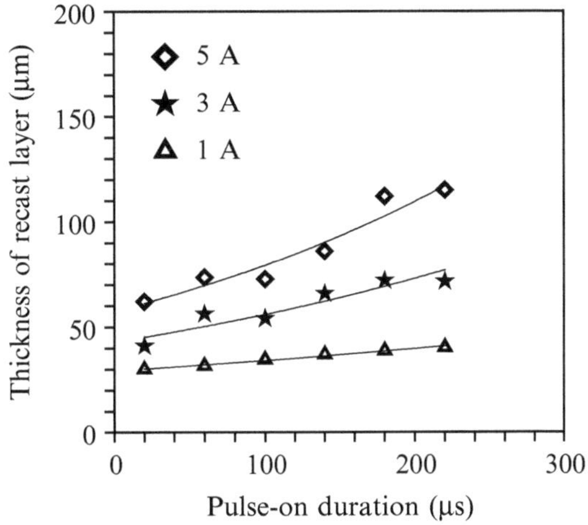

Fig. 2.9 Thickness of the recast layer under various machined conditions [34], reprinted with permission

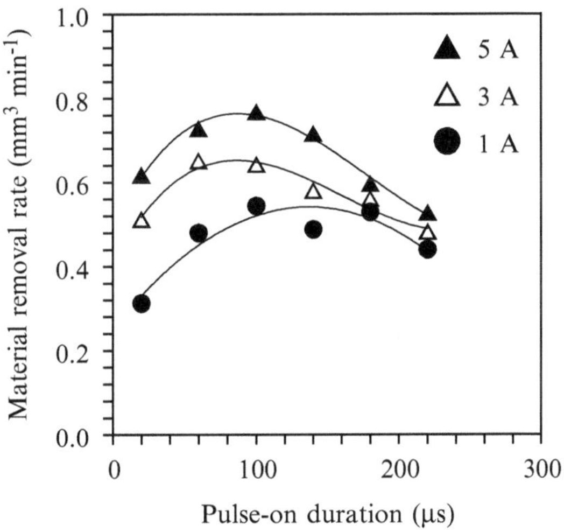

Fig. 2.10 Variation of the material removal rates of the composites during EDM process as functions of pulsed current and pulse-on duration [34], reprinted with permission

$$d_t = 13.696 I_p^{0.534} \tau_{on}^{0.204} \tag{2.3}$$

where the constants depend on electrode materials, the type of dielectric, and the flushing conditions. Equation (2.3) gives average error of 6.3% with respect to the experimental results.

2.2.2.5 Material Removal Rate

Figure 2.10 [34] presents the MRR during the EDM process as a function of pulse-on duration and pulsed current. Clearly, the value of MRR initially increases with pulse-on duration owing to the greater energy input rate. At a small pulse-on duration, the discharge energy is insufficient and the MRR is unable to rise. A pulse-on duration of 100 μs removes material more effectively than other times, but beyond this level the high energy input becomes inefficient and the MRR decreases. Additionally, the surface temperature increases with the pulse-on duration and consequently the melting boundary deepens and widens. This phenomenon prevents the melted material from being completely flushed out by the dielectric fluid and disturbs the gap condition for the next discharge due to the eroded chips filling the gap. The MRR for a larger pulsed current with a higher energy input rate also exceeds that for a smaller pulsed current. A pulsed current of 5 A and duration of 100 μs produces the highest removal rate, of 0.768 $mm^3\ min^{-1}$.

Using regression analysis to reveal correlation between the MRR and machining parameters produces:

$$\mathrm{MRR} = 0.378 I_p^{0.216} \tau_{on}^{0.041} \quad \text{as } \tau_{on} \leq 100\ \mu\mathrm{s} \tag{2.4}$$

$$\mathrm{MRR} = 4.400 I_p^{0.134} \tau_{on}^{-0.428} \quad \text{as } \tau_{on} > 100\ \mu\mathrm{s} \tag{2.5}$$

Equations (2.4) and (2.5) give average errors of 6.2% and 4.8%, respectively, corresponding to the experimental results.

2.2.2.6 Surface Roughness

Figure 2.11 [34] displays the SEM micrographs of the machined surfaces formed at a normal discharge. With a 1 A pulsed current, the surface characteristics are relatively smooth. However, when the discharge energy increases to 5 A, the machined surface shows irregular valleys, and deeper and longer grooves craters are formed by the electrical discharge channel. Carbon fiber-reinforced carbon composites are anisotropic, and thus the strength and thermal conductivity of the composite material depends on the fiber direction. Therefore, crater shape relates strongly to fiber direction and the discharge channel position. If the axial direction of the adjacent fibers runs at a tangent to the region of the discharge crater, the crater is narrow. The narrow crater is a result of less heat conduction along the transverse direction of the fiber. When the fibers are positioned at the crater margin, the delamination propagation occurs more easily along the longitudinal direction than the transverse direction. This phenomenon causes the crater to extend along the fiber direction. Therefore, material surface formation by the EDM process in this composite material differs from that of metals. Notably, the increase of crater depth with the applied pulsed current appears minimal under a small input energy.

EDM erodes surfaces randomly. To determine the effect of the EDM on the surface roughness of composites, this study measured the surface profiles of the topography of the EDM surfaces. Figure 2.12 [34] shows the measurement results. Clearly, a poorer surface finish is obtained with higher pulsed current, since a higher

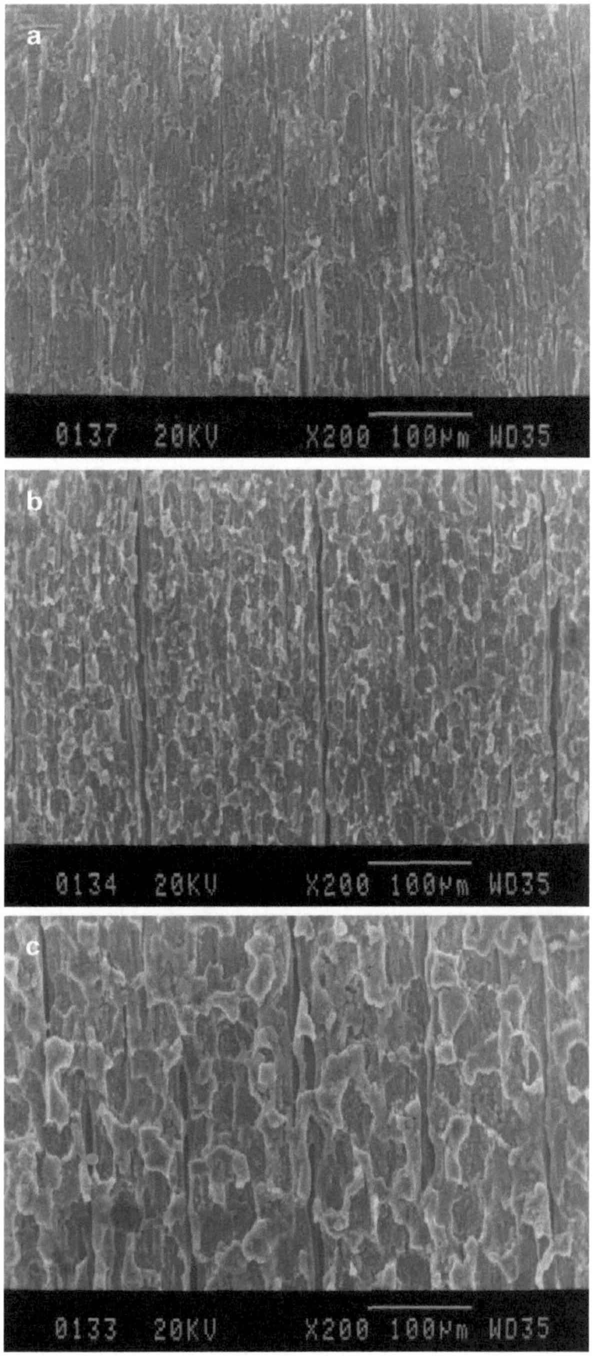

Fig. 2.11 Surface morphology of the faces after EDM with (**a**) 1 A, (**b**) 3 A, and (**c**) 5 A pulsed current, observed by SEM. The pulse-on duration is 100 μs [34], reprinted with permission

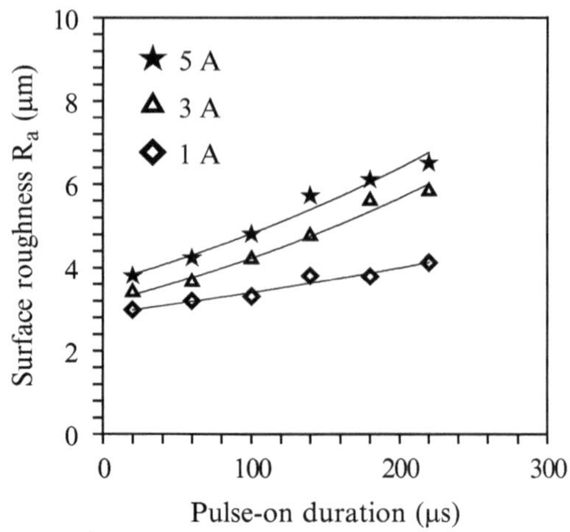

Fig. 2.12 Surface roughness of the EDM surface under various machined conditions [34], reprinted with permission

pulsed current causes more frequent cracking of the dielectric fluid, hence more frequent melt expulsion, and a poorer surface finish. An excellent machined finish can be obtained by setting the machine parameters at low pulsed currents and a small pulse-on duration, but this approach is more time consuming. The surface roughness (R_a) is affected by the process parameters of pulsed current and pulse-on duration. Regression analysis of the experimental data produces the empirical model

$$R_a = 1.326 I_p^{0.224} \tau_{on}^{0.216} \tag{2.6}$$

Equation (2.6) gives average error of 6.1% with respect to the experimental results.

2.3 Effects of Workpiece Rotation on Machinability During Electrical Discharge Machining

2.3.1 Experimental Setup

An electrical discharge machine (Charmc CD-50M) with servo-control was used to conduct the experiments. Figure 2.2 schematically displays the experimental setup of the rotary EDM processes, in which a rotation mechanism was developed and installed on the machining table. A DC motor rotates the workpiece collect at the desired speed. EDM was applied to the central part of the specimen. The copper electrode was connected to positive polarity and the specimen was the negative. The workpiece rotated at 0, 1,200, 2,200, and 5,000 rpm. MRRs for each condition were evaluated by dividing the measured amount of material removal by the machining time. Table 2.2 presents the experimental conditions. The pulsed current, pulse-on duration, and rotation speed were controllable variables. Meanwhile, the tool

Table 2.2 Experimental conditions

Dielectric	Kerosene
Work material	AISI D2 tool steel
Tool material	Copper
Pulsed current (I_p)	1, 5, 10 A
Pulse-on duration (τ_{on})	20, 100, 180, 260 μs
Pulse-off duration (τ_{off})	20 μs
Duty factor $\left(\frac{\tau_{on}}{\tau_{on}+\tau_{off}} \times 100\%\right)$	50, 83, 90, 93%

Table 2.3 Chemical composition of AISI D2 (wt. %)

Element	C	Si	Mn	Mo	Cr	Ni	V	Co	Fe
Weight %	1.5	0.3	0.3	1.0	12.0	0.3	0.8	1.0	Balance

Table 2.4 Mechanical properties of AISI D2

0.2% offset yield strength	1,532 MPa
Tensile strength	1,736 MPa
Hardness (HRC)	56.57

electrode has a radius of 4 mm and a length of 25 mm. Machining was performed on a rod of 8 mm diameter. The specimen is AISI D2 tool steel, which is widely used in the mold industry. The chemical composition of the material is given in Table 2.3. The cold worked AISI D2 tool steels were heated to 1,030°C for 1 h under nitrogen atmosphere and nitrogen-quenched. In the heat treatment, the furnace was heated at a rate of 20°C min^{-1}. After quenching, the specimens were tempered four times at 520°C for 2 h with air cooling. The mechanical properties of the AISI D2 tool steel are given in Table 2.4. The machined surface morphology was examined by SEM. A profilometer was used to measure the surface roughness of the specimens.

2.3.2 *Experimental Findings*

2.3.2.1 Material Removal Rate

The MRR is measured for both conventional and rotary EDM modes. Figure 2.13 [38] shows the relationship between the pulsed current and MRR at different workpiece rotation speeds. According to this figure, the rotary EDM mode is associated with a higher MRR than conventional EDM. At a constant speed, the MRR increases with an increasing pulsed current. At a pulsed current of 1 A, workpiece rotation only slightly affects MRR. Faster workpiece rotation implied more effective flushing of dielectric fluid and, ultimately, a higher MRR.

Figure 2.14 [38] shows the effect of pulse-on duration on MRRs at different rotation speeds. In conventional EDM, the MRR initially increases with increasing

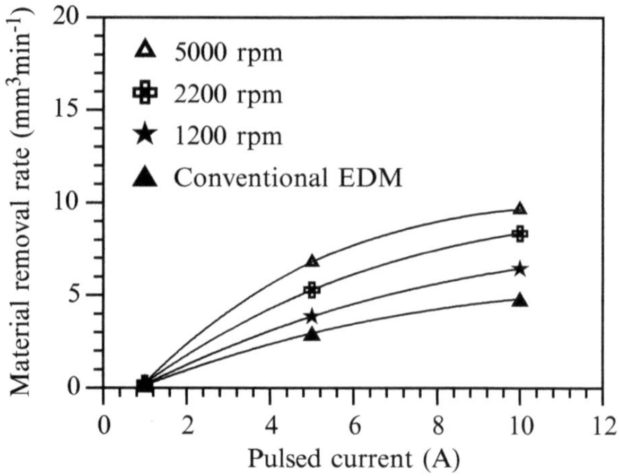

Fig. 2.13 Effect of pulsed current on MRR at different workpiece speeds (pulse-on duration 20 μs) [38], reprinted with permission

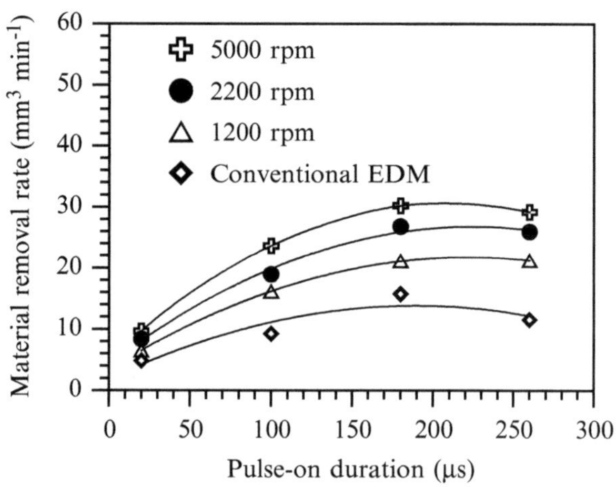

Fig. 2.14 Effect of pulse-on duration on MRR at different workpiece speeds (pulsed current 10 A) [38], reprinted with permission

pulse-on duration owing to the higher energy input rate. A peak value is reached as the pulse-on duration increases beyond a certain value. The material removal in one spark becomes so large that the interelectrode gap condition for the next discharge is disturbed. This discharge instability induced by debris at the gap lowers discharge efficiency. For example, the eroded chips filling the gap will hinder a sound discharge. Experimental results indicate that the maximum removal rate in conventional EDM was 15.7 $mm^3\ min^{-1}$. The maximum MRR has an optimum pulse-on duration of 180 μs. The MRR in rotary EDM is approximately twice as high as in conventional EDM at the same pulse-on duration.

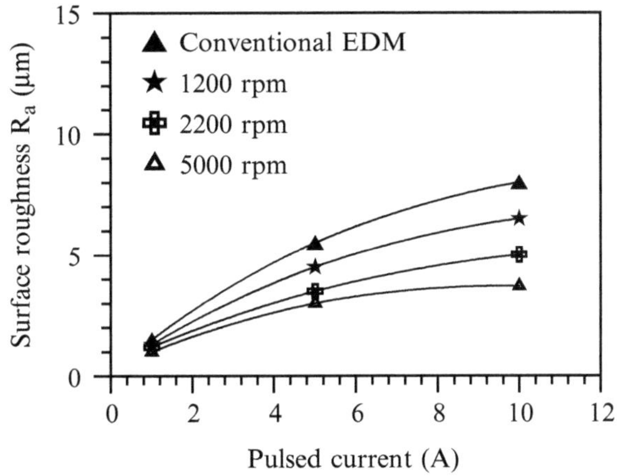

Fig. 2.15 Effect of pulsed current on surface roughness at different workpiece speeds (pulse-on duration 20 μs) [38], reprinted with permission

2.3.2.2 Surface Roughness

EDM erodes surfaces randomly. To determine the effect of machining parameters on the surface roughness (R_a) of tool steel, this study measured the surface profiles of the EDMed surfaces. Figure 2.15 [38] shows the effect of pulsed current on surface roughness for each of the specimen rotation speeds. Higher pulsed current produces a poorer surface finish, while a higher workpiece rotation speed produces a lower surface roughness. On the other hand, the R_a values obtained at the pulsed current of 1 A vary less for various specimen speeds. High pulsed current causes more frequent cracking of the dielectric fluid, causing more melt expulsions and larger tensile residual stresses. These problems in turn result in poor surface finish and serious surface damage. The surface roughness decreases with the rotation speed at the same pulsed current. It can be attributed to the fact that higher rotational speed provides more thorough flushing in the gap, thus the recast layer on the machined surface is more washed off. Furthermore, the temperature distribution during the rotary EDM is more uniform than the conventional EDM; the associated residual stress on the recast layer is less detrimental. Better surface roughness is obtained through these mechanical and thermal effects.

To identify the effect of pulse-on duration on the surface roughness, the pulse-on duration was varied from 20 to 260 μs, while a pulsed current of 10 A is applied throughout the experiments. Figure 2.16 [38] shows the variation in the surface roughness for cases of conventional EDM and rotary EDM. This figure reveals that the surface roughness increases linearly with increased pulse-on duration. The surface roughness for rotary EDM is lower than conventional EDM for the same pulse-on duration. In fact, the roughness value of conventional EDM was almost double that of rotary EDM at 5,000 rpm. Comparing Figs. 2.15 and 2.16 shows that the surface roughness constantly increases with pulsed current as well as pulse-on

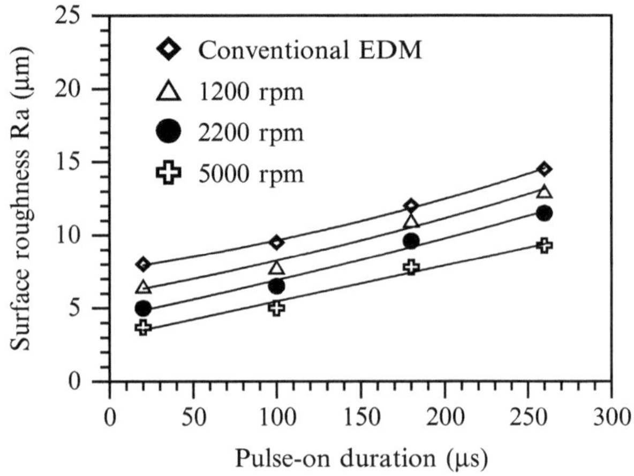

Fig. 2.16 Effect of pulse-on duration on surface roughness at different workpiece speeds (pulsed current 10 A) [38], reprinted with permission

duration. Both quantities deliver thermal energy to the machining location. More amount of molten material is produced, whereas complete flushing by the dielectric fluid becomes more difficult, and a resulted larger recast layer worsens the surface finish. An excellent machined finish can be obtained by setting the machine parameters at a low pulsed current and a small pulse-on duration combined with a high workpiece rotation speed, but this approach is more time consuming. The trends observed for conventional EDM agree with the results reported by previous investigators [28].

The following general conclusion can be made. In conventional EDM, insufficient flushing in the limited space results in the stagnation of the dielectric. The debris particles become suspended in the gap and cause short circuits, reducing machining performance. During rotary EDM, rotation of the workpiece enhances the dielectric fluid circulation and provides a centrifugal force of flow in the gap. Hence, the machining residue in the gap is discharged rapidly. This rapid discharge improves the machining stability, removes more material from the specimen, and produces better surface roughness. Figure 2.17 [38] schematically shows this effect. Furthermore, rotary EDM enhances heat transfer by convection, when the dielectric fluid is carried through the gap. The temperature on the workpiece surface is more uniform compared with the conventional EDM; hence, the resulted residual thermal stress on the machined surface is reduced. These forces produce good surface finish and minor surface damage during rotary EDM.

2.3.2.3 Surface Morphology

The high power density of a single electrical discharge associated with very short interaction time produces a high heating rate within a thin surface layer. The upper surface material in the extremely hot region will be vaporized, while the material

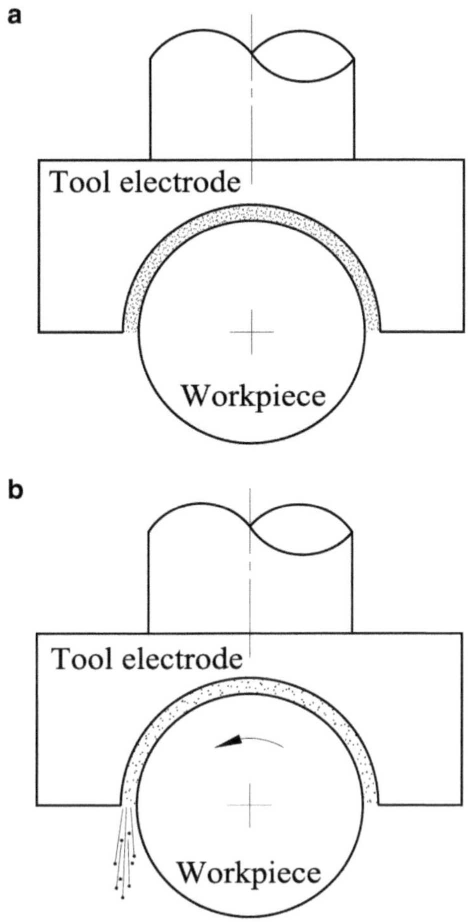

Fig. 2.17 Schematic of machining in conventional and rotary EDM [38], reprinted with permission. (**a**) Insufficient flushing in conventional EDM. (**b**) Enhanced flushing in rotary EDM

beneath will be melted and expelled away by high pressure dielectric turbulence from the electrical discharge. Figure 2.18 [38] shows the scanning electron micrographs of the machined surface obtained in conventional EDM, where I_p is the pulsed current and N is the rotation rate of workpiece. The surface morphology obtained at the pulsed current of 1 A exhibits damage, such as pinholes and micro-voids. These defects will cause stress concentration, eventually leading to deteriorated strength in use. When the discharge energy was increased to 10 A, the machined surface shows a more irregular topography and the defects include voids, debris, spherical particles, and micro-cracks. The spherical shape results from the surface tension. The micro-cracks are the consequence of the thermal stresses. The primary causes of the residual stress in the machined surface are the

Fig. 2.18 Micrographs of machined surfaces in conventional EDM (pulse-on duration 180 μs). (**a**) $I_p = 1$ A, N = 0 rpm (**b**) $I_p = 10$ A, N = 0 rpm [38], reprinted with permission

drastic heating and cooling rate and the nonuniform temperature distribution. The pinholes observed on the surface are due to gas bubbles expelled from the molten material during solidification. Comparing Fig. 2.18a, b reveals that the recast layer thickens with increasing pulsed current, because a high pulsed current causes a steep thermal gradient beneath the melting zone. Hence, some molten and floating material is not flushed out by the dielectric fluid and remains in the electrical discharge gap during EDM. This material will resolidify as an extensive recast layer and will decrease MRR. The ridge-rich surface is formed by the material melted during EDM, which is blasted out of the surface by discharge pressure. However, this material immediately reaches the solidification temperature by cooling in the surrounding working fluid.

Figure 2.19a–d [38] shows the micrographs of surfaces machined by rotary EDM at 1,200 and 5,000 rpm. These micrographs indicate that the number of the micro-voids, pinholes, and appendages formed by the electrical discharge channel decreases with increasing rotation speed. At 5,000 rpm, the machining surface shows fewer appendages. Meanwhile, the micro-voids do not exist at 5,000 rpm during EDM. Better surface finish is obtained. This phenomenon is attributed to the increased centrifugal force and reduced recast layer. Comparing Figs. 2.18 and 2.19, the surface finish obtained by conventional EDM is found to be poorer than for rotary EDM. Workpiece rotation in EDM improves the machining rate and surface finish.

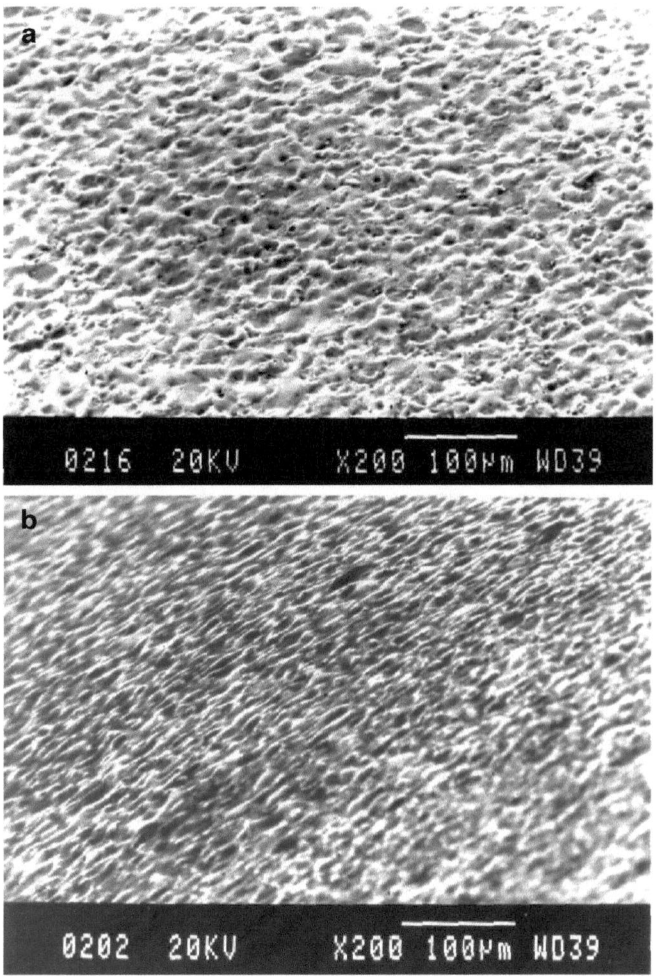

Fig. 2.19 Micrographs of machined surfaces in rotary EDM (pulse-on duration 180 µs). (**a**) $N = 1{,}200$ rpm and $I_p = 1$ A. (**b**) $N = 5{,}000$ rpm and $I_p = 1$ A. (**c**) $N = 1{,}200$ rpm and $I_p = 10$ A. (**d**) $N = 5{,}000$ rpm and $I_p = 10$ A [38], reprinted with permission

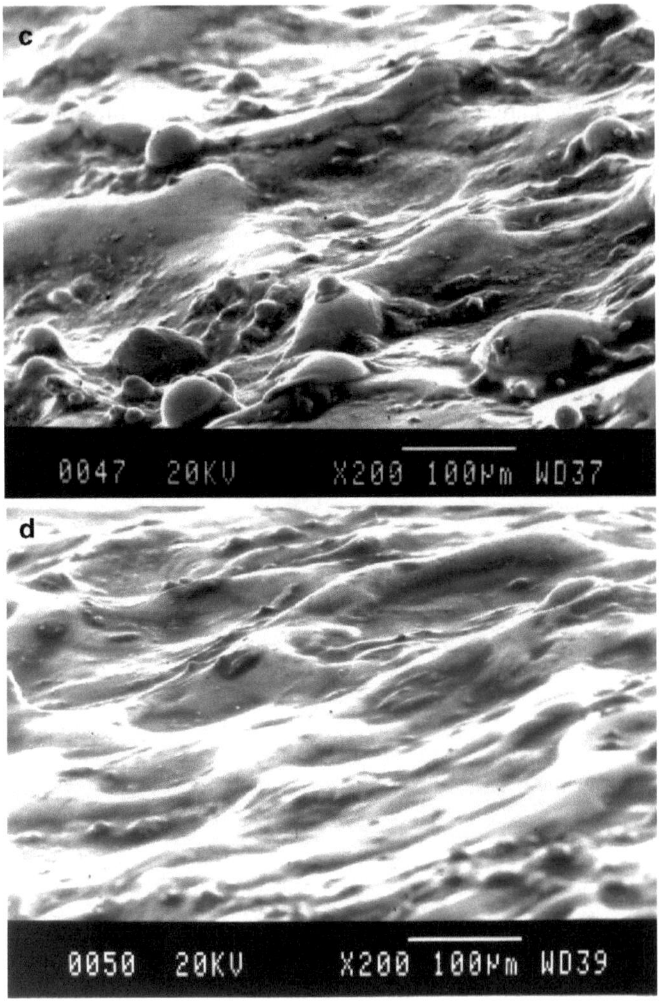

Fig. 2.19 (continued)

2.4 Effect of Electrical Discharge Machining on Surface Characteristics and Machining Damage of AISI D2 Tool Steel

2.4.1 Experimental Setup

AISI D2 tool steel, one of the carbon steels alloyed with Mo, Cr, and V, is widely used for various dies and cutters for its high strength and wear resistance due to the formation of chrome carbides in heat treatment. The base material used for this

Table 2.5 Experimental conditions for EDM

Dielectric	Kerosene
Work material	AISI D2 tool steel
Electrode material	Copper
Pulse current	1, 5, 10 A
Pulse-on duration	20, 100, 180 μs
Pulse-off duration	20 μs

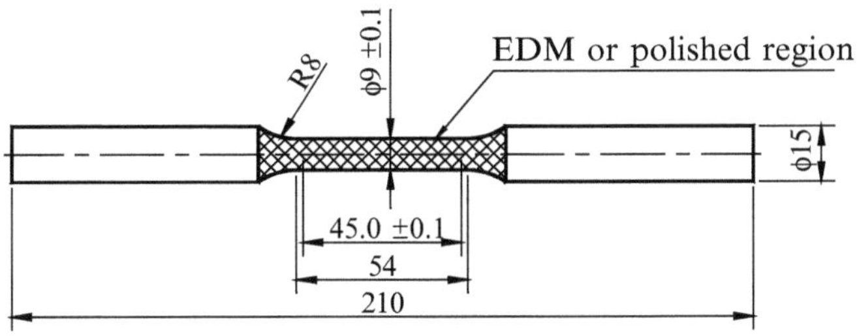

Fig. 2.20 Geometry of the tensile specimen (dimension in mm) [9], reprinted with permission

study was an AISI D2 tool steel. The provided rod of tool steel was prepared by machining leaving a cut depth of 0.4 mm, followed by the heat treatment and subsequent grinding. During the heat treatment, the material was heated to 1,030°C at a heating rate of 20°C/min. Then, the material was held at 1,030°C for 1 h and was quenched. After quenching, the specimens were tempered at 520°C for 2 h and then air cooling. Table 2.3 lists the chemical composition (wt.%) of the material.

There were two types of specimens used in this study. The first kind was the polished specimen and the second kind was the EDM specimen. The mechanical polishing was applied with 3 and 1 μm diamond paste, first in the circumferential direction and finally in the axial direction. To prepare the EDM specimens, the ground specimens were further machined by EDM to remove the unnecessary material at various pulsed currents and pulse-on durations.

An EDM (Charme CD-50M) with servo-control was used to conduct the experiments. Figure 2.2 schematically displays the experimental setup of the EDM processes, in which a rotation mechanism was developed and installed on the machining table. A DC motor rotated the specimen at 1,200 rpm. EDM was applied to the central part of the specimen. The tool electrode geometry was reproduced in the workpiece after the process. The copper electrode was the positive polarity and the specimen was the negative polarity during the EDM process. The processing parameters of the EDM are shown in Table 2.5. Figure 2.20 [9] shows the geometric shape and dimensions of the tensile specimen, as recommended by the American Society for Testing and Materials (ASTM) E 8M-00b [39]. In order to determine the EDM damage, the polished specimen and the EDM specimen were tested in an MTS 810 servohydraulic system under monotonic loading at room temperature. The machined surface and the recast layer were examined using the SEM to determine the nature of the damage. A profilometer

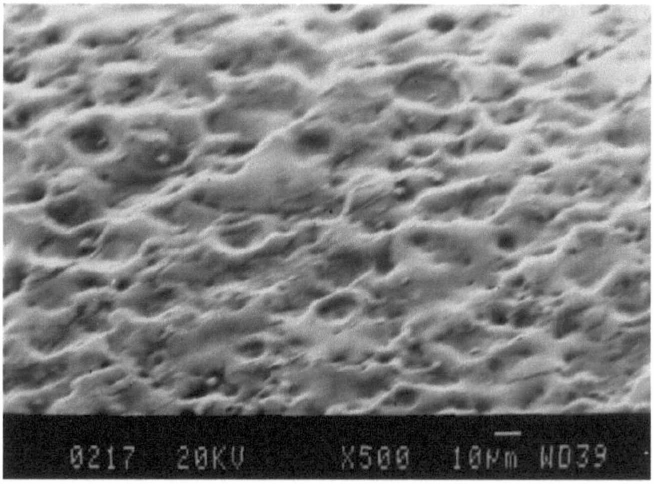

Fig. 2.21 Surface morphology of the faces after EDM with 1 A pulsed current, observed by SEM (pulse-on duration 180 μs) [9], reprinted with permission

was used to measure the surface roughness of the specimens. Residual stress was determined by the X-ray diffractometer technique. The interference line profiles of the (211) planes were measured using $Cr - K_{\beta}$ radiation.

2.4.2 Experimental Findings

2.4.2.1 Surface Roughness

EDM spark erodes surfaces randomly. Figure 2.21 [9] reveals that the EDM process produces damage such as micro-voids and craters on the machined surface. To determine the effect of EDM on the surface roughness of AISI D2 tool steel, the surface profiles of the topography of the EDM surfaces were measured. The cut-off length for each measurement trace was 0.8 mm. Five cut-off lengths were measured on each specimen and their surface roughness was averaged. Figure 2.22 [9] shows the measurement results. The surface roughness on the machined surface varied from 1.3 to 11.0 μm. From these results it is clear that a higher pulsed current causes a poorer surface finish. An excellent machined finish can be obtained by setting the machine parameters at a low pulsed current and a small pulse-on duration, but this approach is more time consuming. The regression analysis of the experimental data yields the semiempirical model

$$R_a = 0.83(I_p)^{0.79}(\tau_{on})^{0.12} \tag{2.7}$$

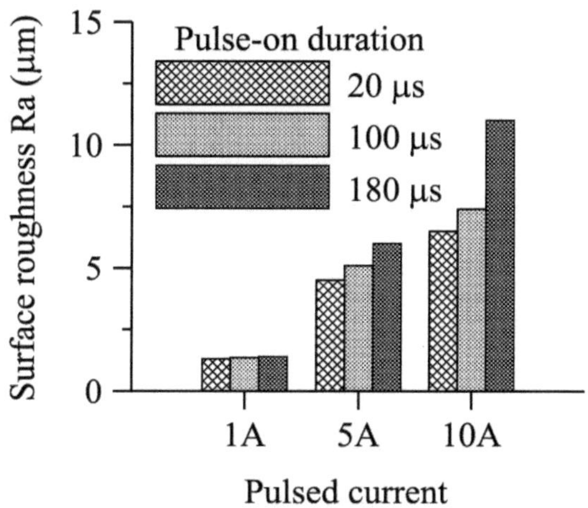

Fig. 2.22 The average surface roughness at various processing conditions [9], reprinted with permission

where R_a is the surface roughness; I_p denotes the pulsed current, and τ_{on} represents the pulse-on duration. Equation (2.7) gives an average error of 9.5% with respect to the experimental results. This semiempirical mode indicates that the pulsed current has a more dominant effect on the surface roughness compared to the pulse-on duration. It can be attributed to the fact that a higher pulsed current may cause more frequent cracking of the dielectric fluid; there is more frequent melt expulsion resulting in the poorer surface finish.

2.4.2.2 Recast Layer

The cross-section profile of the recast layer in the EDM specimen is shown in Fig. 2.23 [9]. Previous investigators [40] showed that the depth of the EDM damage is equal to the thickness of the recast layer. Figure 2.24 [9] shows the dependence of the recast layer thickness on the EDM parameters. Each data point in the figure represents the average thickness of 25 readings measured at various locations on the cross section of the specimen. Sections through a globule appendage were ignored as such reading was an unusual irregularity. The figure shows that the thickness of the recast layer on the machined surface varies from 7 to 31 μm, and grows significantly with the pulsed current and pulse-on duration. The observations are consistent with the experimental results reported by previous investigators [41]. The correlation between the thickness of the recast layer, t, and the machining conditions using regression analysis yields

$$t = 2.93(I_p)^{0.36}(\tau_{on})^{0.29} \tag{2.8}$$

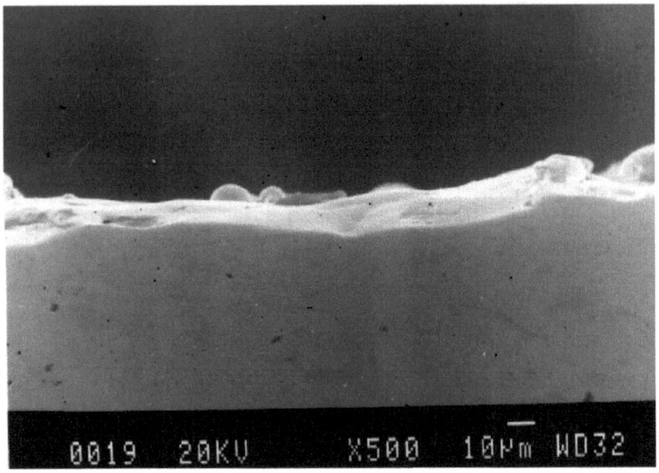

Fig. 2.23 SEM photomicrograph of the recast layer of the tool steel after EDM at 5 A (pulse-on duration 20 μs, white color indicates the recast layer) [9], reprinted with permission

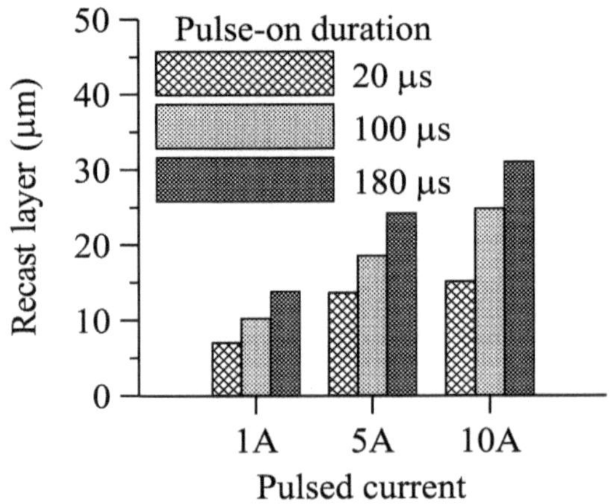

Fig. 2.24 Thickness of the recast layer at various EDM conditions [9], reprinted with permission

Equation (2.8) gives an average error of 4.8% with respect to the experimental results. This model indicates that the thickness of the recast layer is proportional to the pulsed current to a power of about 0.36 and to the pulse-on duration to a power of about 0.29.

2.4.2.3 Residual Stress

The primary cause of residual stress is the fact that the machined surface has been heated and cooled at a high rate and with nonuniform temperature distribution.

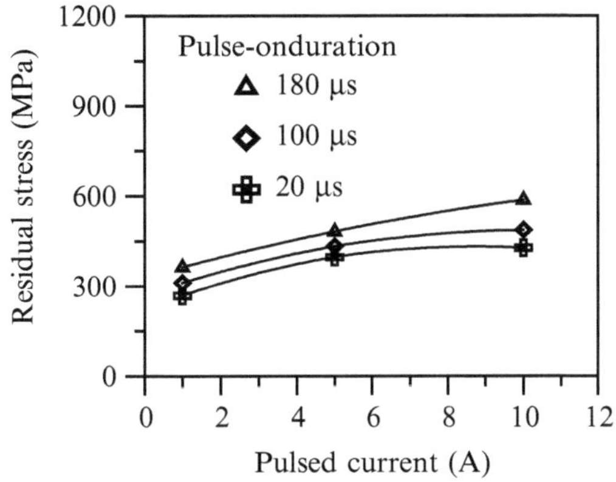

Fig. 2.25 The residual stress at various EDM conditions [9], reprinted with permission

Figure 2.25 [9] shows the surface residual stress of the EDM specimens. Residual tensile stresses of 268–587 MPa were found on the EDM surface. Using regression analysis to determine the correlation between the residual stress, σ_r, and the machining parameters produces

$$\sigma_r = 192.48(I_p)^{0.20}(\tau_{on})^{0.11} \tag{2.9}$$

Equation (2.9) gives an average error of 3.7% corresponding to the experimental results. The equation shows that the magnitude of residual stress depends on the EDM parameters, while the residual tensile stress increases with the increase in pulsed current and pulse-on duration. This is because of the fact that heat supplied to the workpiece due to spark is higher at larger pulsed current and pulse-on duration and hence higher temperature rise is shown. The thermal stress field is tensile and is consistently found on the EDM surface, conforming with results reported by earlier investigators [18].

2.4.2.4 Tensile Strength

The material constitutive response is usually determined through tensile tests that allow determination of the mechanical properties. Figures 2.26 [9] and 2.27 [9] show the engineering stress–strain responses of the polished specimen and the EDM specimen at room temperature, respectively. These curves indicate that the Young's modulus ($E = 140$ GPa) was quite consistent. It is clear that working conditions can influence the tensile strength. The strength of the polished specimen is superior to that of the EDM specimen. A degradation of strength is particularly significant at high pulsed current and long pulse-on duration. This phenomenon occurs because of the discharge energy, which alters the material microstructure, initiates machining damage on the surface, and causes stress concentration,

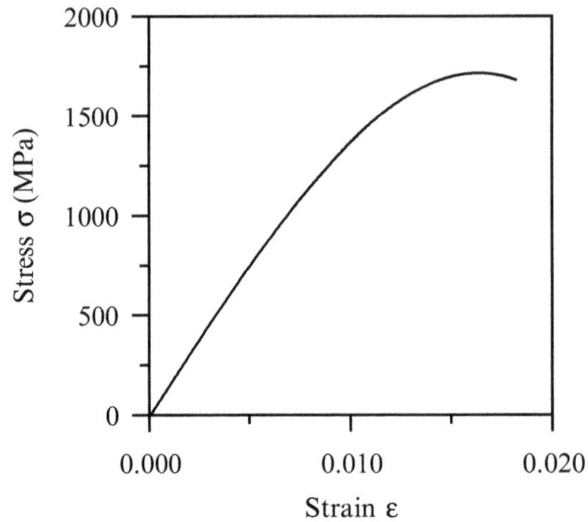

Fig. 2.26 Stress–strain behavior of the polished specimen [9], reprinted with permission

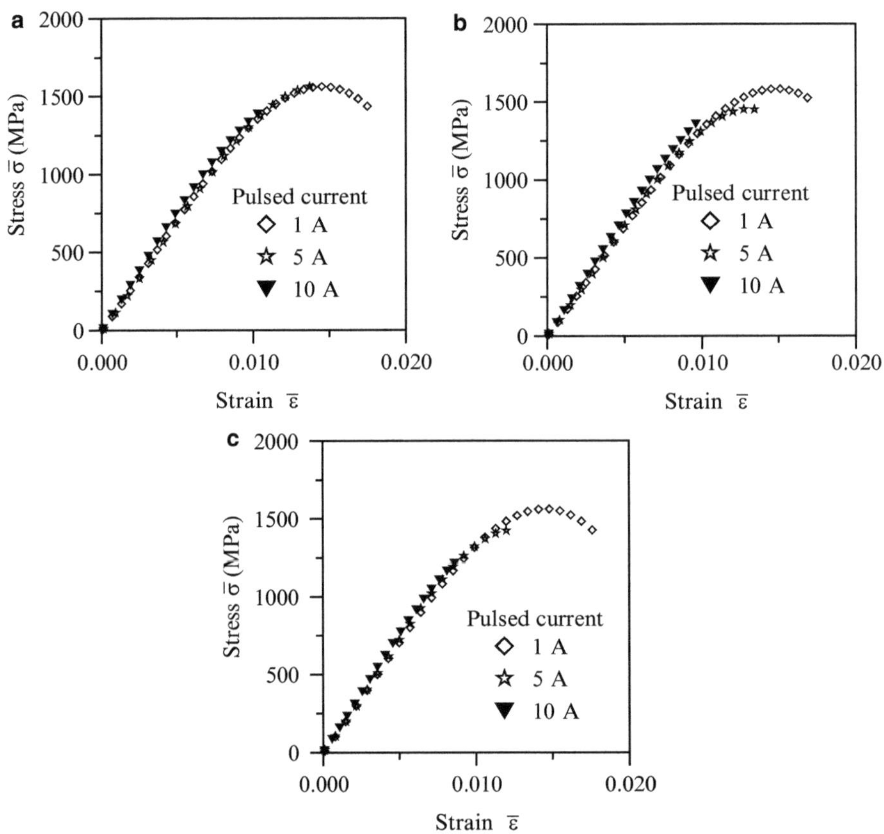

Fig. 2.27 Stress–strain behavior of the EDM specimen (**a**) 20 μs; (**b**) 100 μs; (**c**) 180 μs [9], reprinted with permission

eventually decreasing the service strength. It is therefore considered essential that the surface damage layer induced by EDM process should be removed before they are put to use.

2.4.2.5 Machining Damage

The EDM specimen was composed of the damaged zone occurring at the recast layer and the base metal zone beneath the damaged layer. From the physical point of view, the recast layer can be considered to be the machining damaged zone of microscopic defects, where the strain energy density degradation occurs deteriorating its load carrying capacity. Based on the concept of strain energy density exhaustion, the inherent strain energy density in the material was exhausted after damage. The concept of strain energy density exhaustion was explained in detail by Hibbeler and Fan [42] and Budynas et al. [43]. Furthermore, the ductility is an inherent property of a material, and the conventional measurement of ductility lies in the fracture strain in uniaxial tension. The material fracture strain is nevertheless sensitive to changes of microstructure and mechanical properties so the fracture strain is introduced into damage variable. Under the hypothesis of isotropy, the damage variable, D, can be introduced as

$$D = 1 - \frac{\bar{u}}{u} \tag{2.10}$$

where u is the original strain energy density before damage and represents the strain energy density absorbed by the polished specimen. $\bar{u}$ is the remaining strain energy density of the EDM specimen. From (2.10), if $\bar{u} = u$, then $D = 0$. This corresponds to the undamaged state of the specimen. When the specimen reaches the completely damaged state, the strain energy density of the material is fully exhausted, $D = D_c = 1$. D_c is the critical value of the damage variable. When the strain energy density reaches a critical value, the material will have the damage failure. The strain energy density of the polished specimen, u, was estimated from the available stress–strain curve obtained from the polished specimen. We have, therefore,

$$u = \int_0^{\varepsilon_f} \sigma \, d\varepsilon \tag{2.11}$$

where σ and ε are the stress and strain of the polished specimen, respectively. ε_f is the fracture strain. Using the cubic polynomial equation to fit the data from tensile test of the polished specimen yields

$$\sigma = -9.63 + 1.53 \times 10^5 \varepsilon + 6.44 \times 10^5 (\varepsilon)^2 - 2.16 \times 10^8 (\varepsilon)^3, \quad 0 \leq \varepsilon \leq 0.018 \tag{2.12}$$

Table 2.6 The fracture strain and coefficients of the fitting curves of the EDM specimens

Types of pretreatment		$\bar{\varepsilon}_f$	a_0	a_1	a_2	a_3
EDM	$I_p = 1\,A,\ \tau_{on} = 20\ \mu s$	1.78×10^{-2}	-4.64	1.30×10^5	4.35×10^6	-4.04×10^8
	$I_p = 5\,A,\ \tau_{on} = 20\ \mu s$	1.43×10^{-2}	-2.27	1.23×10^5	5.48×10^6	-4.49×10^8
	$I_p = 10\,A,\ \tau_{on} = 20\ \mu s$	1.07×10^{-2}	0.06	1.50×10^5	2.38×10^6	-3.70×10^8
	$I_p = 1\,A,\ \tau_{on} = 100\ \mu s$	1.72×10^{-2}	-3.85	1.29×10^5	4.27×10^6	-3.88×10^8
	$I_p = 5\,A,\ \tau_{on} = 100\ \mu s$	1.34×10^{-2}	1.85	1.21×10^5	7.06×10^6	-5.99×10^8
	$I_p = 10\,A,\ \tau_{on} = 100\ \mu s$	0.98×10^{-2}	0.06	1.45×10^5	3.97×10^6	-4.52×10^8
	$I_p = 1\,A,\ \tau_{on} = 180\ \mu s$	1.59×10^{-2}	-3.15	1.26×10^5	4.18×10^6	-3.71×10^8
	$I_p = 5\,A\ \tau_{on} = 180\ \mu s$	1.20×10^{-2}	5.25	1.18×10^5	8.60×10^6	-7.17×10^8
	$I_p = 10\,A,\ \tau_{on} = 180\ \mu s$	0.86×10^{-2}	0.06	1.40×10^5	5.64×10^6	-6.33×10^8

The strain energy density of the EDM specimen, $\bar{u}$, was estimated from the available stress–strain curve obtained from the EDM specimen.

$$\bar{u} = \int_0^{\bar{\varepsilon}_f} \bar{\sigma}\, \mathrm{d}\bar{\varepsilon} \tag{2.13}$$

The relationship between the stress and strain of the EDM specimens can be given as follows:

$$\bar{\sigma} = a_0 + a_1\bar{\varepsilon} + a_2(\bar{\varepsilon})^2 + a_3(\bar{\varepsilon})^3, \quad 0 \le \bar{\varepsilon} \le \bar{\varepsilon}_f \tag{2.14}$$

where $\bar{\sigma}$ and $\bar{\varepsilon}$ are the stress and strain of the EDM specimen, respectively. $\bar{\varepsilon}_f$ is the strain at rupture and a_0, a_1, a_2, and a_3 are the fitting coefficients listed in Table 2.6. Substituting (2.11) and (2.13) into (2.10) gives

$$D = 1 - \frac{\int_0^{\bar{\varepsilon}_f} \bar{\sigma}\, \mathrm{d}\varepsilon}{\int_0^{\varepsilon_f} \sigma\, \mathrm{d}\varepsilon} \tag{2.15}$$

This damage variable characterizes the state of deterioration. The substitution of (2.12) and (2.14) into the (2.15) yields the machining damage for the EDM specimens as shown in Fig. 2.28 [9]. The regression analysis of the EDM damage data produces the empirical formula

$$D = 0.07(I_\mathrm{p})^{0.66}(\tau_\mathrm{on})^{0.21} \tag{2.16}$$

Equation (2.16) gives an average error of 13.3% with respect to the experimental results. Notable is the increase in machining damage as the discharge energy is increased. This effect can be explained by the fact that the specimen subjected to EDM develops micro-damages and occurs the stress concentration causing a crack initiation, which deteriorates its carrying capacity of strain energy density.

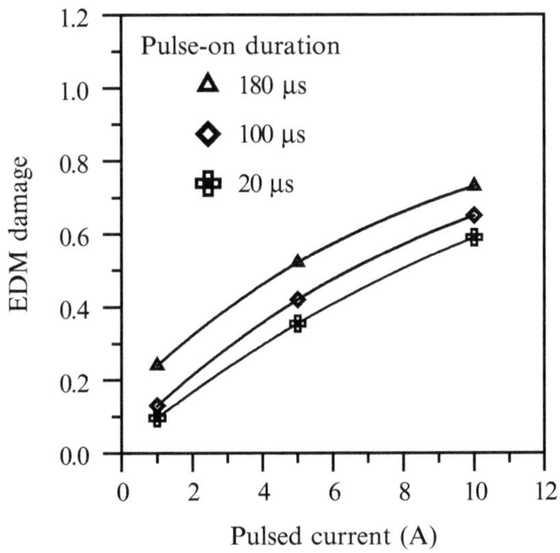

Fig. 2.28 The EDM damage at various machining conditions [9], reprinted with permission

2.5 AFM Surface Imaging of AISI D2 Tool Steel Machined by the EDM Process

2.5.1 Experimental Setup

The specimen was made of the AISI D2 tool steel, which is widely used in the mold industry. The material was heated to 1,030°C at a heating rate of 20°C/min. It was kept at 1,030°C for 1 h and then quenched. After quenching, the specimens were tempered at 520°C for 2 h and then air cool. Table 2.3 lists the chemical composition (wt.%) of the material, while Table 2.4 lists the mechanical properties of the AISI D2 tool steel. The EDM specimens were performed on a die-sinking EDM machine model type Roboform 2000. The experiment was carried out in kerosene dielectric covering the workpiece by 10 mm. A cylindrical copper rod of 3 mm diameter was selected as the tool electrode for drilling the workpiece. The machining depth was 1 mm. The copper electrode was the negative polarity and the specimen was the positive polarity during the EDM process. During EDM, the primary parameters are pulsed current, pulsed voltage, pulse-on duration, and pulse-off duration. From the analysis carried out by Tung [44], Lee and Yur [45], in which the influence of these four parameters was investigated for AISI 1045 carbon steel and H13 die-steel using the Taguchi method, it is known that pulsed current and pulse-on duration are the principal factors which influence surface roughness, thickness of the recast layer, and induced stress. For this reason, this study establishes its full factorial design upon these same two parameters. Table 2.7 shows the EDM conditions. During the EDM process, the pulse-off duration setting

Table 2.7 Experimental conditions for EDM

Dielectric	Kerosene
Work material	AISI D2 tool steel
Electrode material	Copper
Pulsed current	0.5, 1.0, 1.5 A
Pulse-on duration	3.2, 6.4 μs
Pulse-off duration	20 μs

20 μs could effectively control the flushing of the debris from the gap, giving machining stability. Hence, the effect of the pulse-off duration on the machined characteristics was not considered in the present work.

2.5.2 *Experimental Findings*

2.5.2.1 Surface Morphology

During the EDM process, the primary parameters were pulsed current and pulse-on duration, both of which are settings of the power supply. In order to assess the surface measurement results, an AFM study of the surface nanomorphology of the EDM machined surface was conducted. Figure 2.29 shows the three-dimensional AFM images of the machined surface obtained from the EDM specimens, where I_p is the pulsed current, and τ_{on} denotes the pulse-on duration. The darker contrast corresponds to the lower areas of the surface, and the brighter corresponds to the higher. It is clear that the surface microgeometry characteristics include machining damages such as ridge-rich surfaces, micro-voids, and micro-cracks. The ridge-rich surface was formed by material melted during EDM, and blasted out of the surface by the discharge pressure. However, the surface immediately reached the solidification temperature being cooled by the surrounding working fluid. The micro-voids can be attributed to the gas bubbles expelled from the molten material during solidification. The micro-cracks were the result of the thermal stresses. The primary causes of the residual stress in the machined surface were the drastic heating and cooling rate and the nonuniform temperature distribution. In addition, the morphology of the EDM surface was dependent on the applied discharge energy. When applying the smaller pulsed current, 0.5 A, and the pulse-on duration, 3.2 μs (Fig. 2.29a [6]), the surface characteristics have minor hillocks and valleys. When the pulsed current and pulse-on duration increased (Fig. 2.29b [6]), the machined surface exhibited a deeper crack or void and more pronounced defects. The EDM specimens were also analyzed by means of SEM technique. Selected images are shown in Fig. 2.30 [6]. Compared with the AFM method, the SEM technique allows a rapid survey of large sample areas, but it does not reveal the depth of the micro-cracks and the 3D surface textures of the EDM specimen. In the AFM image, it can be seen that the pits have a dark contrast, an observation in accordance with that found in the SEM image.

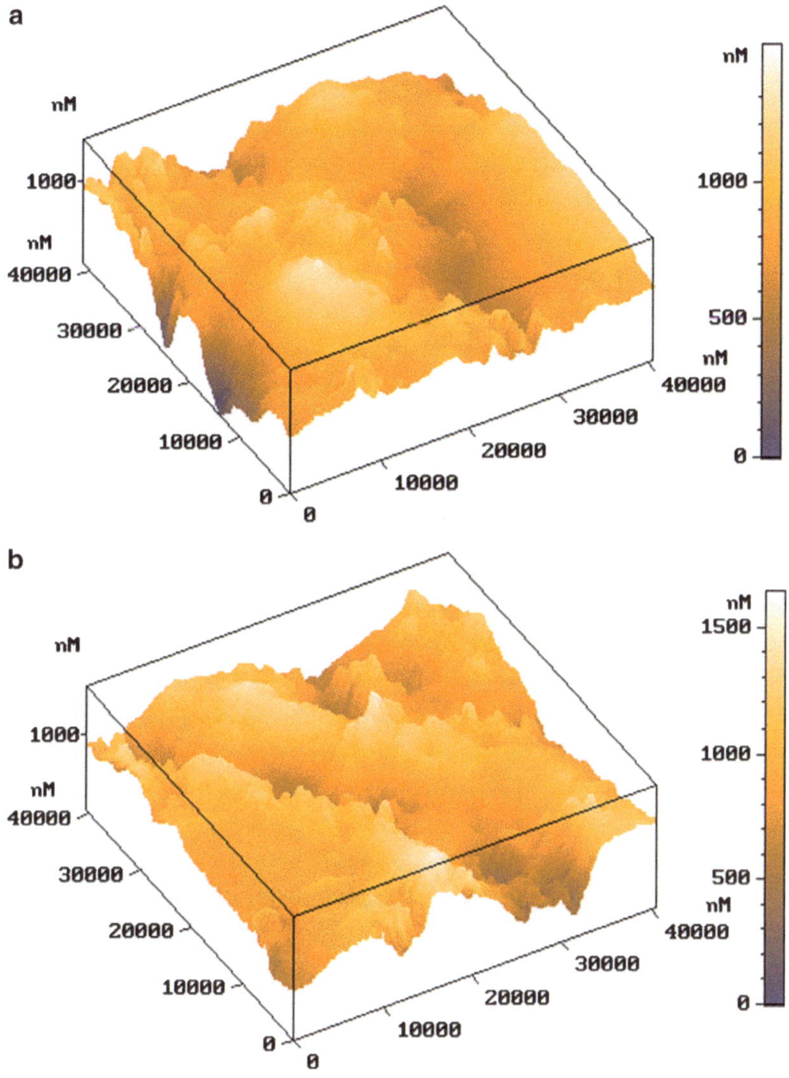

Fig. 2.29 The three-dimensional AFM images of the AISI D2 tool steel after EDM at (**a**) $I_p = 0.5$ A, $\tau_{on} = 3.2$ μs, (**b**) $I_p = 1.5$ A, $\tau_{on} = 6.4$ μs [6], reprinted with permission

2.5.2.2 Surface Roughness

EDM erodes surfaces randomly. To determine the effect of the EDM process on the surface roughness of the tool steel, the surface profiles of the EDM specimens were measured by AFM. The average surface roughness, R_a, of the machined specimen was calculated from the AFM surface topographic data in a scanning area of 40 μm × 40 μm by the equation

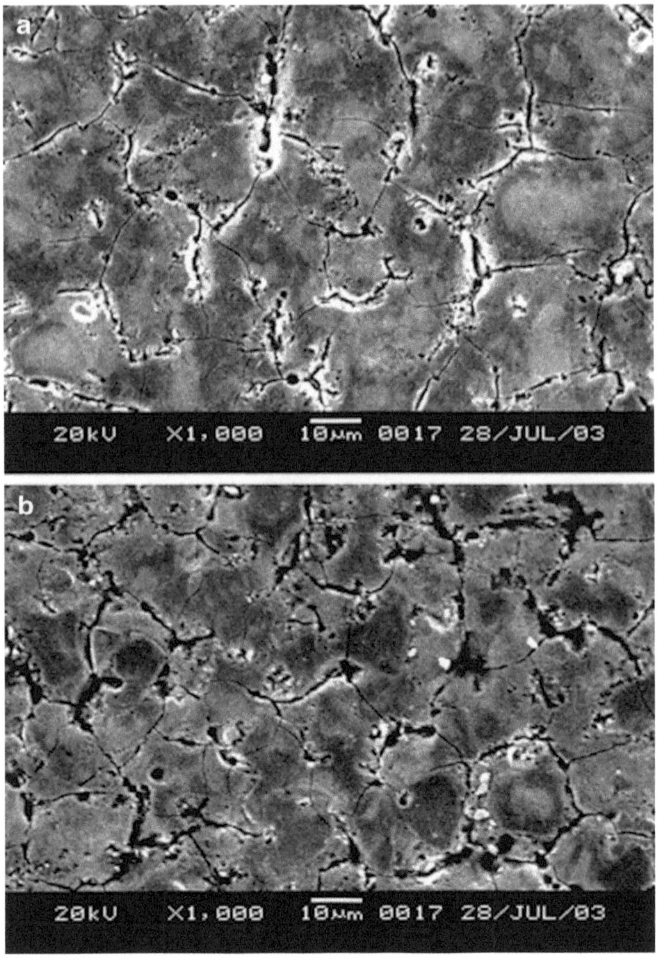

Fig. 2.30 The SEM images of the AISI D2 tool steel after EDM at (**a**) $I_p = 0.5$ A, $\tau_{on} = 3.2$ μs, (**b**) $I_p = 1.5$ A, $\tau_{on} = 6.4$ μs [6], reprinted with permission

$$R_a = \frac{1}{MN}\sum_{i=1}^{M}\sum_{j=1}^{N}\left|Z(x_i, y_j)\right| \tag{2.17}$$

where $Z(x_i, y_j)$ denotes the height of a surface point (x_i, y_j) relative to the mean plane; *MN* is the number of points in an analyzed area. Figure 2.31 [6] shows the measurement results. The surface roughness on the machined surface varied from 103 to 172 nm. From these results it is clear that a higher pulsed current and a longer pulse-on duration cause a poorer surface finish. This can be attributed to the fact that a higher pulsed current and a longer pulse-on duration may cause more frequent cracking of the dielectric fluid, there is also more frequent melt expulsion leading to

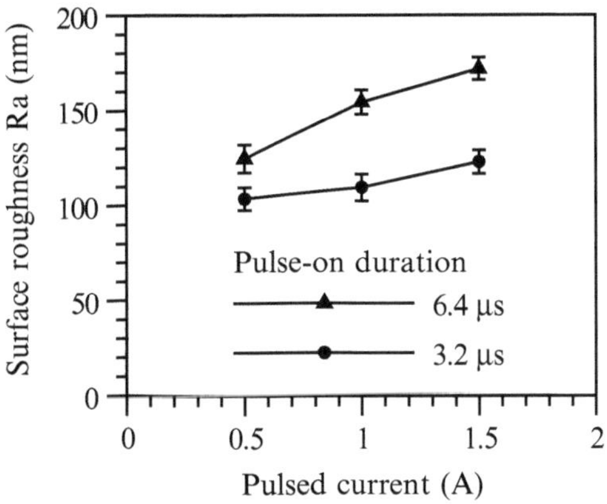

Fig. 2.31 The average surface roughness at various machining conditions [6], reprinted with permission

the formation of deeper and larger craters on the surface of the workpiece. Comparing with the results of Figs. 2.29 and 2.31, we find that an excellent machined finish can be obtained by setting the machine parameters at a low pulsed current and a small pulse-on duration. The trends agree with the results reported by previous investigators [28].

2.5.2.3 Micro-Cracks

In order to measure the maximum depth of the micro-cracks of the EDM specimen, the AFM was used to measure the object generating the surface topography (Fig. 2.32a [6]) and the cross-section profile (Fig. 2.32b [6]). A dotted line in the surface topography shows the position of the cross-section profile. The maximum depth of the micro-cracks can be determined from the distance between the highest peak and the lowest valley. The maximum depth of the micro-cracks is defined as

$$D_{max} = t_{max} - t_{min} \tag{2.18}$$

where D_{max} denotes the maximum depth, t_{max} and t_{min} represent the maximum height and the minimum height of the section profile, respectively. Figure 2.33 [6] shows the dependence of the maximum depth of the micro-cracks on the EDM parameters. The figure shows that the depth of the micro-cracks on the EDM specimen ranges from 1,272 to 1,873 nm increasing significantly with the pulsed current and pulse-on duration. This effect can be explained by the fact that high energy causes a steep thermal gradient beneath the melting zone. Previous investigators [15, 18] have reported that the depth of the surface cracks increases with increasing discharge energy, and the depth of the cracks correlates well with the thickness of machined damage. The machined damages layer generated by the EDM process produces a harmful influence decreasing the service strength and life of the virgin material. This damage layer should be removed before being put to use. It is therefore

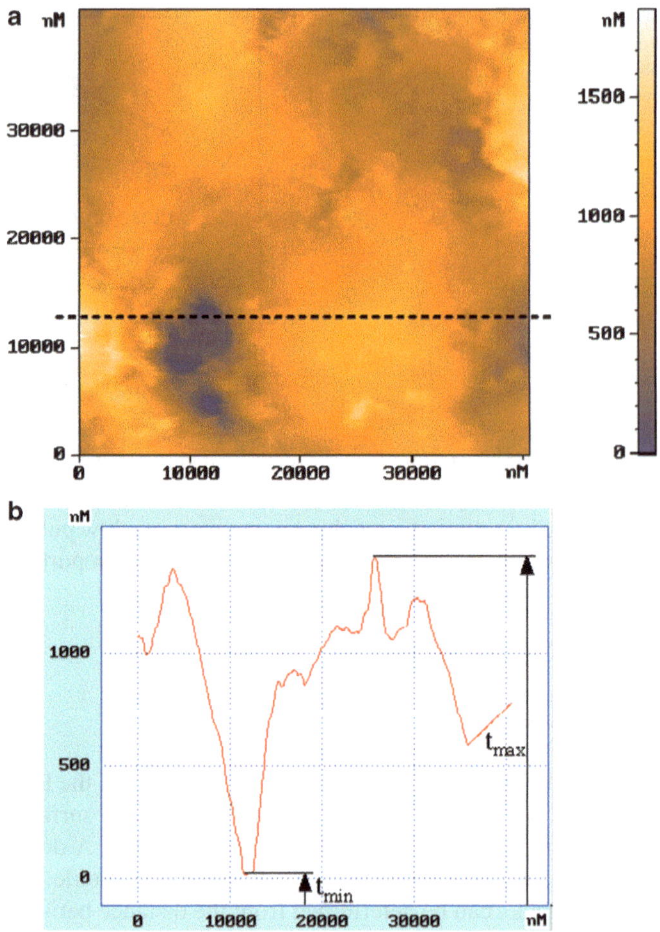

Fig. 2.32 Observation on an EDM surface by AFM (**a**) topography with 1.5 A pulsed current and pulse-on duration 6.4 μs, (**b**) cross-section profile on the morphology signal along the *dotted line* in image (**a**) [6], reprinted with permission

recommended that the EDM specimen should be polished down to at least the maximum depth of the micro-cracks in order to improve its service life.

2.6 Conclusions

Our investigation of carbon fiber-reinforced carbon composite materials yielded empirical formulae for various machining conditions. These findings lead to the following conclusions. Reducing pulse energy can minimize delamination defects around holes in both the upper and lower surfaces of composite materials.

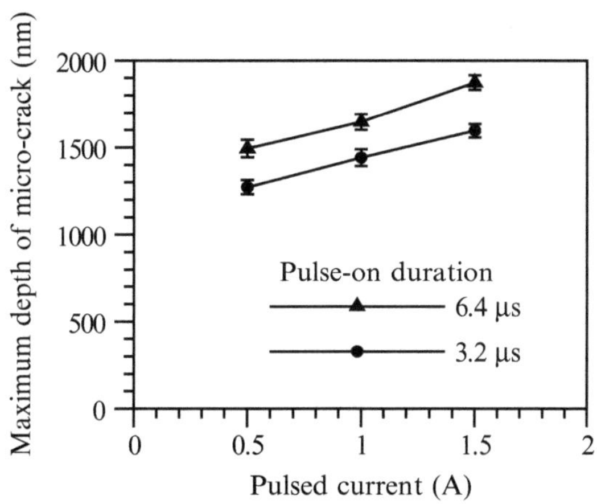

Fig. 2.33 The maximum depth of the micro-cracks on the EDM specimen at various machining conditions [6], reprinted with permission

Moreover, increasing discharge energy causes high temperatures to form, increase surface roughness, create a much larger recast layer, and cause delamination at the cutting edge. The optimal MRR for carbon fiber-reinforced carbon composites occurs at a pulsed current of 5 A and a pulse-on duration of 100 μs. Subsequent melting and vaporization removes the composite material. Our findings demonstrate that EDM is effective in the machining of carbon fiber-reinforced carbon composite materials.

In addition to optimizing electrical process parameters in conventional EDM, workpiece rotation is effective at improving the MRR. Debris in the electrode gap increase discharge instability in conventional EDM. However, in rotary EDM, centrifugal force improves gap flushing and machining efficiency. Rotary EDM MRRs are up to twice those of conventional EDM. In addition, surface roughness decreases with increasing rotation speed; rotation of the sample can reduce the formation of micro-voids and defects on the machined surface. Hence, EDM of an axial-symmetrical die is improved by this technique. This work demonstrates the advantages of providing workpiece rotation in rotary EDM.

Our experimental investigation of AISI D2 tool steel provided empirical formulae for the prediction of various EDM conditions effects on surface roughness, thickness of the recast layer, residual stresses, and EDM damage. The recast layer becomes thicker with increasing pulsed current and pulse-on duration. An excellent machined finish can be obtained by setting the machine parameters to low pulse energy. Lowering the pulse energy reduces the frequency of bursts of dielectric fluid and melt expulsions, and reduces tensile residual stresses. The EDM process causes damage such as pinholes, micro-voids, and craters in the surface layer. Moreover, sample melting increases with increasing peak current, causing severe damage to the surface and subsurface. This reduces the specimen's strain energy density and has a detrimental effect on the strength of the finished product.

EDM is a thermal process to remove material from a workpiece electrode. An AFM study of the surface morphology of the EDM specimen revealed that higher discharge energy results in a poorer surface structure, thus, low discharge energy should be used to avoid excessive damage to the specimen. AFM can provide a three-dimensional image on a nanometer scale to evaluate the depth of micro-cracks formed on the EDM surface. The EDM components should be polished down to at least the maximum depth of the micro-cracks prior to applying this technique.

References

1. Fonda P et al (2008) A fundamental study on Ti-6Al-4V's thermal and electrical properties and their relation to EDM productivity. J Mater Process Technol 202:583–589
2. Clijsters S et al (2010) EDM technology and strategy development for the manufacturing of complex parts in SiSiC. J Mater Process Technol 210:631–641
3. Yan BH et al (2000) Machining characteristics of Al_2O_3/6061Al composite using rotary EDM with a disklike electrode. Int J Adv Manuf Technol 16:322–333
4. Masuzawa T et al (1994) Combined electrical machining process for micronozzle fabrication. CIRP Ann Manuf Technol 43:189–192
5. Casas B et al (2006) Fracture and fatigue behavior of electrical-discharge machined cemented carbides. Int J Refract Metals Hard Mater 24:162–167
6. Guu YH (2005) AFM surface imaging of AISI D2 tool steel machined by the EDM process. Appl Surf Sci 242:245–250
7. Chattopadhyay KD et al (2009) Development of empirical model for different process parameters during rotary electrical discharge machining of copper-steel (EN-8) system. J Mater Process Technol 209:1454–1465
8. Satsangi PS, Chattopadhyay KD (2010) A study of machining parameters during electrical discharge machining of steel by a rotary copper electrode. In: International conference on advances in materials and processing technologies, AMPT 2008, Nov 2–5 2008, Manama, Bahrain, pp 756–763
9. Guu YH, Hocheng H, Chou CY, Deng CS (2003) Effect of electrical discharge machining on surface characteristics and machining damage of AISI D2 tool steel. Mater Sci Eng A 358:37–43
10. Lazar M-B, Xirouchakis P (2011) Experimental analysis of drilling fiber reinforced composites. Int J Mach Tools Manuf 51:937–946
11. Ohzeki H et al (2012) Drilling of carbon fiber reinforced plastic composites with feedback control based on cutting force. In: ICMDT 2011—advances in manufacturing, machine design and tribology, Japan, pp 52–64
12. Tsao CC, Chiu YC (2011) Evaluation of drilling parameters on thrust force in drilling carbon fiber reinforced plastic (CFRP) composite laminates using compound core-special drills. Int J Mach Tools Manuf 51:740–744
13. Hocheng H et al (1998) Feasibility analysis of electrical-discharge machining of carbon-carbon composites. Mater Manufact Process 13:117–132
14. Gourgouletis K et al (2011) Electrical discharge drilling of carbon fibre reinforced composite materials. Int J Mach Mach Mater 10:187–201
15. Ghanem F et al (2011) Effects of finishing processes on the fatigue life improvements of electro-machined surfaces of tool steel. Int J Adv Manuf Technol 52:583–595
16. Lin Y-C et al (2008) Effects of electrical discharge energy on machining performance and bending strength of cemented tungsten carbides. J Mater Process Technol 206:491–499

17. Lai LC et al (2010) Influence of electrical discharged machining and surface defects on the fatigue strength of electrodeposited nanocrystalline Ni. Int J Fatigue 32:584–591
18. Zeid OAA (1997) On the effect of electrodischarge machining parameters on the fatigue life of AISI D6 tool steel. J Mater Process Technol 68:27–32
19. Guu YH, Hocheng H (2001) Improvement of fatigue life of electrical discharge machined AISI D2 tool steel by TiN coating. Mater Sci Eng A 318:155–162
20. Liebowitz H (1989) Computational fracture mechanics, vol 3. Pergamon Press, Oxford
21. Anderson TL (1995) Fracture mechanics: fundamentals and applications, 2nd edn. CRC Press, London
22. Lemaitre J, Dufailly J (1987) Damage measurements. Eng Fract Mech 28:643–661
23. Kachanov LM (1986) Introduction to continuum damage mechanics. Martinus Nijhoff Publishers, Dordrecht
24. Yoshida I, Tsukada T (2006) Uncertainty of wavelength limitation due to stylus tip radius for engineering surface texture based on wavelength and amplitude by FFT. Wear 261:1225–1231
25. Gonzalez-Gonzalez A et al (2010) Morphology analysis of Si island arrays on Si(001). Nanoscale Res Lett 5:1882–1887
26. Jurecka S et al (2010) On the influence of the surface roughness onto the ultrathin SiO2/Si structure properties. Appl Surf Sci 256:5623–5628
27. Cho S-J et al (2011) Three-dimensional imaging of undercut and sidewall structures by atomic force microscopy. Rev Sci Instrum 82:023707
28. Lee LC et al (1988) Quantification of surface damage of tool steels after EDM. Int J Mach Tools Manuf 28:359–372
29. Lee HT, Tai TY (2003) Relationship between EDM parameters and surface crack formation. J Mater Process Technol 142:676–683
30. Ramasawmy H, Blunt L (2002) 3D surface characterisation of electropolished EDMed surface and quantitative assessment of process variables using Taguchi Methodology. Int J Mach Tools Manuf 42:1129–1133
31. Ramasawmy H, Blunt L (2004) Effect of EDM process parameters on 3D surface topography. J Mater Process Technol 148:155–164
32. Ekmekci B (2009) White layer composition, heat treatment, and crack formation in electric discharge machining process. Metall Mater Trans B 40:70–81
33. Guu YH, Hocheng H (2001) High cycle fatigue of electrical-discharge machined AISI D2 tool steel. Int J Mater Prod Technol 16:642–657
34. Guu YH, Hocheng H, Tai NH, Liu SY (2001) Effect of electrical discharge machining on the characteristics of carbon fiber reinforced carbon composites. J Mater Sci 36:2037–2043
35. Fu CT, Li AK (1994) The dependence of surface damage induced by electrical-discharge machining on the fracture strength of Al_2O_3-Cr_3C_2 composites. Mater Chem Phys 39:129–135
36. Fu CT et al (1994) The effect of electrodischarge machining on the fracture strength and surface microstructure of an Al_2O_3-Cr_3C_2 composite. Mater Sci Eng A 188:91–96
37. Yeo SH et al (2008) Critical assessment and numerical comparison of electro-thermal models in EDM. J Mater Process Technol 203:241–251
38. Guu YH, Hocheng H (2001) Effects of workpiece rotation on machinability during electrical discharge machining. Mater Manufact Process 16:91–101
39. American Society for Testing and Materials (2000) Annual book of ASTM standards. Designation: E8M-00b. ASTM, Philadelphia, p 1–27
40. Merdan MR, Arnell RD (1989) The surface integrity of a die steel after electrodischarge machining: 1. Structure, composition and hardness. Surf Eng 5:158–164
41. Lim LC et al (1991) Solidification microstructure of electrodischarge machined surfaces of tool steels. Mater Sci Technol 7:239–248
42. Hibbeler RC, Fan SC (2008) Mechanics of materials, 7th edn. Prentice Hall, Englewood Cliffs, NJ
43. Budynas RG et al (2008) Shigley's mechanical engineering design, 8th edn. McGraw-Hill, New York

44. Tung WH (1998) The relationship between white layer and residual stress induced by EDM hole-drilling. Department of Mechanical Engineering National Cheng Kung University, Tainan, Taiwan
45. Lee HT, Yur JP (2000) Characteristic analysis of EDMed surfaces using the Taguchi approach. Mater Manufact Process 15:781–806

Chapter 3
Electrochemical Machining

P.S. Pa and H. Hocheng

Abstract Electrochemical machining has attracted increasing attention for micro-machining applications. The first section discusses a process to erode a hole of hundreds of microns diameter in a metal surface using a moving electrode. The discussion provides a method to predict the enlargement of the produced hole and to taper under the applied machining conditions. A computational model illustrates how the machined profile develops over time and as the electrode gap changes. The analysis is based on Faraday's laws of electrolysis and the mathematical integral describing a tool. The effectiveness of the model is tested by experiments that apply several electrode movement schemes.

This chapter discusses the surface roughness of several common die materials produced by traditional machining, whereby the internal and external cylindrical surface are electropolished by different electrode designs. Electropolishing efficiency of die materials and parts should be high to improve surface roughness in the shortest amount of time possible, thereby reducing surface residual stresses. The study aims to identify an optimal electrode design, which will help broaden electromachining applications in the future. For electropolishing of internal holes, completely inserted feeding electrodes are supplied with both continuous and pulsed direct current. In the external electropolishing studies, we consider the design of the turning tool electrode, arrowhead electrode, ring-form electrode, and disc-form electrode. For internal electropolishing, an electrode featuring a helix discharge flute performs better than that without a flute or with a straight flute. The borer type electrode performs better an electrode with a lip on the leading edge. Pulsed direct current can improve the polishing, but the machining time and

P.S. Pa (✉)
Department of Digital Content Design, Graduate School of Toy and Game Design,
National Taipei University of Education, Taipei, Taiwan, ROC
e-mail: myhow@seed.net.tw

H. Hocheng
Department of Power Mechanical Engineering, National Tsing Hua University,
Hsinchu, Taiwan, ROC

H. Hocheng and H.-Y. Tsai (eds.), *Advanced Analysis of Nontraditional Machining*,
DOI 10.1007/978-1-4614-4054-3_3, © Springer Science+Business Media New York 2013

costs are increased. In the case of external electropolishing, a smaller nose radius or end radius produces greater current density and provides a faster feed rate and better polishing. Ultrasonic-aided electropolishing improves the polishing effect with no increase in machining time, thus improving efficiency and reducing costs.

Keywords Electrochemical machining • Moving electrode • Hole • Electrode gap • Faraday's laws • Surface roughness • Die materials • Internal cylindrical surface • Electrode design • Electropolishing • Electrobrightening • Pulsed direct current

3.1 Introduction

The electrochemical machining (ECM) was first introduced in 1929 by Gusseff; later it was found that ECM is particularly advantageous for high-strength and high-melting point alloys. The industrial applications have been extended to electrochemical drilling, electrochemical deburring, electrochemical grinding, and electrochemical polishing [1]. An electrical current is made to pass through an electrolyte solution between a cathode tool and an anode workpiece. The workpiece is eroded in accordance with the laws of electrolysis. The tool design determines the machined dimensions on the anode workpiece [2].

ECM has been used widely in manufacturing semiconductor devices and thin metallic films, where high-strength alloys are frequently employed [3–5]. ECM processes were also recognized in the aerospace and electronic industries for shaping and finishing operations of a variety of parts with opening cuts that are a few microns in dimension [6].

Micro-machining refers to the material removal that ranges from several microns to millimeters in dimension. The advanced micro-machining may employ various ultra-precision machining techniques to very small and thin workpieces. A few research attempts at precision micro-ECM have been made by the electromechanical consumer products industries [7]. However, except for electrochemical jet machining, electrochemical etching, and wire electrochemical grinding, ECM research yet falls behind other processes such as EDM in the precision and micro-machining fields [8–10]. It is desired to predict the final shape formed by the ECM process once the machining parameters are given. A prediction also prevents from material wasting and saves time on design. The cost of production trial-and-error for the micro-machining is much higher than the conventional cutting process; thus there is a need for a utility to simulate the final shape of workpiece before the micro-ECM work. A correction factor method has been proposed for tool design of ECM [11]; it proposes an analytical model of electrochemical erosion to predict the final machined diameter of the workpiece. The material removal can be predicted based on the current density distribution semi-empirically obtained from preliminary tests [12]. The boundary element method was used which required large amount of calculation [13]. A graphical method considering local effects can achieve better dimensional accuracy [14]. The current paper proposes a simple model to describe the development of the erosion profile as a function of machining time and the changing electrode gap. The results are compared with experimental work.

The workpiece of electrochemical machining (ECM) is made the anode of an electrolytic cell, and the electrodes are connected to the DC source. The tool is made the cathode and is maintained at a gap of tenths of millimeter from the anode workpiece. An electrolyte is pumped between the gap while a large direct current is passed across, and the material of workpiece is removed by electrolytic dissolution of the anode until it is shaped [15]. The metal removal in ECM depends on the current density between the electrodes, and the tolerances of workpiece depend on knowledge and control of this current distribution [16]. Metals in the state of ions under electrochemical reaction enter into the electrolyte as *MOH*, and the reaction dregs and heat are flushed off the gap by the speedily moving electrolyte, making the reaction of ECM continue until the machining process is completed. The reaction in anode during the electrochemical machining is [17, 18]:

$$M^{+n} + nOH^- \rightarrow M(OH)_n \downarrow \tag{3.1}$$

The produced *MOH* is cleaned away via a filtrating system. The reaction of the cathode is:

$$2H^+ + 2e^- \rightarrow H_2 \uparrow \tag{3.2}$$

The produced hydrogen is let out outdoors via the system of exhausting pipes. Water in the electrolyte is consumed:

$$H_2O \rightarrow 2H^+ + OH^- \tag{3.3}$$

Hence regular supply is required for constant concentration of the electrolyte.

Electrochemical machining is a very complicated machining process. The variables are mainly the electric current, electric voltage, feed speed of electrode, and the electrolyte temperature and flow rate. Because the investment of the original equipment and the tool electrodes is quite expensive, how to effectively choose the best parameters of machining speed and polishing effect is a major concern [19]. In addition, the full understanding and control of the process will help with the analysis of the acquired shapes of workpieces [20].

The theoretical material removal rate of ECM can be predicted by the Faraday's law under the following conditions:

1. The atomic number of ion from ECM has been affirmed.
2. The dissolving response of metal on anode has only one reaction, that is, without any other secondary reactions.
3. Metal is only removed by dissolving atoms one by one, not by the atomic group leading to the falling of chunk. However, the chunk effect is often found [17].

3.1.1 The Volumetric Removal Rate

The amount of metal dissolving on the electric pole of workpiece is $W = KIt$. Then the volume of metal removed is

$$V = \eta\frac{W}{\rho} = \eta\frac{K}{\rho}It \tag{3.4}$$

Let

$$\omega = \frac{K}{\rho} \tag{3.5}$$

then

$$V = \eta\omega It \tag{3.6}$$

where η is current efficiency, ρ *is* density of workpiece, ω is the mass of metal removed per unit of the quantity of electricity, called electrochemical equivalent.If η, ω, I are constants, then the volumetric removal rate is

$$V_m = \frac{V}{t} = \eta\omega I \tag{3.7}$$

3.1.2 The Removal Rate of Depth

When the tube-shaped negative pole is insulated on outer walls, the area in machining process is constant. Let the machined area is A; the machined depth is h; the amount of volume removal $V = Ah$, and the removal rate of depth is

$$v_l = \frac{h}{t} = \frac{V}{At} = \frac{\eta\omega I}{A} = \eta\omega I_A \tag{3.8}$$

where $I_A = I/A$ is the current density.

When the machined areas are changed, the current density can be kept constant, thus (3.8) can be applied. In the balanced state, the removal rate of depth is indeed equal to the feed rate of cathode. When the current efficiency η is constant, the current density is in direct proportion to the feed rate. Once the electrolyte and the machined metals are chosen (i.e., η, ω are constant), the feed rate can be changed to adjust with the current density. Electrochemical machining (ECM) has been known as a highly effective metal removal method for machining hardened metals at high removal rates and little residual stress and tool wear [21]. Since then, electrochemical machining won more attention in industrial application, and the process became an important method for metal removal [22]. Gusseff first filed a patent on electrochemical machining (ECM) in 1929 and found that ECM is suitable for high-strength and high-melting point alloys. It encouraged the application of ECM [18]. It is well suitable for difficult-to-machine materials. Plastic or press dies, wire-drawing dies, optical and electric parts can apply this technique as well [23]. More industrial applications were realized throughout the decades, such as electrochemical drilling, electrochemical grinding, electrochemical deburring, and electropolishing [24]. ECM is suitable for the manufacturing of workpiece with complex shapes. The process is able to shape the workpiece to a profile approximately complementary to the tool, and the complicated shapes of cathode can be machined from a soft material [25]. The technique can also be applied to the repairing of molds, which is hard to cut. It produces good surface roughness and the machined part has no residual stress [26]. However, the early expectations for the potential of the method have not been fully met since a long time. The application of ECM would increase substantially if the dimensional tolerances can be improved [27].

In 1966, Kashcheev et al. proposed that high-density pulse direct current improves the precision of workpiece [28]. The pulse mode provides the efficient method for electrochemical machining of metals; several investigations aimed at improving the parts accuracy and the localization of anodic dissolution [29]. In using pulse direct current, the main controlled factors are peak current, functioning time, and pulse-off time. This can well replace the polishing using continuous direct current and extend the applications. Pulse direct current is supposed to use lower average current density, but can save the setup of high flow rate of electrolyte, since the heat of electrolyte and the electrolytic products can be discharged during the pulse-off time. It meanwhile results in lower removal rate [30]. High-density pulse direct current improves the precision of workpiece. The results of the experiments show lower pulse frequency or flow rate of electrolyte cannot form uniform polishing of workpieces in use of the non-mating electrode [31]. Laboda used water solution $NaC1O_3$ as electrolyte to replace the conventional salt water in order to increase the degree of precision. Due to the concern of burning, $NaClO_3$ was replaced by $NaNO_3$ as electrolyte [30]. Using the electrolyte of NaCl, metal is removed at 100% current efficiency, and the current efficiency is nearly independent of the current density over the anode surface. While the $NaNO_3$ and $NaClO_3$ electrolytes decrease the current efficiency with current density, good dimensional control can be achieved [32]. Noto et al. put forward the study of the electrode gap using sodium nitrate solution [33].

Different materials possess various current efficiency in electrochemical machining, also resulting in different polishing effects. Take medium carbon steel and cast iron as an example. The former has higher current efficiency and polishing rate; the latter is obviously poorer. The major reasons for differences are the carbon ratio in materials, because carbon is the least active element in electrolytic reactions. In electrochemical machining, it will become the non-dissolving dregs of anode, which will remain on the workpieces and be flushed away by flowing electrolyte. Therefore, the higher the carbon amount is, the poorer are the current efficiency and the speed of electrochemical machining. The materials with high ratio of carbon, such as cast iron, are not easy to be processed [17]. In addition, the grain structures and the grain size of the material will affect the removal rate of electrochemical machining and the polishing effect [34]. Opitz et al. put forward the study of the electrode gap. It suggested the control of workpiece geometry in the electrochemical machining [35]. According to the experimental results of Mileham and others, the use of changing machining conditions will directly influence the quality of machined appearances. The experimental results of Mileham et al. showed the quality of the machined surface will be influenced by the current density, flow rate of electrolyte, and the gap width [36]. Datta and Shenoy showed that the gap width between electrode and workpiece directly influences the current condition and the discharge dregs of the electrolyte [37]. Bejar et al. changed the machining gap width as well as the concentration of electrolyte to investigate the influence upon current efficiency. They found that current efficiency is raised with the increase of current density and electrolytic concentration [38]. Bannard proposed the theory that correlates the current efficiency with current density and flow rate of electrolyte. The maximum efficiency varies with the type of electrolyte [39]. Overcut increases with voltage as a result of the increase of current passing between anode and cathode [40]. Larbraga and others use water solution of phosphoric acid and acetic acid electrolytes for carbon steel and aluminum with alternative current, discussing the influence of the current frequency and the current wave on the surface roughness of workpiece. The result shows that the best polishing can be achieved in the range of certain frequency, but the current wave will be less influential [41]. Shen used $NaNO_3$ as the electrolyte to proceed the electropolishing on die surface. The result showed that the surface roughness of workpieces decreases with the increase of current density, flow rate, and concentration of electrolyte. Moreover, polishing with pulse direct current is found better than continuous direct current [42]. Rajurkar et al. obtained the minimum gap width based on Ohm's law, Faraday's law, and the equation of conservation of energy, beyond which the electrolyte will be boiled in electrochemical machining. An on-line monitoring system was proposed [43].

Electrochemical machining is a complicated machining process. Since the investment of the original equipment and the tool electrodes is very expensive, how to effectively choose the best parameters will become the main concern [44]. Acharya et al. proposed that the effective production of the rupture change of workpieces is not easy to be attained. And the heat and bubbles during the machining process are associated with the nature of electrolyte. However, profit and quality can be obtained when operated at optimal conditions [45]. Phillips found the major factors in the electrochemical grinding affecting the metal removal rate are the conductivity of the workpiece, the rate of decomposition, the current capacity of power supply, and the ingredients, concentration, and temperature of electrolyte. The electrochemical grinding has been successfully used for difficult-to-machine alloys [23]. The material removal rate per unit area (feed rate) is proportional to the current density multiplied by the current efficiency, and the feed rate of electrode [46]. Grodzinskii found the increase of the electric voltage elevates the production efficiency when machining heat-resisting metals such as Zhs6K and ZhS6U. The increase of axial feed rate can be achieved by the increase of electric current [47]. The application of electropolishing is yet limited, because the electrode of the cathode is difficult to design, and the geometrical shape of the anode is subject to inaccuracy. The electrochemical drilling (ECD) often makes the holes with taper [20]. Jain and others proposed that the lack of the understanding of removal mechanism as well as the electrode design limits the wide application of ECM. They proved that lowering the height of the naked tips of tools or alienating the sides of the tips of tool will decrease the overcutting [2]. Electropolishing the inner holes of workpieces can not only control the accuracy of the size of the holes but also elevate the corrosion-resistance [48]. The choice of electrode material is also important. It requires good electric conduction, stiffness to resist the electrolyte flushing, and resistant to long-term erosion of electrolyte. Brass and copper are commonly used for electrode material [24].

The design of the cathode shape to achieve a required anode shape is a primary concern in ECM. The basic "cosine rule" has been used for solution [17]. Machining facilities, cost analysis production management, and a host of good intentions are worthless if one does not possess the competence to design ECM tooling. It is often called an "art," since much information has evolved by empirical development of tooling; "tricks" of tool design do work but are not always supported by logical explanation. An "art" implies skills acquired by tedious trial and error, as such it is hardly compatible with the pace of modern industry. However, how to effectively reach the standard of the technology and art is virtually difficult. Even with a precise design guideline, the success still depends on special knowledge and experiments [24]. Tool design in ECM depends on the prediction of anode (workpiece) shape obtained from a tool under specified conditions of machining. The produceable shape by ECM under ideal conditions can be computed from the corresponding equilibrium gap. The results are influenced by a large number of parameters, such as the presence of an anodic film, the current density, the electrode feed rate, change in valency of the work material during machining, the type of electrolyte and the role of additives, the electrolyte pressure distribution, and the electrolyte throwing power [2]. The non-mating electrode is developed. It uses three-axis motion of the electrode to attain the machining of workpieces [49].

In 1927, Wood put glass bars on the glass panel under high-frequency vibration and found glass bars can drill through the glass panel. After 1950, many scholars and experts have devoted to the research of ultrasonic machining. Until the early 1960s, the number of the published papers on ultrasonic machining reached 300. Rozenberg et al. collected the research papers and published a book [50]. In the same period, Markov described the basic machining principles of ultrasonic machining and practical machining systems in 1966. More industrial applications were realized in the areas of material, chemistry, engineering, machining, welding, metallurgy, cleaning, measurement, and communications [51]. The ultrasonic machining comes from the compounding phenomenon of magnetism or electricity via ultrasonic oscillator, which transforms the low-frequency electricity to high-frequency, providing high-frequency mechanical energy to the abrasive grains [52].

It refers to a mixture of abrasive and water; one strikes the abrasives via the vibrating tool (10–15 μm, 15–30 kHz), and the abrasives in turn strike the workpiece, cutting it by finely breaking it. On the aspect of the mechanism of material removal of ultrasonic machining, Shaw's hammering action is mainly the theory of removal mechanism [53]. Rozenberg thinks that the mechanism of material removal is mainly composed of hammering action, free impact action, and cavitation erosion. Through the observation of high-speed photography, the hammering action is found to be the main mechanism of material removal, and the role of the other two is very limited [50]. Dikushin and Barke derived the machining efficiency based on the conservative law of energy, and made experiments to prove that machining efficiency is in direct proportion to the removal rate of materials [54]. The larger ratio of H/E (hardness/elastic modulus) is, the higher is the removal rate of materials. In addition, the use of hard tool materials can result in better roughness [55]. In Komaraiah's opinion, the removal rate is directly related to pressure, the size of grains, the density of grains, and the size of tools [56]. Komaraiah and Narasimha Reddy also found that the removal rate increases with the hardness and breaking intensity of workpiece [57]. Khairy considers ultrasonic machining as a linear moving system and the new tools and abrasives as the inputs, the machined surface as the outputs, and the consumed tools and the abrasives as feedback. A dynamic analysis of ultrasonic machining is derived. Weller divides the consumed area of tools during the machining process into two kinds: (1) the consumption under the hammering action making tools shorter and (2) the consumption under the circulation of mixed liquids and the free impact action making taper tools [58]. Wang takes into account the factor of time. He predicts that the removal rate is in direct proportion to the size of grains, the frequency, and the vibration amplitude. The experiments prove the accuracy is higher than the theories of Shaw [59]. Gilmore compares various non-traditional machining methods of machining ceramic, including grinding, ultrasonic, electric discharge machining, laser beam machining, and water jet machining. He finds that ultrasonic machining has more advantages than those of other machining methods [60]. Ceramic or composite materials have been machined by ultrasonics recently [61]. Hocheng conducts ultrasonic drilling on Carbon/Epoxy and Carbon/PEEK to explore the cutting nature, cutting mechanism, and economic effect, and finds that ultrasonic drilling can avoid the splitting and burrs of the composite materials [62]. In ultrasonic cleaning, the ultrasonic energy is transmitted to the medium of liquid, and the object is put into the washing tank via soaking to execute the cleaning function. The ultrasonic cleaning usually uses high-frequency vibration (the range of frequency is between 20 and 50 kHz) in the cleaning liquid to remove the filth rapidly and to achieve the function of cleaning. In comparison with the traditional cleaning methods, it not only shortens the cleaning time, but has an excellent capability of erasing pollution [63]. The cavitation formed by the ultrasonic energy results in the air bubbles bursting and makes the polluted layers get rid of the surface of solid objects. In addition, other groups of cavitation bubbles invade the ruptures, with the sonic pressure change, the air bubbles will repeatedly shrink and swell, and then the polluted layer is stripped off. The ultrasonic cleansing is widely applied to families, stores, and factories [64].

Either hand polishing or machine polishing often results in nonuniform residual stress due to the contact between tool and workpiece. Surface crack and microvoids are often induced and deteriorate the service life. The electropolishing can efficiently produce workpieces of good surface finish, and the electropolishing can efficiently produce workpieces free of the above-mentioned shortcomings [23]. The potential for electropolishing is yet to explore, and the major difficulty of electropolishing is the cost and the design of tool electrode. The authors will develop the electrode geometry of inserted and feeding electrodes for electropolishing of the internal cylindrical holes of several die materials after drilling and reaming. The electropolishing is used after the hole is drilled, while the electrobrightening is applied after the hole is drilled and reamed. Among various factors affecting the electropolishing and electrobrightening, the design of tool electrode is majorly discussed. However, since the amount of material removal in electropolishing is quite limited, the electrode design in the present study is concerned with the influence of the various forms, instead of the shape compensation, on the attained surface roughness. The study also discusses the effect of tool design in electropolishing of the external cylindrical surface when turning and drawing cannot achieve the desired fine surface finish. After rough turning or drawing, the average surface roughness of common die materials is about 3.0–6.3 μm, and better surface finish (0.8–1.6 μm) can be obtained later through fine turning or grinding [65]. When surface finish better than 0.8 μm is required, subsequent conventional techniques such as polishing by hand or machine are applied. However, they heavily depend on the sophisticated experience. An efficient electropolishing process using low-cost electrodes is discussed, and various design of electrode will be presented.

Various electrodes of internal or external cylindrical surface are developed based on the following considerations [66–73]:

3.2 The Dimension of Workpiece

3.2.1 The Size of the Hole

For medium and small holes, a completely inserted electrode can be used. The whole surface of the cylindrical walls is polished at the same time; hence the production cycle is kept short. For large holes, higher electrical current rating is required. To reduce the cost of the power supply, the feeding electrode should be employed at the expense of the increased cycle time. Nevertheless, the discharge of the electrolytic product is more advantageous in this case, which is important for large-area electrochemical machining.

3.2.2 The Dimension of the Cylinder

As to the short external cylindrical surface, one can use the mate-electrodes of the same length to shorten the production cycle. In the case of longer workpiece or unsuitable polishing by counterpart electrodes, one needs to use the feeding electrodes. The polishing by feeding electrodes can not only save the investment cost of large DC power supply but also contributes to better discharge of dregs than that of the mate-electrodes. It is a significant point of the design of electrodes.

3.3 Increase of Current Density to Provide Fast Feed of Electrode

Since the electropolishing requires a sufficient electrical current density, a good electrode design should meet this requirement. The rotation of electrode can cut a portion of the working area along the circumferential direction. Small end radius or sharp inner edge rounding of feeding electrodes leads to high current density and feed rate.

3.4 More Effective Discharge of Electrolytic Products

The discharge of the electrolytic products out of the gap is crucial for the polishing. The electrode rotation, the open space, straight and helix flute, large rake angle, and smaller cross section will be incorporated into the design.

3.5 Reduction of Secondary Machining

To ensure the dimensional and geometrical accuracy of the polished hole or external cylindrical surface, the secondary overcut induced from the working gap should be eliminated as possible.

3.5.1 Pulse Effect

Pulse current has been found advantageous in electropolishing considering the improved discharge of the electrolytic product during the off-time. A power supply with the function of pulse current is more expensive. Through the design of electrode, an equivalent effect of pulse current can be achieved. The rib and flute width on the electrode edge or pin-type will be incorporated into the design.

3.6 Ultrasonic-Aided Dregs Discharge

Using vibrating energy of ultrasonic in electrolyte is beneficial to discharge dregs of the electropolishing and to make the reaction process smoothly, which will improve the polishing quality. The ultrasonic-aided electropolishing is also a low-cost setup and the polishing time is not raised.

3.7 Cost and Manufacturability of the Electrodes

The design of the electrode will not require unusual manufacturing technique.

The study includes the experiment on process parameters consists of electropolishing of holes and external cylindrical surface; electrode design in electropolishing of holes including completely inserted electrode and feeding electrode; design of electrobrightening of holes; electrode design in electropolishing of external cylindrical surface consisting of turning tool electrode, disc-form electrode, arrow-head electrode, and ring-form electrode; and ultrasonic-aided electropolishing of holes and external cylindrical surface.

3.8 Generation of Erosion Profile of Through Hole in Electrochemical Boring Analytical Approach

3.8.1 Three-Dimensional Method

Figure 3.1 shows the scheme of the three-dimensional ECM. The electrode is a cylinder electrode of diameter d, and is placed above the workpiece at gap δ.

Coulomn's law states the electric field intensity (E) applied on the workpiece made by a point charge $P(x')$ on electrode is inversely proportional to the square of distance (R) between $P(x')$ and the considered point on workpiece $Q(x)$.

$$E \propto \frac{q}{R^2} \tag{3.9}$$

where q is the charge density, and

$$R = \sqrt{x^2 + x'^2 - 2xx' \cos(\theta - \alpha) + \delta^2} \tag{3.10}$$

x and α are the radial and angular coordinates on the workpiece, respectively, while x' and θ are the angular coordinates on the electrode.

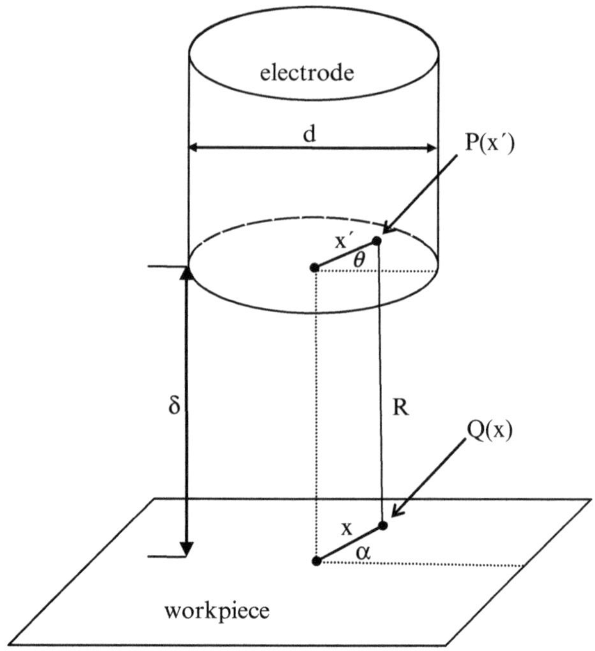

Fig. 3.1 Scheme of the 3D model [74], reprinted with permission

The amount of material erosion on the workpiece caused by a single charge on the electrode is assumed proportional to the electrical intensity [12]. In the model, the cathode end surface is composed of differential point charge sources. Considering one point on the anode workpiece, one finds that it is influenced by all point charges on the cathode end. To add up all the erosion effects, a double integral over the charge sources on the entire end surface of cathode is carried out. The actual eroded depth at the considered point on anode produced by the cathode at each time increment can be calculated by an iteration

$$y_1(x,\alpha) = \delta$$

$$y_{i+1}(x,\alpha) = y_i(x,\alpha) + \Delta t \cdot m_i(x,\alpha) \quad (i \geq 1),$$

where

$$\begin{aligned} m_i(x,\alpha) &= \int_0^{d/2}\int_0^{2\pi} \frac{c_1}{R^2}\,\mathrm{d}\theta\,\mathrm{d}x' \\ &= -\int_0^{d/2}\int_0^{2\pi} \frac{c_1}{x^2 + x'^2 - 2x \cdot x' \cos(\theta - \alpha) + y_i(x,\alpha)^2}\,\mathrm{d}\theta\,\mathrm{d}x'. \end{aligned} \tag{3.11}$$

C_1 is a constant of electric efficiency determined by preliminary experiment, which differentiates the effects of the applied voltage and electrolyte concentration.

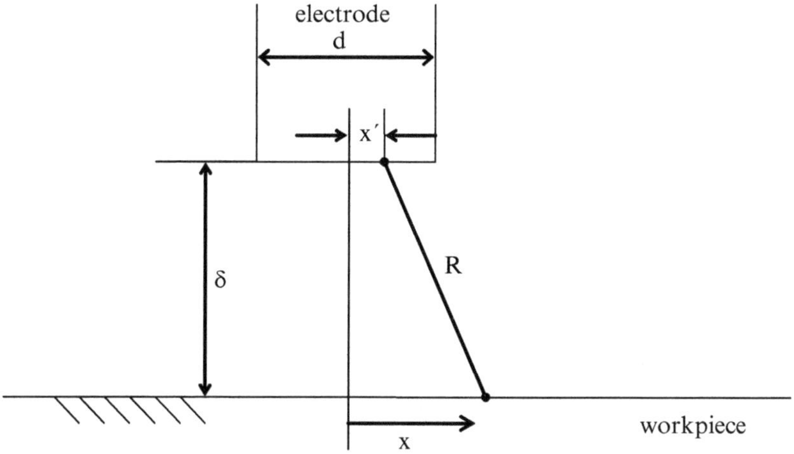

Fig. 3.2 Scheme of the 2D model [74], reprinted with permission

3.8.2 Equivalent Simplified Two-Dimensional Method

The result of the inner integral of $m_i(x, \alpha)$ in (3.11) is found

$$\int_0^{2\pi} \frac{c_1}{x^2 + x'^2 - 2x \cdot x' \cos(\theta - \alpha) + \delta^2} \mathrm{d}\theta = \frac{2c_1\pi}{\mathrm{x}^2\beta + \mathrm{x}'^2\beta + 2\mathrm{x} \cdot \mathrm{x}'\beta + \delta^2\beta} \tag{3.12}$$

where

$$\beta = \sqrt{\frac{x^2 + x'^2 - 2x \cdot x' + \delta^2}{x^2 + x'^2 + 2x \cdot x' + \delta^2}} \tag{3.13}$$

The results show that $m_i(x, \alpha)$ is actually independent of α, namely $m_i(x, \alpha) \equiv m_i(x)$. The erosion profile on anode is thus found axis-symmetric as can be conceived. Furthermore, when $\theta = 0$, the three-dimensional model is reduced to a two-dimensional model as shown in Fig. 3.2. A simplified model considers the erosion profile along a diametral line on workpiece produced by all point charges on a diametral line on the electrode. $m_i(x)$ is reduced to $m'_i(x)$ at $\theta = 0$ as follows,

$$\begin{aligned} m'_i(x) = m_i(x)|_{\theta=0} &= -\int_{-d/2}^{d/2} \frac{c_2}{x^2 + x'^2 - 2x \cdot x' \cos(0) + y_i(x)^2} \mathrm{d}x' \\ &= -\int_{-\mathrm{d}/2}^{\mathrm{d}/2} \frac{\mathrm{c}_2}{(\mathrm{x} - \mathrm{x}')^2 + \mathrm{y}_\mathrm{i}(\mathrm{x})^2} \mathrm{d}\mathrm{x}' \end{aligned} \tag{3.14}$$

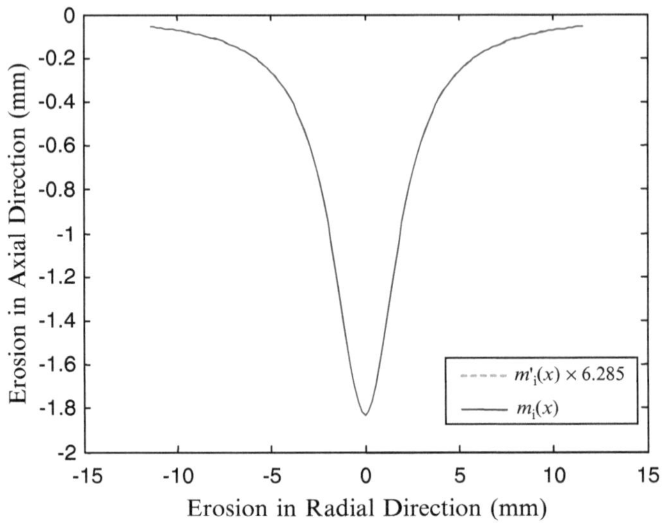

Fig. 3.3 Constant difference between $m_i(x)$ and $m'_i(x)$ [74], reprinted with permission

Figure 3.3 shows the numerical results of $m_i(x)$ and $m'_i(x)$. It clearly demonstrates $m_i(x)$ and $m'_i(x)$ are identical except the difference by a constant of 6.285. This constant is considered as the product of C_1 in (3.11) and C_2 in (3.14). Hence the simple integral of $m'_i(x)$ along a diametral line electrode is used to replace the double integral of $m_i(x)$ over the entire cathode end, and the whole scheme of the model can be then reduced to a two-dimensional equivalent model. The difference of constant in between can be found by the preliminary experiment, which is also required when the three-dimensional simulation would be used. Based on (3.11) and (3.14), the eroded depth produced by the equivalent line electrode at each time increment can be calculated by the following iteration equivalent to (3.11).

$$y_1 = \delta$$

$$y_{i+1} = y_i + \Delta t \cdot m_i \quad (i \geq 1),$$

where

$$m_i = -\int_{-d/2}^{d/2} \frac{c}{R^2} \mathrm{da} = -\int_{-d/2}^{d/2} \frac{\mathrm{c}}{(\mathrm{x}-\mathrm{x}')^2 + \mathrm{y_i}^2} \mathrm{da} \tag{3.15}$$

At the first step of the machining, the equation draws an initial eroded profile on the workpiece. When the first iteration is done, the workpiece was eroded into an indent so that the gap distance across the electrode and the new generated surface of workpiece has changed; thus each point on the new surface of workpiece has to

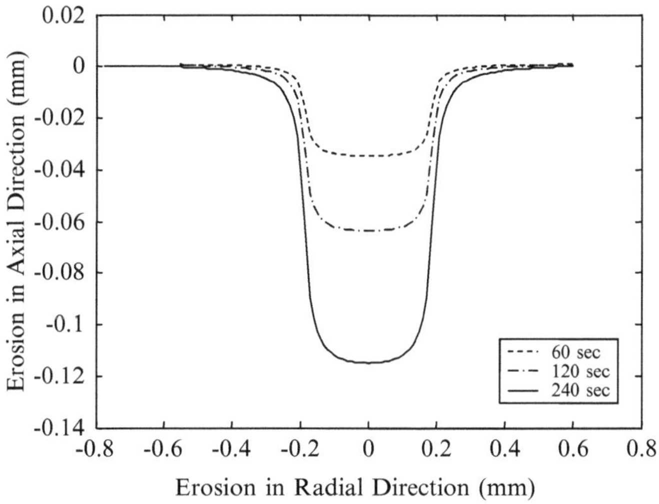

Fig. 3.4 Development of the predicted erosion profile [74], reprinted with permission

renew its gap value after the first time step. Substituting the renewed gap into the model for the second iteration, one can redraw the eroded profile. In use of the iterations continuously, the model describes the development of the erosion profile on workpiece during the ECM process. Figure 3.4 shows the formation of the erosion indent calculated by the iterations [74].

3.8.3 Boring-Through Model

Once the eroded profile touches the bottom plane of the workpiece, a hole will be bored through. Figure 3.5 is a photograph of the bored hole. One identifies the eroded area of diameter $2q$ on top surface and an opening through the bottom surface at early stage of diameter 2 m. The previous method of calculating the material removal cannot be continued from the moment of boring-through, since there is no material to be eroded underneath. The electrical current before and after the formation of through hole was monitored and showed approximately on the same level (see Fig. 3.6), which indicates that the same amount of material is removed. Hence, the electric charge will be consumed to increase the side material removal for wider opening of hole. The concept is presented in Fig. 3.7. If the hole were not bored through, as shown in Fig. 3.7a, the electric charge will distribute quite evenly along all the machined surface for material removal. The machined profile can be predicted by the iteration model. When the hole is bored through the bottom of the workpiece, there is no material beneath the bottom and the electric charges will be distributed to the side wall of the hole, as shown in Fig. 3.7b.

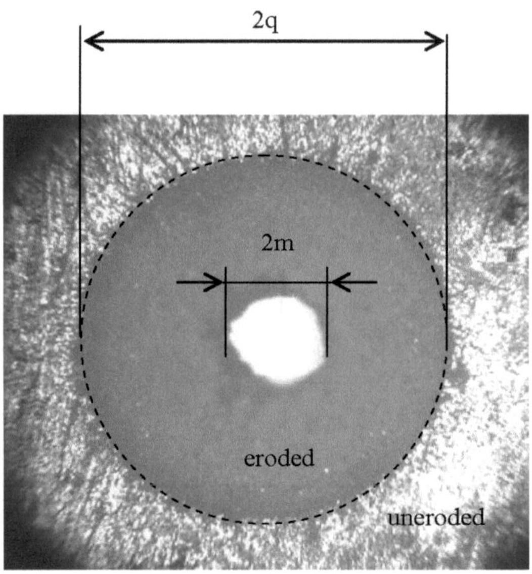

Fig. 3.5 Top view of the eroded hole after boring through [74], reprinted with permission

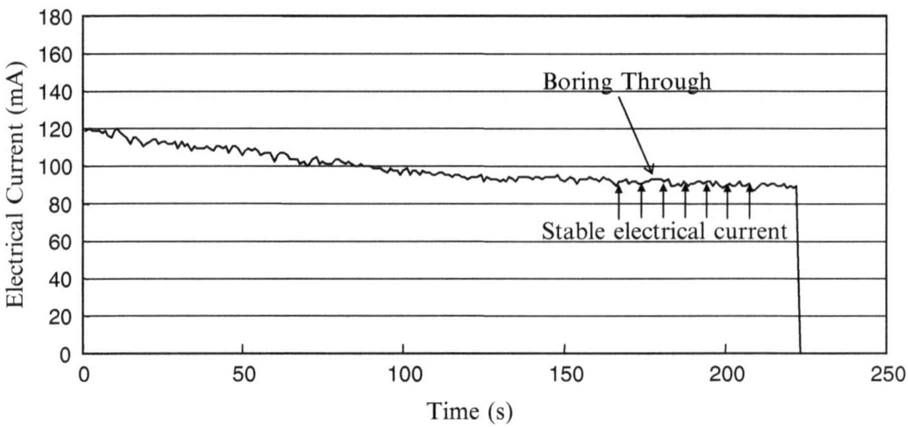

Fig. 3.6 Electrical current rating before and after boring through (voltage = 8 V, electrolyte concentration = 5 M, initial gap = 0.25 mm, electrode diameter = 0.3 mm)

The hole will be therefore enlarged more than previously predicted. The method considering the bottomless hole formation is introduced as follows.

When the calculated profile reaches beyond the actual workpiece thickness (h), as described by curve y_i and the shaded area A in Fig. 3.8, the erosion profile will be modified based on the principle of constant amount of material removal, as elaborated below.

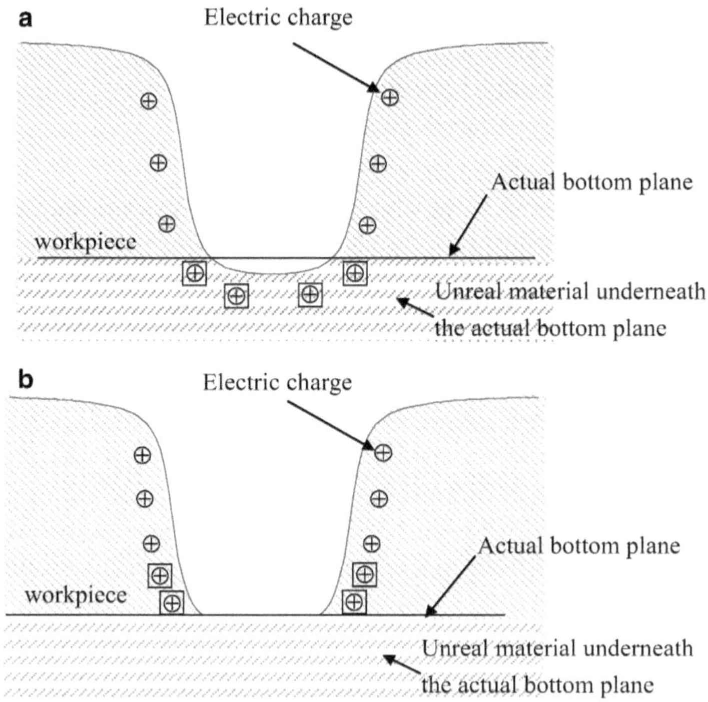

Fig. 3.7 Schematic redistribution of electrical charge when boring through [74], reprinted with permission. (**a**) If hole were not bored through. (**b**) Hole is bored through

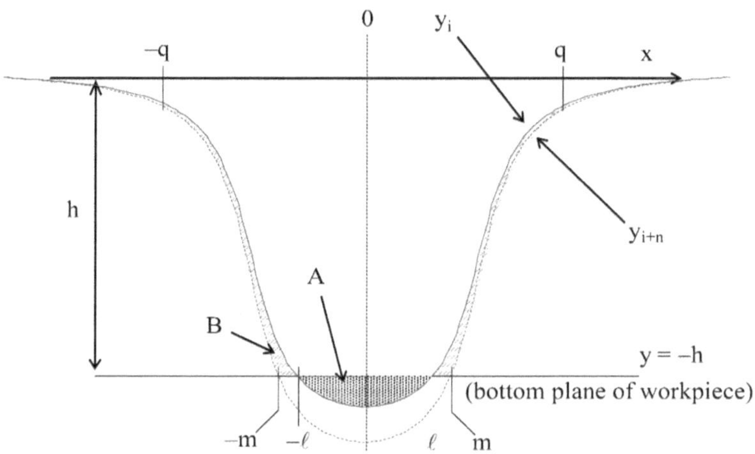

Fig. 3.8 Scheme of the converted material removal [74], reprinted with permission

The area A is considered a virtually removed amount of material between the bottom line and the calculated profile

$$A = \int_{-\lambda}^{\lambda} (-h - y_i)\,\mathrm{dx} \tag{3.16}$$

This part will be converted to the part in workpiece (B) to be removed, namely $A = B$. To find B, one conducts the calculation for another profile, y_{i+n}, n time steps beyond y_i. In the calculation, y_{i+n} is set limited between 0 and $-h$. B is the area between y_i, y_{i+n} and $y = -h$, the converted part of material removal.

$$B = 2\left[\int_{-q}^{-m} (y_i - y_{i+n})\,\mathrm{dx} + \int_{-\mathrm{m}}^{-\ell} (\mathrm{y_i} - (-\mathrm{h})\,\mathrm{dx}\right] \tag{3.17}$$

where n is determined by setting the area B equal to area A. Namely,

$$A = \int_{-\lambda}^{\lambda} (-h - y_i)\,\mathrm{dx} = 2\int_{-\mathrm{q}}^{-\mathrm{m}} (\mathrm{y_i} - \mathrm{y_{i+n}})\,\mathrm{dx} + 2\int_{-\mathrm{m}}^{-\lambda} (\mathrm{y_i} - (-\mathrm{h})\,\mathrm{dx}$$
$$= \mathrm{B} \tag{3.18}$$

The profile y_{i+n} is considered the actually machined one, and the opening at bottom is 2 m instead of 2 l. To calculate the bored profile at the next time step, this new profile y_{i+n} is replaced as y_i into (3.15). The calculation procedure is shown in Fig. 3.9.

3.9 Experimental Design

The experiment is set up on a desktop CNC machine for control of the electrode movement. The gap between electrode and workpiece can be controlled by Z-axis of the CNC machine within 2 μm of tolerance. The electrode is a commonly used CuZn73 alloy machined to uniform diameter and flat end by special technique of micro-EDM. The prepared end of electrode is shown in Fig. 3.10. The workpiece is SK5 stainless steel of 0.2 mm in thickness polished to the surface roughness of Ra = 0.1 μm to ensure the constant initial condition of electrochemical machining. It is clamped in a tank filled with $NaNO_3$ electrolyte at room temperature. The circulation of the electrolyte was controlled carefully to maintain the stability of the electrical current monitored by a current meter. The flow rate of the electrolyte is 6 cm^3/s. The electrical power supply is a stable voltage source which can be adjusted from 0 to 15 V. The machined hole diameter is measured by both optical microscope and WYKO high-resolution surface profiler. The experimental parameters in the tests include machining time and electrode gap, while the voltage is set 8 and 9 V, the electrolyte is 5 M $NaNO_3$, and the diameter of the electrode is

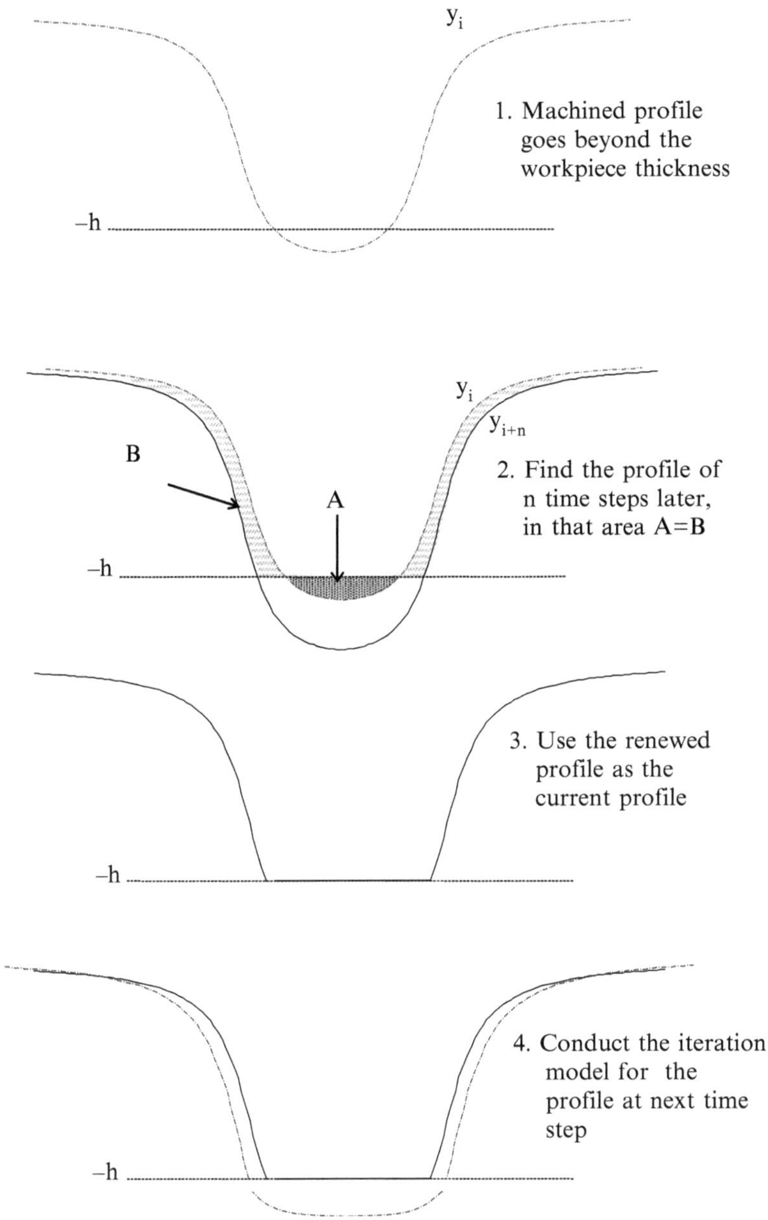

Fig. 3.9 Calculation of machined profile after boring through [74], reprinted with permission

0.3 mm. The experiment includes two parts with the stationary electrode as well as the moving electrode. Four schemes of electrode motion A to D with rough and fine moving pace are designed to investigate the eroded hole profile and compared to the prediction of the model. The four schemes of electrode movement are shown

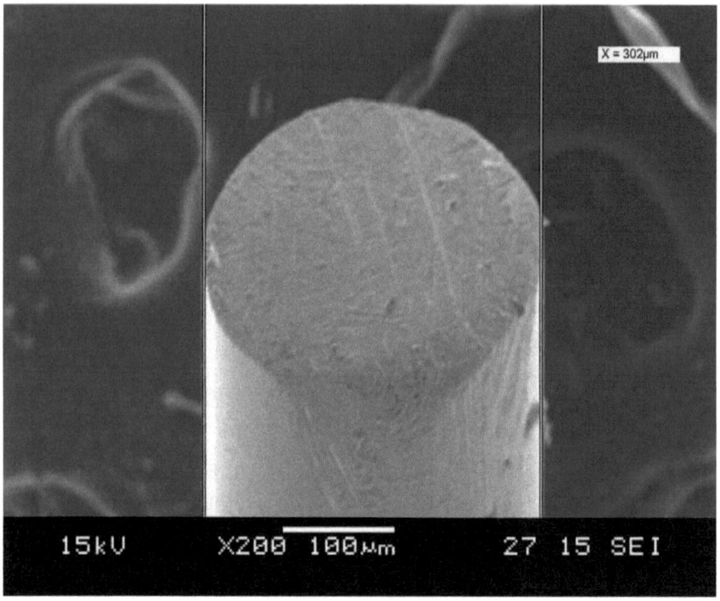

Fig. 3.10 SEM of electrode tip [74], reprinted with permission

in Fig. 3.11. The movement of the electrode starts moving first after 30 s in consideration of the transient beginning of machining, and stops moving after 90 s. The gap between electrode and workpiece is kept larger than 0.15 mm throughout the experiment, or the electrical arc will occur.

3.10 Results and Discussions

3.10.1 Stationary Electrode

Figure 3.12 is an example of the experimental results of the stationary electrode [78]. One notices that the boring-through occurs at around 200 s of machining time. The prediction agrees fairly with the experiment by the method described in Sect. 8.2. The opening at bottom enlarges from the initial boring-through to a stable size near to 1 mm around 300 s, while the opening at top surface settles at its final size of 1 mm earlier than bottom opening does. Further observation shows a taper of the hole forms during this development period. A desired hole taper can be achieved by careful exercise of this process with the aid of the proposed model. In ECM, the material erosion occurs beyond the transpassive zone. If the electric intensity acting on part of the workpiece is below a threshold value, no material removal occurs and the machining profile of that part of workpiece will not change. Such a consideration

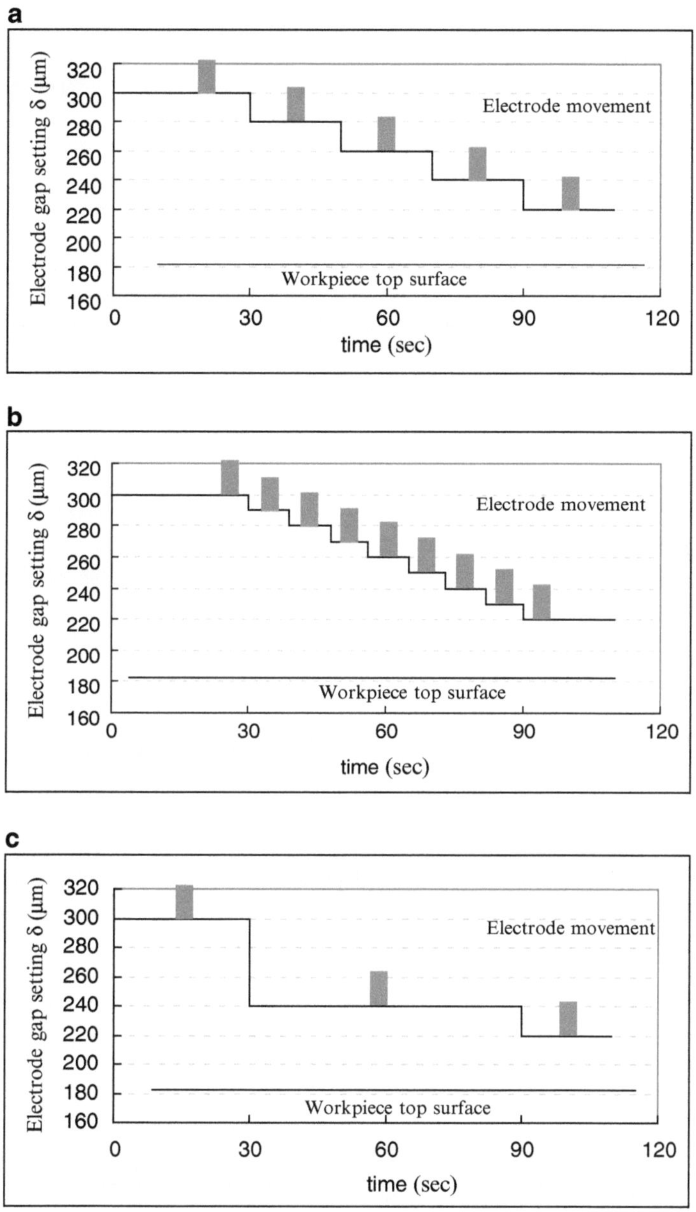

Fig. 3.11 Moving schemes of the electrode in the experiment [74], reprinted with permission. (**a**) Moving scheme A. (**b**) Moving scheme B. (**c**) Moving Scheme C. (**d**) Moving Scheme D

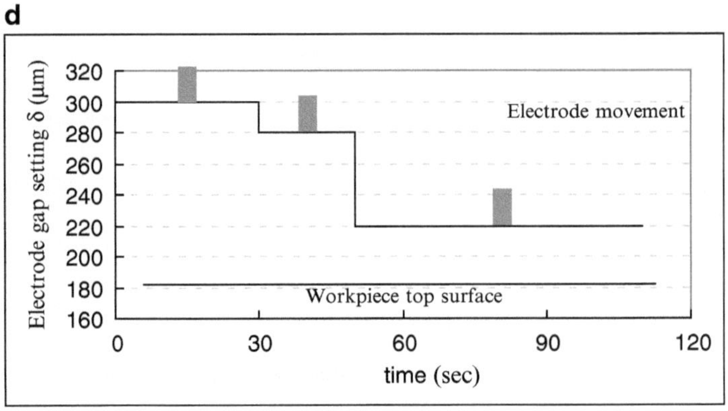

Fig. 3.11 (continued)

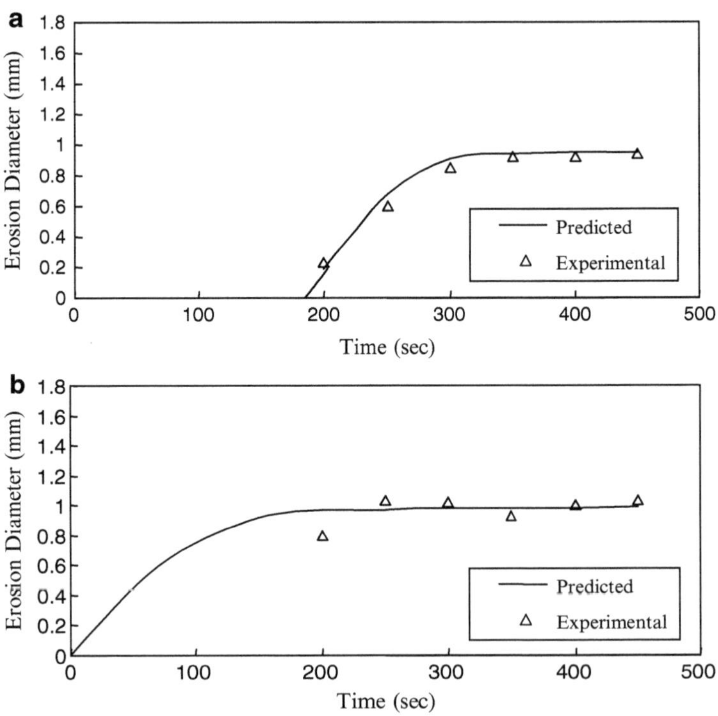

Fig. 3.12 Predicted and experimental diameter of opening (voltage = 8 V, electrolyte concentration = 5 M, initial gap = 0.25 mm, electrode diameter = 0.3 mm) [74], reprinted with permission. (**a**) Diameter of bottom opening. (**b**) Diameter of top opening

is applied to the hole edge of the workpiece. A calculated erosion profile is shown in Fig. 3.6. The points of y_i in (3.17) less than the predefined C_3/V lie between the x-coordinates $-q$ and q, where V is the voltage between electrode and workpiece, C_3 is a constant of 0.05 determined in preliminary experiments. $2q$ is correspondingly considered the machined hole diameter on top surface of workpiece.

3.10.2 Moving Electrode

Once the experiment using stationary electrode proves the effectiveness of the proposed model preliminarily, the experiment using moving electrode is conducted. Figures 3.13, 3.14 and 3.15 show the experimental results obtained from different

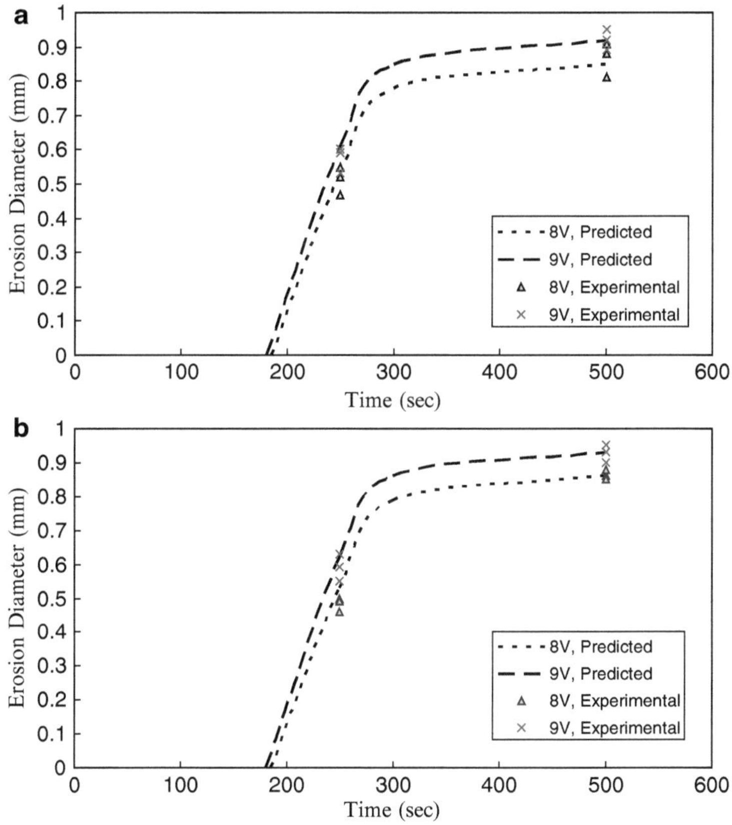

Fig. 3.13 Predicted and experimental bottom hole diameter [74], reprinted with permission. (**a**) Moving scheme A. (**b**) Moving scheme B. (**c**) Moving scheme C. (**d**) Moving scheme D

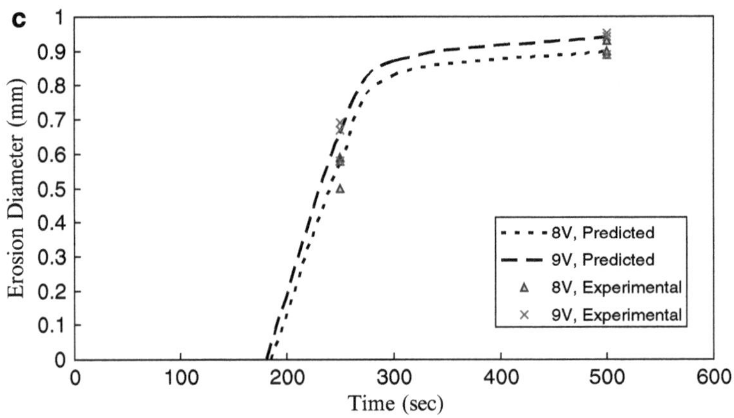

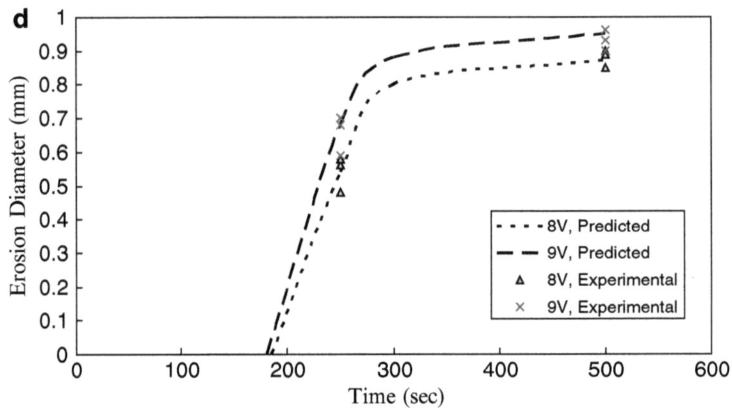

Fig. 3.13 (continued)

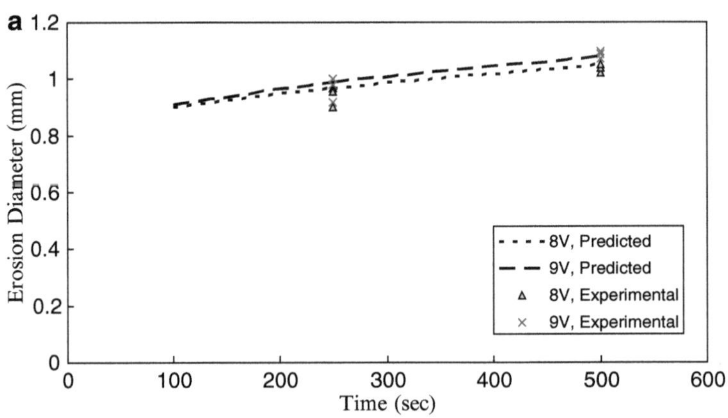

Fig. 3.14 Predicted and experimental top hole diameter [74], reprinted with permission. (**a**) Moving scheme A. (**b**) Moving scheme B. (**c**) Moving scheme C. (**d**) Moving Scheme D

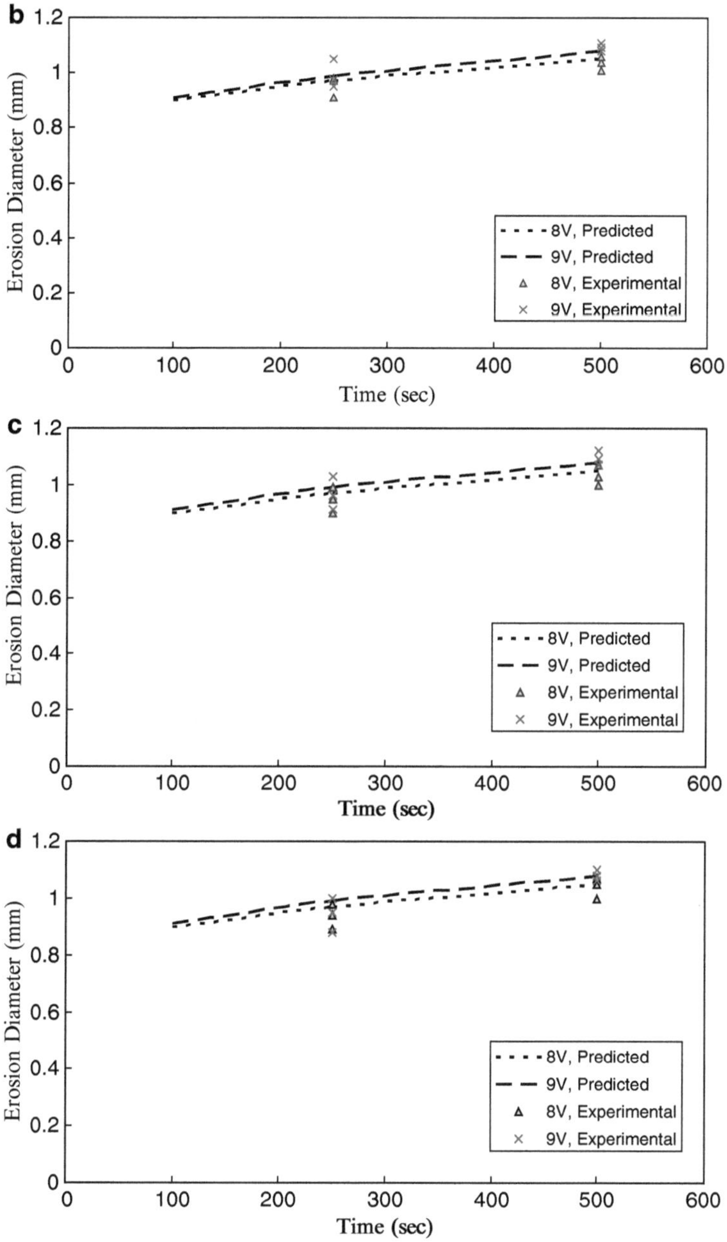

Fig. 3.14 (continued)

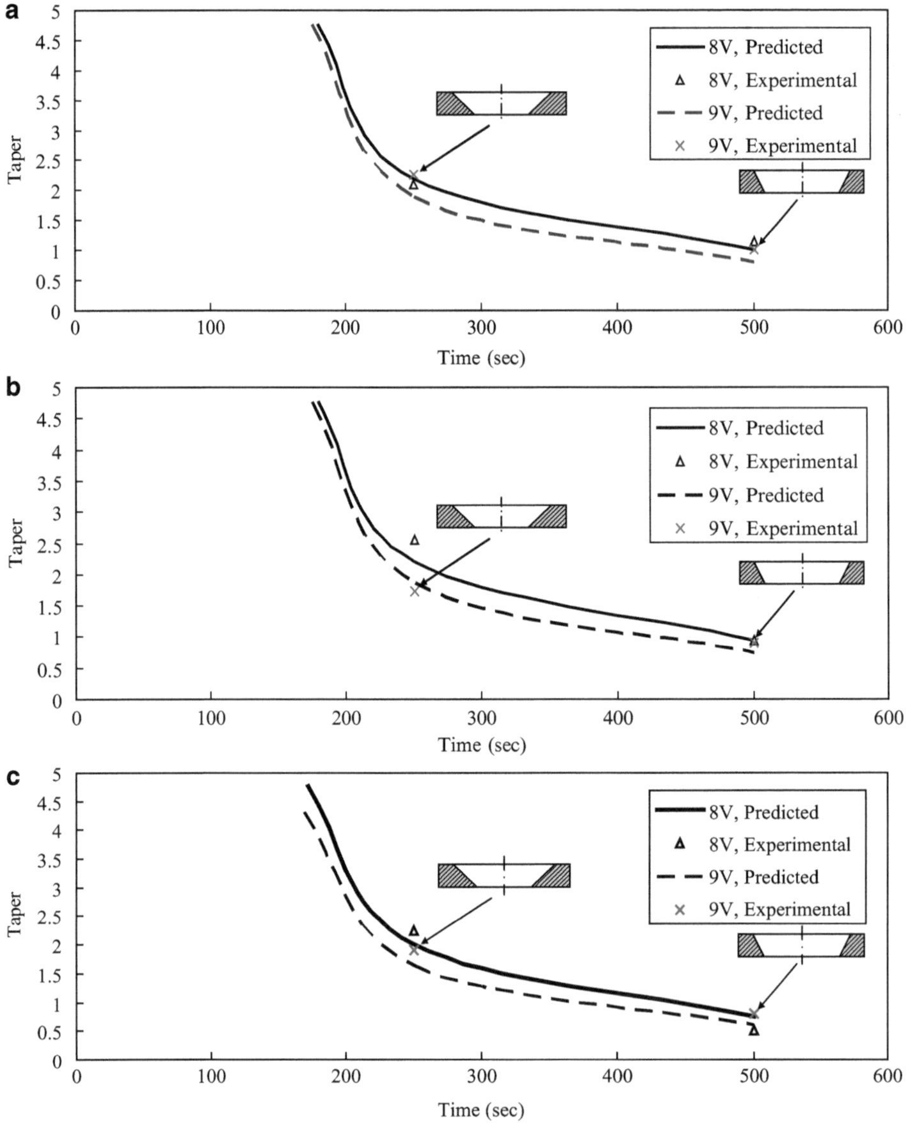

Fig. 3.15 Taper prediction of moving electrode machining [74], reprinted with permission. (**a**) Moving scheme A. (**b**) Moving scheme B. (**c**) Moving scheme C. (**d**) Moving scheme D

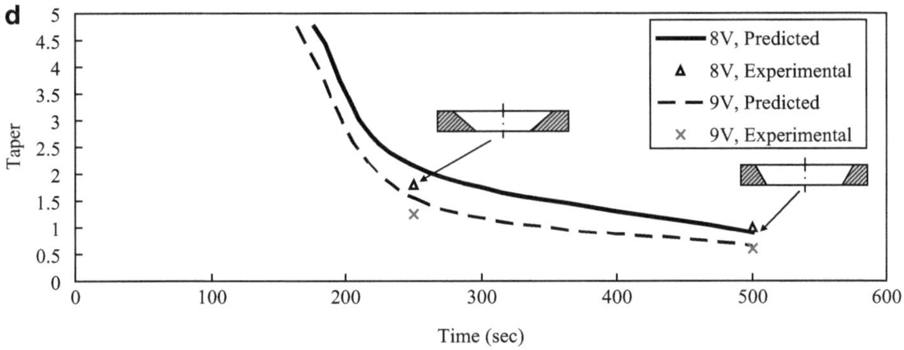

Fig. 3.15 (continued)

moving schemes of the electrode compared with the model predictions. Each experimental condition was tested three times. The comparisons of the predicted and experimental erosion opening at bottom and top are shown in Figs. 3.13 and 3.14, respectively. The experimental results lie on the predicted route of hole enlargement developing with time in all moving schemes. Higher voltage produces faster and larger erosion as expected. The opening at bottom enlarges drastically after boring-through at around 200 s lasting for 50 s, until the diameter approaches to 0.8 and 0.9 mm for 8 and 9 V, respectively. The top diameter increases at constant pace to the extent slightly wider than the bottom opening. The maximum errors between the experimental and predicted bottom opening of moving schemes A to D are 15, 12, 17, and 18%, respectively. As to the top opening diameter, the maximum errors are 7, 7, 8, and 10%, respectively. The errors on the top are smaller than the bottom. Since the top opening is eroded ahead of the bottom opening, it grows to a stable dimension while the bottom opening is yet developing after being bored through. The experimental results show the proposed model is simple for use with acceptable accuracy.

Figure 3.15 shows the predicted eroded taper in the model and the average of the measured taper. The taper is defined as the difference between top and bottom diameter per unit workpiece thickness.

$$\text{Hole Taper} = \frac{\text{top opening diameter} - \text{bottom opening diameter}}{\text{workpiece thickness}} \tag{3.19}$$

As observed in Figs. 3.13 and 3.14, the top erodes earlier than the bottom until boring-through; thus the taper first drops at that moment and keeps reducing for more than 200 s approaching to the value of one, namely a straight-wall through hole. The average error of the model prediction is 8.9 and 12% for the condition of 8 and 9 V, respectively. It illustrates the proposed model is able to predict not only the dimensions of opening but also the taper of the hole during the electrochemical

Table 3.1 Comparison of polishing methods

	Manpower	Mechanic	Electropolishing
Power	Man	Grinding or polishing machine	Dissolved by electrolytic
Tool	Polisher, Polishing-wheel et al.	Honing-head, Polishing-wheel et al.	Electrode (cathode)
Tool aid	Polishing cloth, Polishing past et al.	Polishing past	Non
Automatic process	No	Yes	Yes
Cutting way	Contact	Contact	No contact
Tooling wear	Yes	Yes	No
Relative to workpieces	Yes	Yes	No
Polish hour	Long	Long	Very short
Surface roughness	Direct proportion to polishing technique and polishing hour, about 0.1~0.8 μm	Direct proportion to polishing technique and polishing hour, about 0.1~0.8 μm	0.2~1.6 μm
Even ratio of polishing	Depends on labor's technique and experience	Good	Good
Residual stress	Yes	Yes	Non
Over polishing stress	Surface crack and microvoids	Surface crack and microvoids	Non
Surface layer of transformation	Little	Little	Non
Cost of equipment	Low	Medium	High
Labor cost	High	Medium	Low
Limit of polishing area	Yes	Non	Non

boring process. With careful design of the process, one can produce a small and tapered hole of a wide selection of taper.

The model assumes no undercut or abrupt shape changes on electrodes, with a stable flow of electrolyte in between. The dimension of the specified ECM area is small that the effects of the nonuniform electrolyte concentration, voltage, and electrolyte flow rate are considered limited.

3.11 Electropolishing

ECM can effectively deal with the difficult-to-machine materials, and the results are good surface roughness without residual stress [26]. Table 3.1 shows the characteristics of electropolishing. There are wide applications of the hole electropolishing, such as the inner wall of guns and artillery, the hole of wire-drawing, extrusion or deep-drawing dies, the hardened inner wall, the hole of

Table 3.2 Comparison of common hole-machining methods

	Drilling	Reaming	EDM	ECM
Cutting mechanism	Removed by drill lip	Removed by reamer	Gasified by electric discharge	Dissolved by electrolytic
Cutting way	Contact	Contact	Noncontact	Noncontact
Wearing-type of tool	Mechanic	Mechanic	Electric discharge	Non
Workpieces	Non-limited	Non-limited	Conductor	Conductor
Relative to workpieces	Yes	Yes	No	No
Dimensional accuracy	0.1–0.5 mm	0.02–0.1 mm	0.01–0.05 mm	0.01–0.1 mm
Surface roughness	8–25 μm	2–6.3 μm	1.6–6.3 μm	0.2–3.2 μm
Residual stress	Yes	Yes	Yes	Non
Surface layer of transformation	Mechanical stress and heat stress	Mechanical stress and heat stress	Remelted of quench	Non
Removal rate (mm^3/min)	1,000–7,000	20–600	0.25–25	2–200
Cost	Low	Low	High	High

instruments, or the hole from electric discharge machining. The processes for hole making are compared in Table 3.2. ECM can be successfully applied to cylinder machining. Table 3.3 illustrates the comparison between several processes. ECM can be successfully applied to cylinder machining. Table 3.3 illustrates the comparison between several processes.

Among traditional techniques of hole wall or cylindrical surface machining, reaming or fine turning is often used for the close dimensional tolerance. Subsequent conventional techniques such as polishing by hand or machine are often employed to achieve fine surface finish. However, polishing by hand is heavily dependent on the sophisticated experience, and either hand or machine polishing will result in non-uniform residual stress due to the contact between tool and workpiece. Surface crack and microvoids are often induced and deteriorate the service life of parts. In case that the internal hole is produced by electric discharge machining, the brittle surface layer due to the carburation and quench in the process creates additional difficulty for the following conventional polishing process. The electropolishing can efficiently produce workpieces free of the above-mentioned shortcomings. It is well suitable for difficult-to-machine materials. Plastic or press dies, wire drawing dies, optical and electric parts can apply this technique as well [23].

In terms of the application of electropolishing technology in Taiwan, it is limited to the cleaning of the surface of stainless steel, which has been mechanically machined. Meanwhile, the workpieces being electrochemically cleaned increase the ability of preventing erosion. In recent years, the application of electrochemical machining has been increasing. The Die Institute of the Tai-Lien Tech University in mainland China is using the erosion method of electrochemical machining to polish the surface of dies. It not only decreases manpower requirement but also increases

Table 3.3 Comparison of common cylinder machining methods

	Turning	Rolling	Drawing	Extrusion	ECM
Machining mechanism	Removed by drill lip	Formed by roller	Formed by mode	Formed by mode	Dissolved by electrolytic
Contact	Yes	Yes	Yes	Yes	No
Wearing type of tool	Mechanic	Mechanic	Mechanic	Mechanic	Non
Workpieces	Non-limited	Melt	Melt	Melt	Conductor
Relative to workpieces	Yes	Yes	Yes	Yes	No
Dimensional accuracy	0.01~0.13 mm	0.21~0.33 mm	0.13~0.21 mm	0.21~0.33 mm	01~0.1 mm
Surface roughness	2~25 μm	0.8~6.3 μm	0.8~6.3 μm	0.8~6.3 μm	0.2~3.2 μm
Residual stress	Yes	Yes	Yes	Yes	Non
Surface layer of transformation	Mechanical stress and heat stress	Mechanical stress and heat stress	Mechanical stress and heat stress	Mechanical stress and heat stress	Non
Removal rate (mm^3/min)	1,000–7,000	Non	Non	Non	2–200
Cost	Low	Low	Low	Low	High

Table 3.4 Chemical composition of workpieces

(Wt %)	Fe	C	Si	Mn	P	S	Cr	Mo	Al	V	Cu	Ni
SKD61	90.70	0.38	0.96	0.43	0.29	0.03	5.31	1.08	–	0.82	–	–
SKD11	88.65	1.40	0.40	0.30	0.02	0.03	8.20	0.80	–	0.20	–	–
NAK80	92.06	0.13	0.60	1.50	–	–	–	0.25	1.12	–	1.24	3.1
SNCM8	96.48	0.39	0.30	0.90	0.02	0.03	0.80	0.25	–	–	0.03	2.0

the accuracy control of dies. For over 20 years, Japan has contributed significantly in the development and application of electrochemical machining. For instance, Gin-Kan Company of Japan has produced the electropolishing machine of good quality. The polishing of workpieces still depends on mate-electrodes, where the design and production of electrode requires high cost. Hence the application of electropolishing is limited. In terms of the insufficiency of the application of electropolishing and the shortcomings of the current polishing technology, it is necessary to advance the technology of electropolishing, particular to improve the design of the polishing electrodes. In this study, the electropolishing is applied to the surface of holes and cylinder. In fact, electropolishing of internal surface includes two processes, namely the electropolishing and electrobrightening according to the amount of the material removal in the polishing process. ECM after drilling or rough turning is called electropolishing, while after reaming or fine turning is called electrobrightening. The study is aimed to design the electrodes, to undergo the electropolishing of inner hole and cylinder. The electrodes are expected low cost, effective in discharging dregs, high efficiency in polishing, and widely applicable for the industry.

3.11.1 Experimental Setup

3.11.1.1 Preparation of Workpiece

The materials of workpiece in this study are SKD11, SKD61, NAK80, and SNCM8, representing four common die and mold materials; their chemical compositions are shown in Table 3.4. The experiment of the polishing of internal cylindrical surface is divided into two parts. The first is the electropolishing after the workpiece is drilled to ϕ8.0 mm. The second is the electrobrightening after the workpiece is further reamed to ϕ7.8 mm. All workpieces are required to be examined by a caliper (accuracy of 0.01 mm) before undergoing electropolishing or electrobrightening. The tolerance of the inner holes must be controlled under 0.07 mm after drilling and 0.02 mm after reaming. The dimension of the workpiece of the experiment of the polishing of external cylindrical surface is 10.2 mm in diameter and 50 mm in length.

All workpieces after electropolishing are measured by the surface roughness measurement (Hommel T500, the accuracy is within ±5% after standard correction). The surface roughness is characterized by Ra, where the length of cut-off is 0.8 mm, and the measuring direction is perpendicular to the tooth mark. The measuring data are chosen from two different locations at least. The initial average surface roughness value (Ra) of the workpiece after drilling or rough turning is 10–15 μm. The removal amount of the diameter after electropolishing is 0.2 ± 0.03 mm, and the amount after electrobrightening is about 0.03 ± 0.01 mm. After turning, drilling, or reaming, all workpieces are put to ultrasonic-cleaning for 5 min before electropolishing or electrobrightening.

3.11.1.2 Experimental Parameters

(a) Electropolishing or electrobrightening of holes

The electrolyte is $NaNO_3$ of 25 wt.%. The temperature of the machining is maintained at 25 ± 5° C. Beside the polishing time and the current wave (continuous or pulse direct current), the side gap width between the electrode and hole wall varies at 0.2, 0.3, 0.4, 0.5, and 0.6 mm; the rotational speed of electrode is 100, 200, 400, 600, 800, 1,000, and 1,200 rpm; and the flow rate of electrolyte is 1, 2, 3, and 4 L/min. In this study, the power supply uses the design of fixed current, while the working voltage is confined to 10 V. The current density is 15, 30, 45, and 60 A/cm^2, and the current rating is 5, 10, 15, and 20 A. The axial feed rate for the feeding electrodes ranges from 0.5 to 3.5 mm/min.

(b) Electropolishing of external cylindrical surface

The controlled factors of electropolishing of external cylindrical surface are the electrolyte of $NaNO_3$ of 25 wt.%, the flow rate of 4 L/min, and the machining temperature of 25 ± 5°C. The gap width between electrode and workpiece is set 0.3 mm. The main parameters include the geometry of electrode, die material, the rotational speed of workpiece, and the electrode feed rate, as shown in Table 3.5. The gap between the disc electrode and workpiece is open when the flute on the disc edge encounters with the workpiece. It provides the current-off effect; hence the on-off pulse instead of continuous direct current can be applied. The on-off ratio is set by the rib and flute width on the electrode in the present study, which ranges from 1 to 5.

(c) Ultrasonic-aided electropolishing of holes and external cylindrical surface

The experiment is divided into two parts. The first selects six completely inserted electrodes and feeding electrodes for electropolishing of holes. The second uses the other six types of mate-electrode and six types of electrode for electropolishing of external cylindrical surface. The experiment includes electropolishing (EP), pulse-electropolishing (PEP), ultrasonic-aided electropolishing (UEP), and ultrasonic-aided pulse-electropolishing (UPEP). The material of workpiece is SKD61. The primary parameters include the frequency and the power of ultrasonic, current density, current rating, the

Table 3.5 Experimental parameters in electropolishing of external cylindrical surface

(a) Turning-tool electrodes			
Rotation speed of workpiece (rpm)		200, 400, 600, 800, 1,000, 1,200, 1,400	
Current rating (A)		5, 10, 15, 20	
Electrode feed rate (mm/min)		0.5, 1, 1.5, 2, 2.5, 3, 3.5, 4	
Back rake angle (degree)		−20, −10, 0,10, 20, 30, 40	
Side rake angle (degree)		0, 10, 20, 30, 40	
End clearance angle (degree)		0, 10, 20, 30, 40	
Side clearance angle (degree)		0, 10, 20, 30, 40	
End cutting edge angle (degree)		0, 10, 20, 30, 40	
Side cutting edge angle (degree)		0, 10, 20, 30, 40	
Nose radius (mm)		0.1, 0.2, 0.3, 0.4, 0.5, 1, 1.5, 2, 2.5, 3, 3.5	
Cross section (mm^2)		5×5, 7.5×7.5, 10×10, 12.5×12.5, 15×15	
(b) Arrow-head electrodes			
Rotation speed of workpiece (rpm)		200, 400, 600, 800, 1,000, 1,200	
Rotation speed of electrode (rpm)		0, 200, 400, 600, 800, 1,000, 1,200	
Current rating (*A*)		5, 10, 15, 20	
Electrode feed rate (mm/min)		0.5, 1, 1.5, 2, 2.5, 3, 3.5, 4	
End radius (mm)		0.5, 1, 1.5, 2, 2.5, 3, 3.5, 4, 4.5, 5, 5.5, 6, 6.5, 7, 7.5, 8, 8.5, 9, 9.5, 10	
Taper angle (degree)		30, 45, 60, 75, 90, 105	
Back rake angle (degree)		0, 10, 20, 30, 40	
Side rake angle (degree)		0, 10, 20, 30, 40	
Tail angle (degree)		0, 5, 10, 15, 20, 25	
Flat electrode thickness (mm)		5, 6, 7, 8	
Flute depth (mm)		4, 6, 8, 10	
(c) Ring-form electrodes			
I. Material			
1	Workpiece	SKD11, SKD61, NAK80, SNCM8	Ar, Br, Cr, Dr, Er, Fr
II. Power supply			
2	Current rating (A)	5, 10, 15, 20, 25, 30A	Ar, Br, Cr, Dr, Er, Fr
3	Pulse on/off time (ms/ms)	100/100, 100/200, 100/300, 100/400, 100/500	Ar, Br, Cr, Dr, Er, Fr
III. Motion control			
4.1	Rotational speed of workpiece (rpm)	200, 400, 600, 800, 1,000, 1,200	Ar, Br, Cr, Dr, E, Fr
4.2	Rotational speed of workpiece (rpm) (with electrode rotation)	0	Ar, Br, Cr
4.3	Rotational speed of workpiece (rpm) (without electrode rotation)	0	Ar, Br
5.1	Rotational speed of electrode (rpm)	0, 200, 400, 600, 800, 1,000, 1,200	Ar, Br, Cr
5.2	Rotational speed of electrode (rpm)	0	Dr, Er, Fr
6	Feed rate of electrode (mm/min)	0.5, 1, 1.5, 2, 2.5, 3, 3.5, 4	Ar, Br, Cr, Dr, Er, Fr

(continued)

Table 3.5 (continued)

IV. Electrode design			
7	Thickness of electrode (t) (mm)	2, 3, 4, 5	Ar, Br, Dr, Er
8	Inner taper angle (θ) (degree)	0, 10, 20, 30, 40	Ar, Br, Dr, Er
9	Inner edge radius (r) (mm)	0.25, 0.5, 0.75, 1, 1.25,1.5	Ar, Br, Dr, Er
10	Pin end radius (ρ) (mm)	0.25, 0.5, 0.75, 1, 1.25,1.5	Cr, Fr
11	Pin length (L) (mm)	2.5, 2.75, 3, 3.25, 3.5, 3.75, 4	Cr, Fr
12	Diameter of pin (d) (mm)	3, 3.5, 4, 4.5, 5	Cr, Fr
13.1	Pin number	1, 2, 3, 4	Cr
13.2	Pin number	1	Fr
14	Arc extension of partial ring (S) (mm)	1.3, 2.3, 3.3, 4.3, 5.3	Dr
15	Inner radius (R) (mm)	6, 9, 12, 15	Er
16	Height of partial ring (H) (mm)	6, 9, 12, 15	Er, Fr

(d) Disc-form electrodes	
Rotation speed of workpiece (rpm)	200, 400, 600, 800, 1,000, 1,200
Rotation speed of electrode (rpm)	0, 200, 400, 600, 800, 1,000, 1,200
Current rating (A)	5, 10, 15, 20
Electrode feed rate (mm/min)	0.5, 1, 1.5, 2, 2.5, 3, 3.5, 4
Disc thickness (mm)	2, 3, 4, 5
Disc edge radius (mm)	0.25, 0.5, 0.75, 1.0, 1.25, 1.5
Taper angle (degree)	0, 10, 20, 30, 40
Back rake angle (degree)	0, 10, 20, 30, 40
Side rake angle (degree)	0, 10, 20, 30, 40
Flute depth (mm)	2.5, 3, 3.5, 4
Flute width ratio	1/1, 1/2, 1/3, 1/4, 1/5
Pin end radius (mm)	0.25, 0.5, 0.75, 1.0, 1.25, 1.5
Pin length (mm)	2.5, 2.75, 3, 3.25, 3.5, 3.75, 4

current wave (continuous or pulse direct current), pulse period, rotational speed of tool electrode, the electrode feed rate, and the geometry of electrodes. The rotational speed of workpiece is 600 rpm. The rotational speed of tool electrode is 100, 200, 400, 600, 800, 1,000, and 1,200 rpm. The frequency and power of ultrasonic is 46 kHz/50 W, 46 kHz/80 W, 120 kHz/80 W. The current density is 15, 30, 45, 60 A/cm^2. The polishing time is 30 s. The current rating is 5, 10, 15, and 20 A. The axial feed rate for the feeding electrodes ranges from 0.5 to 4.0 mm/min. The pulse period (on/off time) is 100 ms/100 ms, 100 ms/200 ms, 100 ms/300 ms, 100 ms/400 ms, and 100 ms/500 ms. Their settings are shown in Table 3.6.

(d) Electropolishing of spherical surface

The material of workpiece is NAK80. The amount of reduction in the workpiece's surface after electropolishing is 10 μm, which is designed in the processes for the dimensional control of parts. The electrolyte is $NaNO_3$

Table 3.6 Experimental parameters

Frequency and power of ultrasonic (kHz/W)	46/50, 46/80, 120/80
Current rating (*A*)	5, 10, 15, 20
Current density (A/cm^2)	15, 30, 45, 60
Polishing time (s)	30
Axial feed rate for the feeding electrodes (mm/min)	From 0.5 to 4.0
Pulsed period (on/off time) (ms/ms)	100/100, 100/200, 100/300,100/400, 100/500
Rotational speed of electrode (rpm)	200, 400, 600, 800, 2,000, 1,200
Different finish process	EP, PEP, UEP, UPEP

Table 3.7 Experimental parameters

Frequency and power of ultrasonic (kHz/W)	120/80
Gap width between electrodes (mm)	0.3, 0.4, 0.5, 0.6
Current density (A/cm^2)	10, 15, 20, 25, 30, 35
Flow rate of electrolyte (L/min)	5, 10, 15, 20, 25
Thickness of wire electrode (mm)	5, 6, 7, 8, 9, 10
Pulsed period (on/off time) (ms/ms)	100/100, 100/200, 100/300,100/400, 100/500
Rotational speed of electrode (rpm)	200, 400, 600, 800, 2,000, 1,200
Different finish process	EP, PEP, UEPWR, UPEPWR

of 20 wt.%. The temperature of the electrolyte is maintained at 35 $\pm$ 5°C. The flow rate of the electrolyte is 15 L/min. The current density is 10, 15, 20, 25, 30, and 35 A/cm^2. The gap width between the electrode and workpiece is 0.3, 0.4, 0.5, and 0.6 mm. The diameter of the workpiece is 40 mm. The thickness of the wire electrode is 5–10 mm. The rotational speed of the magnetic-aid system is 100, 200, 300, 400, 500, and 600 rpm. The pulsed period (on/off time) is 100 ms/100 ms, 100 ms/200 ms, 100 ms/300 ms, 100 ms/400 ms, and 100 ms/500 ms. The different features for the finishing process include electropolishing (EP), pulsed electropolishing (PEP), ultrasonic-aided electropolishing with workpiece rotation (UREP), and ultrasonic-aided pulsed electropolishing with workpiece rotation (URPEP). Their settings are shown in Table 3.7.

3.11.1.3 Experimental Equipment

The equipment of electropolishing of internal or cylindrical surface includes DC power supply, pulse generators, pump, flow meter, electrolytic tank, and filter. The experimental setup is illustrated in Figs. 3.16 and 3.17. The experimental setup for ultrasonic-aided electropolishing of internal or cylindrical surface adds the ultrasonic generators and ultrasonic tank, as illustrated in Figs. 3.18 and 3.19.

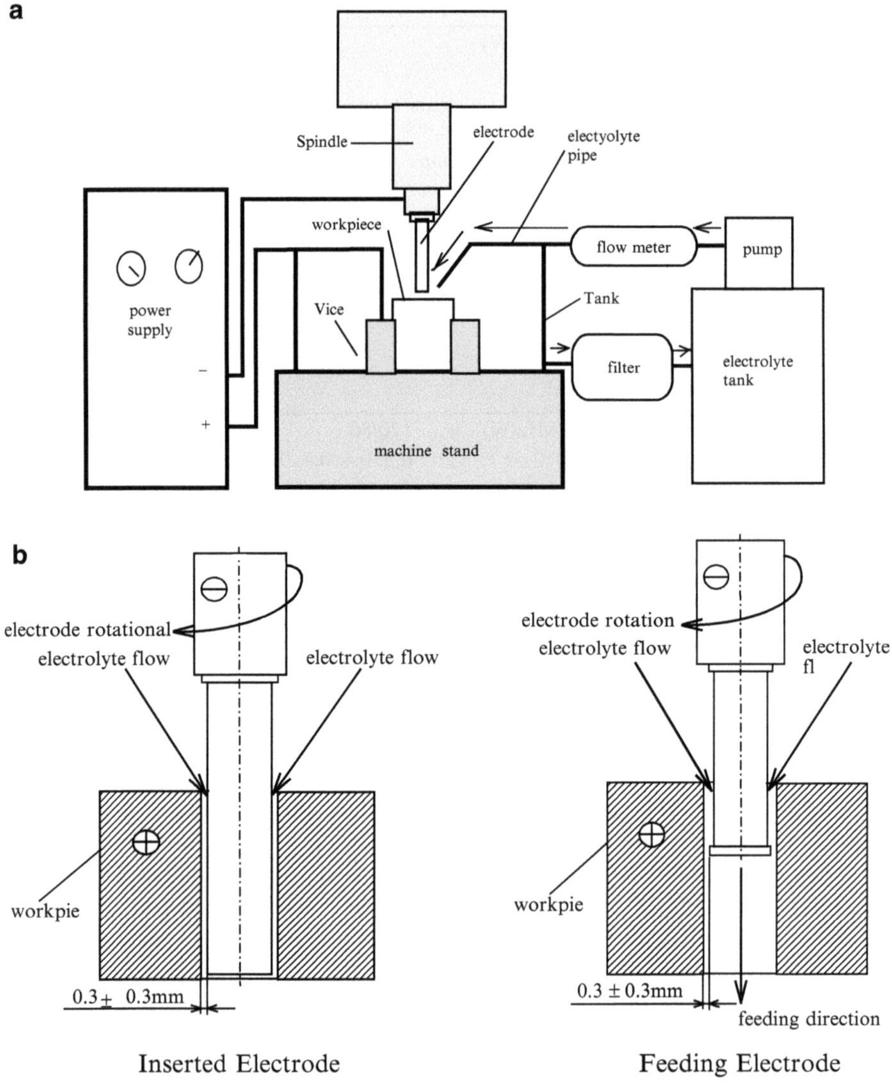

Fig. 3.16 Experimental setup of electropolishing of internal cylindrical surface. (**a**) System schematics ([66, 67], reprinted with permission). (**b**) Configuration of tool and workpiece ([67], reprinted with permission)

The experimental setup for electropolishing of spherical surface is as illustrated in Fig. 3.20. The first selects six completely inserted electrodes and feeding electrodes for electropolishing of holes, as illustrated in Fig. 3.21. The electrode design of external cylindrical surface is illustrated in Fig. 3.22. The electrode for ultrasonic-aided electropolishing of internal surface is shown in Fig. 3.23. The second uses the other six types of mate-electrode and six types of electrode for electropolishing of external cylindrical surface for ultrasonic-aided electropolishing as shown in Fig. 3.24. The third uses one type of wire electrode for electropolishing of spherical surface; the electrode is illustrated in Fig. 3.25. The steps of centering the internal cylindrical surface with the electrode are as follows: (a) Use the vice with *V* flute in

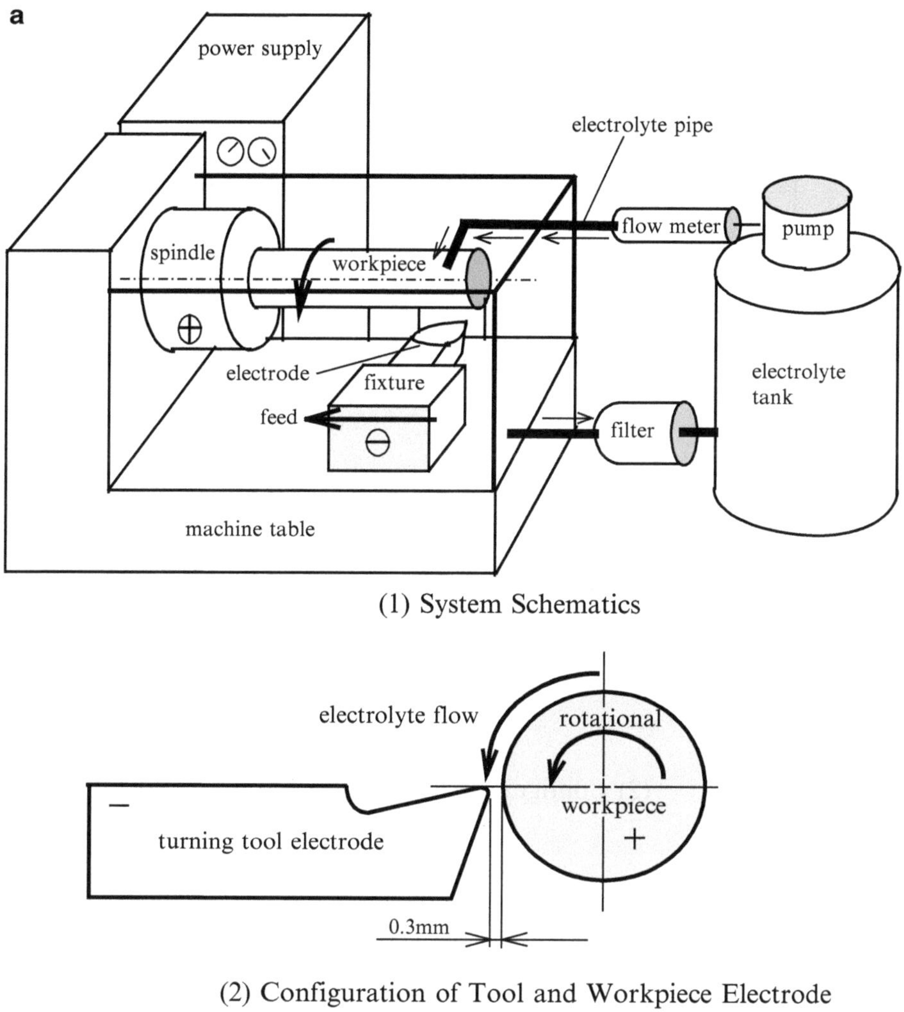

Fig. 3.17 Experimental setup of electropolishing of external cylindrical surface. (**a**) Turning tool electrode ([68], reprinted with permission). (**b**) Arrow-head electrode ([69], reprinted with permission). (**c**) Ring-form electrode ([70], reprinted with permission). (**d**) Disc-form electrode ([71], reprinted with permission)

b

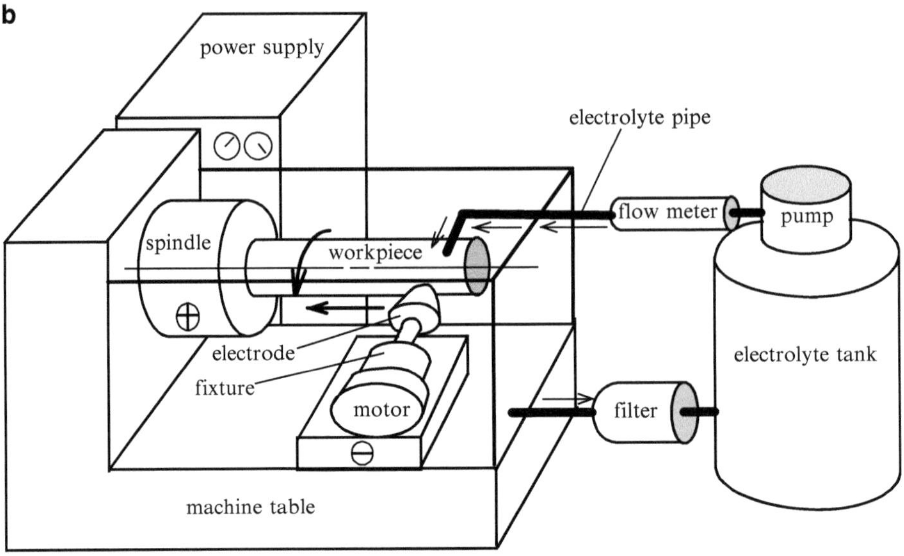

(1) System Schematics

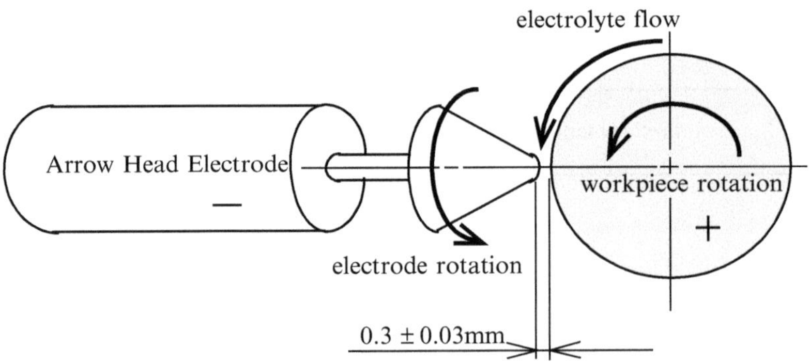

(2) Configuration of Tool and Workpiece

Fig. 3.17 (continued)

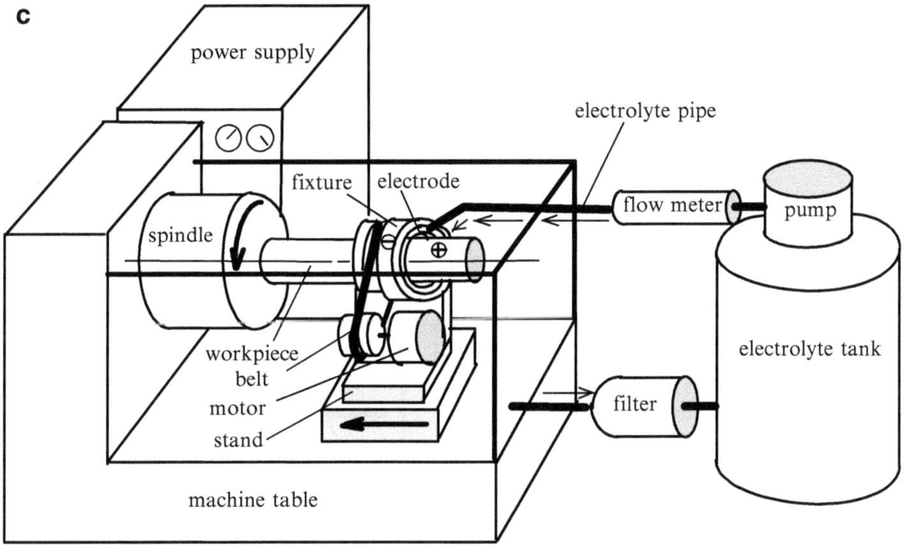

(1) System Schematics

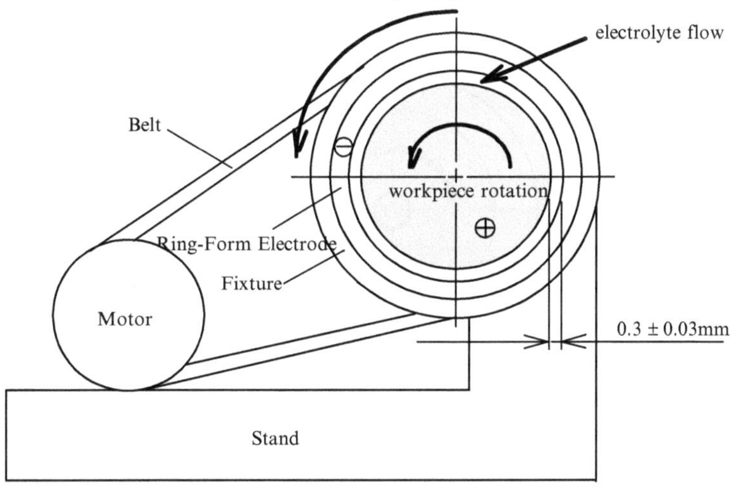

(2) Configuration of Tool and Workpiece

Fig. 3.17 (continued)

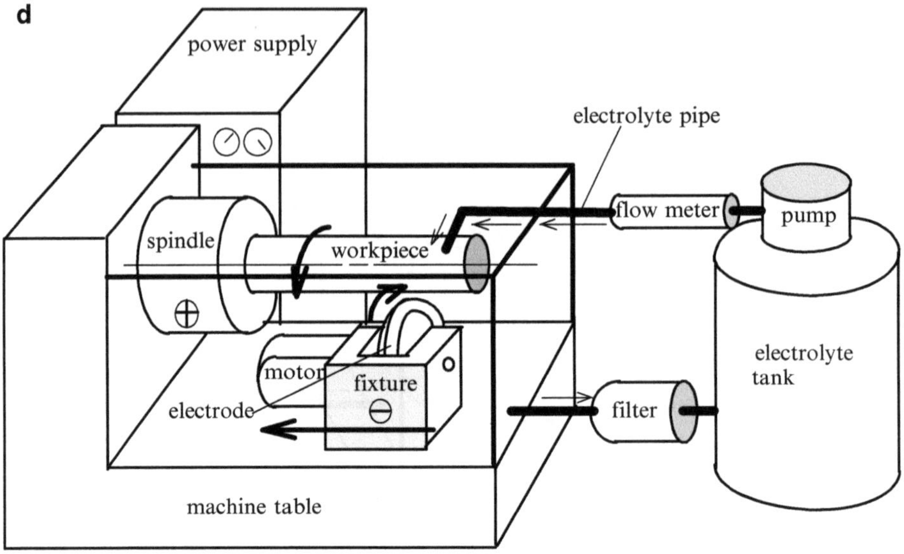

(1) System Schematics

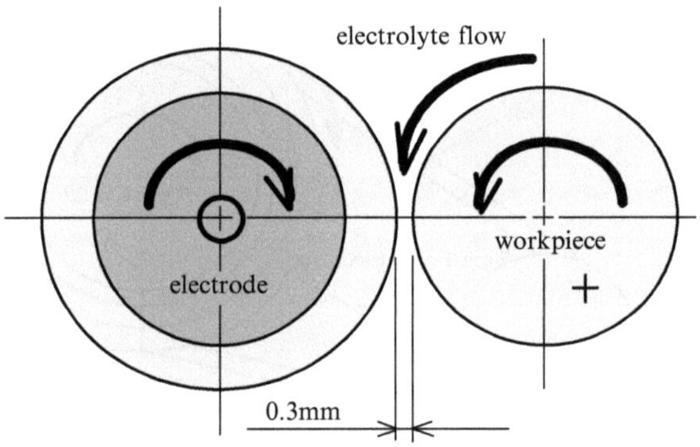

(2) Configuration of Tool and Workpiece Electrode

Fig. 3.17 (continued)

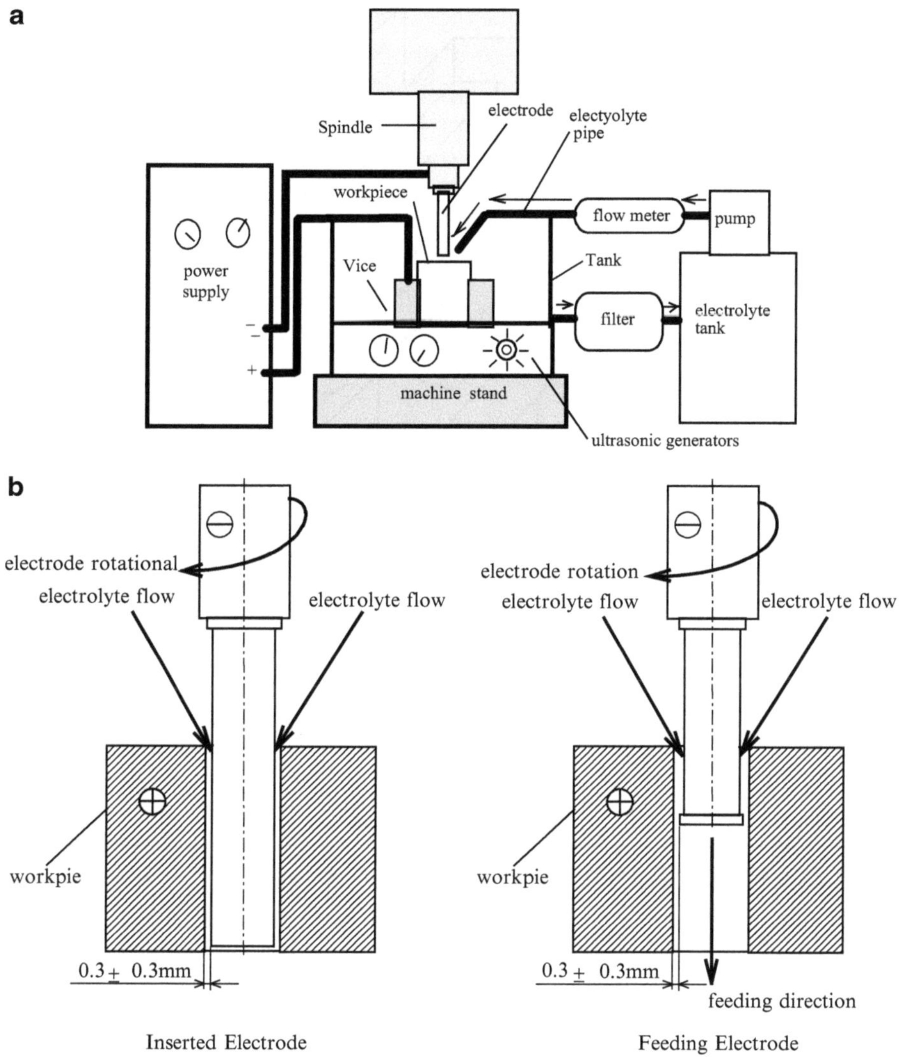

Fig. 3.18 Experimental setup of ultrasonic-aided electropolishing of holes. (**a**) System schematics. (**b**) Configuration of tool and workpiece ([66, 67], reprinted with permission)

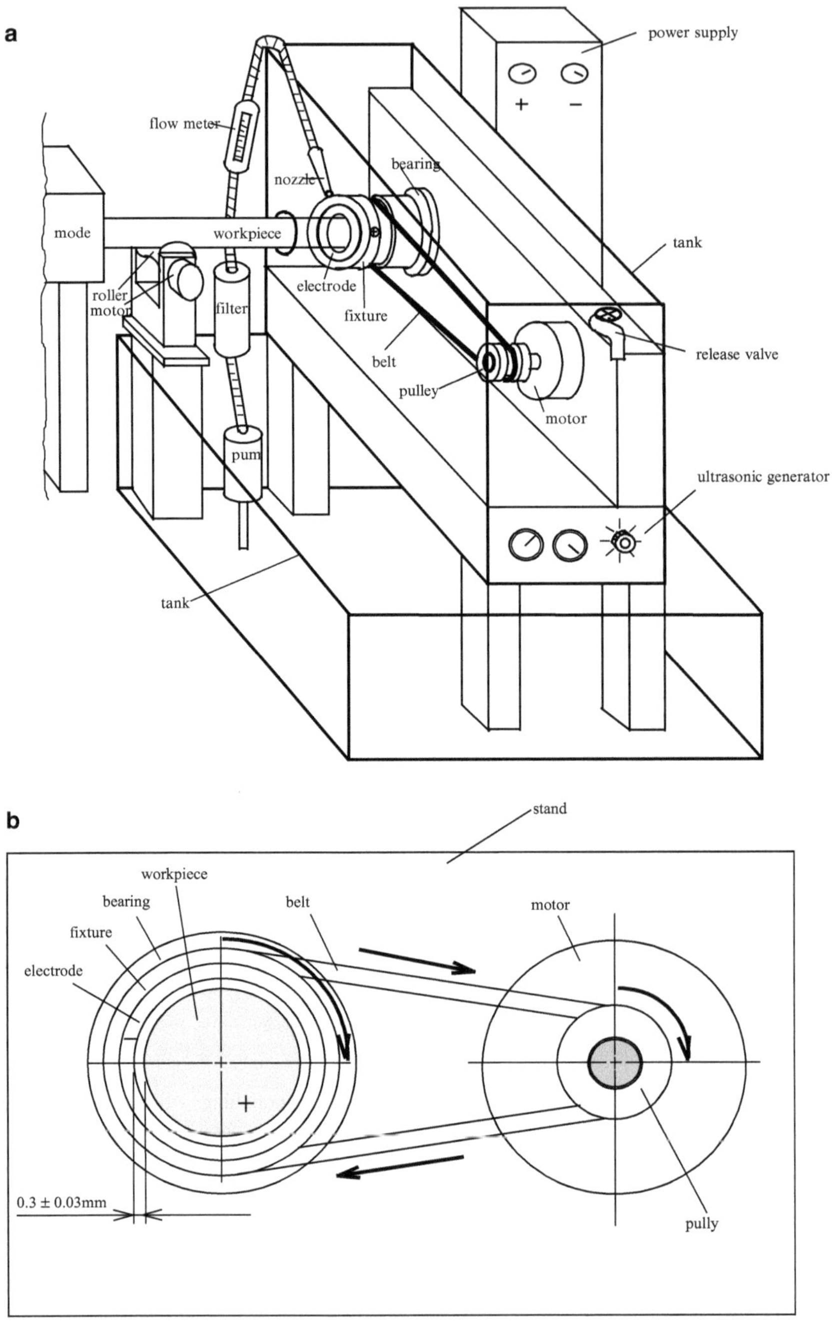

Fig. 3.19 Experimental setup of ultrasonic-aided electropolishing of external cylindrical surface (in the example of ring-form electrode) ([72], reprinted with permission). (**a**) System schematics. (**b**) Configuration of electrode and workpiece

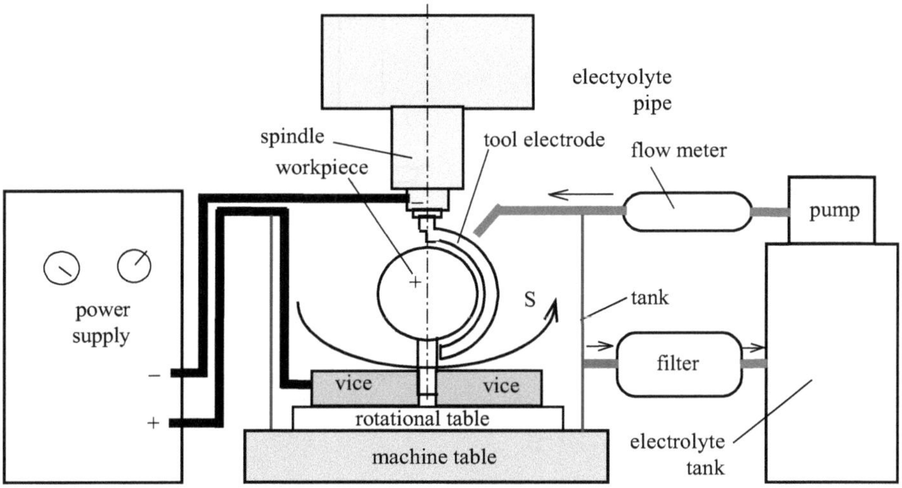

Fig. 3.20 Experimental setup

the jaw to hold the cylindrical surface of workpiece. (b) Fix the tool electrode on the machine spindle. (c) Move the tool electrode into the hole of workpiece. (d) Remove the tool electrode to touch one side of the hole wall. (e) Move the tool electrode 0.3 mm toward the other side of the hole and make it straight toward the center of workpieces. (f) Switch on the power to revolve the tool electrode; use electric guiding instrument to examine whether the tool electrode is short-circuit. The electrode and workpiece for the electropolishing of external cylindrical surface are set up as follows: (a) Use fixtures to hold the tool electrode in intended position. (b) Hold the workpiece on the machine spindle. (c) Move the tool electrode near workpiece until slight contact. (d) Remove the tool electrode 0.3 mm off the workpieces.

3.11.2 Form Design of Electrode for Smoothing of Holes

The experimental study is divided into three stages. The first stage is the parametric study using the basic electrode forms (namely A_i and A_f). The set of process parameters giving the finest surface polishing found in the first stage and is used in the following experiments. The second stage is the design and evaluation of the various inserted electrodes. The best design and its functions will be presented. The third stage compares the electropolishing with electrobrightening.

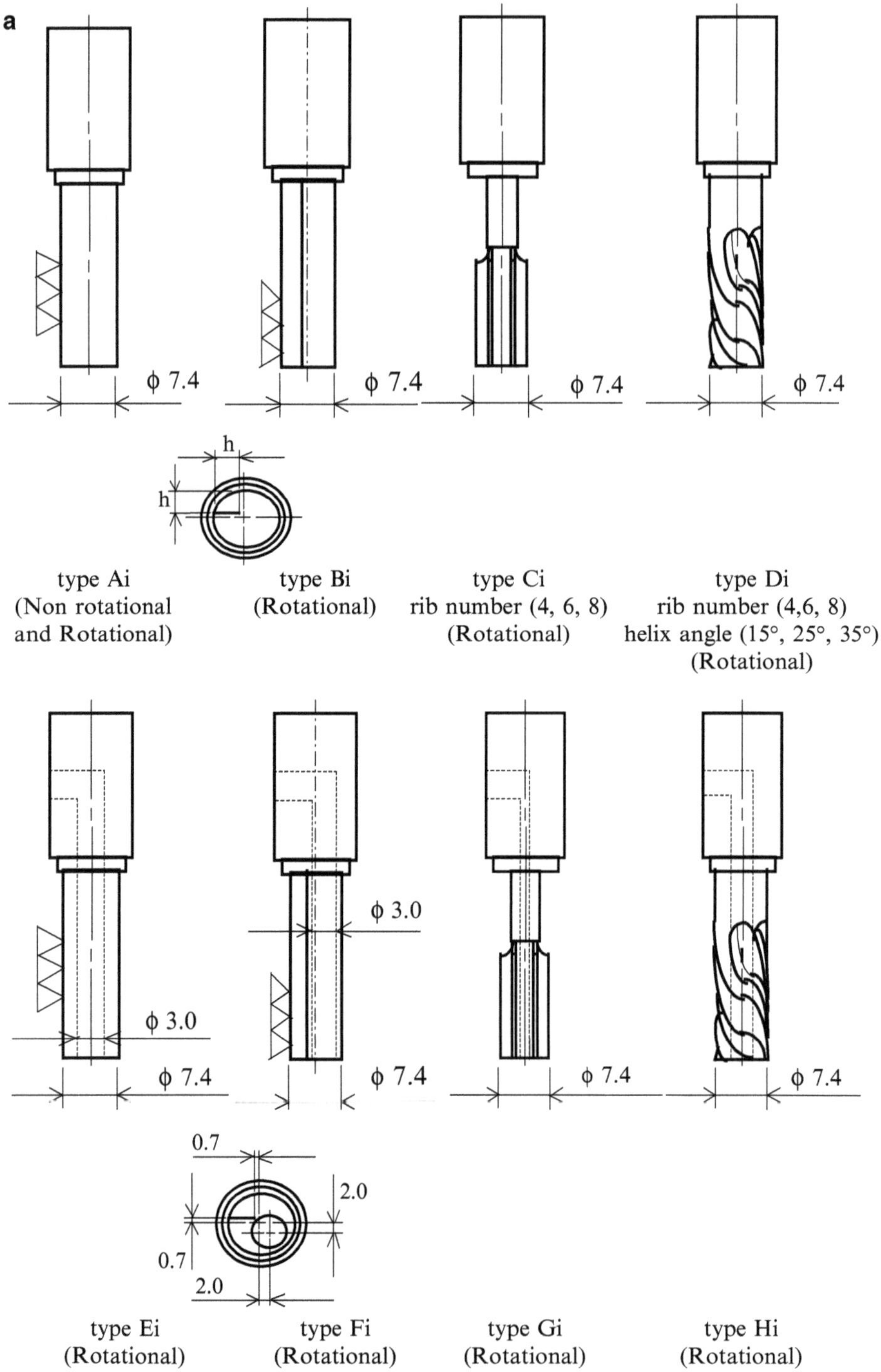

Fig. 3.21 Design of electrodes for internal cylindrical surface ([73], reprinted with permission). (**a**) Inserted electrodes. (**b**) Feeding electrodes

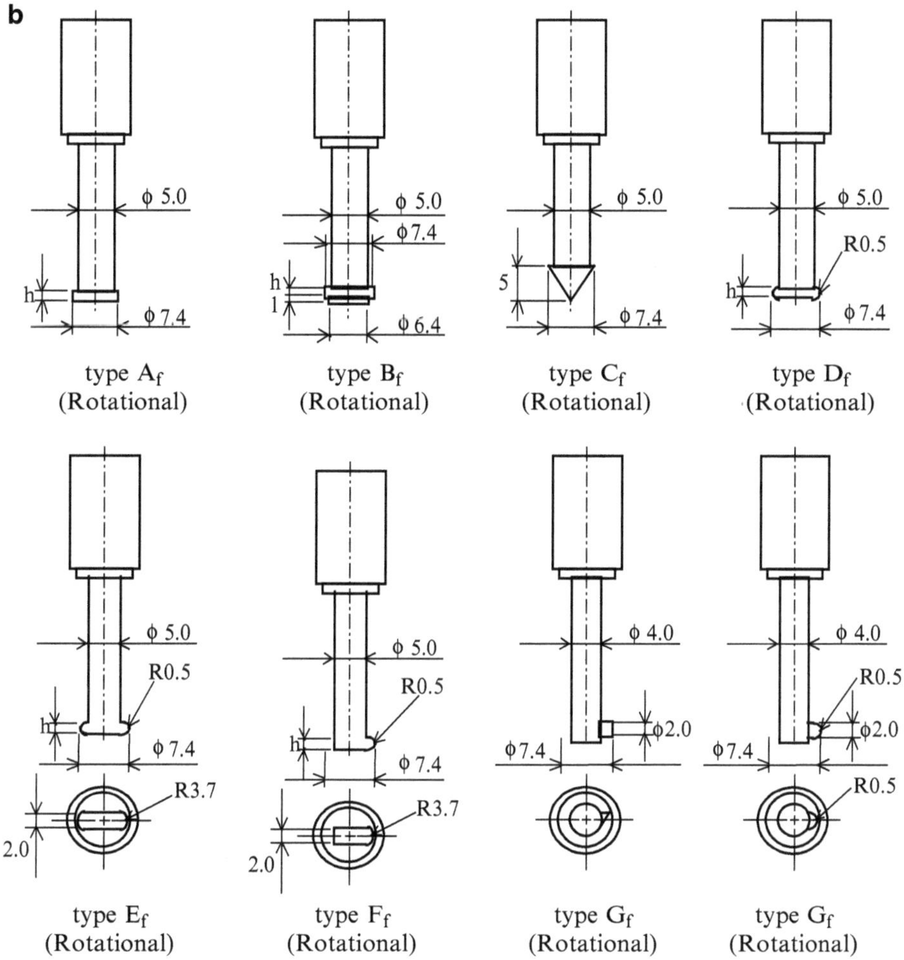

Fig. 3.21 (continued)

3.11.2.1 Experiment on Process Parameters

Figure 3.26 shows that tighter side gap between the electrode and hole wall produces smoother surface. However, gap width down to 0.2 mm tends to cause unstable operation. The electrolyte flushing also becomes more difficult and it is easy to lead to short-circuit. Thus the side gap width of 0.3 mm is suggested for the following test of electrode design. The results also show the polishing of SKD61 is the best, followed by SKD11, NAK80, and SNCM8. As to the effect of electrolytic flow rate, Figure 3.27 shows that the larger the flow rate is, the more effective is the polishing, since the electrolytic products and heat can be brought away more rapidly. Therefore, under the condition that the current equipment can bear the flowing pressure of the electrolyte, the electrolytic flow rate of 4 L/min is used. Figure 3.28(a) tells the polishing effect of completely inserted electrodes is better between 30 and 45 A/cm^2. Although 45 A/cm^2 produces faster removal rate, the control stability of electric instruments is worse than 30 A/cm^2. The electropolishing of 15 A/cm^2 is very mild and the polishing time is long. The condition of 45 A/cm^2 up to 60 A/cm^2 takes less time for the same amount of material removal. However, the discharge of electrolytic products from the gap is more difficult; thus the operation stability becomes a concern. Similarly, Fig. 3.28 (b) suggests the operation at the current rating between 10 and 15 A for the feeding electrode.

The effect of electrode rotation is shown in Fig. 3.29. The range between 400 and 800 rpm produces better polishing effect. Below 200 rpm, the rotational flow energy is insufficient for effective flushing. High-speed rotation often causes the run-out of electrode, and the associated centrifugal force often leads to the counter flow of the electrolyte. It affects the stability of gap width; thus the polishing effect is worsened. The effective rotational flow energy at the proper rotational speed of the electrode will help to discharge the reacting products of electropolishing and thus improve the polishing effect of the surface roughness of workpieces. The chemical compositions of materials are shown in Table 3.8. The ideal removal

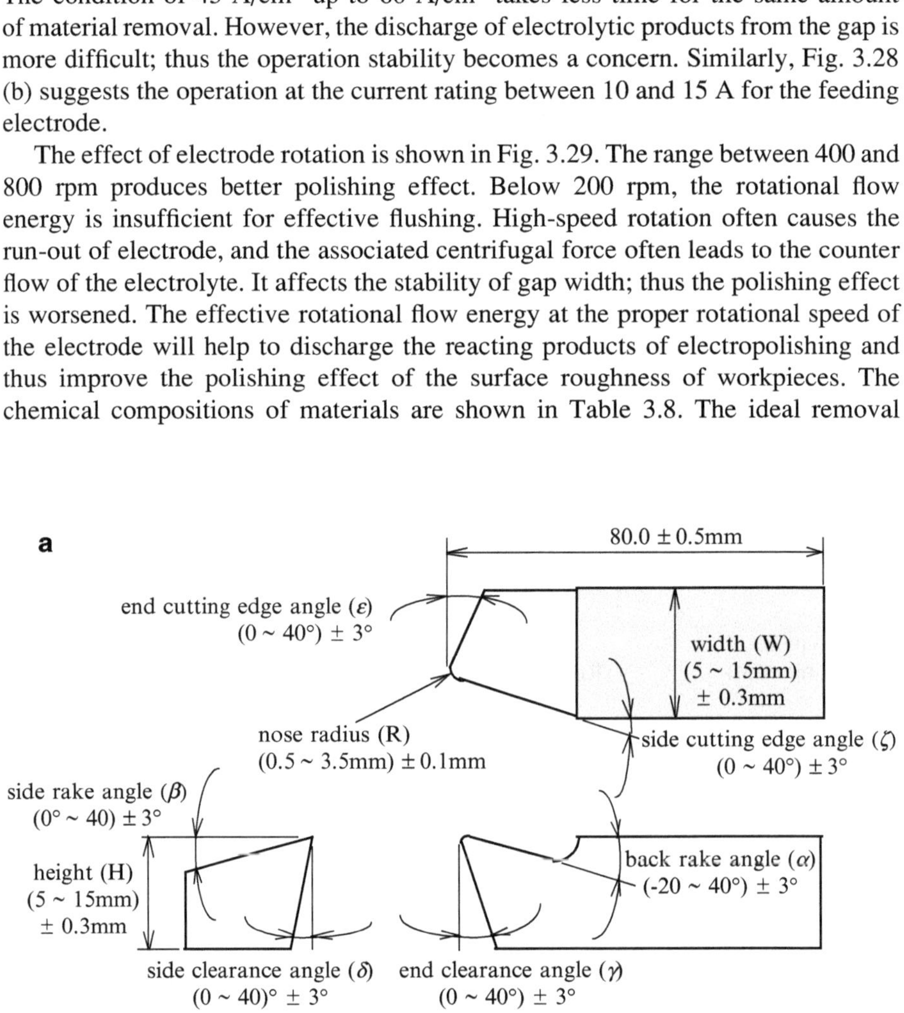

Fig. 3.22 Design of electrodes for external cylindrical surface. (**a**) Turning tool electrodes ([68], reprinted with permission). (**b**) Design of arrow head electrode ([69], reprinted with permission). (**c**) Design of ring electrode for cylindrical workpiece ([70], reprinted with permission). (**d**) Disc-form electrodes

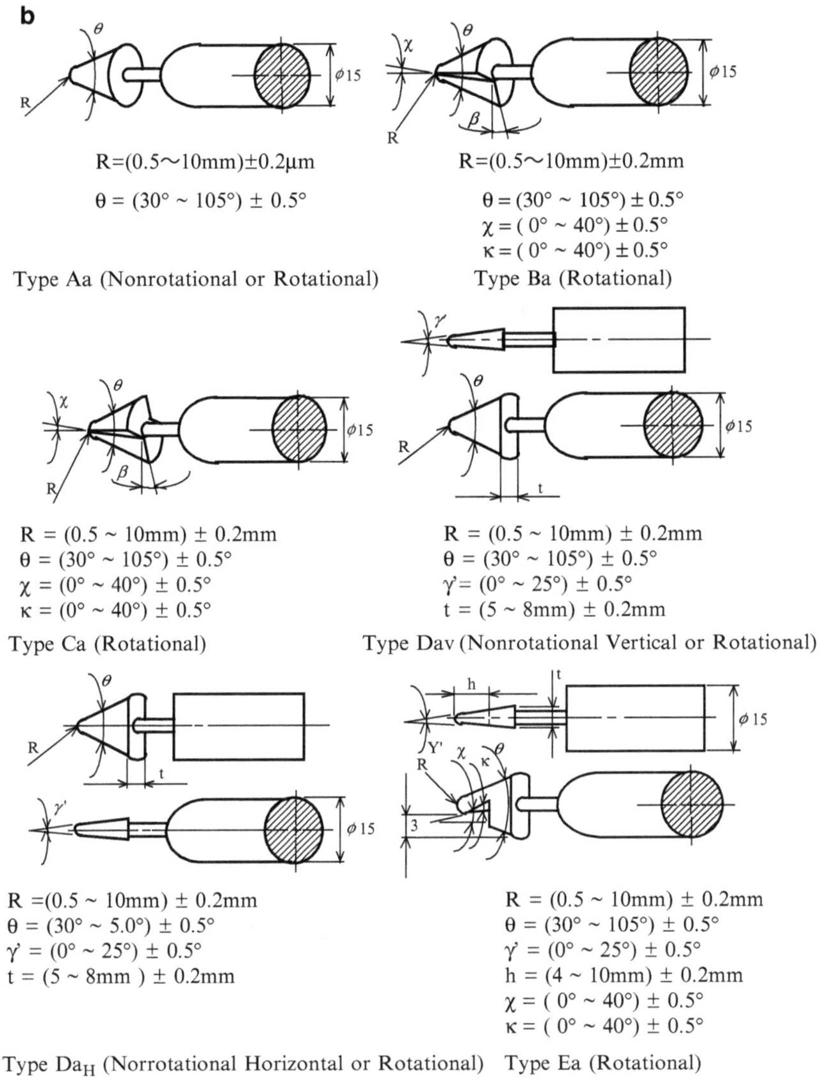

Cone Arrow Head	A	B	C	χ: back rake angle κ: side rake angle	R: end radius h: flute depth
Flat Arrow Head	D_V	D_H	E	γ': tail angle θ: taper angle	t: plate thickness

Fig. 3.22 (continued)

c

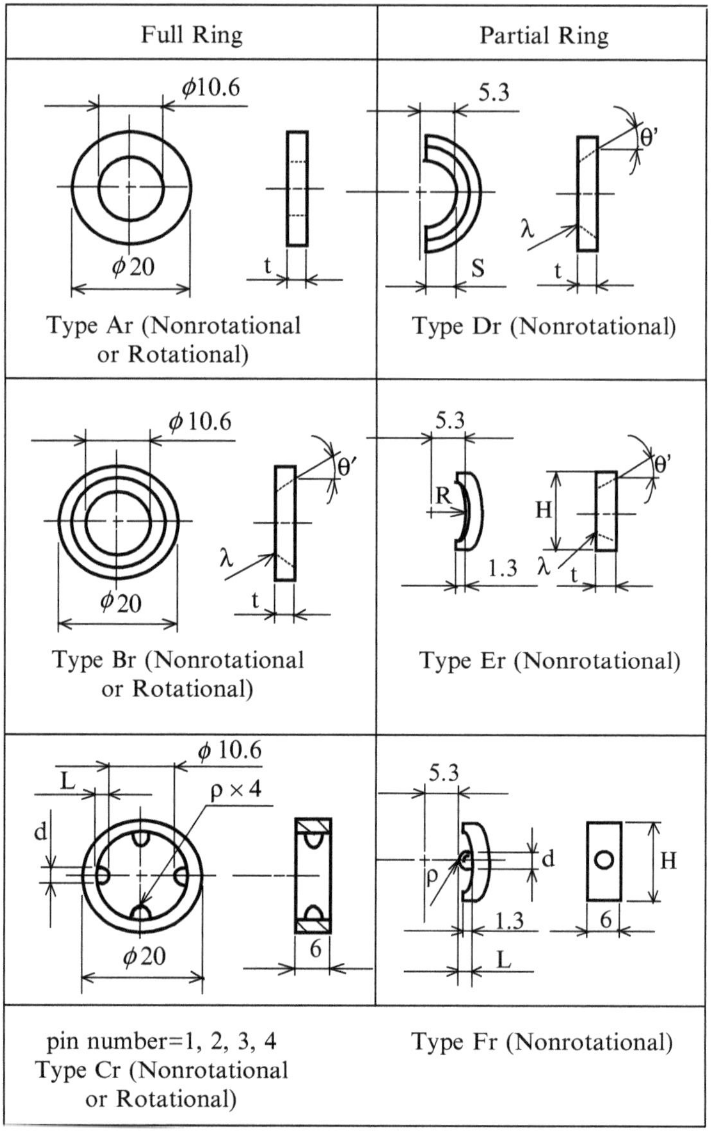

H: height of partial ring d: pin diameter θ': inner taper angle
L : pin length R: inner edge rounding ρ: pin end radius
λ': inner radius t: thickness of electrode
S : arc extension of partial ring

Fig. 3.22 (continued)

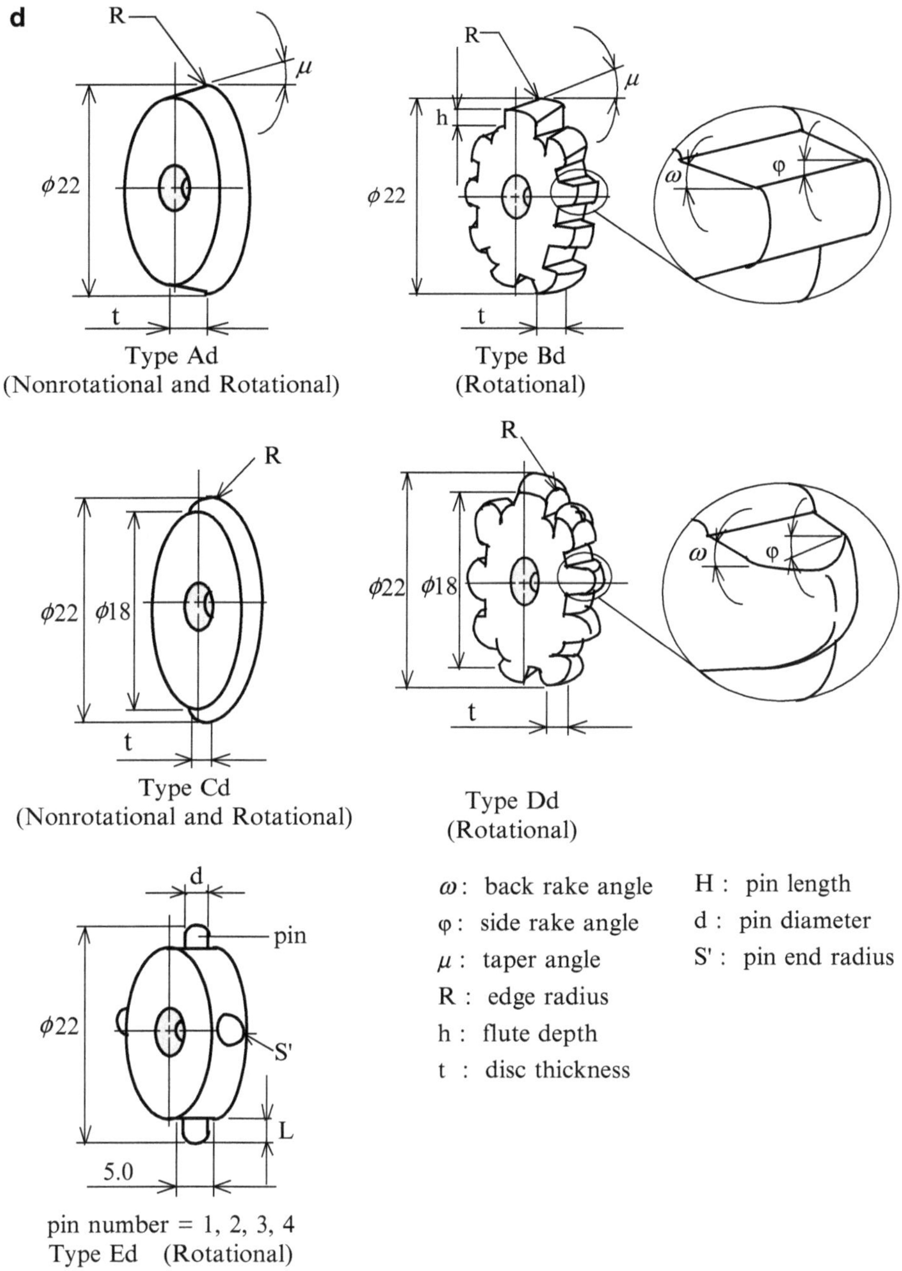

Fig. 3.22 (continued)

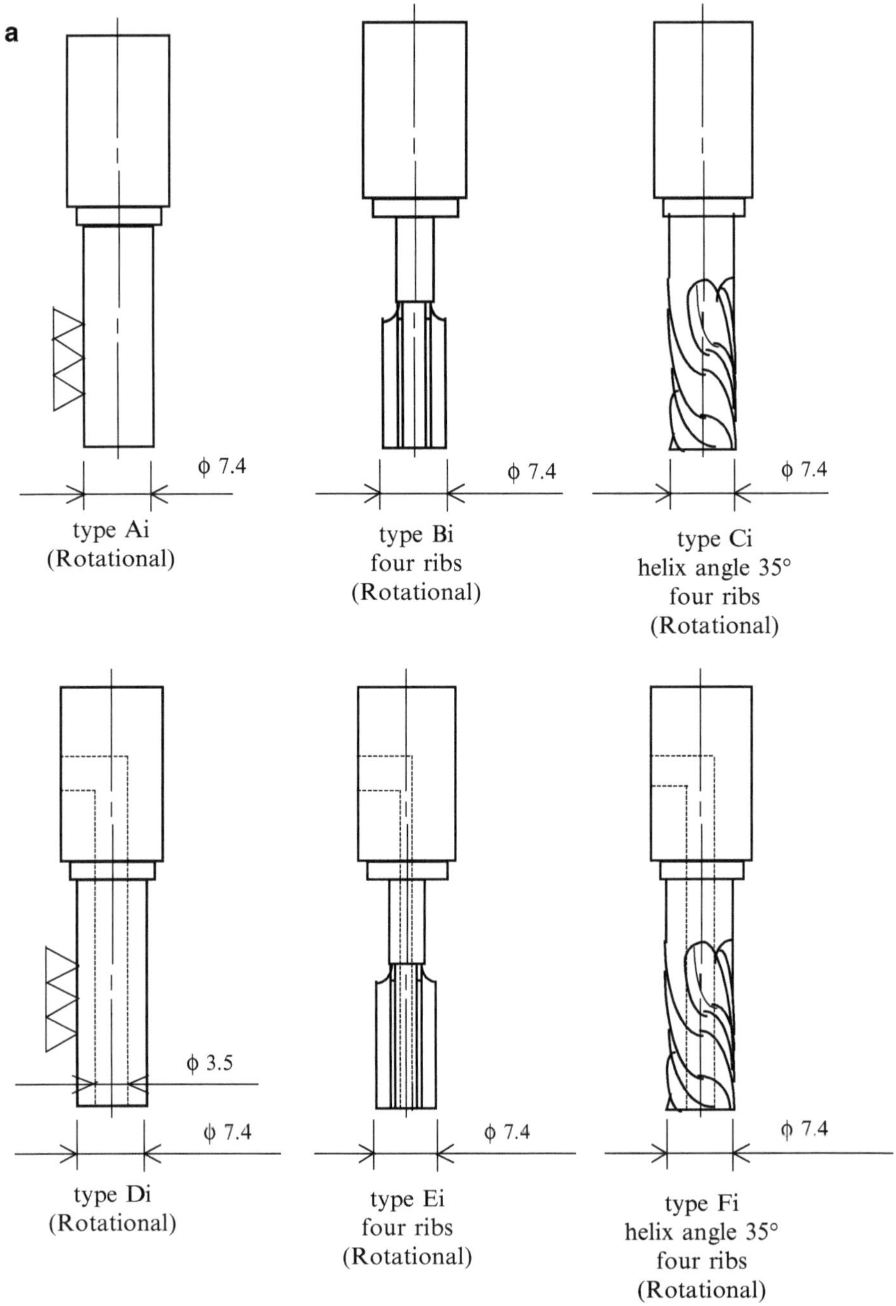

Fig. 3.23 Electrodes in ultrasonic-aided electropolishing of holes. (**a**) Inserted electrodes ([67], reprinted with permission). (**b**) Feeding electrodes

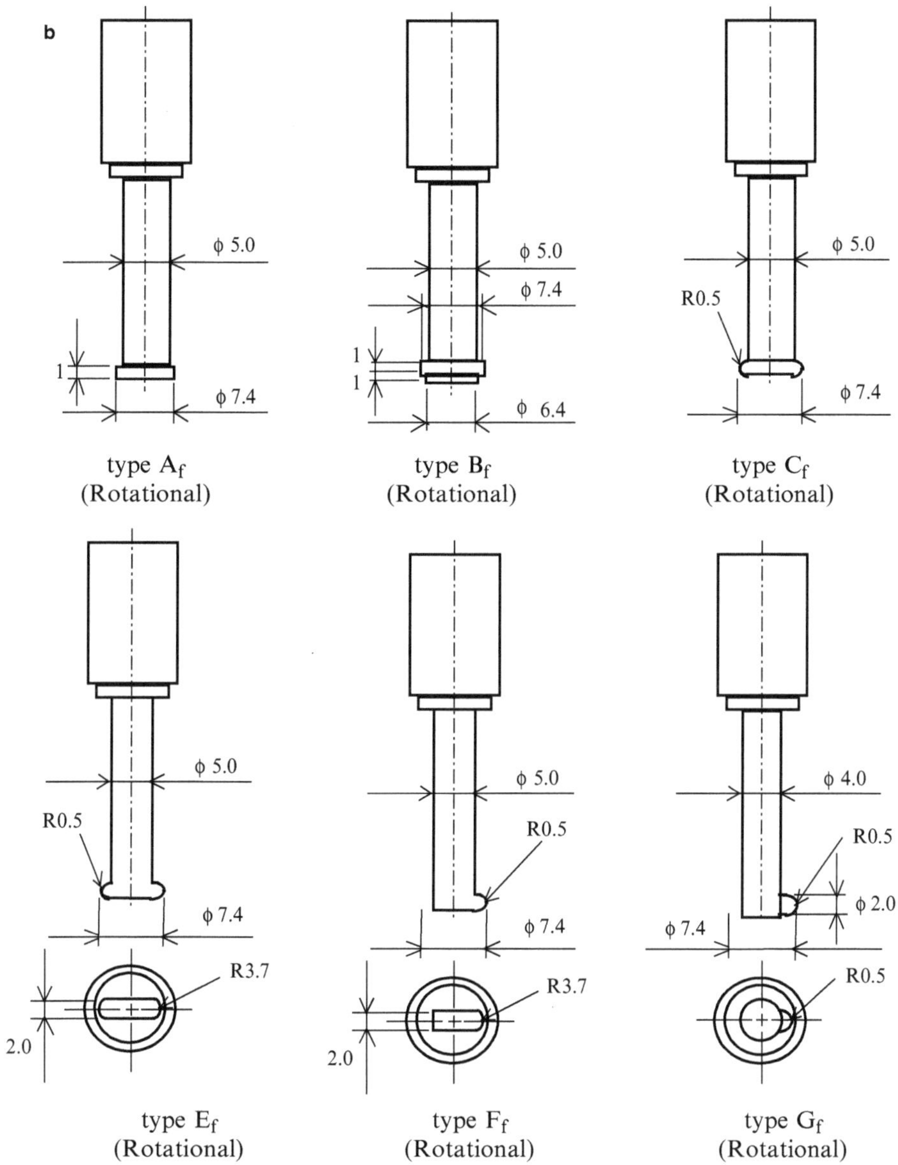

Fig. 3.23 (continued)

rate of different materials at the same machining time (30 s) and current rating (10 A) is calculated: SNCM8 (0.1274 g) >NAK80 (0.12488 g) >SKD61 (0.1242 g) >SKD11 (0.1213 g). Comparing with the results from the experiment: SNCM8 (0.073 g), NAK80 (0.092 g), SKD61 (0.122 g), and SKD11 (0.105 g), one obtains the current efficiency (η) [17]:

$$\eta = \frac{\text{actural material removal rate}}{\text{theoretical material removal rae}} \tag{3.20}$$

The results are shown in Table 3.8. The author uses SKD61 (showing the best current efficiency) to investigate the effects of eight electrode forms on polishing. To reach the same amount of material removal of 0.2 mm in diameter, the required time for different materials varies. SKD61 takes 30 s, SKD11 takes 35 s, NAK80 takes 40 s, and SNCM8 takes 50 s (see Fig. 3.30). SKD61 requires the shortest time, and its polishing effect is the best. The suggested process parameters for the following study are as follows: side gap width between the electrode and hole

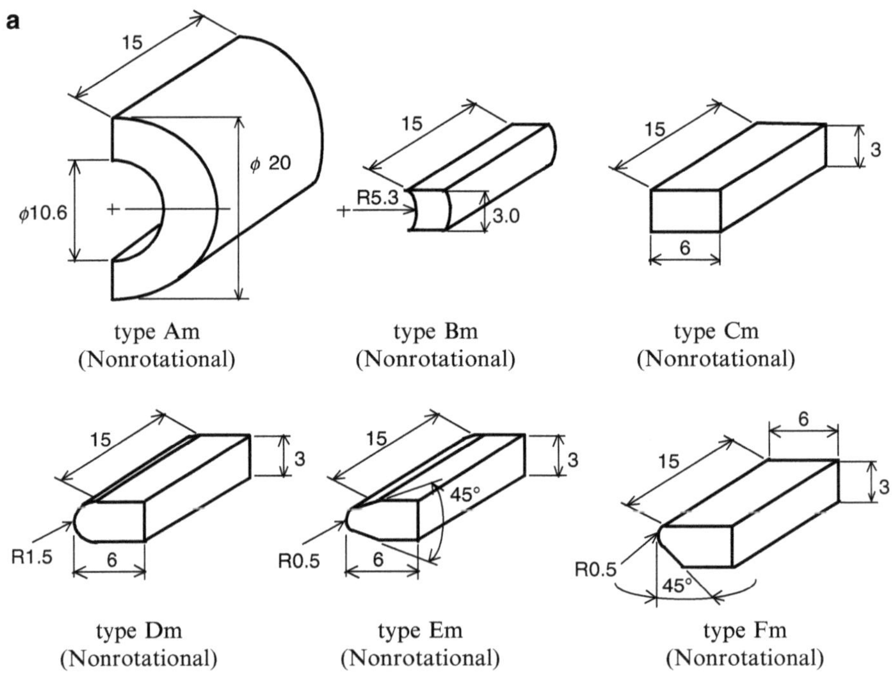

Fig. 3.24 Electrodes in ultrasonic-aided electropolishing of external cylindrical surface. (**a**) Mate-electrodes. (**b**) Feeding electrodes ([71], reprinted with permission)

b

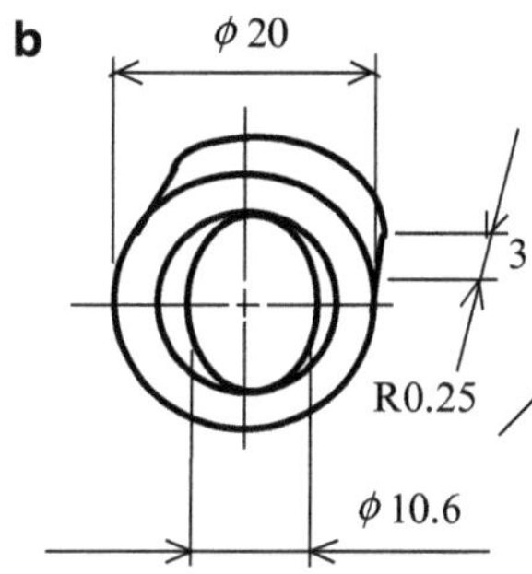

type Au
(Rotational)

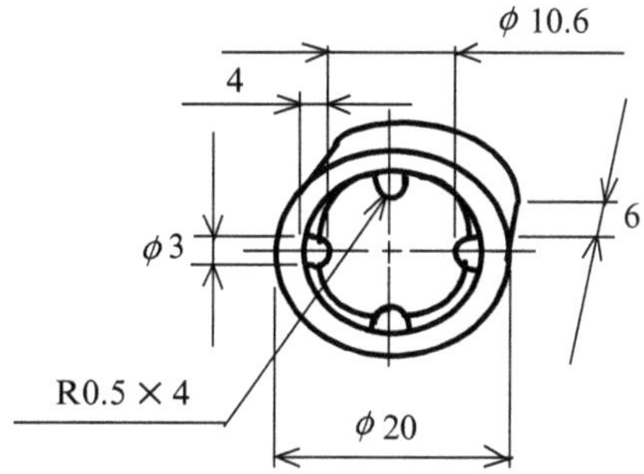

type Bu
(Rotational)

([72], reprinted with permission)

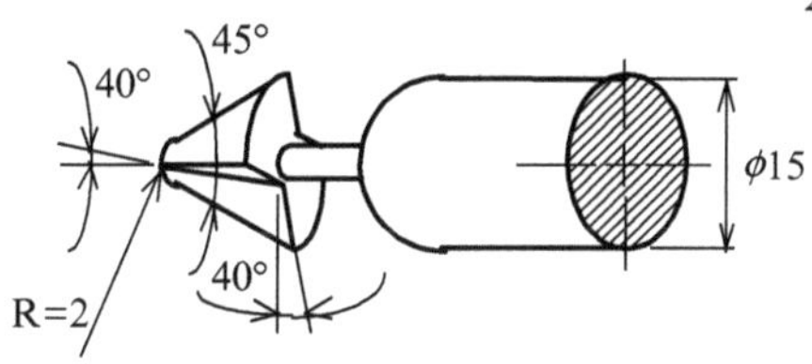

type Cu
(Rotational)

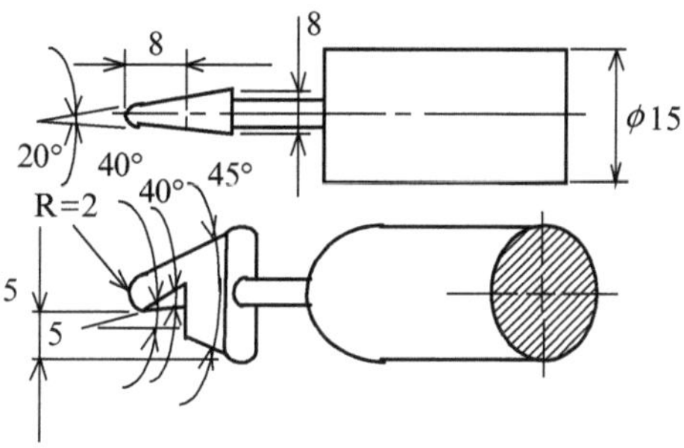

type Du
(Rotational)

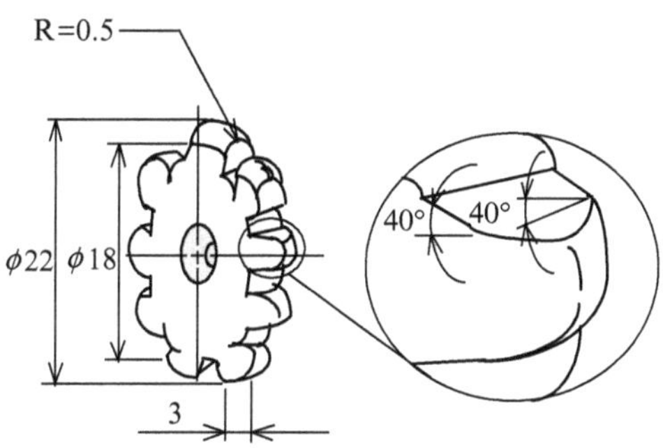

type Eu([69], reprinted with permission)
(Rotational)

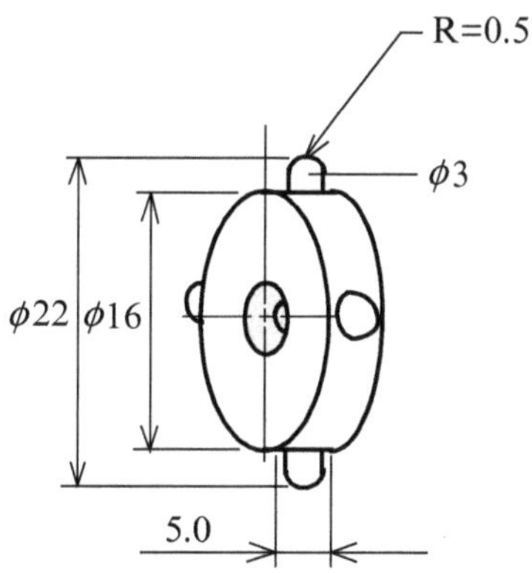

type Fu([69], reprinted with permission)
(Rotational)

Fig. 3.24 (continued)

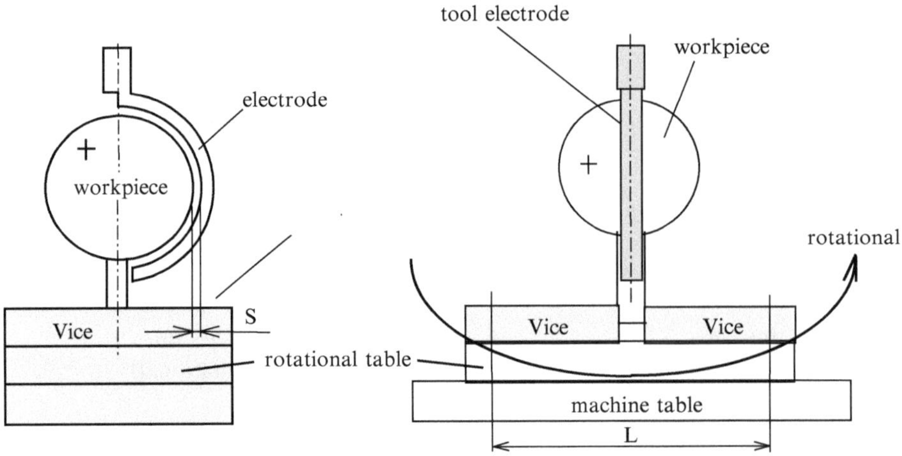

Fig. 3.25 Configuration of wire electrode for electropolishing of spherical surface

wall of 0.3 mm, the electrode rotational speed of 600 rpm, the electrolytic flow rate of 4 L/min, the current density of 30 A/cm^2, and the current rating of 10 A. They are adopted for the second-stage experiment on the electrode design.

3.11.2.2 Experiment on Electrode Design

Inserted Electrode

The drilled hole of 7.8 mm in diameter was electropolished to 8.0 mm in diameter using various electrodes. The results of surface finish developed with time are shown in Fig. 3.31. The hole is polished within 5–30 s. Type H_i with helix flute and water hole performs the best among the eight electrodes. Type A_i of the simplest form provides the mediocre effect. One finds that the flute on the electrode elevates the polishing effect. And the deep flute depth also provides mild improvement of the polishing (see Fig. 3.32). It is closely related to its capability of discharging electrolytic dregs, and the effect varies with different types of flute. The polishing effect of electrode with discharge flute is significantly improved compared to the simple cylinder electrode, and the flute of more straight ribs (C_i) is better than the single straight flute (B_i). Besides, helix flute (D_i) is better than straight flute (C_i), the polishing effect very slightly improves when the rib number is increased (see Fig. 3.33), and larger helix angle is better up to 45° (see Fig. 3.34). Based on the mechanics of a slope, one can explain the effects of the helix angle. For a constant rotational speed of the electrode, larger helix angle produces higher flow velocity of the electrolyte along the flute; thus the discharge is more effective. The electrode with water hole (H_i) helps the electrolyte circulation and the flushing

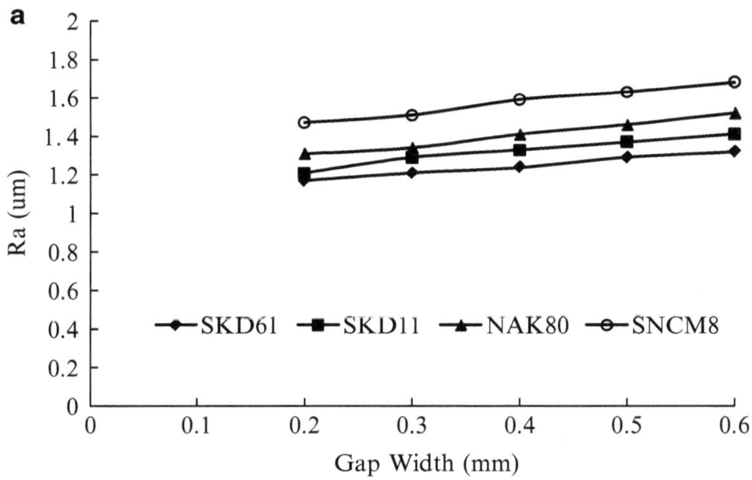

Type A_i/ nonrotational, 30A/cm^2, 30sec

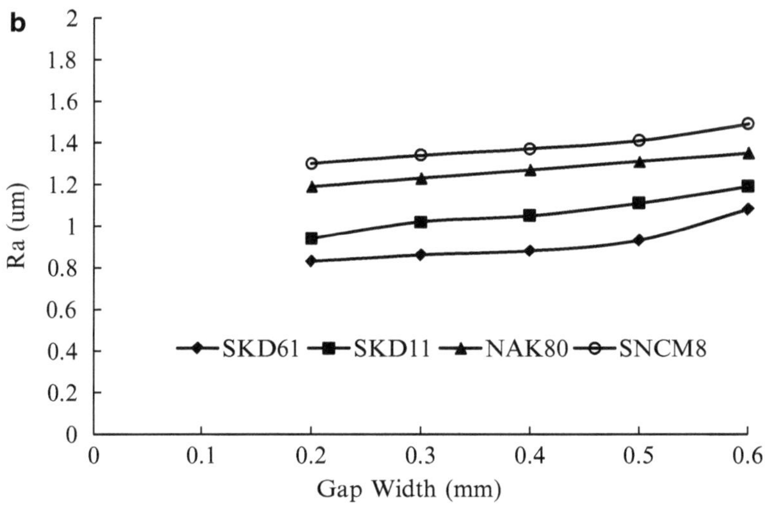

Type Af / nonrotational, 10A, 2mm/min

Fig. 3.26 Electropolishing at different gap width between electrode and hole wall (4 L/min, Continuous DC)

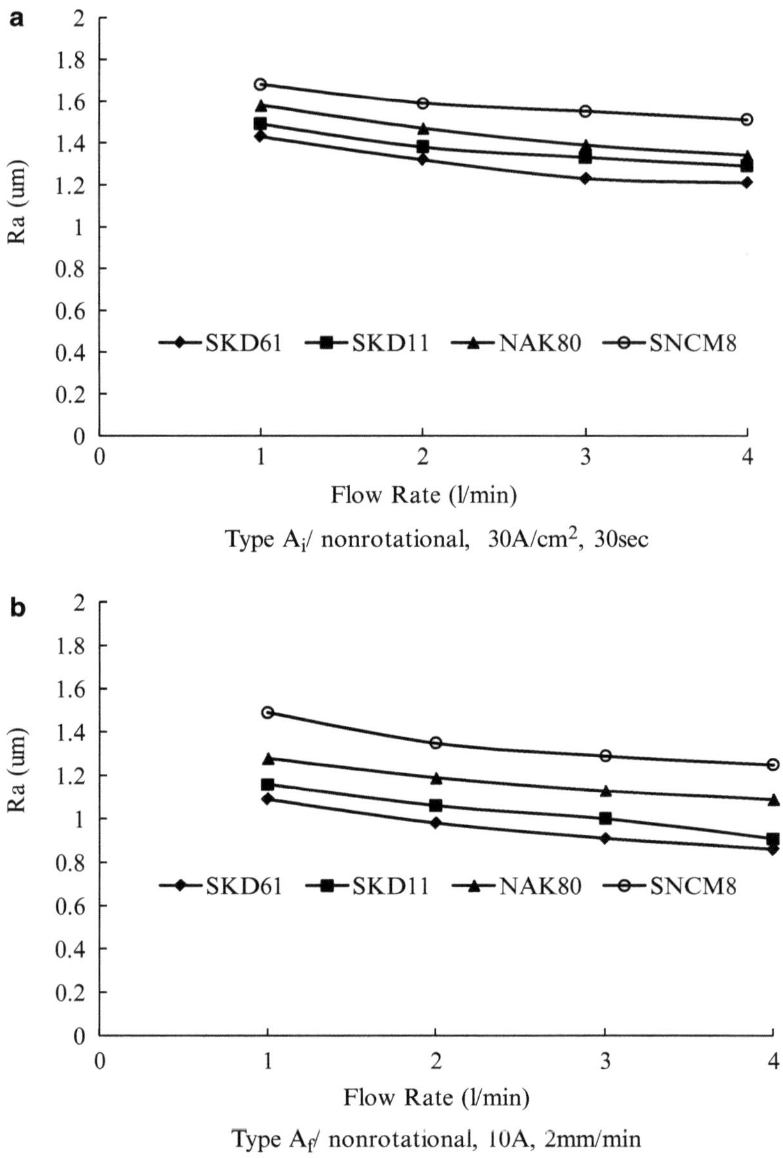

Fig. 3.27 Electropolishing at different flow rate of electrolyte (4 L/min, continuous DC)

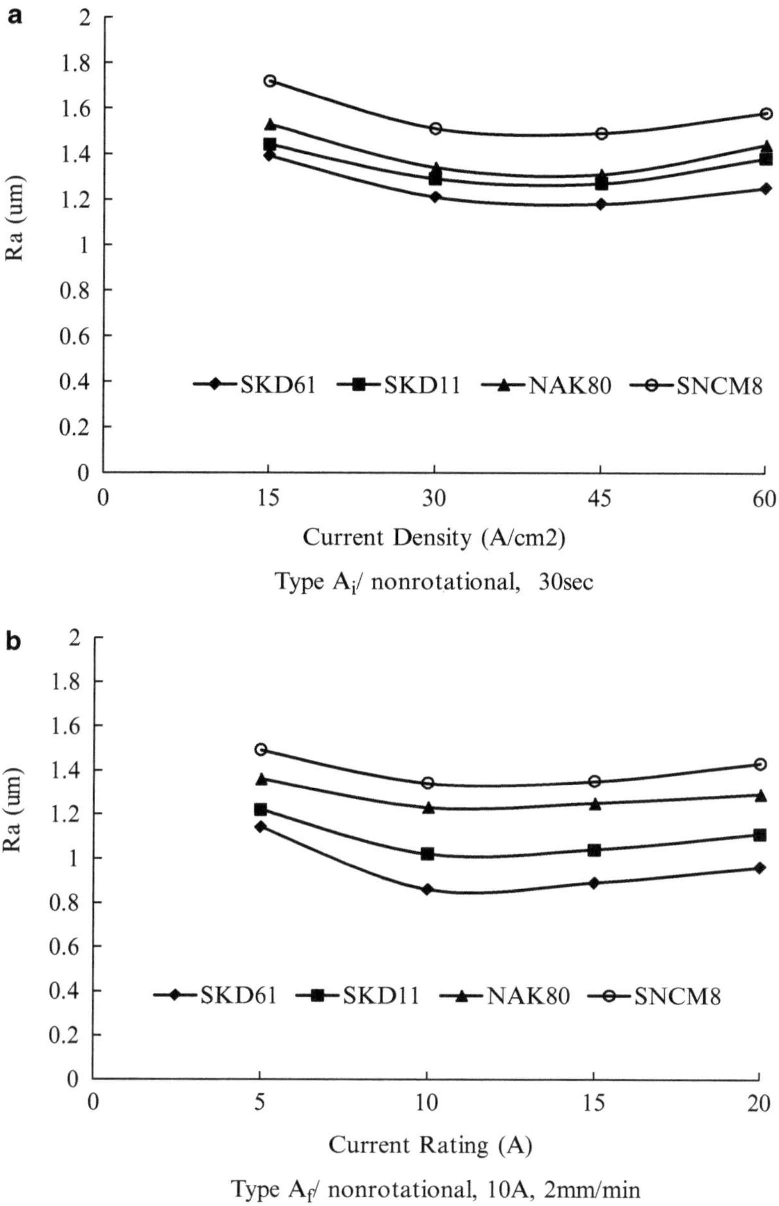

Fig. 3.28 Electropolishing at different current density (4 L/min, continuous DC)

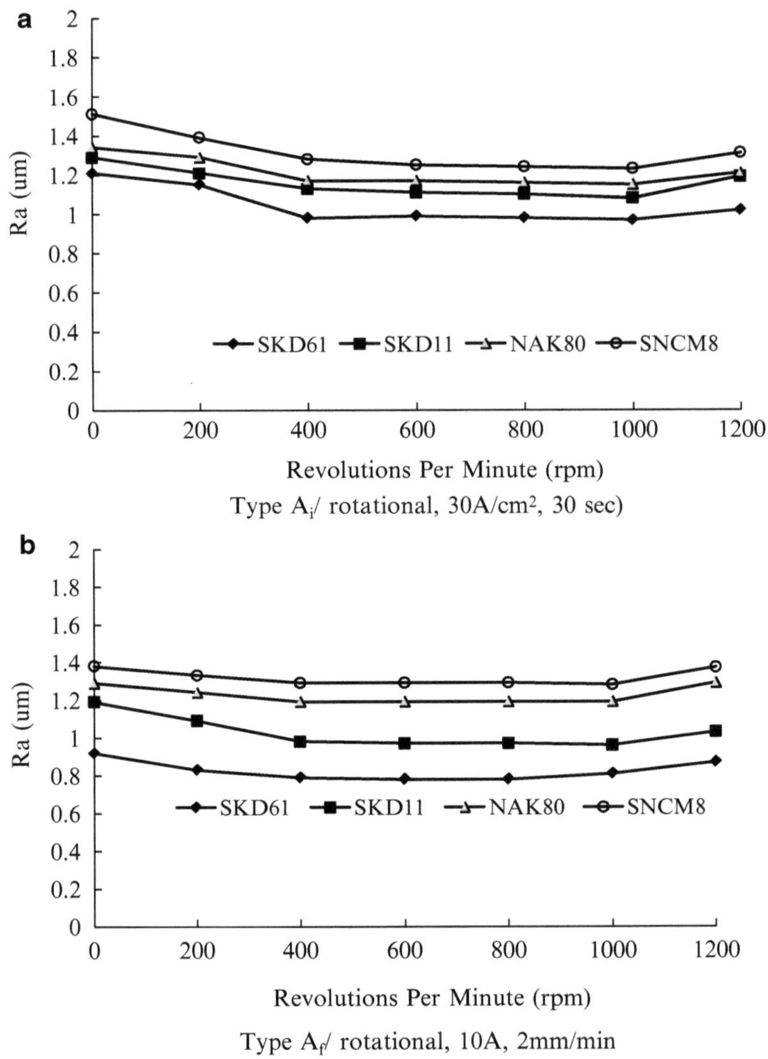

Fig. 3.29 Electropolishing at different rotational speed of electrode (4 L/min, continuous DC)

Table 3.8 The current efficiency of different materials

SKD 61	SKD 11	NAK 80	SNCM 8
98.20%	86.60%	73.70%	57.30%

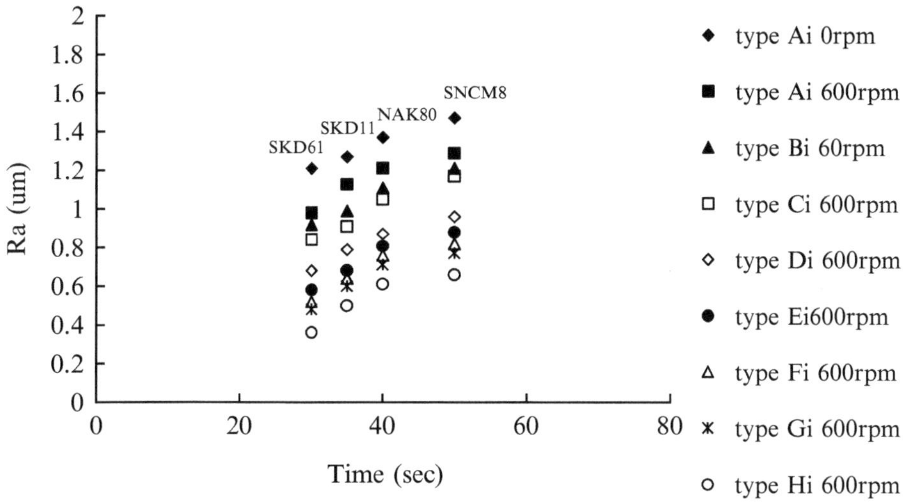

Fig. 3.30 Electropolishing with different types electrode at the same polishing time (4 L/min, continuous DC, 30 A/cm^2) ([67], reprinted with permission)

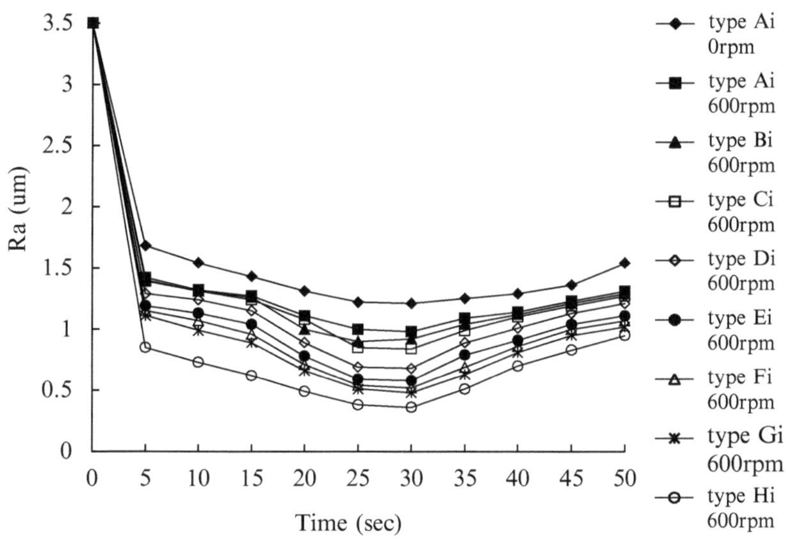

Fig. 3.31 Electropolishing with different types of electrode (SKD61, 4 L/min, continuous DC, 30 A/cm^2) ([67], reprinted with permission)

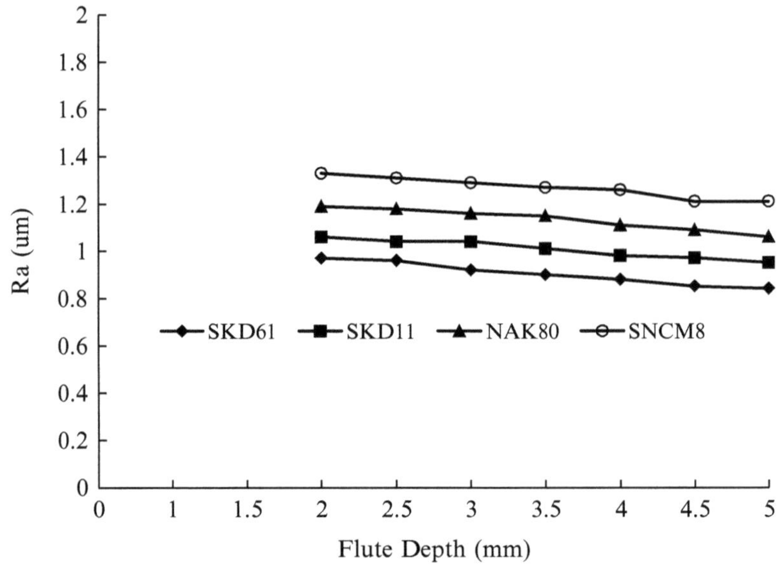

Fig. 3.32 Electropolishing with different flute depth of electrode (Type Bi, 600 rpm, 4 L/min, continuous DC, 30 A/cm^2, 30 s)

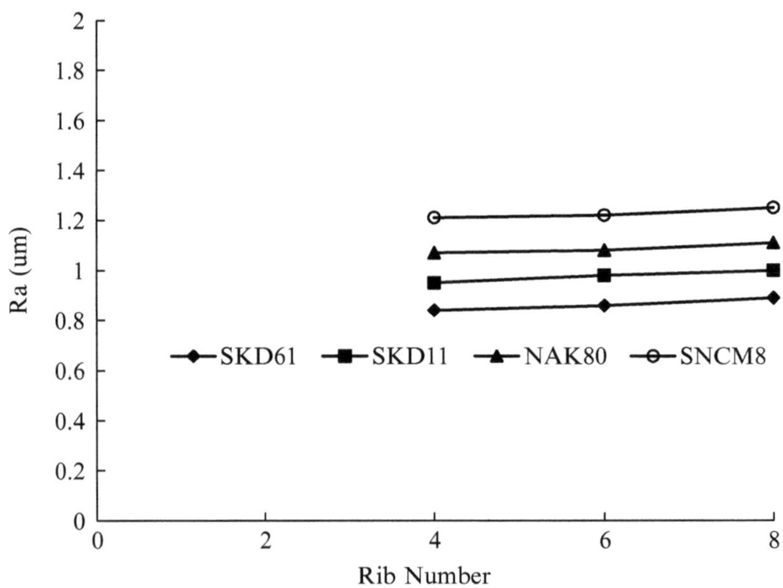

Fig. 3.33 Electropolishing with different rib number of electrode (Type Ci, 600 rpm, 4 L/min, continuous DC, 30 A/cm^2, 30 s)

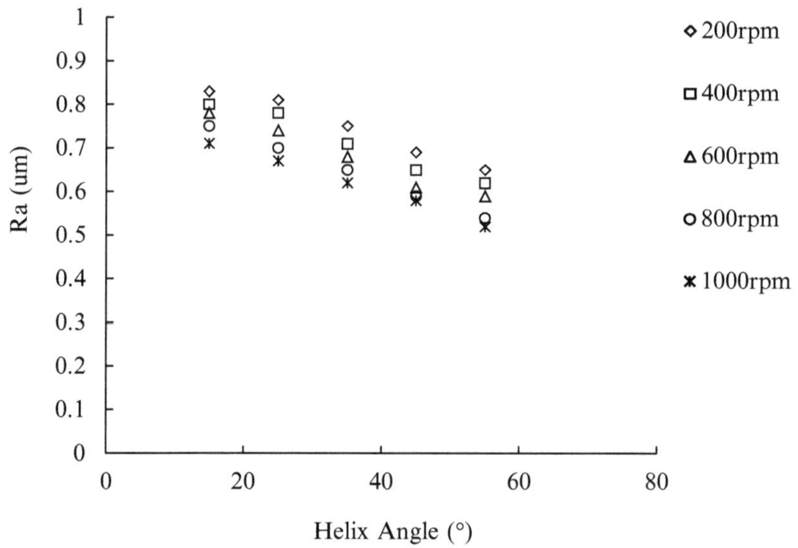

Fig. 3.34 Electropolishing compared with different helix angle and rotational speed of electrode (SKD61, Type C_i, 4 L/min, continuous DC, 30 A/cm^2)

of dregs, since it provides guided flow of the electrolyte with higher velocity. Hence H_i performs the best among the eight electrodes. Figure 3.35 shows the downstream electrolyte associated with electrode rotation in helix direction gives the best flushing. The effect of the pulse direct current using the A_i electrode is shown in Fig. 3.36. Longer off-time is more advantageous, because the discharge of polishing dregs during the off-time is more thorough. However, the total machining time and cost will increase with the prolonged off-time. One also finds that the D_i electrode with helix flute performs the best in this case as shown in Fig. 3.37. The application of pulse current (100 ms/500 ms) to the helix flute electrode can only improve the surface roughness by 0.05 μm (from 0.68 to 0.63 μm); thus the use of pulse direct current is not as effective as the design change of electrode from D_i to H_i. The ratio of surface roughness improvement using different completely inserted electrodes is shown in Fig. 3.38; the improvement obtained by H_i (helix flute with water hole/ rotational) is 70 and 74% at continuous and pulse current, respectively. Figure 3.39 shows the distribution of the surface roughness improvement obtained by H_i through the application of pulse (6%), the electrode rotational (24%), the helix flute (31%), and the water hole (39%). Therefore, the electrode design of water hole and helix flute contributes the most to the polishing, while the use of pulse current is of limited advantage, particularly when the increased polishing time and cost is considered.

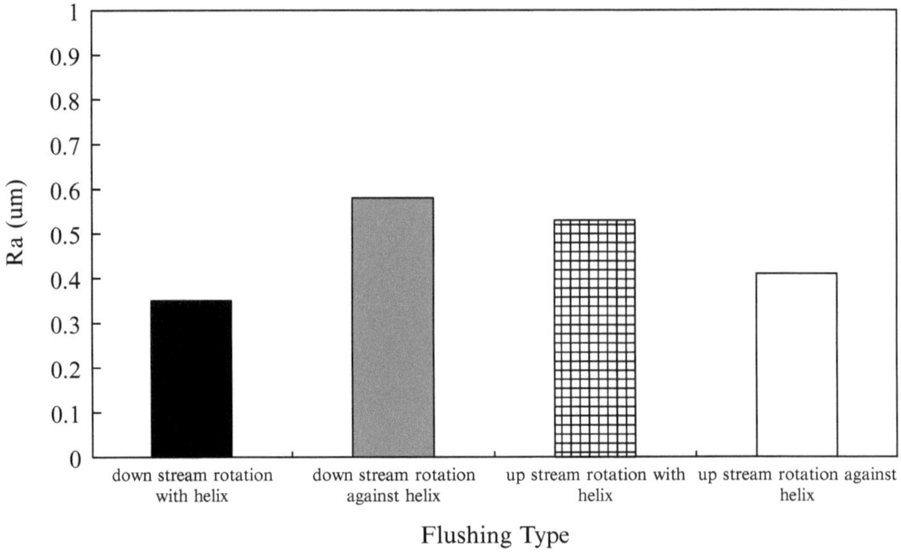

Fig. 3.35 Electropolishing with different flushing types (Type H_i, SKD61, 4 L/min, continuous DC, 30 A/cm^2, 30 s) ([67], reprinted with permission)

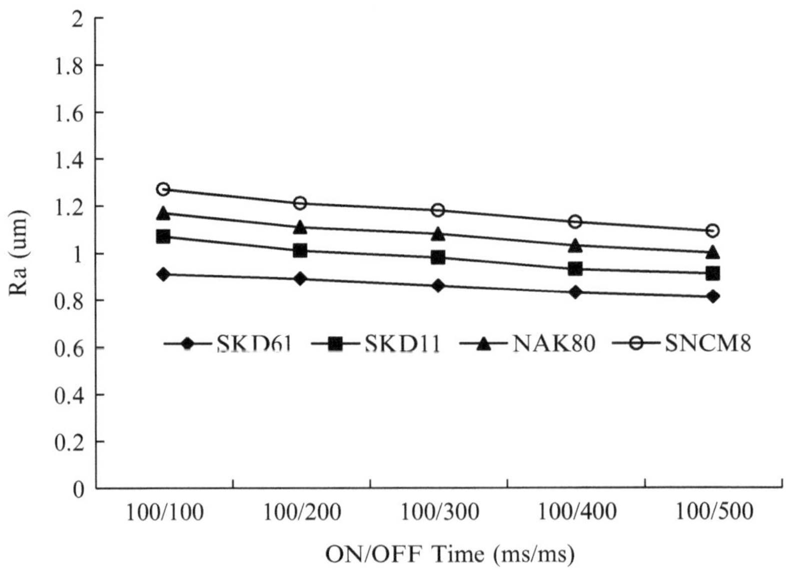

Fig. 3.36 Electropolishing through pulse direct current with rotational electrode Type Ai (4 L/min, pulse DC, 30 A/cm^2, ON Time 30 s)

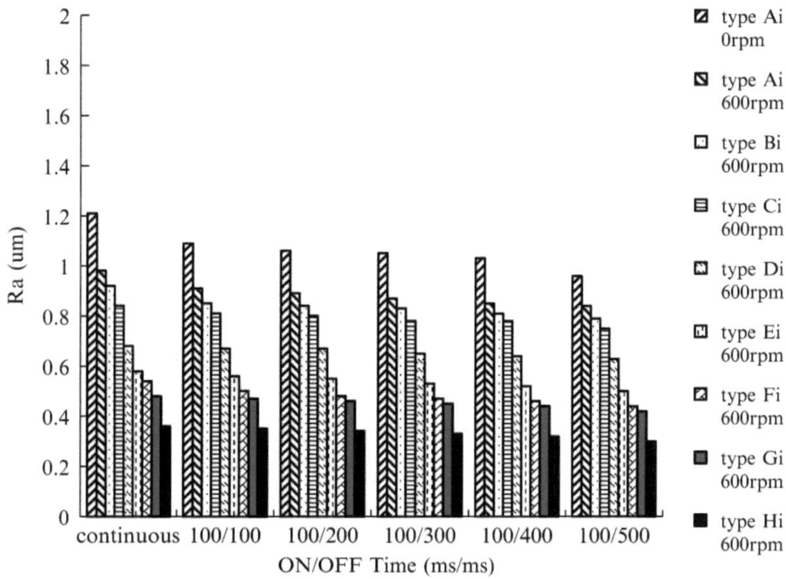

Fig. 3.37 Electropolishing with different types of electrode at continuous and pulse direct current (SKD61, 4 L/min, DC, 30 A/cm^2, ON Time 30 s) ([67], reprinted with permission)

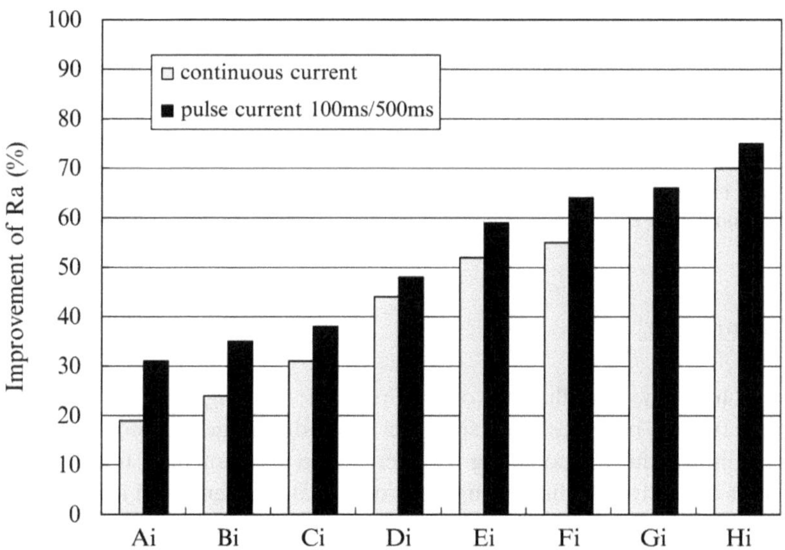

Fig. 3.38 The surface roughness improvement using different completely inserted electrodes (SKD61, 4 L/min, 600 rpm, 30A/cm^2, 30 s)

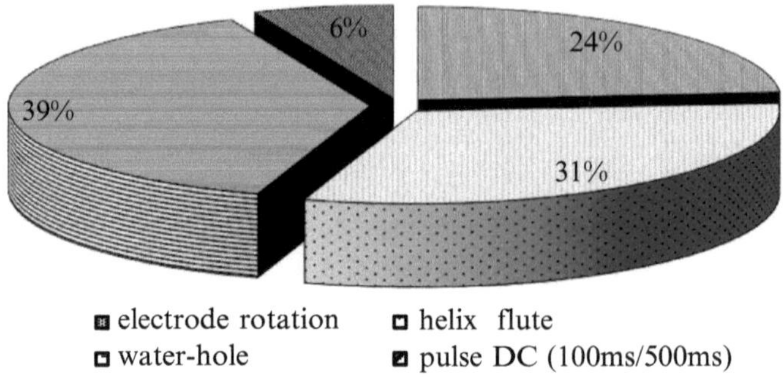

Fig. 3.39 The contribution pie of surface roughness improvement of electrode H_i (SKD61, 4 L/min, 600 rpm, pulse DC, 30 A/cm^2, 100 ms/500 ms)

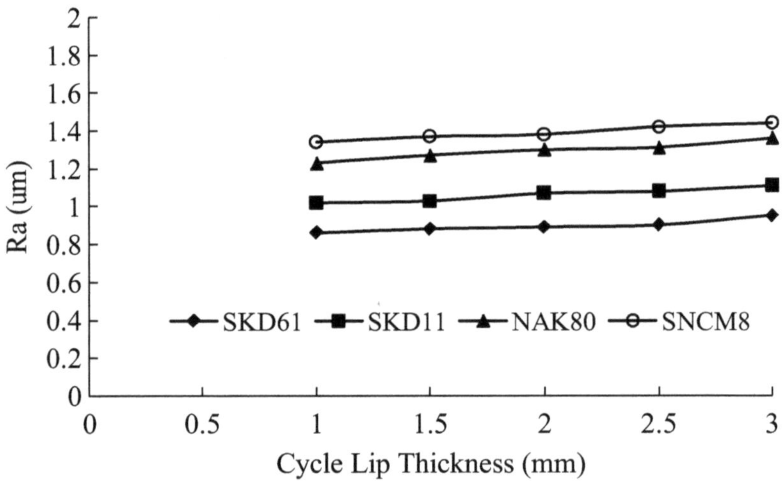

Fig. 3.40 Electropolishing with different cycle lip thickness of electrode (Type A_f, 0 rpm, 4 L/min, continuous DC, 10 A, 2 mm/min)

Feeding Electrode

Figure 3.40 shows that the thin circle lip on the leading edge of the cylinder of electrode type A_f provides more space for dregs discharge and reduces the secondary machining; thus the polishing is better. In the experiment of feeding electrode, the results of electropolishing using electrode A_f at different axial feed rate is shown in Fig. 3.41. Low feed rate is associated with more electrochemical reaction per unit area and more dregs in the side gap; thus the polishing effect is reduced. On the other hand, high feed rate does not provide enough time for the polishing effect to be fully developed. The results show that the feed rate of 2.0 mm/min for

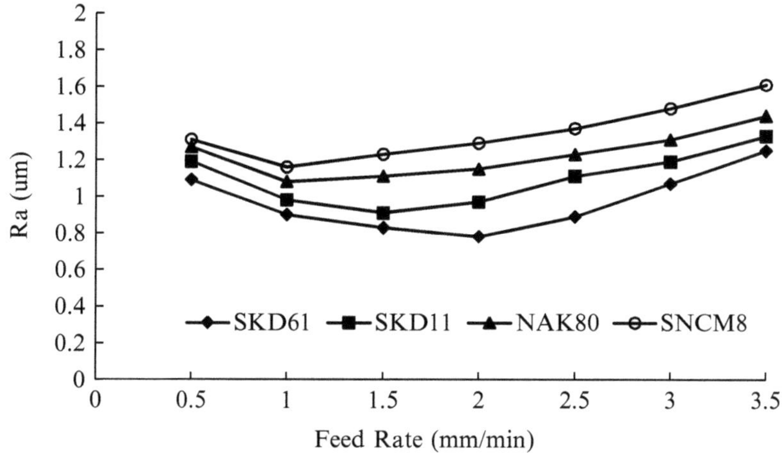

Fig. 3.41 Electropolishing with different feed rate of electrode (Type A_f, 600 rpm, 4 L/min, continuous, DC, 10 A) ([67], reprinted with permission)

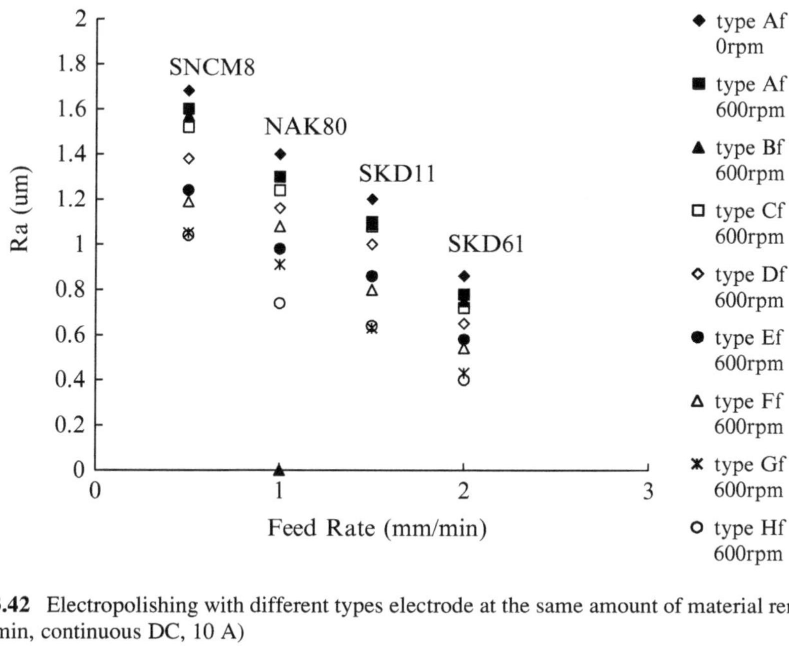

Fig. 3.42 Electropolishing with different types electrode at the same amount of material removal (4 L/min, continuous DC, 10 A)

electropolishing SKD61 is optimal. The polishing of SKD61 is the best, followed by SKD11, NAK80, and SNCM8, as shown in Fig. 3.42. Figure 3.43 shows the same good polishing can be achieved by adequate combination of current rating with electrode feed rate, such as 5 A with 1 mm/min, 10 A with 2.0 mm/min, 15 A

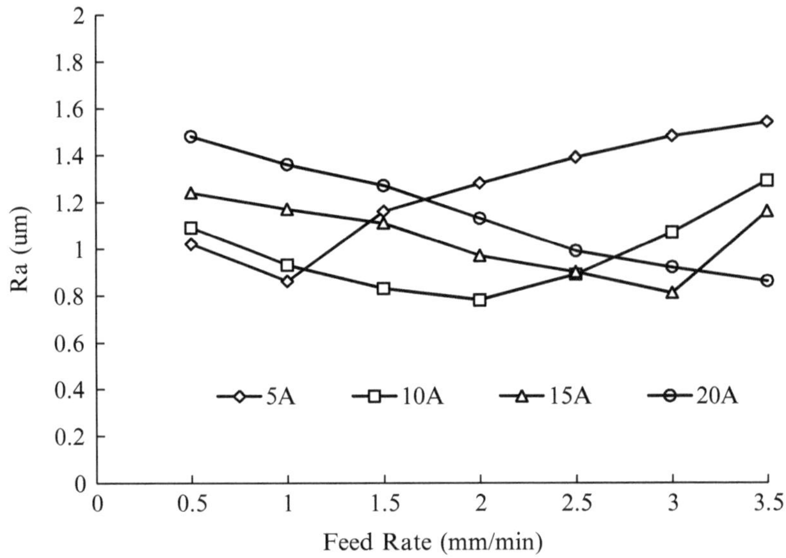

Fig. 3.43 Electropolishing at different feed rate of electrode using different current rating (SKD61, Type A_f, 600 rpm, 4 L/min, continuous DC, 10 A)

with 3.0 mm/min, and 20 A with 3.5 mm/min. These combinations imply the concept of an optimal amount of electrochemical energy input per unit polished area. Lower values do not produce effective polishing, while higher values often cause problems of dregs discharge. Figure 3.44 compares the electropolishing of SKD61 with different current rating through different feeding electrodes. The results also show that among the eight types of electrode, the electrode H_f gives the best surface finish at all current ratings. Figure 3.45 also verifies this result for different workpiece materials. The electrode A_f has a circle lip on the leading edge of the cylinder. The polishing results from the rotating electrode are better than the non-rotating one, since the rotation helps discharge dregs by means of the centrifugal force. Electrode B_f provides two circle lips, and the second lip produces second-step machining, which is slightly more advantageous over the A_f. Electrode C_f has a circle lip in triangular form, which provides finer stepwise secondary polishing. Electrode D_f can actually operate at higher current density, since the edge of the lip has a retreat instead of a full front. Electrode E_f, F_f, G_f, and H_f are of borer type with much more space for dregs discharge, the polishing effect is then better than the four electrodes with full circle lip. The electrode H_f of a globular borer tip performs the best among all electrodes. Figure 3.46 also shows the improvement by the use of H_f, compared to A_F, reaches nearly 60%. Figure 3.47 shows the contribution of surface finish improvement obtained by H_f, through the electrode rotation (17%), borer leading edge (66%), and the globular borer geometry (17%).

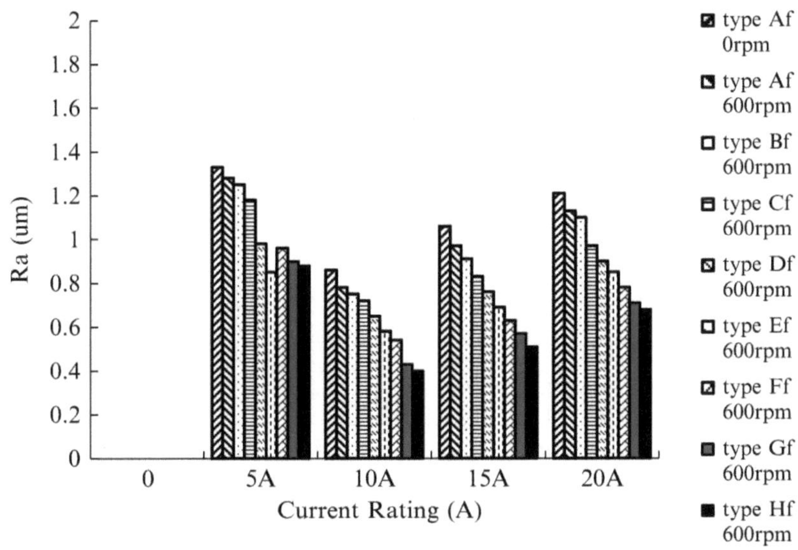

Fig. 3.44 Electropolishing at different current rating (SKD61, 4 L/min, continuous DC, 2 mm/min) ([67], reprinted with permission)

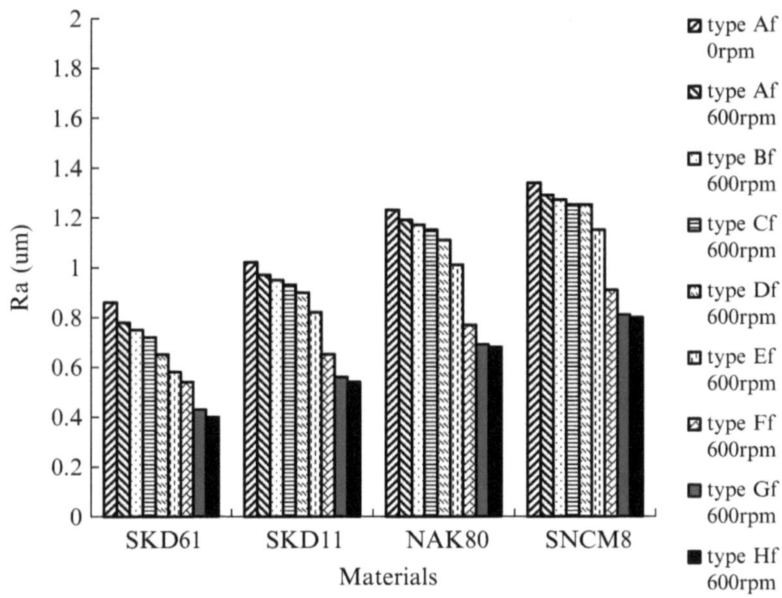

Fig. 3.45 Electropolishing for various materials (4 L/min, continuous DC, 10 A, 2 mm/min)

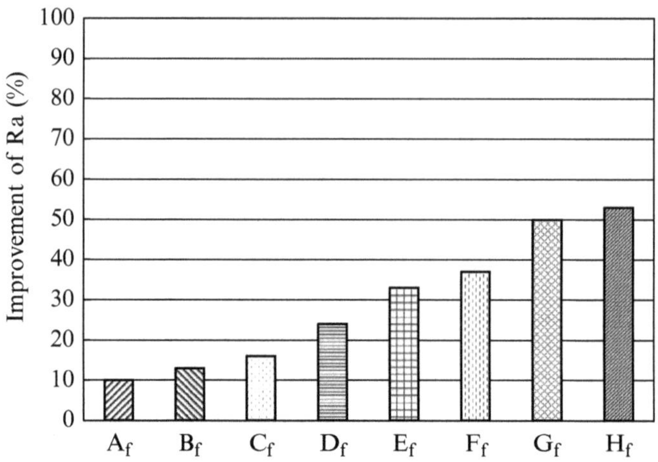

Fig. 3.46 The surface roughness improvement through electropolishing with different feeding electrode (SKD61, 4 L/min, 600 rpm, continuous DC, 10 A, 2 mm/min)

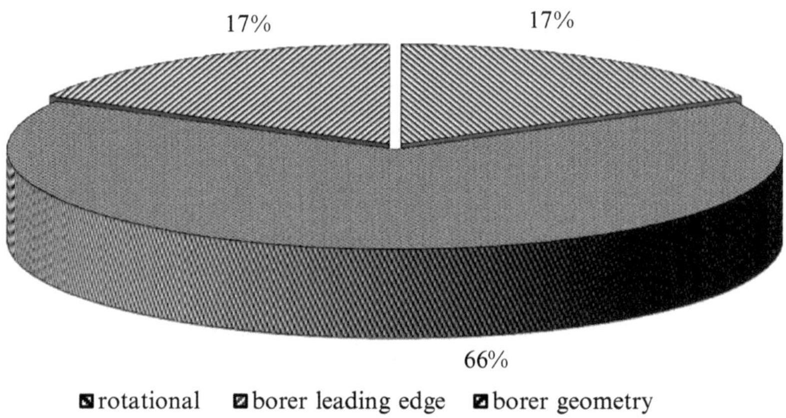

Fig. 3.47 The contribution pie of surface roughness improvement of electrode H_f (SKD61, 4 L/min, 600 rpm, continuous DC, 10 A, 2 mm/min) ([67], reprinted with permission)

Applications of Inserted and Feeding Electrodes

The inserted and feeding electrodes possess various characteristics. Table 3.9 lists the major items for comparison. In summary, both methods of electropolishing can be realized without major difficulties, and the inserted electrodes require larger power supply to polish faster, which are suitable for smaller holes, while the feeding electrodes can polish larger holes in moving manner using smaller power supply.

Table 3.9 Comparison of hole polishing using inserted and feeding electrodes

	Inserted electrode	Feeding electrode
Polishing cycle time	Short (stationary)	Long (traveling)
Size of hole	Small/medium	Medium/large
Power of supply	Large	Small
Electrode cost	Low	Low
Selection of processing parameter	Wide	Wide
Integration with the preceding machine	Yes	Yes

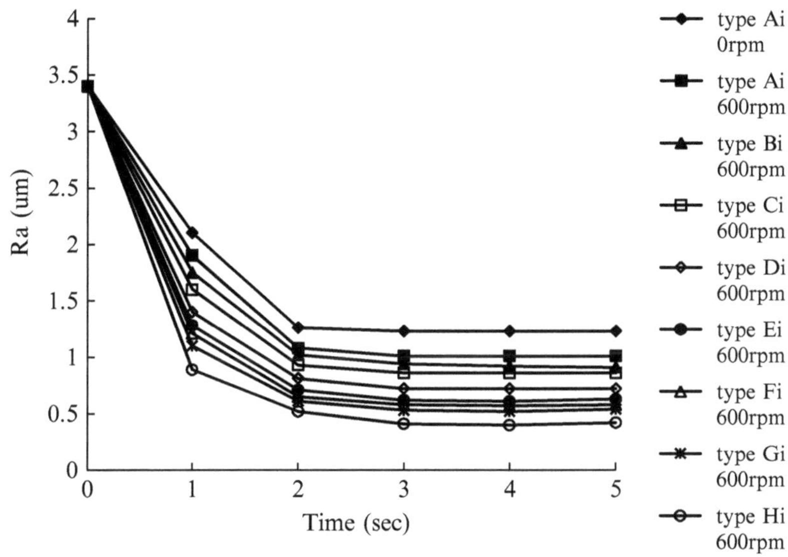

Fig. 3.48 Electrobrightening with inserted electrodes (SKD61, 4 L/min, continuous DC, 30 A/cm^2) ([67], reprinted with permission)

3.11.2.3 Electrobrightening

After the hole is drilled to 7.8 mm and further reamed to 8.0 mm in diameter, the electrobrightening is carried out. It differentiates from the electropolishing in that the amount of material removal of the latter is much larger than the former, and the average surface roughness after reaming is around 3.5 μm instead of 10 μm after drilling. When expressed in the machining depth in the radial direction in the current case, it is about 200 and 2 μm, respectively. A comparison of these two processes is of interest in this dissertation. Figure 3.48 illustrates the surface finish against the working time for various electrodes. The obtained surface finish and the influence of the electrodes are similar to electropolishing (see Fig. 3.31), while the time required for finishing is much less (within 1–3 s). Since the electrobrightening takes very short time, the electrolytic products are few. It greatly reduces the

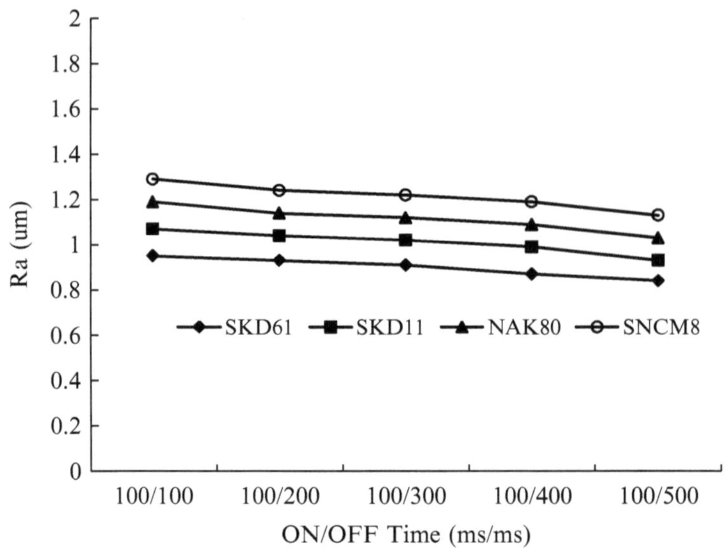

Fig. 3.49 Electrobrightening through pulse direct current with rotational electrode Type Ai (4 L/min, Pulse DC, 30 A/cm^2, ON Time 3 s)

interference of the electrolytic products in the polishing process. However, the preceding reaming should be considered in the total cycle time. Figures 3.49, 3.50, 3.51, 3.52, 3.53, and 3.54 further show the results of electrobrightening; one concludes that its technological characteristics are basically the same as electropolishing (see Figs. 3.37, 3.38, 3.41, 3.43, 3.44, and 3.45).

The helix-flute electrode with water hole also produces the best effect among the eight electrodes. The selection of process between these two alternatives will depend on the production time and cost, where the electrobrightening is faster and requires reaming in advance, while the electropolishing does not.

3.11.2.4 Ultrasonic-Aided Electropolishing

The experiments are carried out by four methods: electropolishing (EP), pulse-electropolishing (PEP), ultrasonic-aided electropolishing (UEP), and ultrasonic-aided pulse-electropolishing (UPEP). The comparisons among six types of completely inserted electrodes and among six types of feeding electrodes are discussed. Figure 3.55 shows that the use of ultrasonics helps to elevate the polishing quality, higher frequency and greater power produces better polishing. The electrolyte via vibrating energy of high frequency indeed can discharge dregs off the small machining gaps rapidly. Figure 3.56 shows the appropriate rotation (400–800 rpm) of electrode is advantageous for electropolishing, while the ultrasonic-aided electropolishing requires lower rotational energy to achieve the

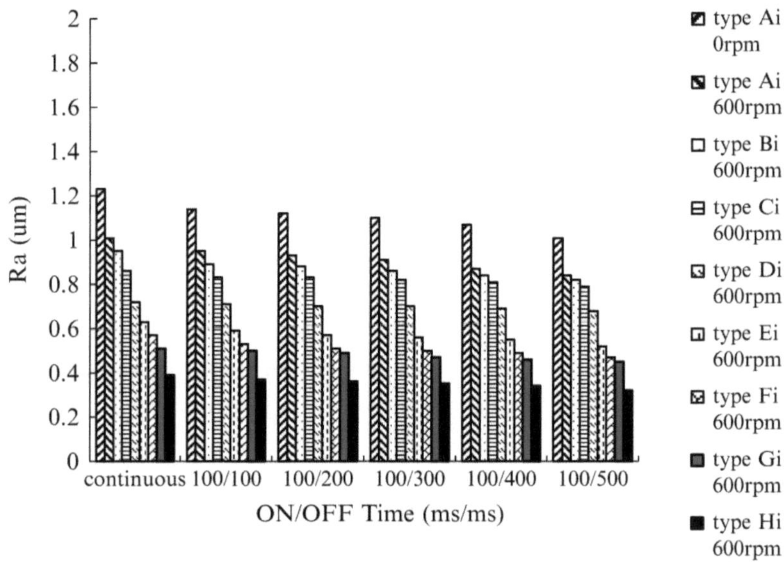

Fig. 3.50 Electrobrightening with different types of electrode using continuous and pulse direct current (SKD61, 4 L/min, DC, 30 A/cm^2, ON Time 3 s) ([67], reprinted with permission)

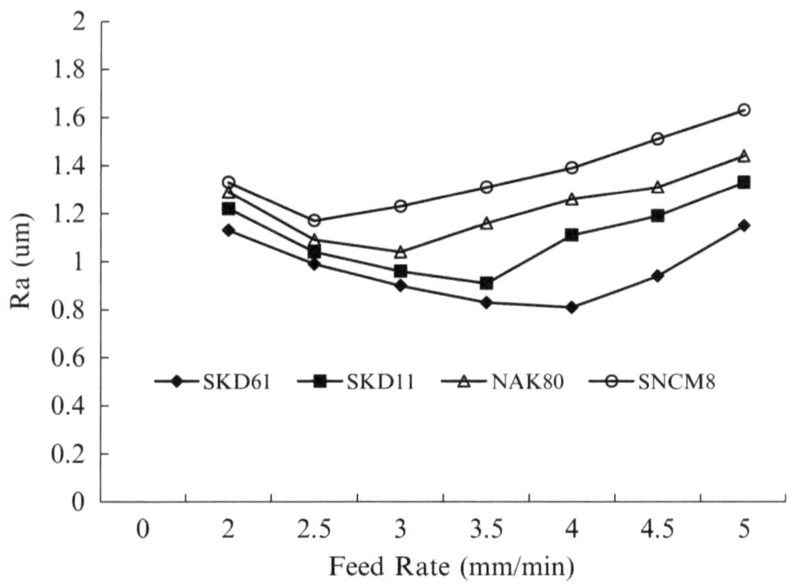

Fig. 3.51 Electrobrightening with different feed rate of electrode (Type A_f, 600 rpm, 4 L/min, continuous, DC, 10 A)

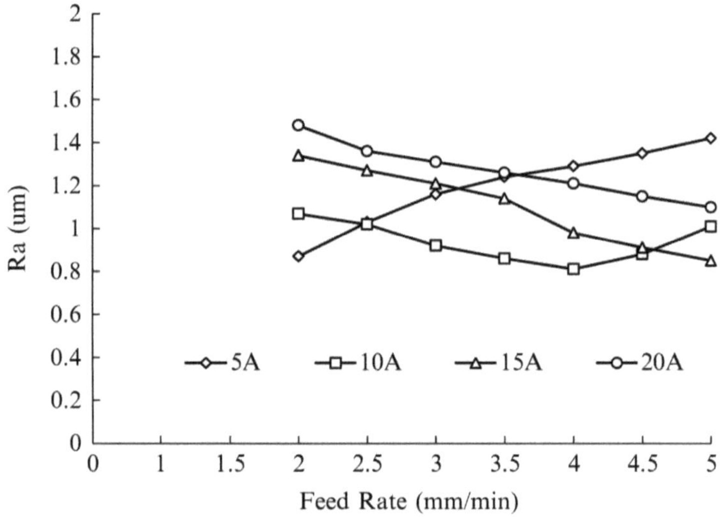

Fig. 3.52 Electrobrightening at different feed rate of electrode using different current rating (SKD61, Type A_f, 600 rpm, 4 L/min, continuous DC)

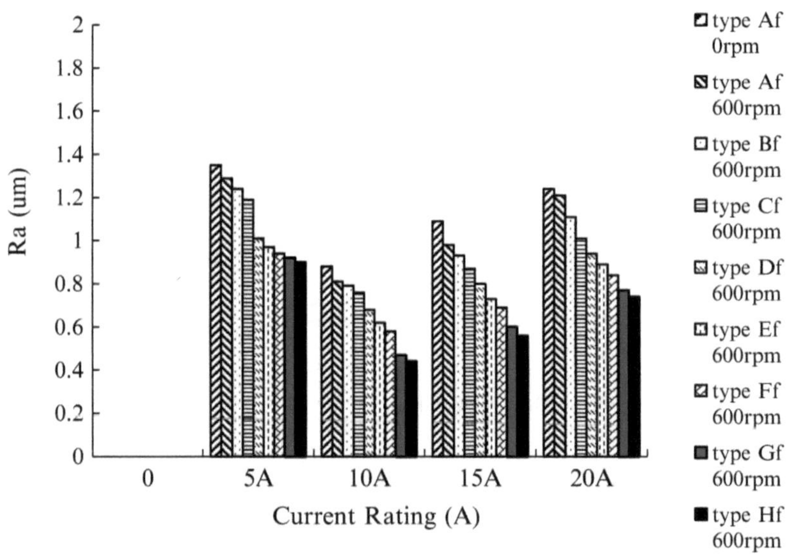

Fig. 3.53 Electrobrightening with different current rating through different type of feeding electrode (SKD61, 4 L/min, continuous DC, 4 mm/min)

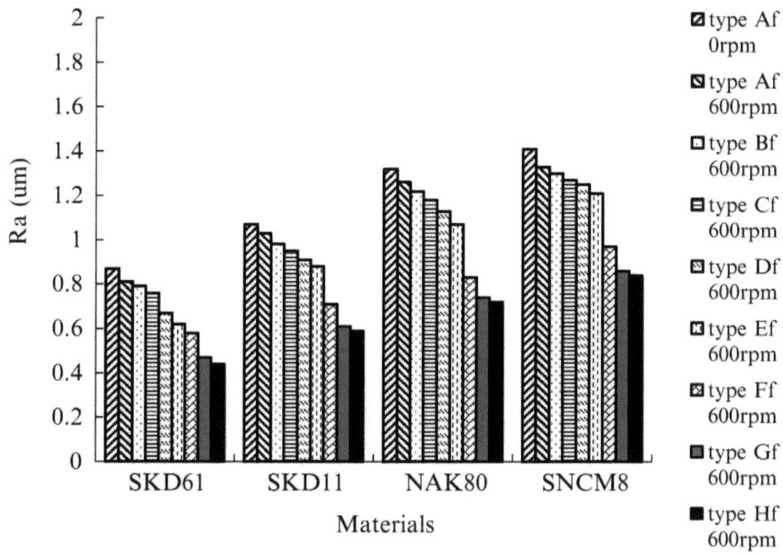

Fig. 3.54 Electrobrightening with different types of electrode (4 L/min, continuous DC, 10 A, 2 mm/min)

same effect. Namely, the ultrasonic is more beneficial at higher rotational speed. Among the six types of inserted electrode, the best polishing comes from the electrode F_i with large discharge space and off-center round pin (see Fig. 3.57). The comparison of improvement of surface roughness in electropolishing using different electrodes is shown in Fig. 3.58. The best result in electropolishing is obtained by the electrode type Fi (70%), followed by Ei (59%), Di (51%), Ci (38%), Bi (25%), and Ai (17%). In case of ultrasonic-aided pulse-electropolishing, it elevates the polishing effect on average of 19%. The improvement ratios are Fi (72%), Ei (61%), Di (53%), Ci (40%), Bi (28%), and Ai (19%). The improvement of ultrasonic-aided pulse-electropolishing only increases 3% compared to the ultrasonic-aided electropolishing. The average contribution of surface finish improvement obtained through pulse (16%), ultrasonic (19%), the electrode rotation (12%), and the electrode geometry (53%) is shown in Fig. 3.59. As mentioned above, the electrode type Fi with ultrasonic-aided pulse-electropolishing can achieve the best polishing effect. Since the resting period (pulse-off time) for dregs discharge takes more working time, one should use ultrasonic-aided electropolishing to save the polishing time.

Figure 3.60 shows that the best polishing comes from the electrode F_f with large discharge space and borer type. Figure 3.61 shows the improvement ratios of surface roughness by ultrasonic-aided electropolishing using different electrodes are F_f (69%), E_f (59%), D_f (55%), C_f (31%), B_f (28%), and A_f (15%). The ultrasonic-aided pulse-electropolishing elevates the polishing effect on average at to 26%. The improvement ratios are F_f (72%), E_f (63%), D_f (59%), C_f (31%),

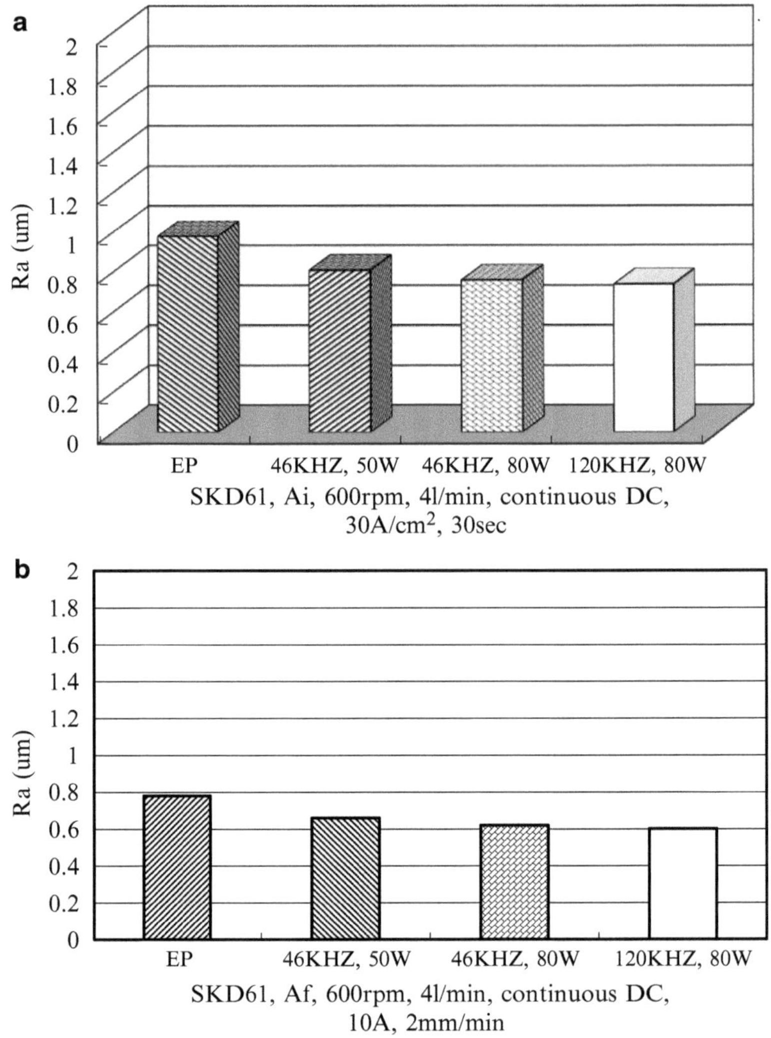

Fig. 3.55 Compared the frequency and power of ultrasonic in electropolishing

B_f (32%), and A_f (19%). It also shows the polishing effect of ultrasonic-aided pulse-electropolishing only increases 5% than the ultrasonic-aided electropolishing. Since the pulse-off time prolongs the polishing time and cost as mentioned above, the use of ultrasonic-aided electropolishing is worthy. In summary, the average contribution of surface finish improvement obtained through pulse (16%), ultrasonics (21%), electrode rotation (12%), and electrode geometry (52%) is shown in Fig. 3.62.

a

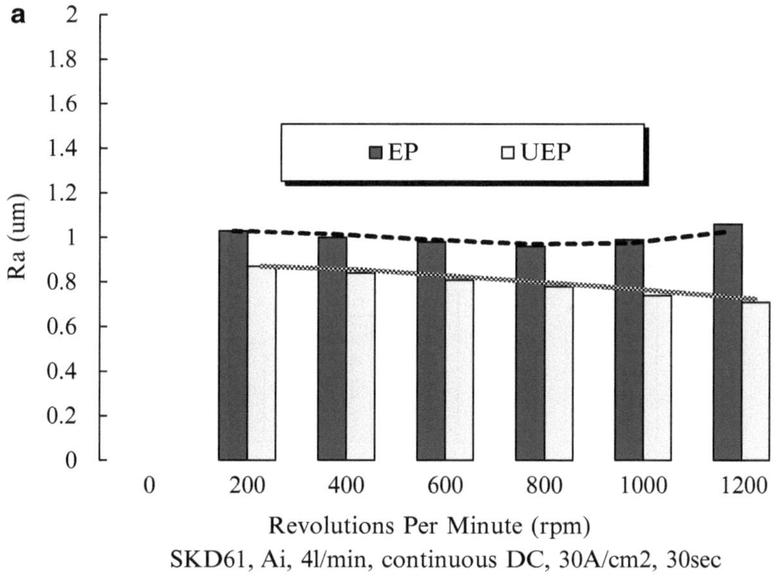

b

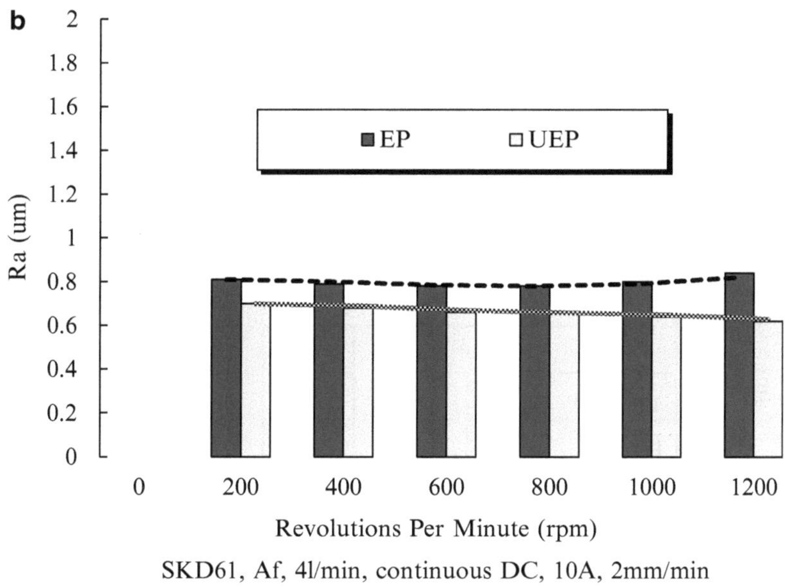

Fig. 3.56 Effects of rotational speed of electrode

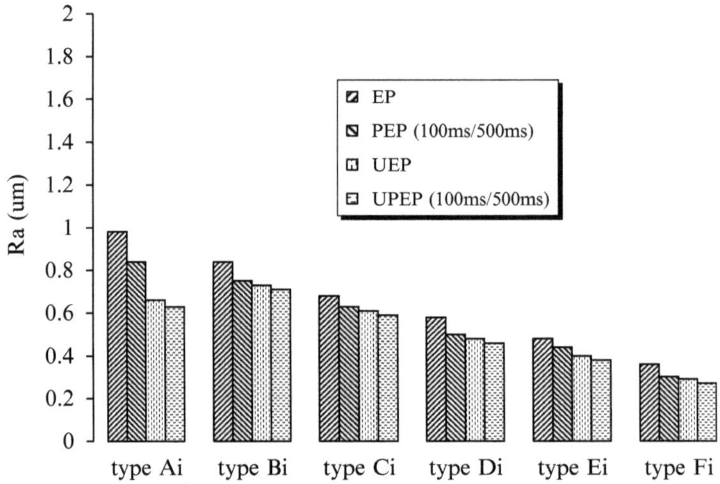

Fig. 3.57 Comparison of electropolishing with inserted electrodes (SKD61, 600 rpm, 4 L/min, 30 A/cm^2, 30 s)

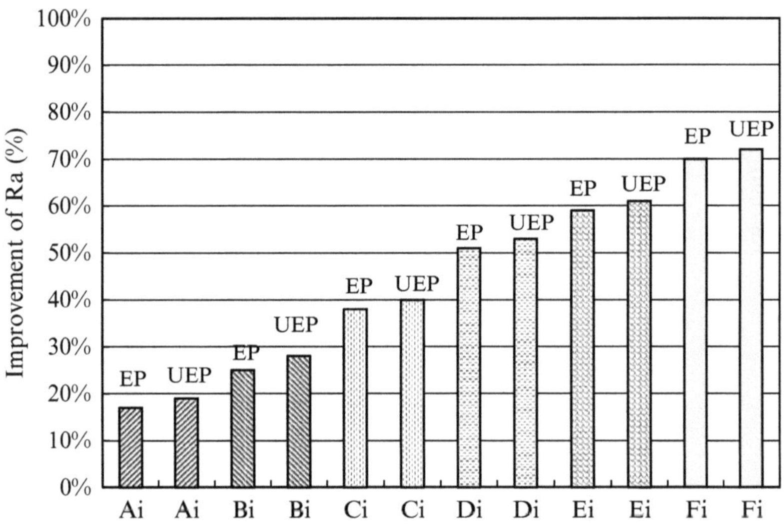

Fig. 3.58 The surface roughness improvement of ultrasonic-aided electropolishing for different inserted electrodes (SKD61, 600 rpm, 4 L/min, continuous DC, 30 A/cm^2, 30 s)

Evaluation of Manufacturability and Cost for Different Electrodes

It contributes the lowest cost if the electrode is only through turning for the electropolishing of internal cylindrical surface. However, in order to increase discharge space and effectively promote the mobility of discharge dregs, it can be

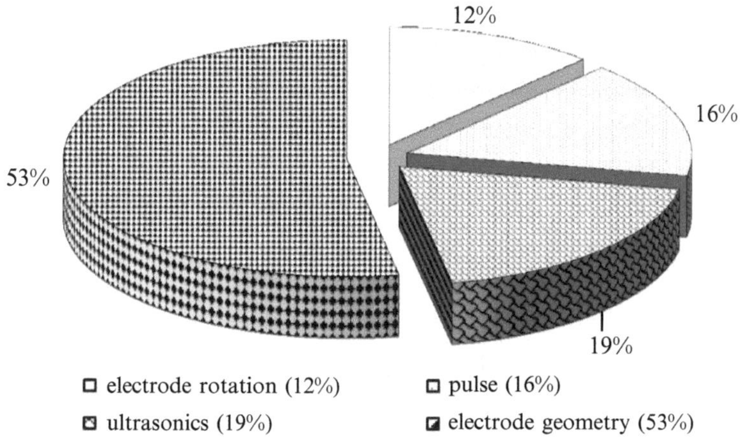

Fig. 3.59 The contribution pie of surface finish improvement using inserted electrodes (SKD61, 600 rpm, 4 L/min, DC, 30 A/cm^2, 30 s)

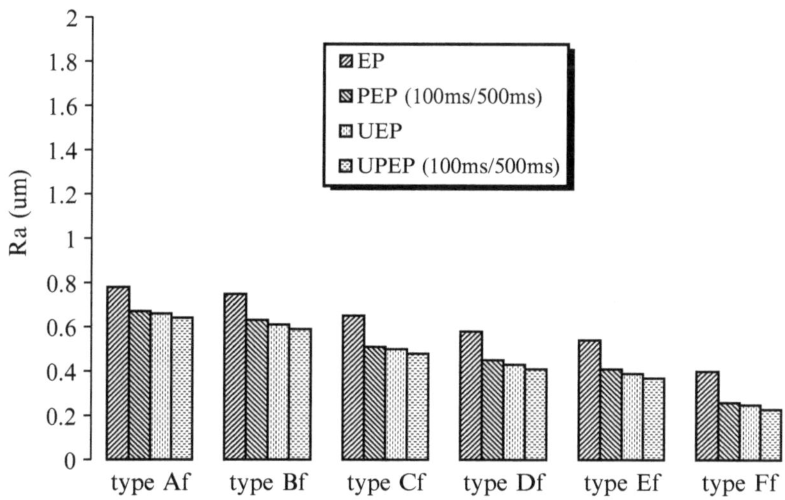

Fig. 3.60 Comparison of electropolishing with feeding electrodes (SKD61, 600 rpm, 4 L/min, 10 A, 2 mm/min)

taken into consideration to add discharge flute or water holes on the inserted electrodes, but the machining cost increased. According to the previous experiments, it can be proved that the electrode with discharge flute contributes better polishing effect. Though the cost of electrode is increased, but the characteristic of electropolishing has noncontact between electrode and workpieces and electrode itself with nonconsumption, the electrode can be used long time if the

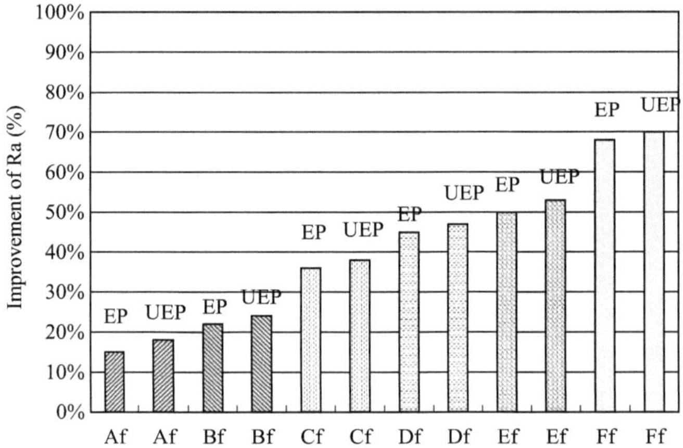

Fig. 3.61 The surface roughness improvement of ultrasonic-aided electropolishing for different feeding electrodes (SKD61, 600 rpm, 4 L/min, continuous DC, 10 A, 2 mm/min)

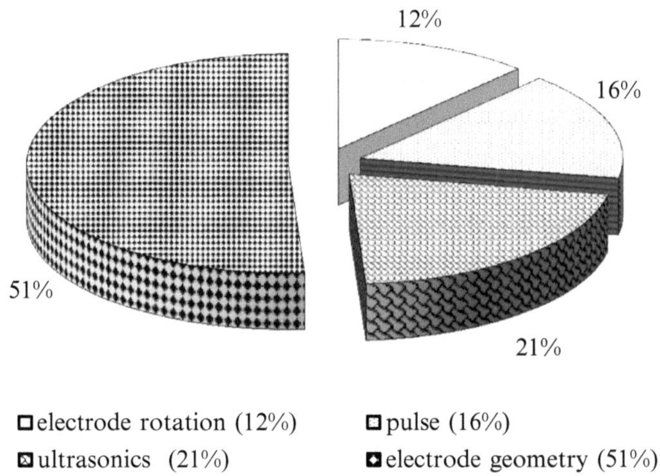

Fig. 3.62 The contribution pie of surface finish improvement using feeding electrodes (SKD61, 600 rpm, 4 L/min, DC, 2 mm/min)

process without short-circuit. Authors recommend the fluted electrodes when consider the overall effect and benefit. Likewise, the feeding electrodes of thin cross section are beneficial to polishing and the cost also increased. The borer electrode not only has the widest discharge space but also has the advantage of low cost. It deserves our recommendation among feeding electrodes.

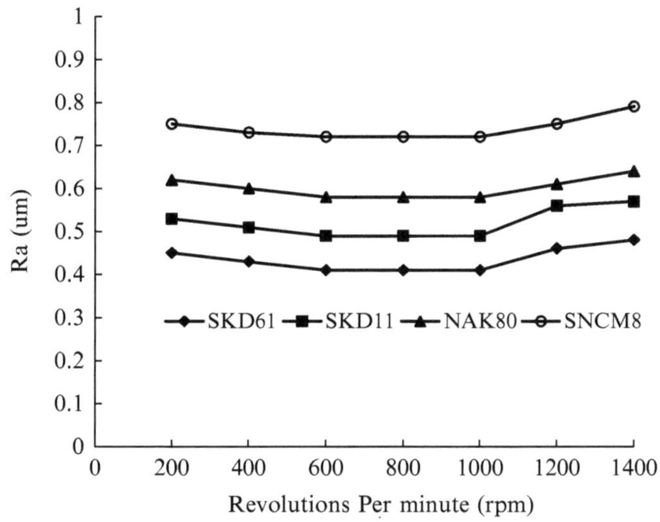

Fig. 3.63 Electropolishing at different rotational speed of workpiece (4 L/min, continuous DC, 10A, 2 mm/min, $R = 2$ mm, $\alpha, \beta, \gamma, \delta, \varepsilon, \zeta = 40°$, cross section = 100 mm^2) ([68], reprinted with permission)

3.11.3 Form Design of Electrode in Electrochemical Smoothing of External Cylindrical Surface

The processing parameters of SKD61, the electrolytic flow rate of 4 L/min, and the gap width of 0.3 mm determined in Sect. 3.2.2 are used in this chapter. Four types of feeding electrodes are designed and studied. The effects of ultrasonic are discussed in the last section.

3.11.3.1 Turning Tool-Form Electrodes

The mild effect of workpiece rotation is shown in Fig. 3.63. The range between 200 and 1,000 rpm is suggested. Below 200 rpm, the workpiece rotation contributes little to effective flushing, while high-speed rotation often causes the run-out of electrode, which will affect the stability of gap width. Figure 3.64 suggests an adequate range of tool feed in electropolishing. High feed rates produce insufficient polishing, while low feed rates worsen the surface finish by excessive materials removal, as found in electropolishing for too long. Figure 3.65 suggests the operation at the current rating between 5 and 15 A for the feeding electrode in the present case. The process parameters for the following experiment on the geometry design of turning tool electrode are the workpiece rotation of 600 rpm, the electrode feed rate of 2 mm/min, and the current rating of 10 A.

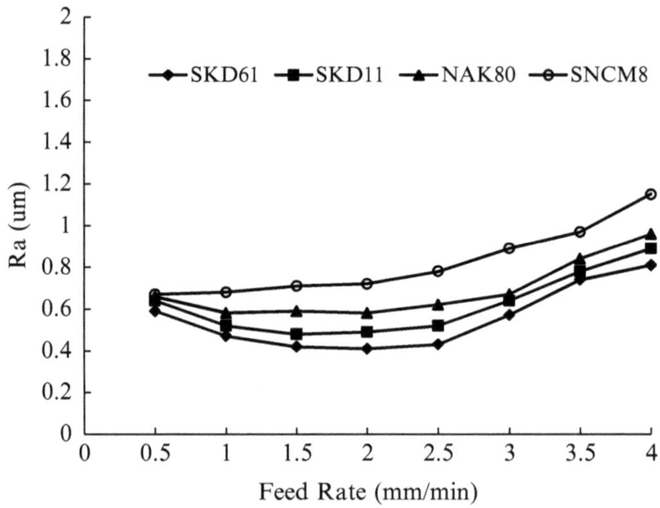

Fig. 3.64 Electropolishing at different feed rate of electrode (4 L/min, continuous DC, 10A, $R = 2$ mm, α, β, γ, δ, ε, $\zeta = 40°$, cross section $= 100$ mm^2, workpiece 600 rpm)

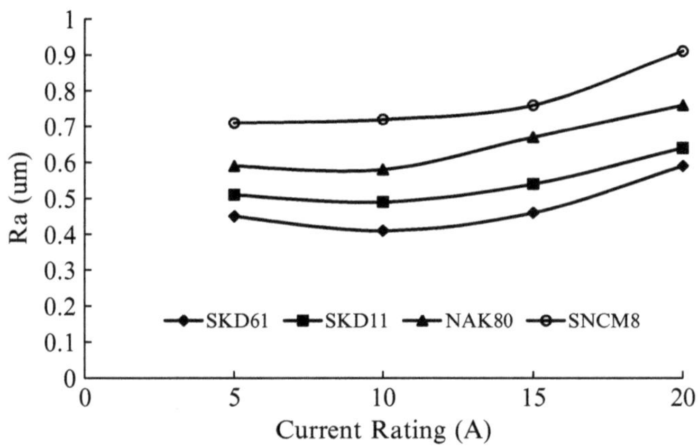

Fig. 3.65 Electropolishing at different current density (4 L/min, continuous DC, 2 mm/min, $R = 2$ mm, α, β, γ, δ, ε, $\zeta = 40°$, cross section $= 100$ mm^2, workpiece 600 rpm)

Figure 3.66 shows that the larger the back rake angle is, the more effective is the polishing, since it produces the inclined plane which allows the higher flow velocity of the electrolyte leading to effective flushing along the inclined plane. In the meantime, the electrolytic products and heat can be brought away more rapidly. Figure 3.67 tells that the effects of the side rake angle are similar, since the electrolytic products can be also transported away more effectively from the side

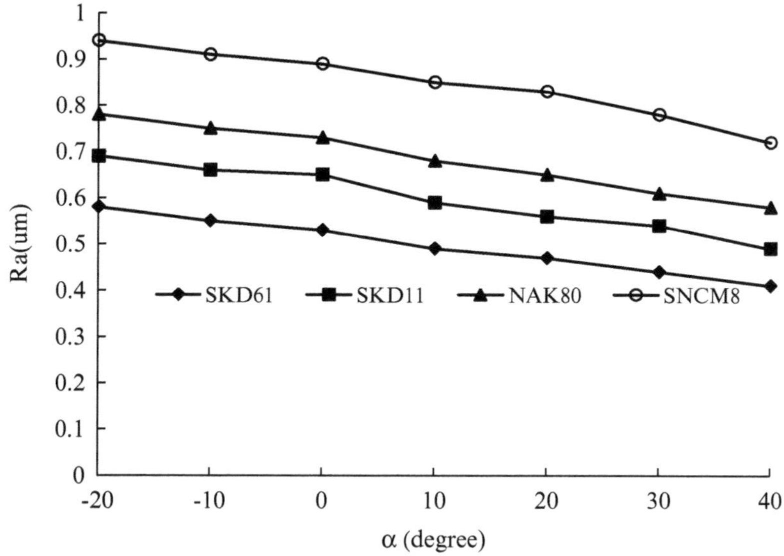

Fig. 3.66 Electropolishing at different back rake angle (4 L/min, continuous DC, 10 A, 2 mm/min, $R = 2$ mm, $\beta, \gamma, \delta, \varepsilon, \zeta = 40°$, cross section = 100 mm^2, workpiece 600 rpm)

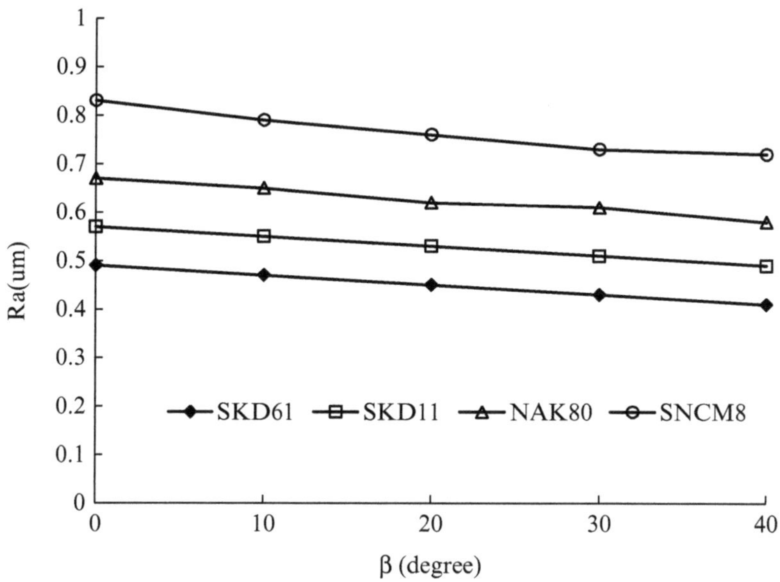

Fig. 3.67 Electropolishing at different side rake angle (4 L/min, continuous DC, 10 A, 2 mm/min, $R = 2$ mm, $\alpha, \gamma, \delta, \varepsilon, \zeta = 40°$, cross section = 156.25 mm^2, workpiece 600 rpm)

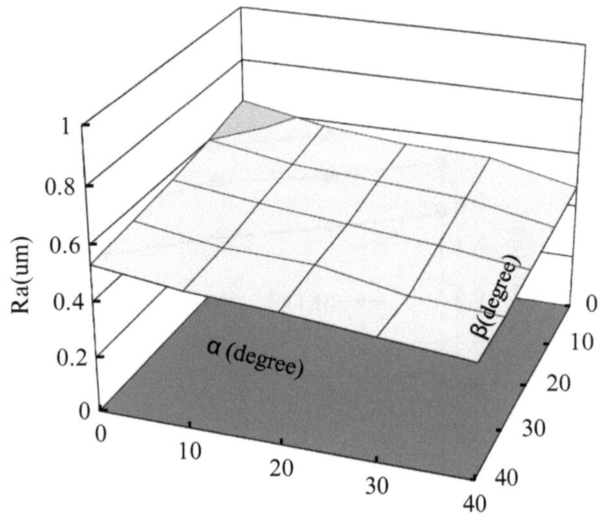

Fig. 3.68 Electropolishing through different back rake angle and side rake angle (SKD61, 4 L/min, continuous DC, 10A, 2 mm/min, $R = 2$ mm, $\gamma, \delta, \varepsilon, \zeta = 40°$, cross section = 100 mm^2, workpiece 600 rpm) ([68], reprinted with permission)

of tool surface; thus the polishing effect is better. Figure 3.68 shows the effects of these two angles on the polishing of SKD61. Figures 3.69 and 3.70 show the larger end and side clearance angle provide wider space beyond the gap between workpiece and tool nose, which is advantageous for effective flushing. The effects of the former are more significant than the latter. Figure 3.71 illustrates the effects of two clearance angles on the polishing of SKD61. The secondary machining can be eliminated effectively with larger end cutting edge angle. More space for dregs discharge is also obtained; thus the polishing effect is better, as shown in Fig. 3.72. However, the increase of side cutting edge angle causes no improvement on polishing. Since the turning tool does not contact with the workpiece, the side cutting angle in front of the polishing site does not bring noticeable improvement in metal turning, as shown in Fig. 3.73. Figure 3.74 shows that the smaller the nose radius is, the more effective is the polishing. However, if the nose radius is too small (<0.4 mm), the current density becomes too large causing violent electrolytic reaction and nonuniformity of the surface finish (see Fig. 3.75). Decreased cutting edge width and height are associated with larger electrolyte flow and discharge space, which is advantageous for the polishing, as shown in Figs. 3.76 and 3.77. Figure 3.78 shows that the improvement of polishing effect is related to the average reduction ratio of the cross section of tool electrode. The reduction ratio (κ) is calculated based on the maximum cross-sectional area (15 × 15 mm^2):

$$\kappa = \left(1 - \frac{A_n}{15 \times 15\,\mathrm{mm}^2}\right)\% \tag{3.21}$$

where A_n is the cross-sectional area of 12.5 × 12.5, 10 × 10, 7.5 × 7.5, and 5 × 5 mm^2; the resulted ratios are 31%, 56%, 75%, and 89%, respectively.

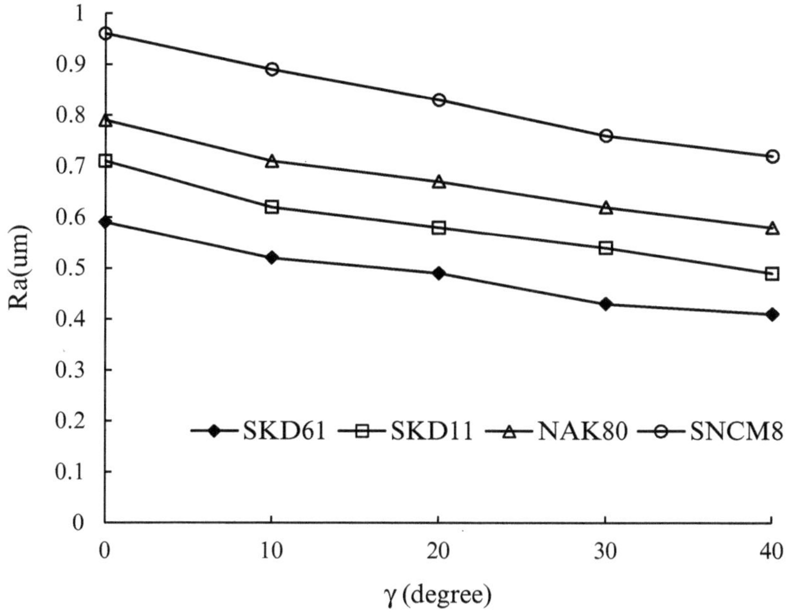

Fig. 3.69 Electropolishing at different end clearance angle (4 L/min, continuous DC, 10A, 2 mm/min, $R = 2$ mm, $\alpha, \beta, \delta, \varepsilon, \zeta = 40°$, cross section = 100 mm^2, workpiece 600 rpm)

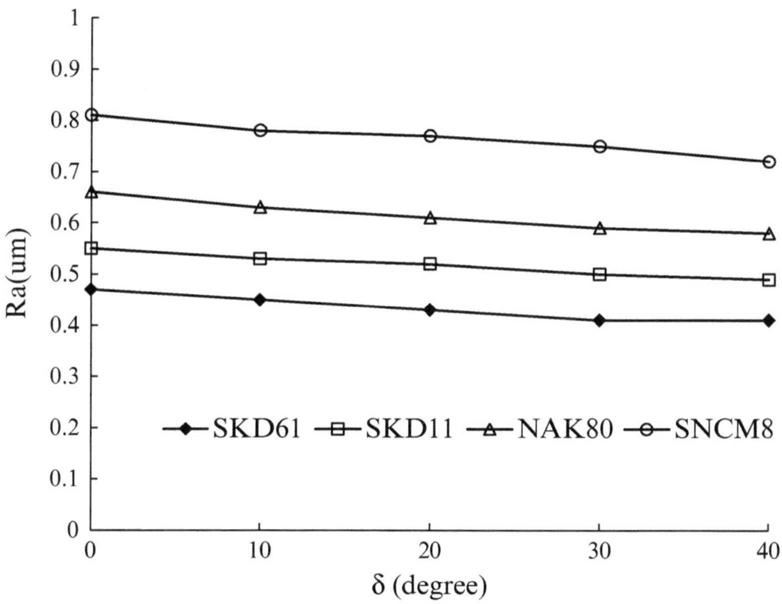

Fig. 3.70 Electropolishing at different side clearance angle (4 L/min, continuous DC, 10 A, 2 mm/min, $R = 2$ mm, $\alpha, \beta, \gamma, \varepsilon, \zeta = 40°$, cross section = 100 mm^2, workpiece 600 rpm)

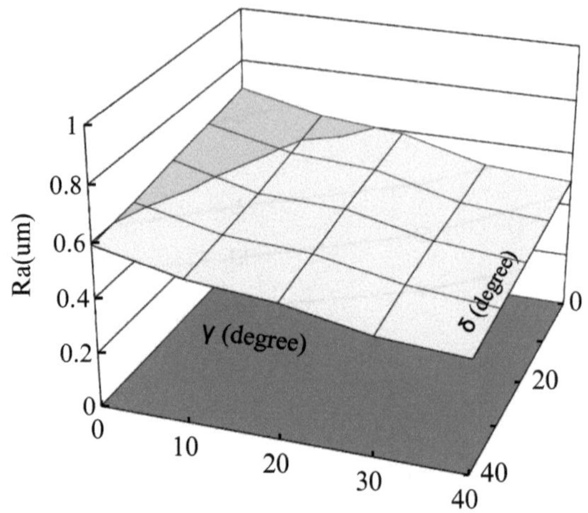

Fig. 3.71 Electropolishing through different end clearance angle and side clearance angle (SKD61,4 L/min, continuous DC, 10 A, 2 mm/min, $R = 2$ mm, α, β, δ, ε, $\zeta = 40°$, cross section $= 100\ \mathrm{mm}^2$, workpiece 600 rpm) ([68], reprinted with permission)

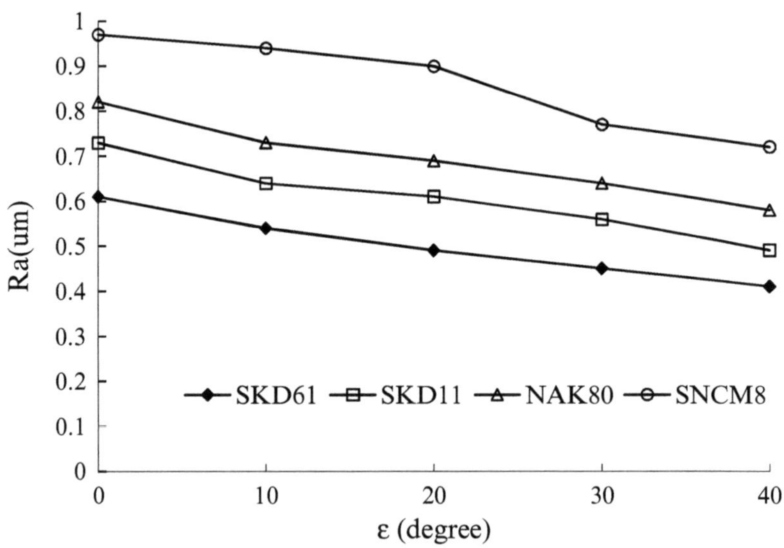

Fig. 3.72 Electropolishing at different end cutting edge angle (4 L/min, continuous DC, 10 A, 2 mm/min, $R = 2$ mm,α, β, γ, δ, $\zeta = 40°$, cross section $= 100\ \mathrm{mm}^2$, workpiece 600 rpm)

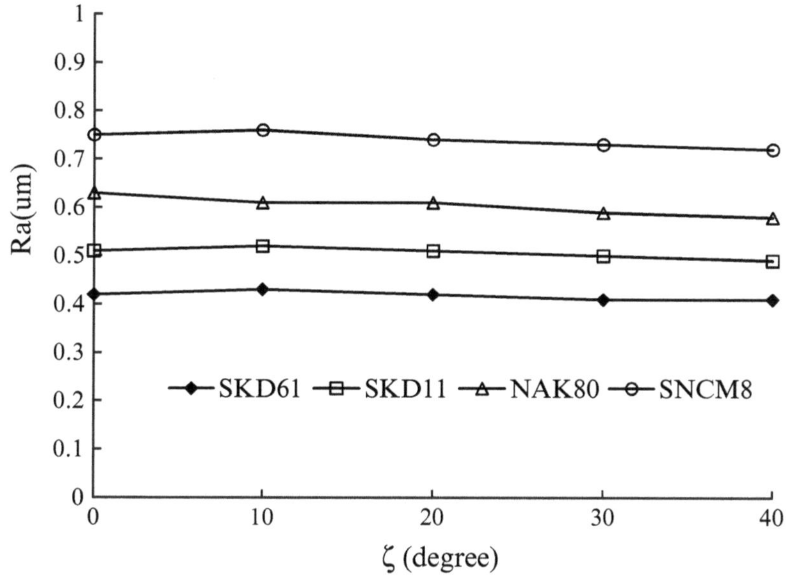

Fig. 3.73 Electropolishing at different side cutting edge angle (4 L/min, continuous DC, 10A, 2 mm/min, $R = 2$ mm, α, β, γ, δ, ε=40°, cross section=100 mm^2, workpiece 600 rpm)

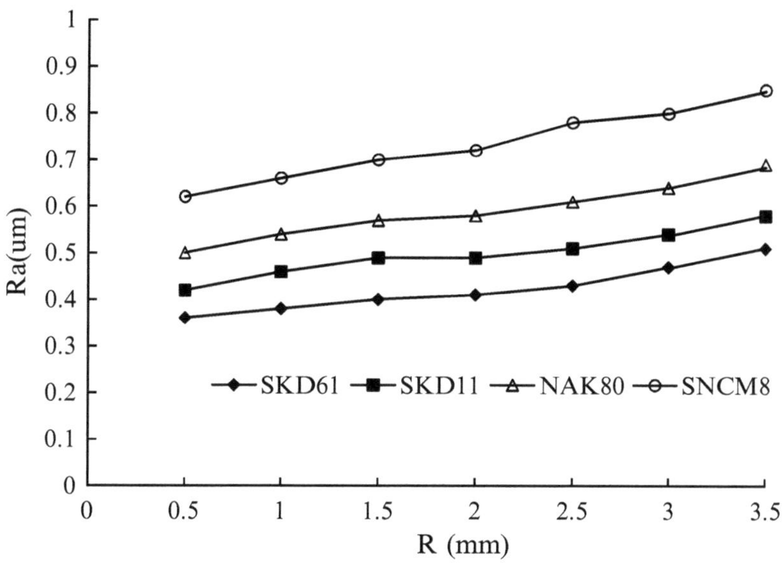

Fig. 3.74 Electropolishing at different nose radius (4 L/min, continuous DC, 10 A, 2 mm/min, α, β, γ, δ, ε, ζ = 40°, cross section = 100 mm^2, workpiece 600 rpm)

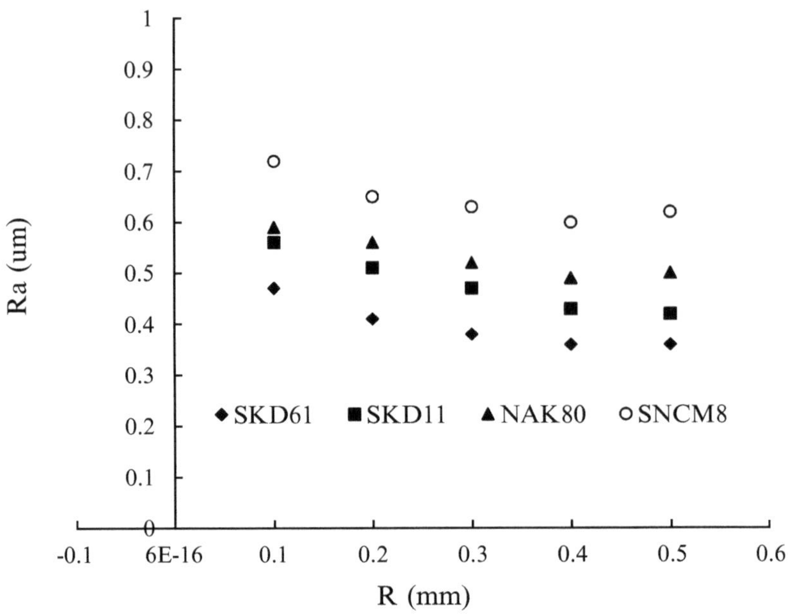

Fig. 3.75 Electropolishing at different nose radius (4 L/min, continuous DC, 10 A, 2 mm/min, α, β, γ, δ,ε, ζ = 40°, cross section = 100 mm^2, workpiece 600 rpm)

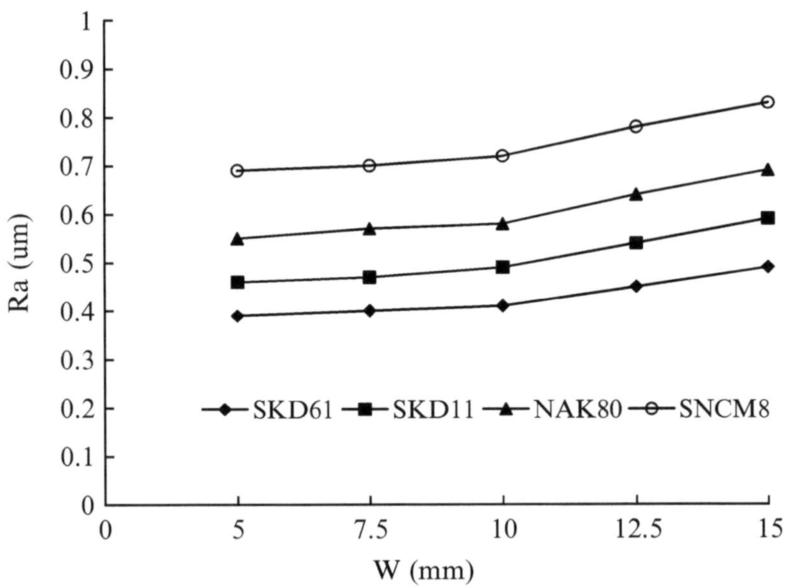

Fig. 3.76 Electropolishing at different width of cutting edge (4 L/min, continuous DC, 10 A, 2 mm/min, R = 2 mm, α,.β, γ, δ, ε, ζ = 40°, H = 10 mm, workpiece 600 rpm)

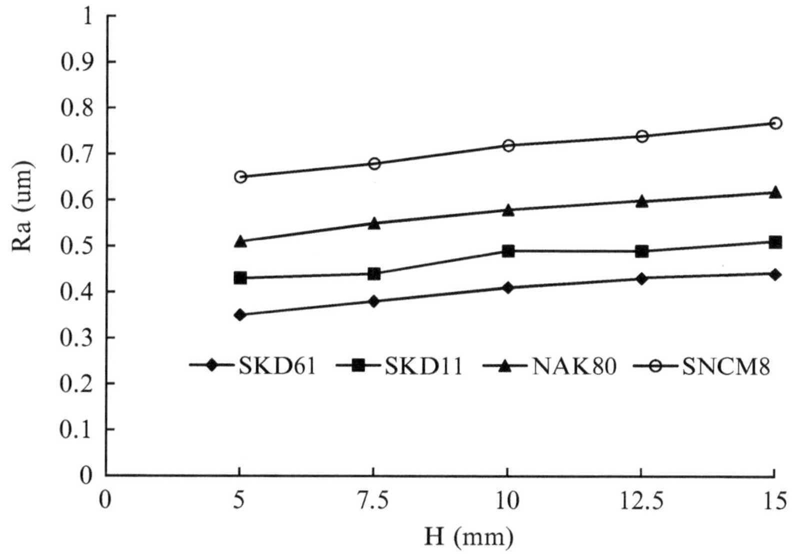

Fig. 3.77 Electropolishing at different height of cutting edge (4 L/min, continuous DC, 10 A, 2 mm/min, $R = 0.5$ mm, α, β, γ, δ, ε, $\zeta = 40°$, $W = 10$ mm, workpiece 600 rpm)

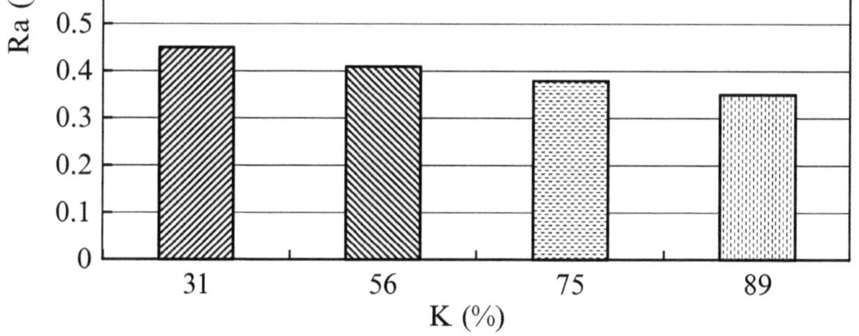

Fig. 3.78 The surface roughness improvement using reduction of the cross section of cutting edge (SKD61, 4 L/min, continuous DC, 10 A, 2 mm/min, $R = 2$ mm, α, β, γ, δ, ε, $\zeta = 40°$, workpiece 600 rpm)

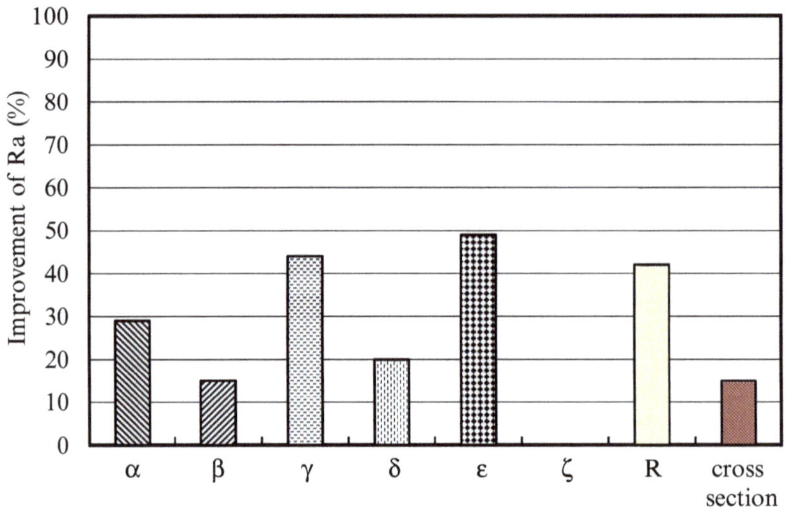

Fig. 3.79 The significance of surface roughness improvement using the designed geometry of tool electrode during electropolishing (SKD61, 4 L/min, continuous DC, 10 A, 2 mm/min, workpiece 600 rpm)

The effects of the tool angle, nose radius, and the size of tool cross section on surface roughness in electropolishing are illustrated in Fig. 3.79. It shows that the end cutting edge angle is the most significant factor to be considered, followed by end clearance angle, nose radius, and back rake angle, while the side cutting edge angle has no effects.

At the same current rating of 10 A, the smaller nose radius results in higher current density; thus the feed rate of the tool electrode can be increased, and the surface roughness can be improved, as shown in Fig. 3.80. However, too small nose radius cause unstable electropolishing. Figure 3.81 shows the amount of diameter reduction is inversely proportional to the feed rate for all the tested materials, while the level of variation is different. Figure 3.82 suggests the same good polishing can be achieved by adequate combination of current rating and electrode feed rate, such as 5 A and 1 mm/min, 10 A and 2 mm/min, 15 A and 3 mm/min, and 20 A and 4 mm/min. Figure 3.83 shows the contribution of surface finish improvement obtained through the tool angle (34%), the nose radius (16%), the tool cross section (16%), the workpiece rotation (13%), and the electrode feed rate (21%). The design of tool, including various angles, nose radius, and cross-sectional area, leads to 66% of the improvement, compared to 34% of the operation parameters, i.e., workpiece rotational and tool feed rate. Figure 3.84 shows the surface roughness of die materials after electropolishing using a common turning tool. The tool angles are different from those discussed above. Comparing Figs. 3.64 and 3.84, one notices that the polishing effect of the designed turning tool electrode is better than that of the typical tool. Figure 3.85 shows the comparison of the surface roughness using

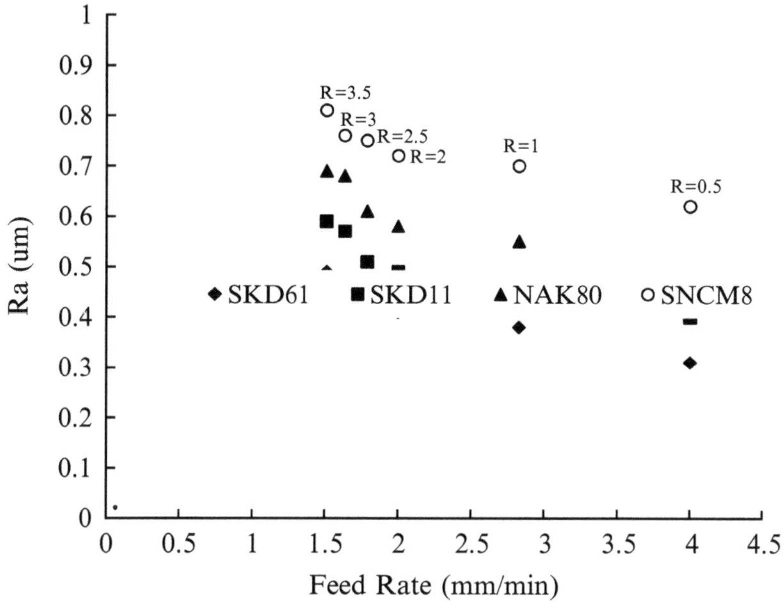

Fig. 3.80 Electropolishing at different nose radius and feed rate (4 L/min, continuous DC, 10 A, $R = 2$ mm, $\alpha, \beta, \gamma, \delta, \varepsilon, \zeta = 40°$, cross section = 100 mm^2, workpiece 600 rpm)

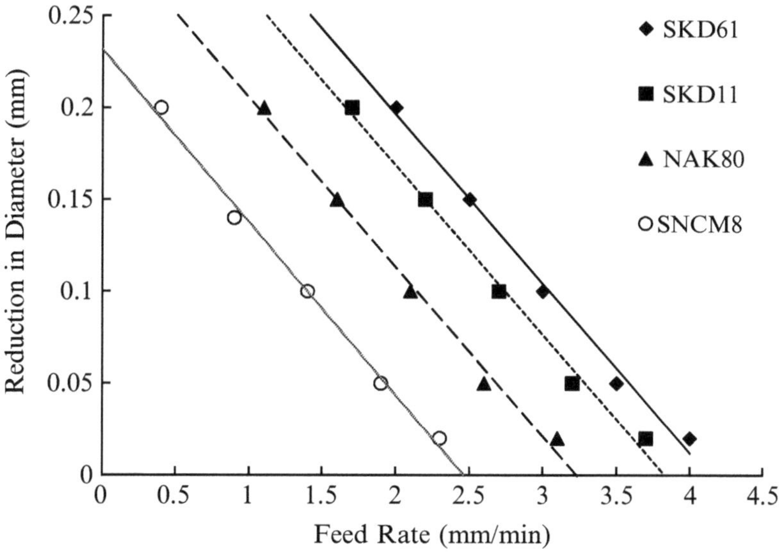

Fig. 3.81 Correlation between cutting depth and feed rate (4 L/min, continuous DC, 10 A, $R = 2$ mm, $\alpha, \beta, \gamma, \delta, \varepsilon, \zeta = 40°$, cross section = 100 mm^2, workpiece 600 rpm) ([68], reprinted with permission)

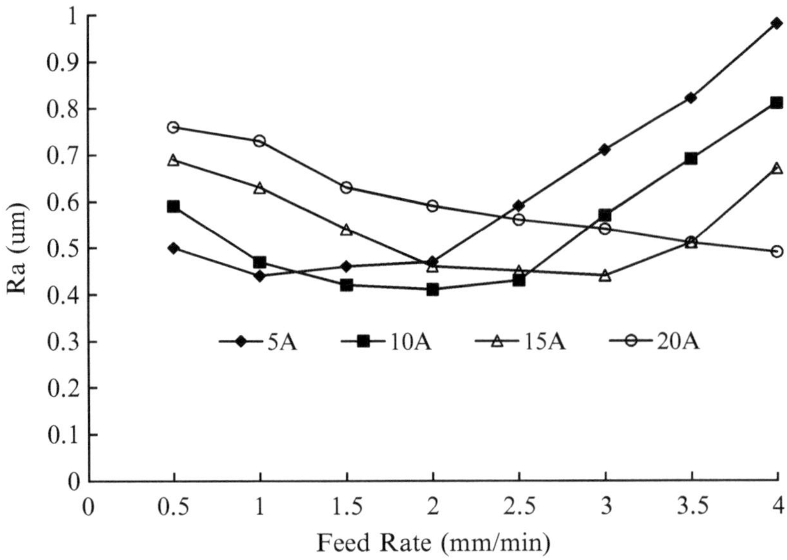

Fig. 3.82 Electropolishing at different feed rate and current rating (SKD61, 4 L/min, continuous DC, $R = 2$ mm, α, β, γ, δ,ε, $\zeta = 40°$, cross section = 100 mm^2, workpiece 600 rpm) ([68], reprinted with permission)

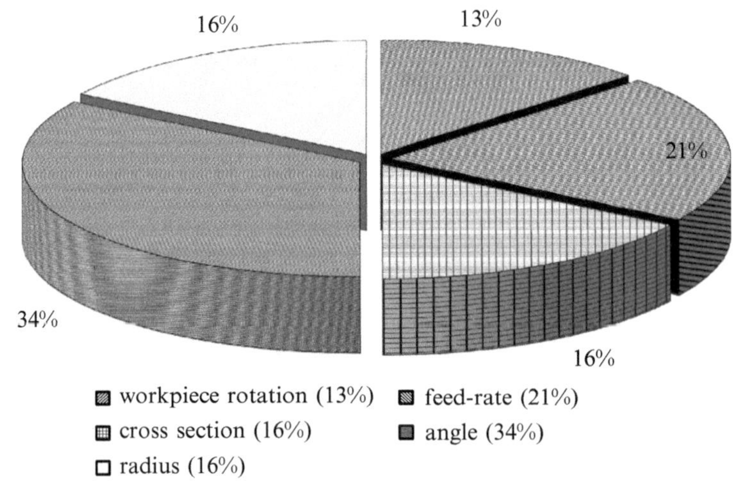

Fig. 3.83 The factors pie of surface roughness improvement (SKD61, 4 L/min, continuous DC, 10 A, 2 mm/min, workpiece 600 rpm)

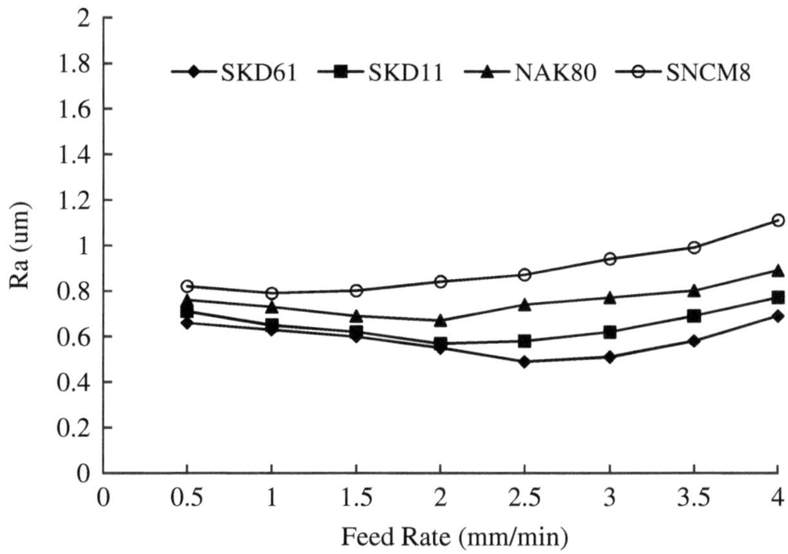

Fig. 3.84 Electropolishing by a common turning tool as (4 L/min, continuous DC, 10 A, $R = 1.0$ mm, $\alpha = 0°$, γ, $\delta = 10°$, $\varepsilon = 20°$, $\zeta = 15°$, workpiece 600 rpm, cross section = 100 mm^2)

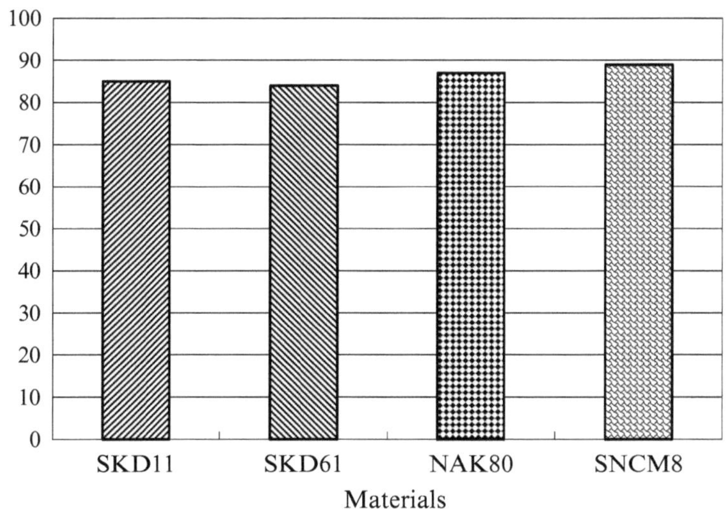

Fig. 3.85 The comparison of the polishing effect through designed tool and common tool (4 L/min, continuous DC, 10 A, workpiece 600 rpm)

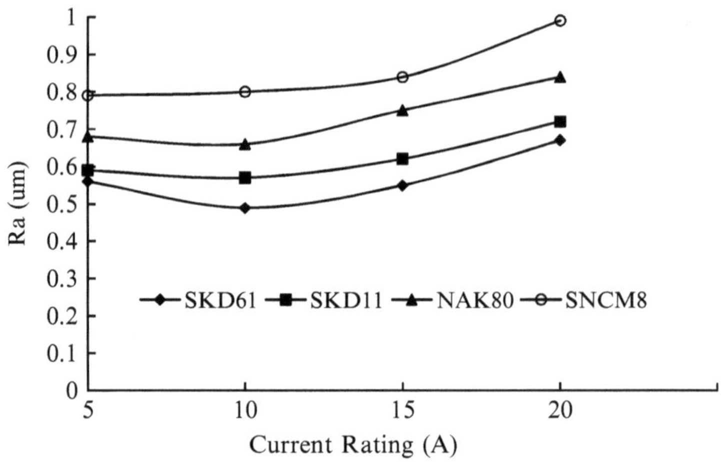

Fig. 3.86 Electropolishing at different current density (Type Aa/nonrotational, 4 L/min, continuous DC, 2 mm/min, $R = 2$ mm, $\theta = 45°$, workpiece 600 rpm) ([69], reprinted with permission)

the designed turning tool and the common turning tool as electrode. The latter achieves 86% of the surface finish in average. It shows a common turning tool can be used in electropolishing; nevertheless, the common turning tool should be slightly modified with increased angles for optimal electropolishing, where the cost for the modified tool is expected inexpensive.

3.11.3.2 Arrow Head-Form Electrode

Figures 3.86 and 3.87 show the polishing should be carried out at the adequate electrode feed rate and current rating. Large feed rate and low current provide insufficient energy density for the electropolishing, while the contrary will remove material fast and rough. Figure 3.87 suggests the material-dependant proper feed rate is: SKD61 1.5–2.5 mm/min, SKD11 1.0–2.0 mm/min, NAK80 0.5–1.5 mm/min, and SNCM8 0.5–1.0 mm/min. Figure 3.88 shows the effects of different arrow head end radius and feed rate. Small radius with fast feed rate are advantageous for polishing, since smaller radius means smaller effective area and higher current density at the same current rating; thus faster feed rate of electrode can be used. Figure 3.89 shows that small end radius is indeed effective during the polishing. Figure 3.90 shows the workpiece rotation is effective for polishing, although the variation causes mild effect only. Figures 3.86, 3.87, and 3.90 also illustrate the polishing results of SKD61 are the best, followed by SKD11, NAK80, and SNCM8. Figure 3.91 shows high rotational speed of electrode is advantageous in the example of type Aa; an average reduction of surface roughness by 6% at the increase of 100 rpm is achieved. One believes the rotational flow energy elevates the discharge mobility, which improves the electropolishing. The moderate

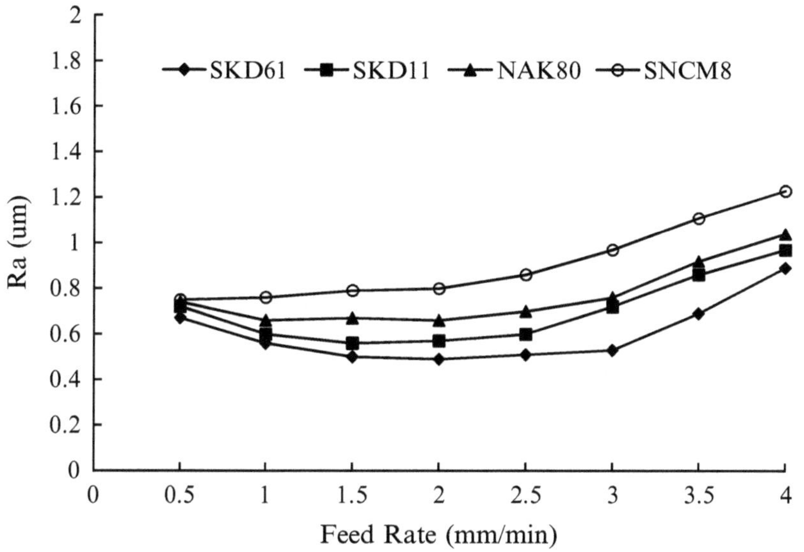

Fig. 3.87 Electropolishing at different feed rate of electrode (Type Aa/nonrotational, 4 L/min, continuous DC, 10 A, $R = 2$ mm, $\theta = 45°$, workpiece 600 rpm) ([69], reprinted with permission)

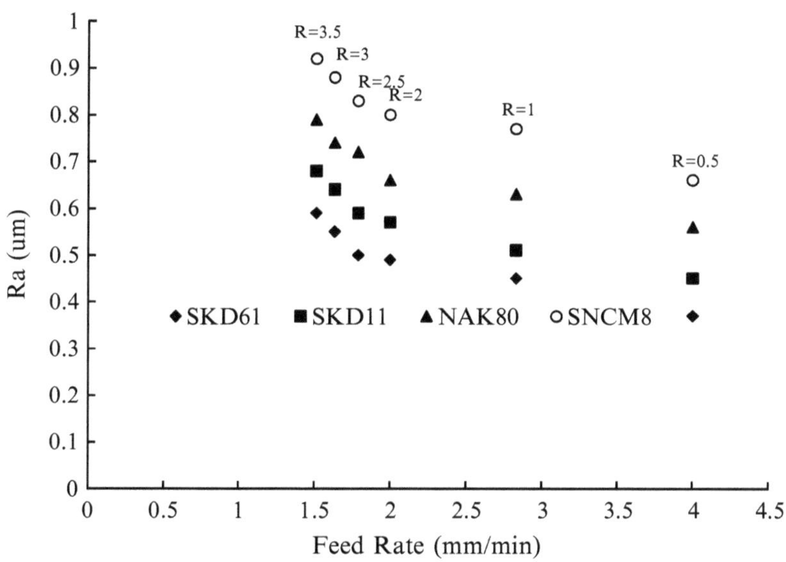

Fig. 3.88 Electropolishing at different end radius and feed rate (Type Aa/nonrotational, 4 L/min, continuous DC, 10 A, $R = 2$ mm, $\theta = 45°$, workpiece 600 rpm) ([69], reprinted with permission)

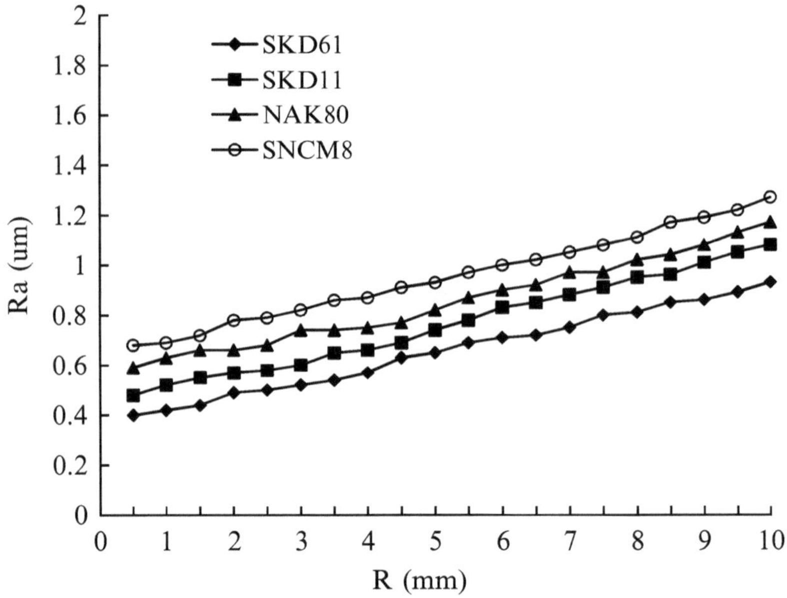

Fig. 3.89 Electropolishing at different end radius (Type Aa/nonrotational, 4 L/min, continuous DC, 10 A, $R = 2$ mm, $\theta = 45°$, workpiece 600 rpm) ([69], reprinted with permission)

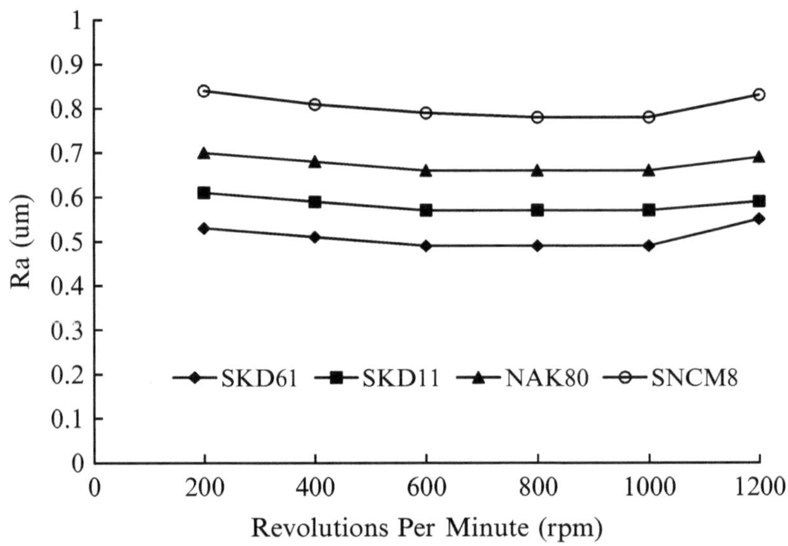

Fig. 3.90 Electropolishing at different rotational speed of workpiece (Type Aa/nonrotational, 4 L/min, continuous DC, 2 mm/min, $R = 2$ mm, $\theta = 45°$) ([69], reprinted with permission)

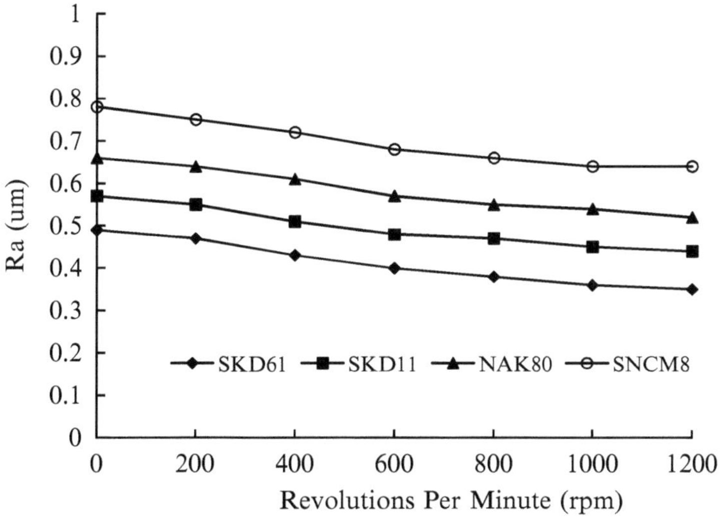

Fig. 3.91 Electropolishing at different rotational speed of electrode (Type Aa, 4 L/min, continuous DC, 10 A, $R = 2$ mm, $\theta = 45°$, workpiece 600 rpm) ([69], reprinted with permission)

rotational speed of workpiece (600–1,000 rpm) associated with high rotational speed of electrode produces good polishing, as shown in Fig. 3.92.

Figure 3.93 shows that reducing the taper angle is effective for the polishing; the effect increases 7% when the taper angle is reduced by 15°. The reduced taper angle is advantageous for dreg flushing and also the reduction of secondary overcutting; the taper of the polished external cylindrical surface is controlled below 0.15° when the taper angle is 45°. Figure 3.94 shows the small taper angle and high rotational speed of electrode should be used. Figure 3.95 illustrates the optimum current rating varies with the feed rate for a material and an electrode. The existence of the optimum is shown in Fig. 3.86. The adequate combinations of current rating and electrode feed rate in this case are 5 A and 1 mm/min, 10 A and 2 mm/min, 15 A and 3 mm/min, and 20 A and 4 mm/min. The design of flute opening can help remove the electropolishing products from the gap. Figure 3.96 shows electrode type Ca with double-discharge flute is better than type Ba (single-flute) and Aa, since the double-flute provides more discharge space and flushing mobility. The average improvement is about 9%. Large flute back rake angle and side rake angle will slightly improve the polishing effect for electrode type Ba and type Ca (see Figs. 3.97 and 3.98). Figure 3.99 shows that large rake angle and high rotational speed of electrode produce better polishing. They both contribute to effective dregs discharge. Figure 3.100 summarizes the contribution of surface finish improvement using cone arrow head electrode type Ca. It is obtained from the electrode rotation (23%), the end radius (32%), the taper angle (22%), the flute back rake angle (7%), the flute side rake angle (4%), and the flute number (12%). One notices the end radius is the most effective factor in electrode design.

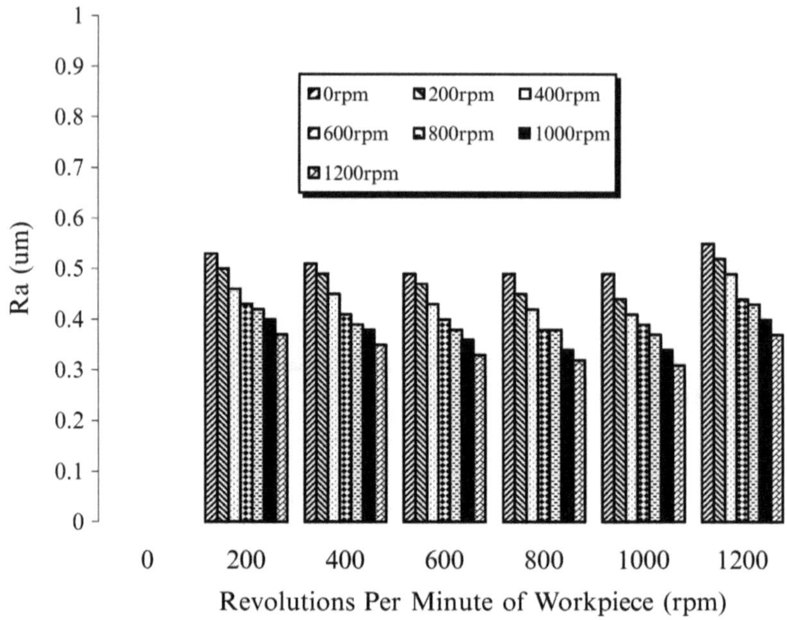

Fig. 3.92 Electropolishing at different rotational speed of electrode and workpiece (Type Aa, 4 L/min, continuous DC, 10 A, 2 mm/min, $R = 2$ mm, $\theta = 45°$, workpiece 600 rpm) ([69], reprinted with permission)

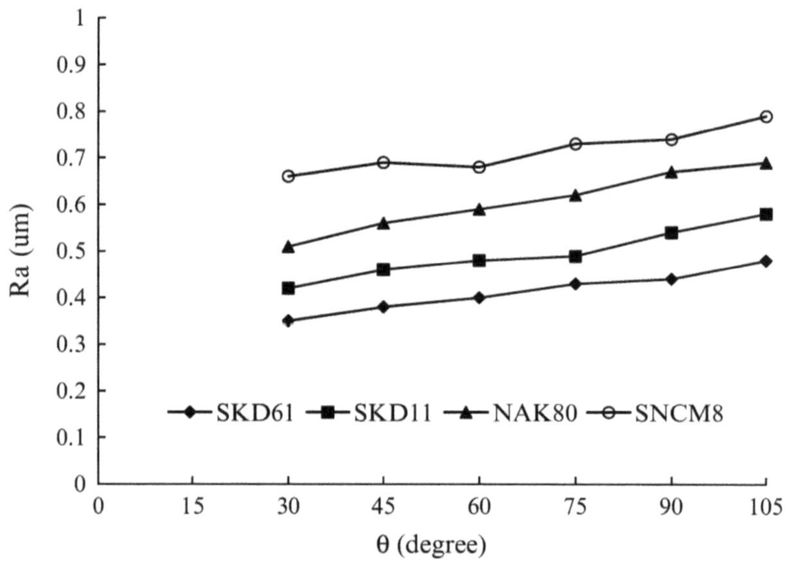

Fig. 3.93 Electropolishing at different taper angle (Type A/800 rpm, 4 L/min, continuous DC, 10 A, 2 mm/min, $R = 2$ mm, workpiece 600 rpm) ([69], reprinted with permission)

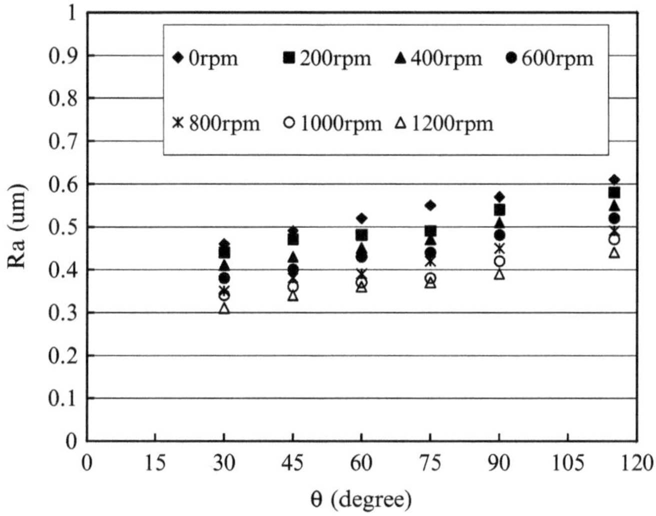

Fig. 3.94 Electropolishing at different taper angle and rotational speed of electrode (Type Aa, 4 L/min, continuous DC, 10 A, 2 mm/min, R =2 mm, workpiece 600 rpm) ([69], reprinted with permission)

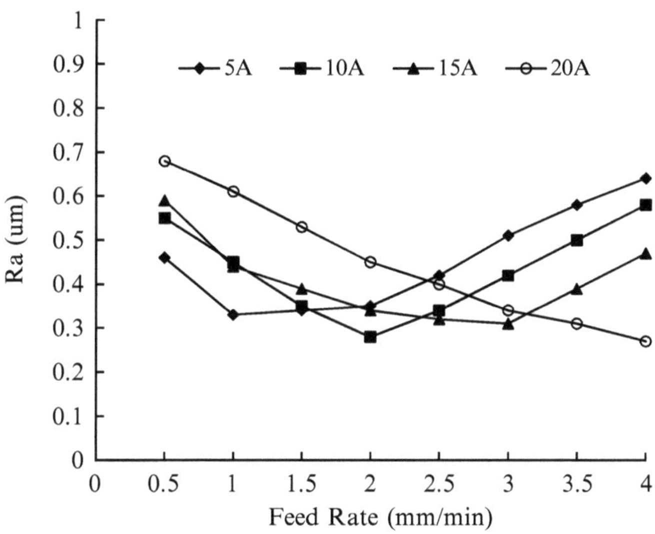

Fig. 3.95 Electropolishing at different feed rate and current rating (SKD61, Type Ca/800 rpm, 4 L/min, continuous DC, R = 2 mm, $\theta = 45°$, χ and $\kappa = 40°$, workpiece 600 rpm) ([69], reprinted with permission)

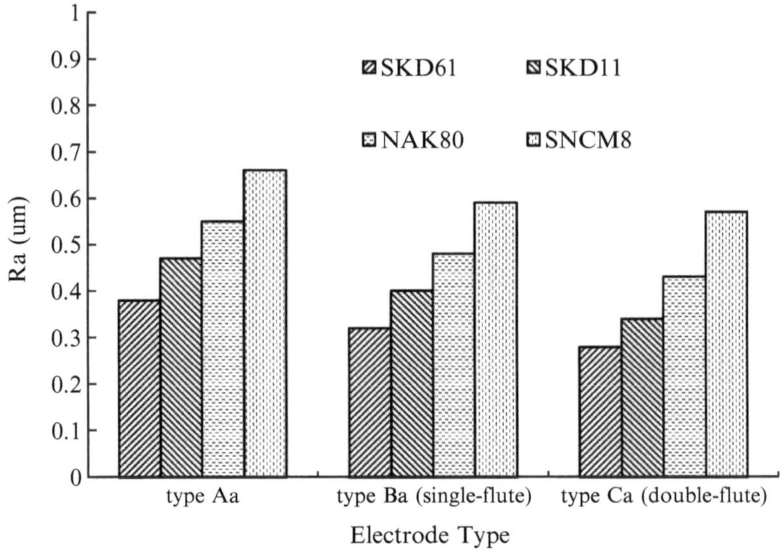

Fig. 3.96 Electropolishing with cone arrow head electrode (800 rpm, 4 L/min, continuous DC, 10 A, 2 mm/min, $R = 2$ mm, $\theta = 45°$, χ and $\kappa = 40°$, workpiece 600 rpm) ([69], reprinted with permission)

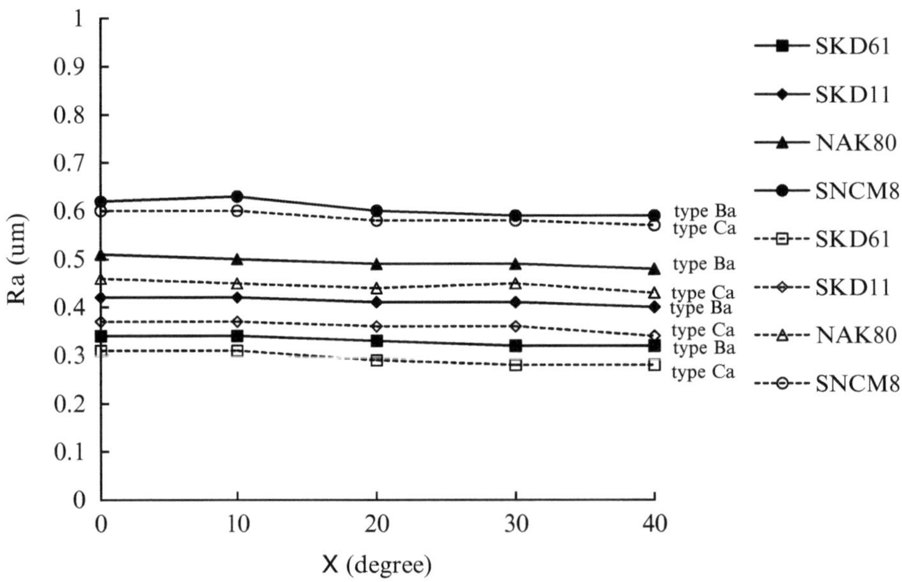

Fig. 3.97 Electropolishing at different back rake angle (800 rpm, 4 L/min, continuous DC, 10 A, 2 mm/min, $R = 2$ mm, $\theta = 45°$, $\kappa = 40°$, workpiece 600 rpm) ([69], reprinted with permission)

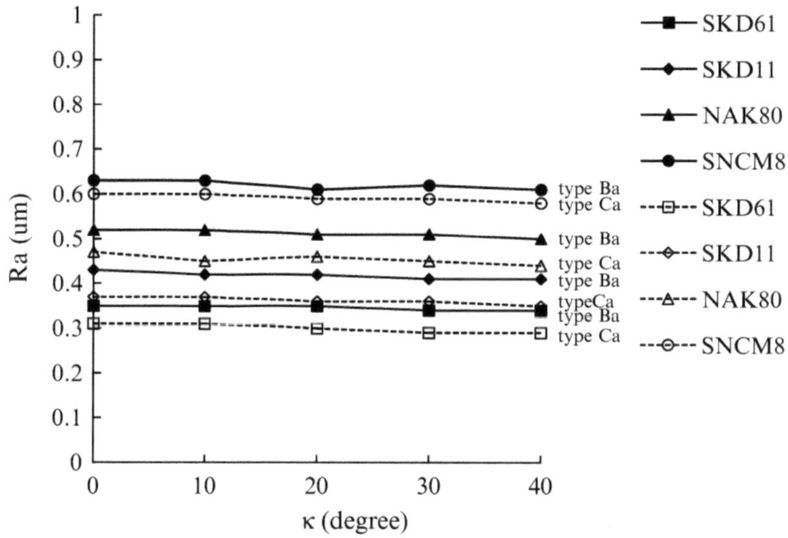

Fig. 3.98 Electropolishing at different side rake angle (800 rpm, 4 L/min, continuous DC, 10 A, 2 mm/min, $R = 2$ mm, $\theta = 45°$, $\chi = 40°$, workpiece 600 rpm) ([69], reprinted with permission)

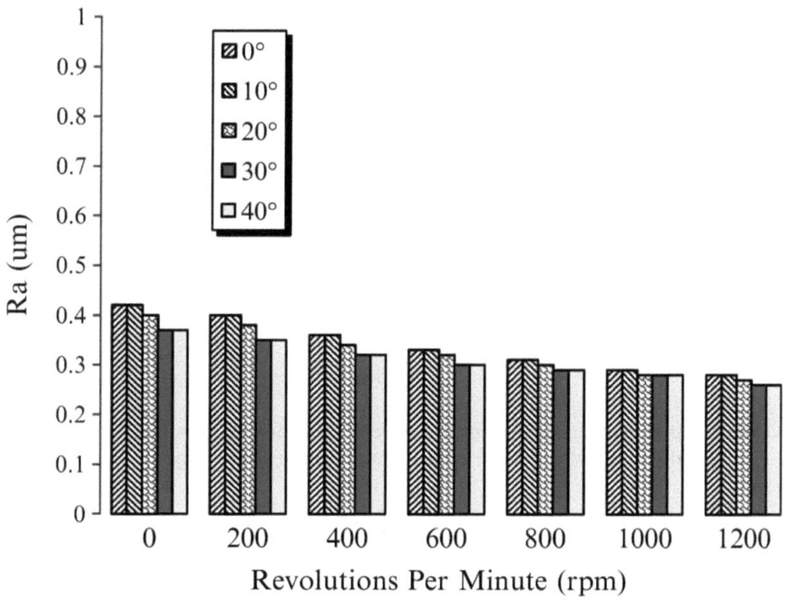

Fig. 3.99 Electropolishing at different electrode rotational speed and rake angles (SKD61, Type Ca, 4 L/min, continuous DC, $R = 2$ mm, $\theta = 45°$, workpiece 600 rpm) ([69], reprinted with permission)

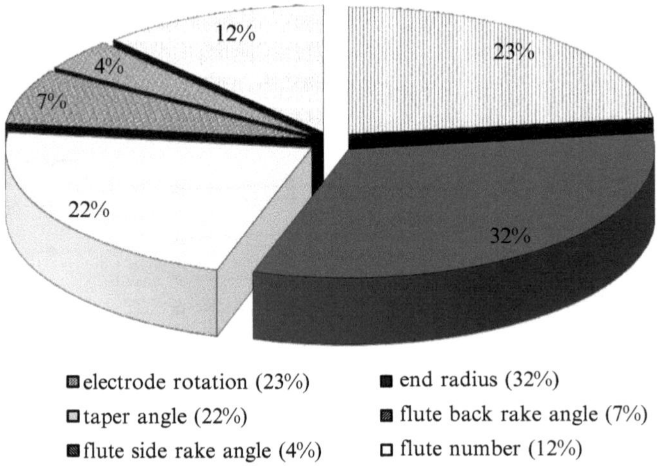

Fig. 3.100 The contribution pie of surface finish improvement of type Ca (SKD61, electrode 800 rpm, 4 L/min, continuous DC, 10 A, 2 mm/min, $R = 2$ mm, $\theta = 45°$, χ and $\kappa = 40°$, workpiece 600 rpm) ([69], reprinted with permission)

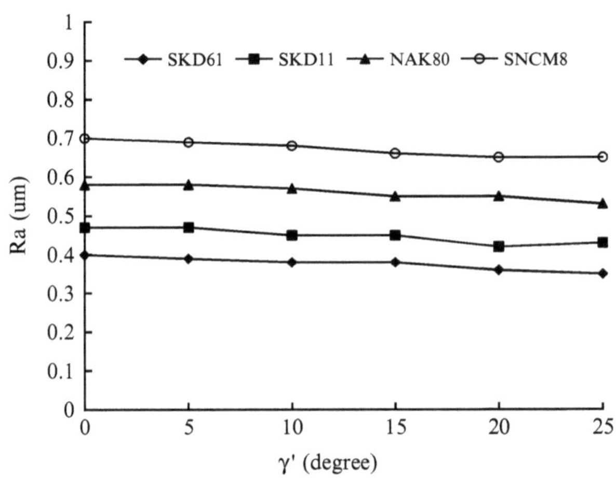

Fig. 3.101 Electropolishing at different tail angle (Type Da_V/800 rpm, 4 L/min, continuous DC, 10 A, 2 mm/min, $R = 2$ mm, $\theta = 45°$, $t = 5$ mm, workpiece 600 rpm) ([69], reprinted with permission)

The arrow head electrode can be also designed flat (electrode type Da_V, Da_H, and Ea). The secondary machining can be reduced effectively with large tail angle. More space for dregs discharge is also obtained; thus the polishing is better, as shown in Fig. 3.101. Thin electrode improves the polishing for the same reason, as

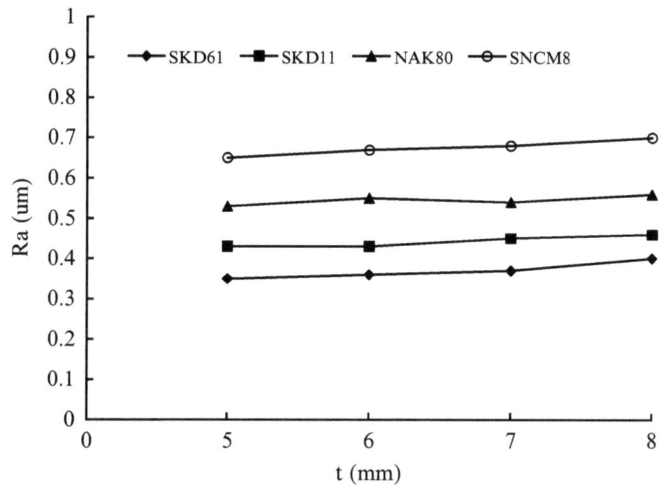

Fig. 3.102 Electropolishing at different electrode thickness (Type Da_V/800 rpm, 4 L/min, continuous DC, 10 A, 2 mm/min, $R = 2$ mm, $\theta = 45°$, $\gamma' = 25°$, workpiece 600 rpm) ([69], reprinted with permission)

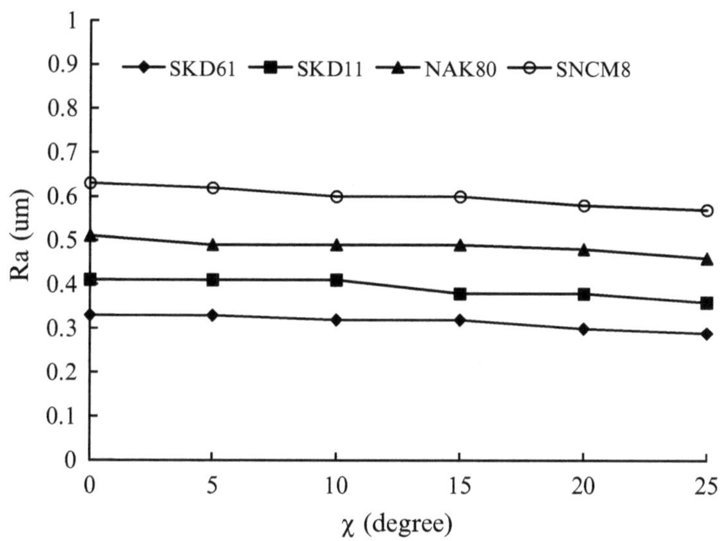

Fig. 3.103 Electropolishing at different back rake angle (Type Ea/800 rpm, 4 L/min, continuous DC, 10A, 2 mm/min, $R = 2$ mm, $t = 5$ mm, $h = 10$ mm, $\theta = 45°$, $\gamma' = 25°$, $\kappa = 20°$, workpiece 600 rpm) ([69], reprinted with permission)

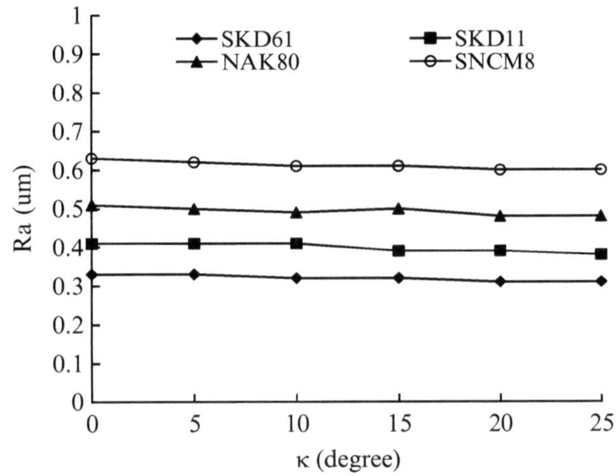

Fig. 3.104 Electropolishing at different side rake angle (Type Ea/800 rpm, 4 L/min, continuous DC, 10 A, 2 mm/min, $R = 2$ mm, $t = 5$ mm, $h = 10$ mm, $\theta = 45°, \gamma' = 25°, \chi = 20°$, workpiece 600 rpm) ([69], reprinted with permission)

seen in Fig. 3.102. Figure 3.103 shows that the larger the flute back rake angle is, the more effective is the polishing. It provides the inclined plane allowing higher flow velocity of the electrolyte for effective flushing. In the meantime, the electrolytic products and heat can be brought away more rapidly. Figure 3.104 tells the similar effects of the side rake angle of electrode. Deep flute depth is desired for the improvement of the polishing, since it enlarges the space for dregs discharge (see Fig. 3.105). The contribution pie of surface finish improvement using the flat arrow head electrode type Ea is shown in Fig. 3.106. Among the electrode rotation (18%), the end radius (24%), the taper angle (17%), the tail angle (9%), the flute back rake angle (8%), the flute side rake angle (6%), the flute depth (11%), and the plate thickness (7%), the end radius remains the most effective factor in electrode design, as in the case of cone arrow head electrode.

Figure 3.107 shows the polishing results from different electrodes and current rating. The current rating of 10 A should be used and type Ca and Ea are recommended in the current study (using SKD61, 4 L/min, continuous DC 10 A, 2 mm/min, $R = 2$ mm, $\theta = 45°$, electrode 800 rpm, workpiece 600 rpm, and the amount of the reduction of diameter after electropolishing is 0.2 mm). Figure 3.108 shows the polishing results of different type electrodes for various materials. SKD61 is the most suitable material and the electrode type Ca and Ea are the best design. SKD61 also achieves the fastest feed in electropolishing, when compared with other materials at the same amount of material removal, as shown in Fig. 3.109. The comparison of improvement of surface roughness by electropolishing using different electrodes rotating at 800 rpm is shown in Fig. 3.110. The best result is obtained by the electrode type Ca (cone arrow head with double discharge flute, 43%) and type Ea (flat arrow head with discharge flute opening and rake angle, 41%), followed by the electrode type Ba (35%), the electrode type Da_V and Da_H (29%), and the electrode type Aa (18%). In summary, the average contribution of surface finish improvement obtained from the factors of

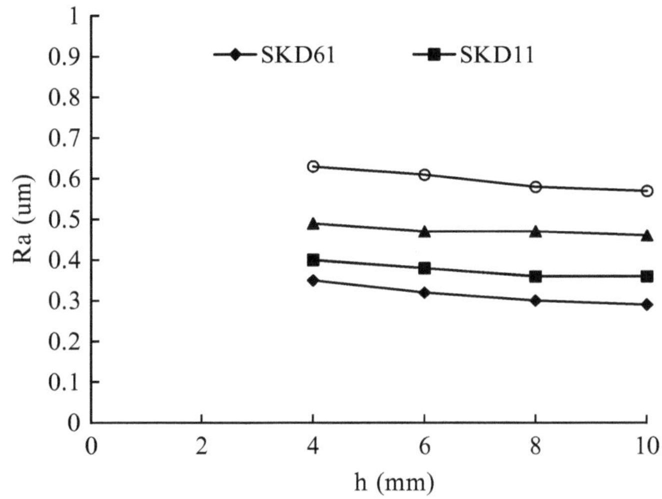

Fig. 3.105 Electropolishing at different electrode flute depth (Type Ea/800 rpm, 4 L/min, continuous DC, 10 A, 2 mm/min, $R = 2$ mm, $t = 5$ mm, $\theta = 40°$, $\gamma' = 25°$, $\chi = \kappa = 20°$, workpiece 600 rpm) ([69], reprinted with permission)

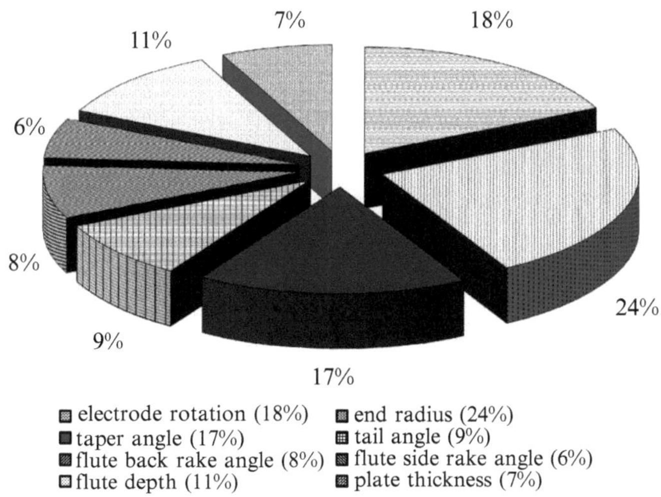

Fig. 3.106 The contribution pie of surface roughness improvement (Type Ea/800 rpm, 4 L/min, continuous DC, 10 A, 2 mm/min, $R = 2$ mm, $t = 5$ mm, $h = 10$ mm, $\theta = 45°$, $\gamma' = 25°$, $\chi = \kappa = 20°$, workpiece 600 rpm) ([69], reprinted with permission)

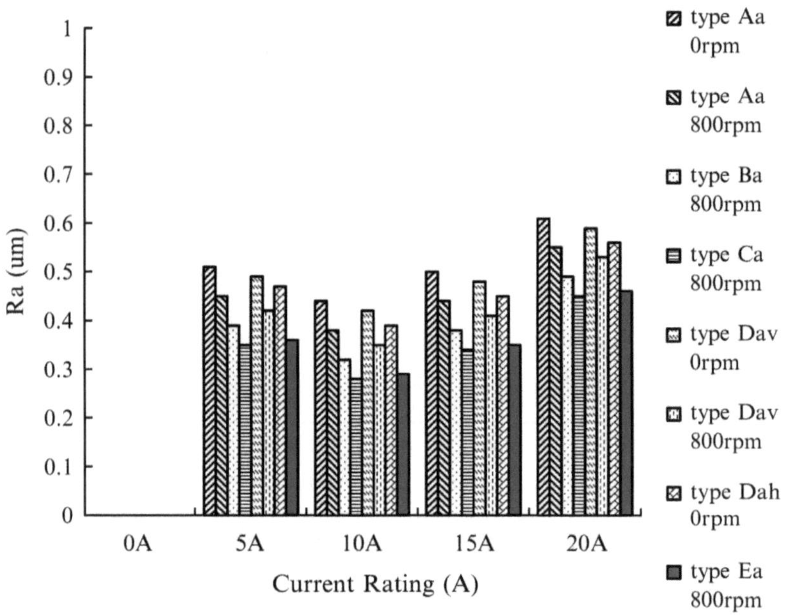

Fig. 3.107 Electropolishing using different current rating and electrode (SKD61, 4 L/min, continuous DC, 2 mm/min, $R = 2$ mm, $\theta = 45°$, workpiece 600 rpm) ([69], reprinted with permission)

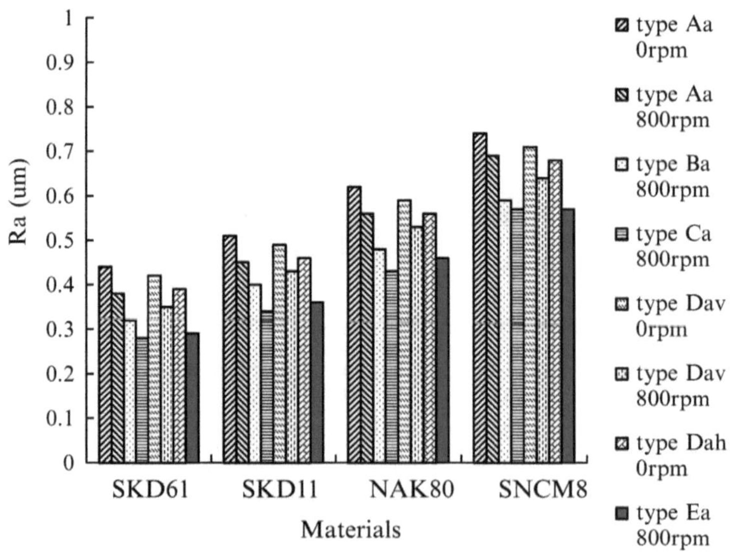

Fig. 3.108 Electropolishing using various material and electrode (4 L/min, continuous DC, 10 A, 2 mm/min, $R = 2$ mm, $\theta = 45°$, workpiece 600 rpm) ([69], reprinted with permission)

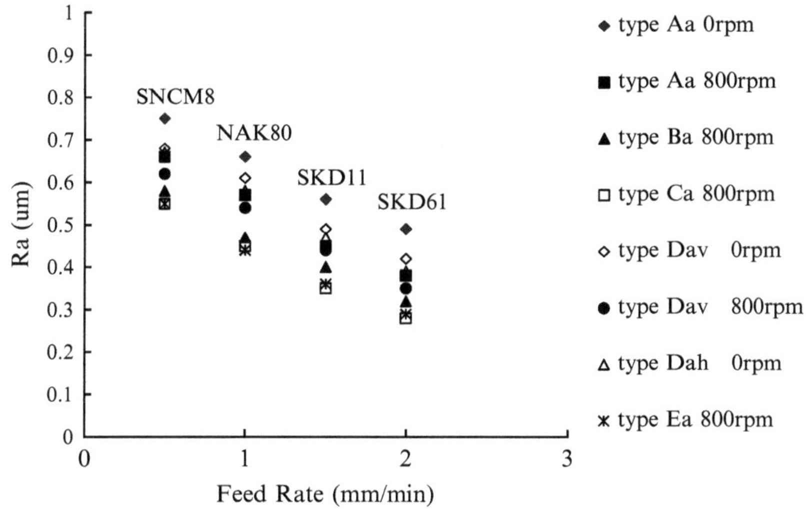

Fig. 3.109 Electropolishing using different material and electrode (4 L/min, continuous DC, $R = 2$ mm,$\theta = 45°$, workpiece 600 rpm) ([69], reprinted with permission)

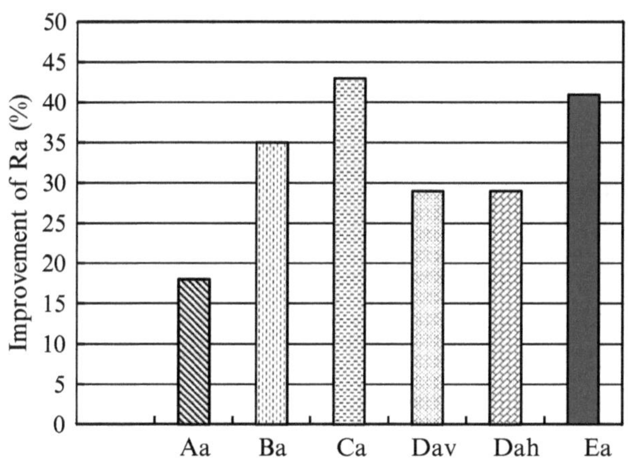

Fig. 3.110 The surface finish improvement obtained from different electrode (SKD61, electrode 800 rpm, 4 L/min, continuous DC, 2 mm/min, $R = 2$ mm, $\theta = 45°$, workpiece 600 rpm) ([69], reprinted with permission)

electrode, i.e., the electrode rotation, the electrode feed rate, and the tool geometry, is shown in Fig. 3.111. The design of tool, including end radius, various angles, thickness, and flute depth, leads to 58% of the improvement, compared to 42% from the electrode rotation and feed rate.

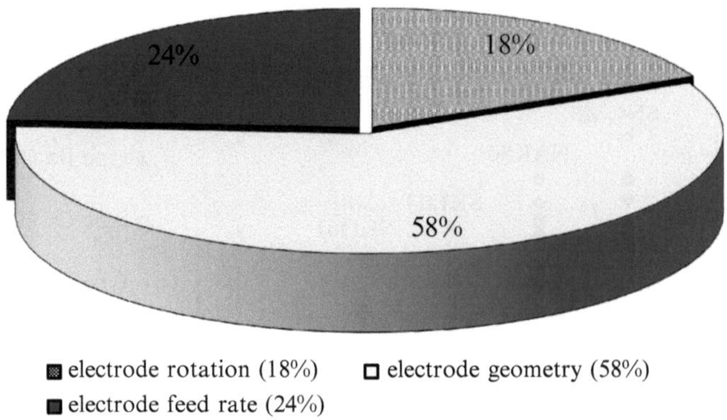

Fig. 3.111 The contribution pie of surface finish improvement (SKD61, 4 L/min, continuous DC, 2 mm/min, $R = 2$ mm, $\theta = 45°$, workpiece 600 rpm) ([69], reprinted with permission)

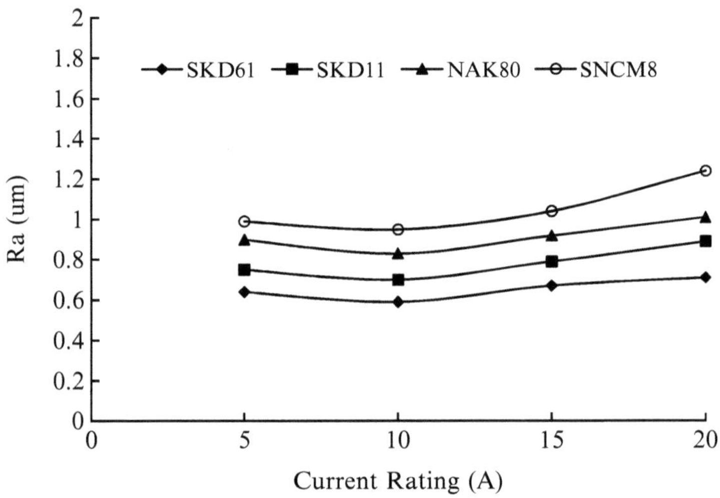

Fig. 3.112 Electropolishing at different current rating (Type Br, 0 rpm, 4 L/min, continuous DC, 2 mm/min, $t = 3$ mm, $\theta = 40°$, $\lambda = 0.5$ mm, workpiece 0 rpm) ([70], reprinted with permission)

3.11.3.3 Ring-Form Electrode

Figure 3.112 suggests there exists an optimal current rating for electropolishing. Excessive or insufficient electrical energy cannot effectively reduce the surface roughness. Under the same machining conditions, the polishing effect of SKD61 is the best, followed by SKD11, NAK80, and SNCM8. Figure 3.113 shows different workpiece materials require their individual suitable feed rates as: SKD61 1.5–2.5 mm/min, SKD11 1.0–2.0 mm/min, NAK80 0.5–1.5 mm/min, and

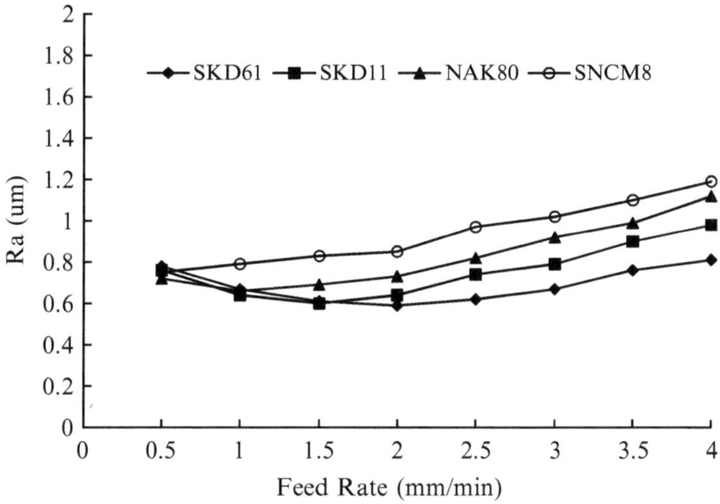

Fig. 3.113 Electropolishing at different feed rate of electrode (Type Br, 0 rpm, 4 L/min, continuous DC, 10 A, $t = 3$ mm, $\theta' = 40°$, $\lambda = 0.5$ mm, workpiece 0 rpm) ([70], reprinted with permission)

SNCM8 0.5–1.0 mm/min. Figure 3.113 shows that good polishing is achieved by adequate combination of electrode feed rate and current rating, as results of Figs. 3.112 and 3.114. The principle lies in the optimal energy density across the gap during electropolishing. The workpiece rotation above 400 rpm improves the polishing, as shown in Fig. 3.115. Similar effect of electrode rotation is shown in Fig. 3.116. Both are attributed to that the rotational energy elevates the discharge mobility of dregs. The rotational speed of workpiece above 600 rpm associated with fast electrode rotation produces good polishing, as shown in Fig. 3.117. Figure 3.118 illustrates the thickness of ring-form electrode plays minor role as compared to workpiece materials. Thin electrode is thus preferred for the reduction of the secondary overcutting. To maintain the stability of the gap along electrode feed, a certain thickness is however needed. Three millimeter is adopted in the current study, and the taper of the polished external cylindrical surface is controlled below 0.15° when the electrode thickness is 3 mm. Figure 3.119 shows that the larger the inner taper angle is, more effective is the polishing. It is more efficient for dregs discharge and concentration of polishing current. The adequate inner edge rounding (about 0.5–0.75 mm radius of curvature) should be taken into consideration, as shown in Fig. 3.120. If the edge radius is too small (<0.5 mm), the current density becomes too large causing violent electrolytic reaction and nonuniformity of the surface finish. On the other hand, large radius results in insufficient polishing energy and rough finish. It is worth noticing that sharp inner edge associated with fast feed rate will keep the current density at the appropriate level; therefore better polishing can be obtained, as shown in Fig. 3.121. Figure 3.122 illustrates that adequate feed rate (1.5–2.5 mm/min) and long pulse-off time is desired for fine finish.

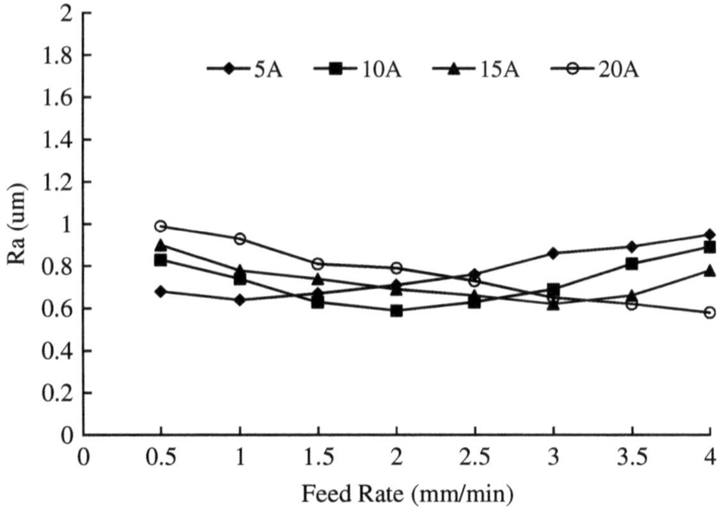

Fig. 3.114 Electrodepolishing at different feed rate of electrode and current rating (SKD61, Type Br, 0 rpm, 4 L/min, continuous DC, $t = 3$ mm, $\theta' = 40°$, $\lambda = 0.5$ mm, workpiece 0 rpm) ([70], reprinted with permission)

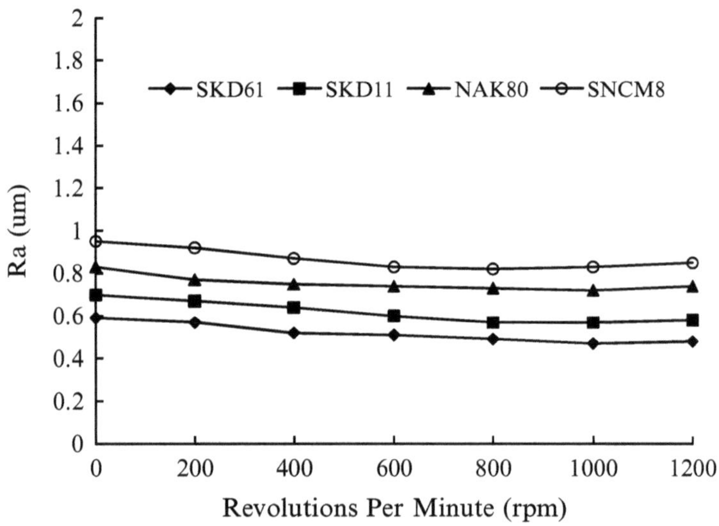

Fig. 3.115 Electropolishing at different rotational speed of workpiece (Type Br, 0 rpm, 4 L/min, continuous DC, 10 A, 2 mm/min, $t = 3$ mm, $\theta' = 40°$, $\lambda = 0.5$ mm) ([70], reprinted with permission)

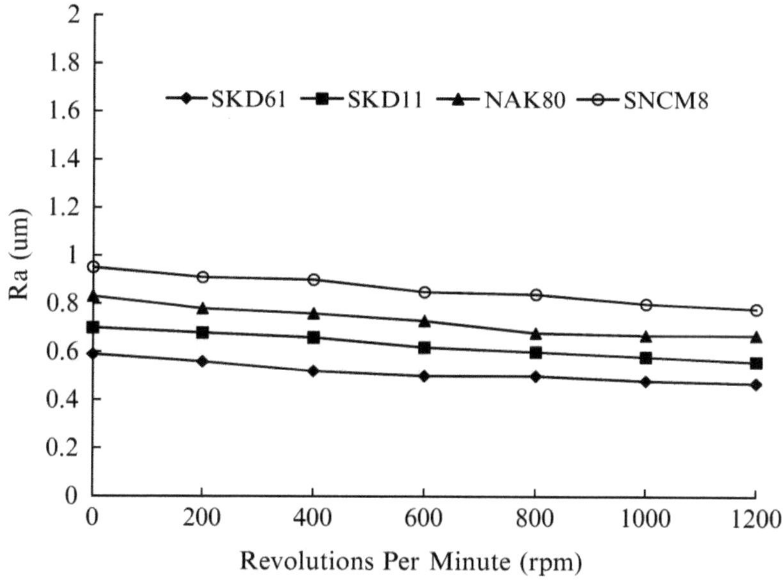

Fig. 3.116 Electropolishing at different rotational speed of electrode (Type Br, 4 L/min, continuous DC, 10 A, 2 mm/min, $t = 3$ mm, $\theta' = 40°$, $\lambda = 0.5$ mm, workpiece 0 rpm) ([70], reprinted with permission)

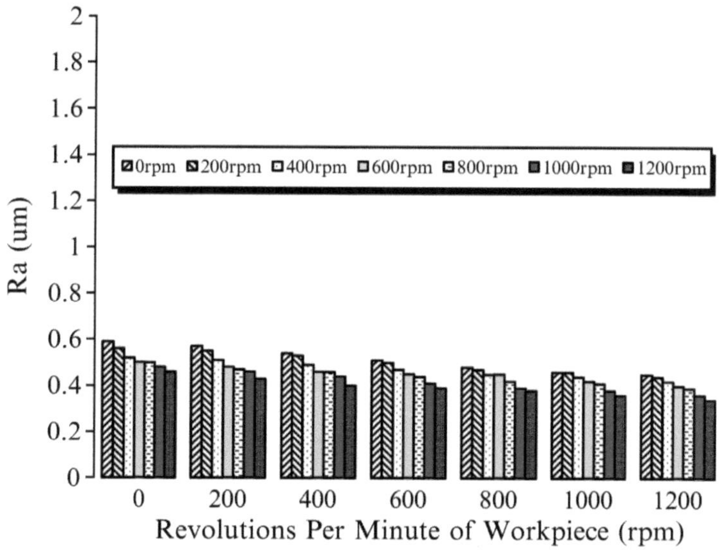

Fig. 3.117 Electropolishing at different rotational speed of electrode and workpiece (SKD61, Type Br, 4 L/min, continuous DC, 10 A, 2 mm/min, $t = 3$ mm, $\theta' = 40°$, $\lambda = 0.5$ mm) ([70], reprinted with permission)

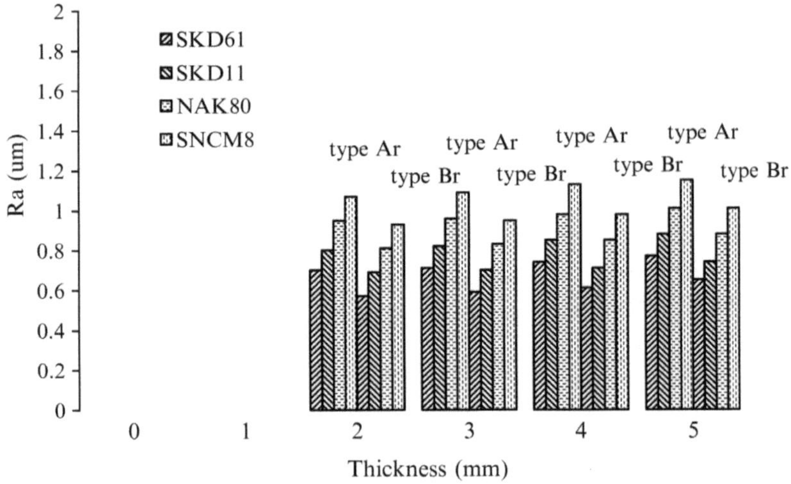

Fig. 3.118 Electropolishing at different electrode thickness (0 rpm, 4 L/min, continuous DC, 10 A, 2 mm/min, $\theta' = 40°$, $\lambda = 0.5$ mm, workpiece 0 rpm) ([70], reprinted with permission)

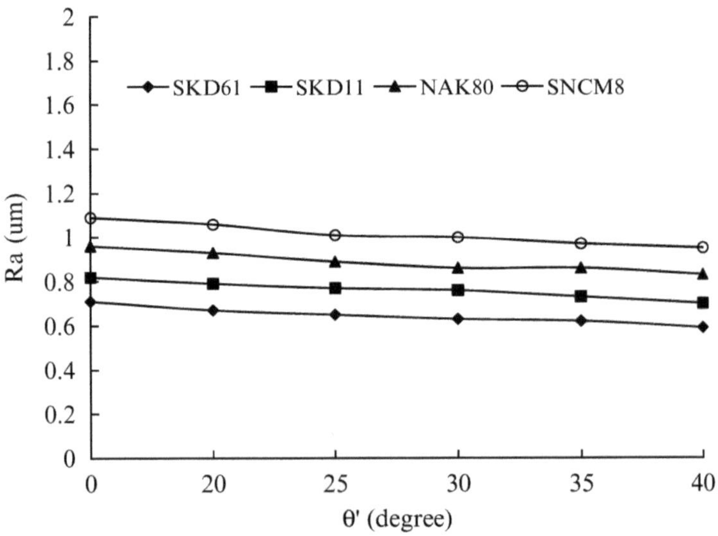

Fig. 3.119 Electropolishing at different taper angle (Type Br, 0 rpm, 4 L/min, continuous DC, 10 A, 2 mm/min, $t = 3$ mm, $\lambda = 0.5$ mm, workpiece 0 rpm) ([70], reprinted with permission)

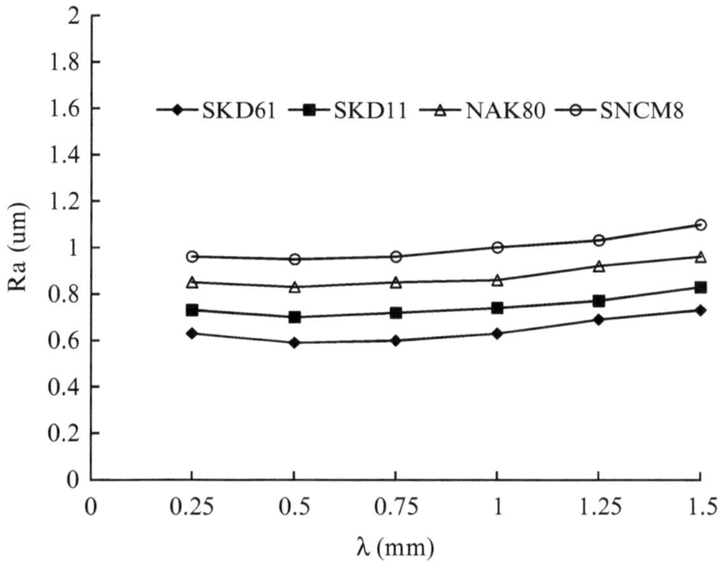

Fig. 3.120 Electropolishing at different inner radius (Type Br, 0 rpm, 4 L/min, continuous DC, 10 A, 2 mm/min, t = 3 mm, θ' = 40°, workpiece 0 rpm) ([70], reprinted with permission)

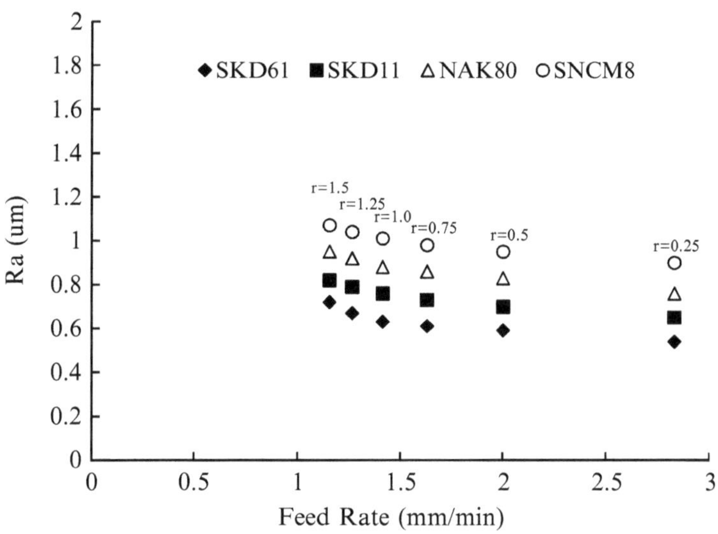

Fig. 3.121 Electropolishing at different inner radius and feed rate (Type Br, 0 rpm, 4 L/min, continuous DC, 10A, workpiece 0 rpm) ([70], reprinted with permission)

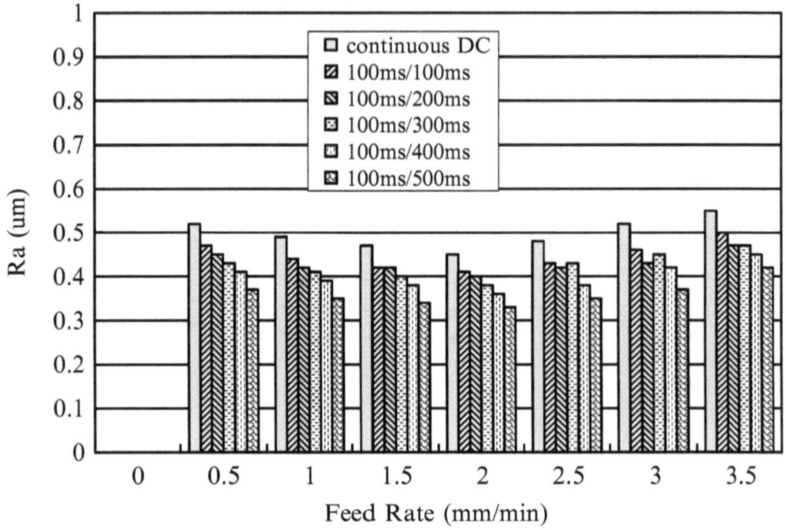

Fig. 3.122 Electropolishing using two techniques at different feed rate (SKD61, Type Cr, four pins, 600 rpm, 4 L/min, 10 A, d = 3 mm, L=3 mm, ρ = 0.5 mm, workpiece 0 rpm)

The former is related to the optimal current density across the gap, while the latter provides time for removal of the dregs out of the gap.

As to the electrode type Ca, there are short pins inside the ring. The discharge space for the electrolytic products is much more increased, when the polishing is done at the pin top. Figures 3.123 and 3.124 show that a long pin and pin end radius of about 0.5 mm produce better finish. Long pins can increase discharge space. Whatever the pin end radius is too large or too small will cause insufficient or excessive current density, respectively. The polishing effect will not be improved when the number of pins increases, as shown in Fig. 3.125, because the dregs discharge is impeded. One also notices that the electrode with only one pin costs the least. Figure 3.126 shows that small diameter of pin is slightly advantageous for increasing the discharge space. The effects of different electrodes and the pulse direct current using different electrodes are shown in Fig. 3.127. Longer off-time is more advantageous, because the discharge of electrolytic dregs during the off-time is more effective. However, the total machining time and cost will increase with the prolonged off-time. The electrode type Cr with pins performs the best among three electrodes. Figure 3.128 compares the use of rotation of workpiece and electrode. Though the workpiece rotates with electrode is the most beneficial to the polishing, other arrangements are used in different occasions (see Table 3.10).

Figure 3.129 shows the contribution to surface finish improvement in use of full-ring electrodes obtained through the workpiece rotation (10%), electrode rotation (11%), electrode feed rate (14%), pulse direct current (12%), electrode thickness (5%), inner taper angle (9%), inner edge radius (11%), pin length (5%), pin end radius (13%), pin number (6%), and pin diameter (4%). The electrode feed rate and

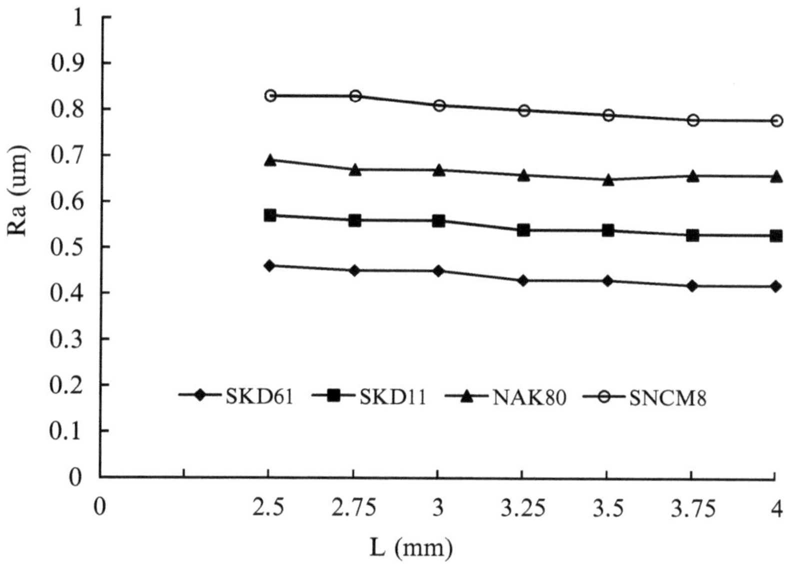

Fig. 3.123 Electropolishing at different pin length (Type Cr, 600 rpm, four pins, 4 L/min, continuous DC, 10 A, 2 mm/min, $d = 3$ mm, $\rho = 0.5$ mm, workpiece 0 rpm) ([70], reprinted with permission)

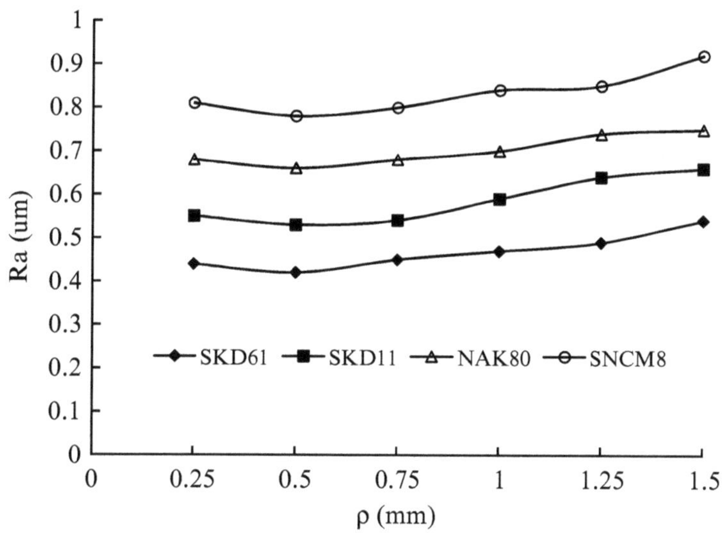

Fig. 3.124 Electropolishing at different pin end radius (Type Cr, 600 rpm, four pins, 4 L/min, continuous DC, 10 A, 2 mm/min, $d = 3$ mm, $L = 3$ mm, workpiece 0 rpm)

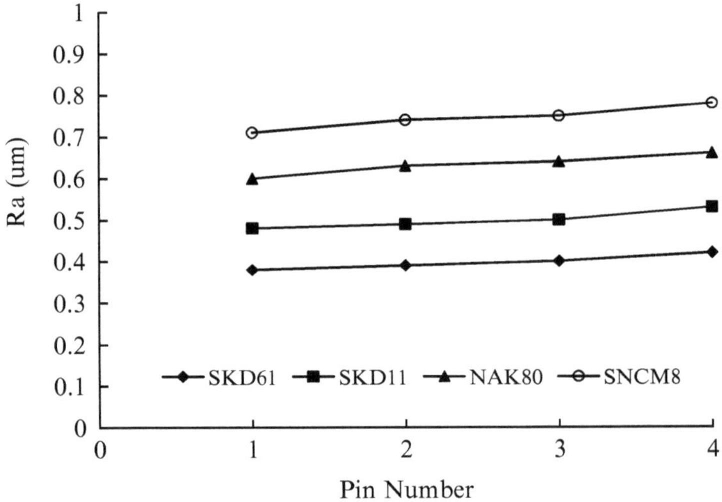

Fig. 3.125 Electropolishing at different number of pins (Type Cr, 600 rpm, 4 L/min, continuous DC, 10 A, 2 mm/min, $d = 3$ mm, $L = 3$ mm, $\rho = 0.5$ mm, workpiece 0 rpm) ([70], reprinted with permission)

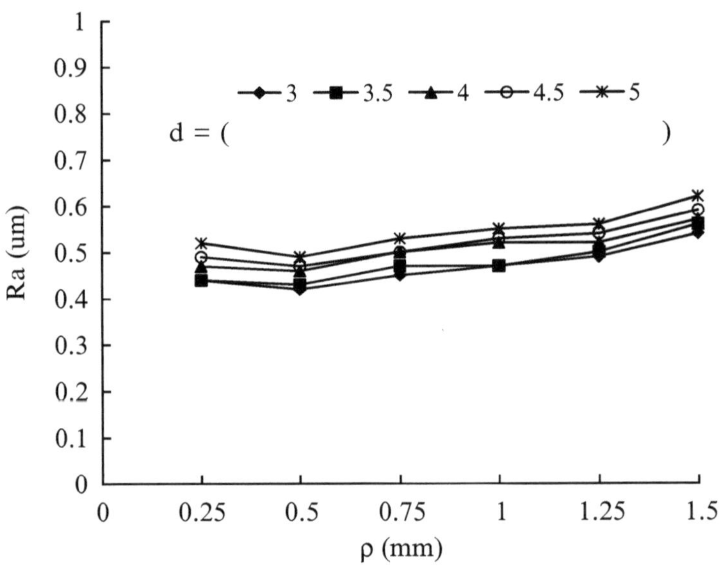

Fig. 3.126 Electropolishing at different pin end radius and pin diameter (Type Cr, 600 rpm, four pins, 4 L/min, continuous DC, 10A 2 mm/min, $d = 3$ mm, $L = 3$ mm, workpiece 0 rpm) ([70], reprinted with permission)

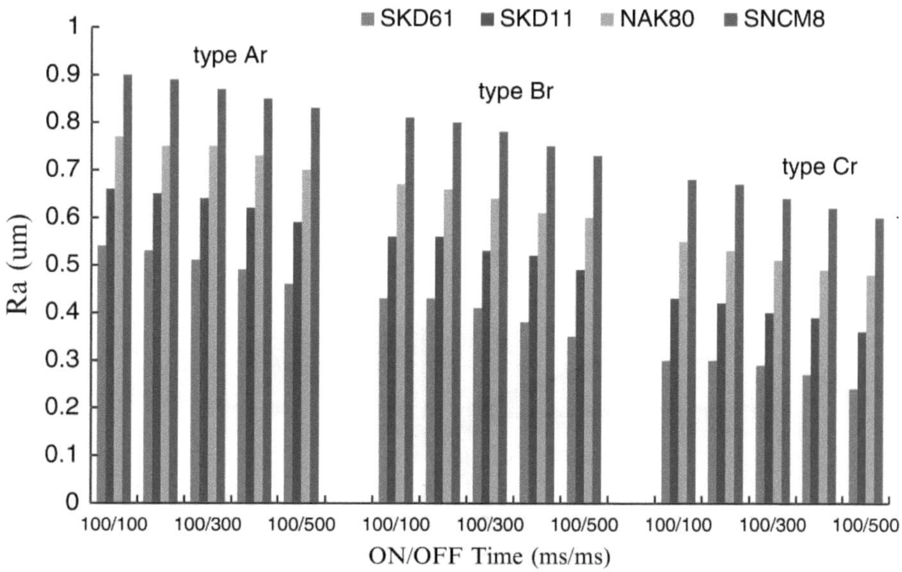

Fig. 3.127 Electropolishing with pulse direct current of different electrodes (600 rpm, 4 L/min, 10 A, 2 mm/min, workpiece 0 rpm)

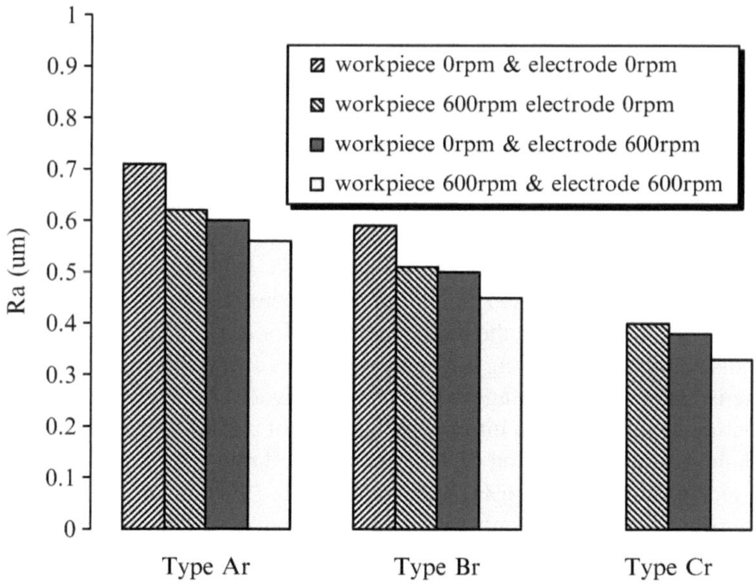

Fig. 3.128 Electropolishing at different arrangement for rotation of workpiece and electrodes ([70], reprinted with permission)

Table 3.10 Applications of ring electrodes

Rotation			
Workpiece	Electrode	Applicable process	Electrode
No	No	Rolling, drawing, extrusion, turning	Ar, Br, Cr
	Yes		
Yes	No	Turning	Ar, Br, Cr, Dr, Er, Fr
	Yes		Ar, Br, Cr

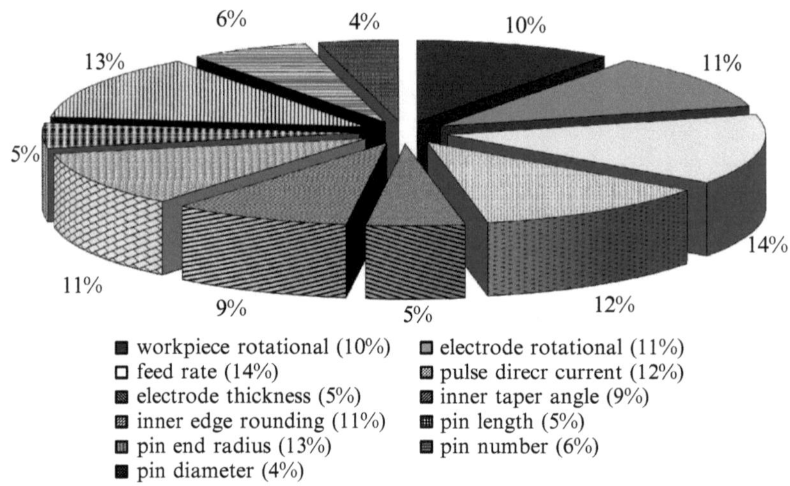

Fig. 3.129 The contribution pie of surface roughness improvement of full ring (SKD61, 600 rpm, 4 L/min, 10A, 2 mm/min, workpiece 600 rpm) ([70], reprinted with permission)

pin end radius contribute the most to the polishing. Decreased arc extension of electrode type Dr is associated with less restricted electrolyte flow and more discharge space, which are advantageous for the polishing, as shown in Fig. 3.130. Large inner radius (*R*) of type Er provides large discharge space hence slightly improves polishing (see Fig. 3.131). It is also advantageous for the polishing in use of decreased height of the electrodes Er and Fr, as shown in Figs. 3.132 and 3.133, since the discharge space for dregs is increased. The effects of the pulse direct current using different electrodes are shown in Fig. 3.134. Longer off-time is also more advantageous, similar to Fig. 3.127. Figure 3.135 shows the contribution to surface finish improvement in use of partial-ring electrodes obtained through the workpiece rotation (9%), electrode feed rate (15%), pulse direct current (12%), electrode thickness (6%), inner taper angle (9%), inner edge radius (11%), arc extension (7%), pin length (5%), pin end radius (12%), pin diameter (4%), electrode height (6%), and inner radius (4%). The electrode feed rate, pulse direct current, and pin end radius contribute the most to the polishing. Figure 3.136 shows the polishing results of electrodes from Ar to Fr for various materials. One finds that SKD61 is the most suitable material for this process and the electrodes Cr (with

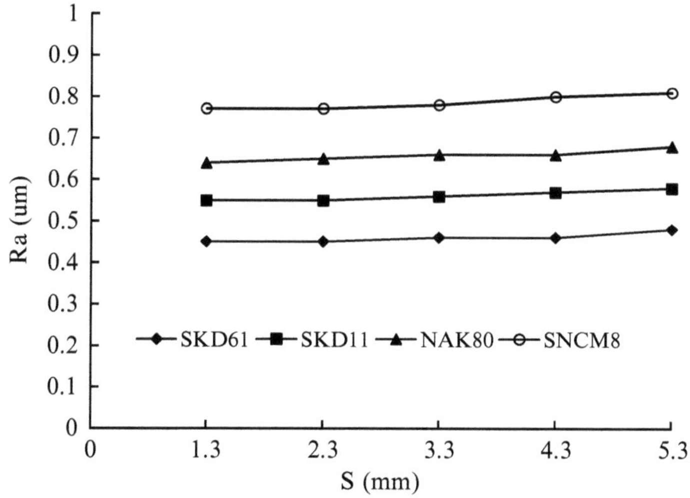

Fig. 3.130 Electropolishing using different arc extension of partial ring (Type Dr, 0 rpm, 4 L/min, continuous DC, 10 A, 2 mm/min, $t = 3$ mm, $\lambda = 0.5$ mm, $\theta' = 40°$, workpiece 600 rpm) ([70], reprinted with permission)

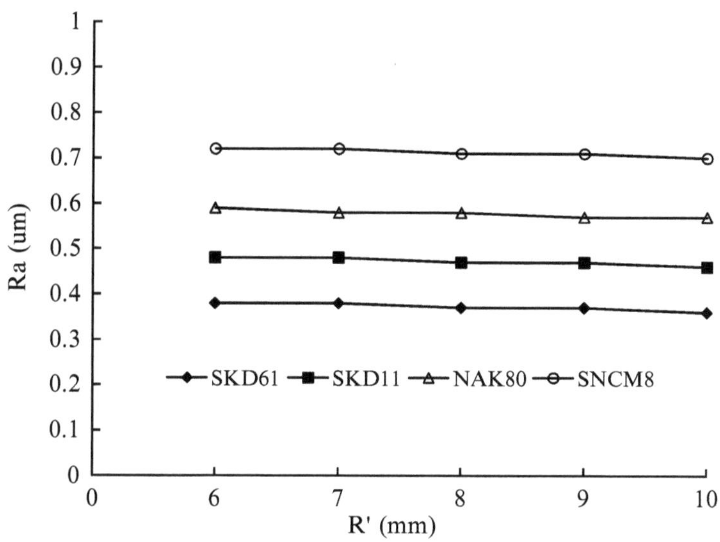

Fig. 3.131 Electropolishing at different inner edge of partial ring (Type Er, 0 rpm, 4 L/min, continuous DC, 10 A, 2 mm/min, $t = 3$ mm, $\lambda = 0.5$ mm, $\theta' = 40°$, $H = 6$ mm, workpiece 600 rpm)

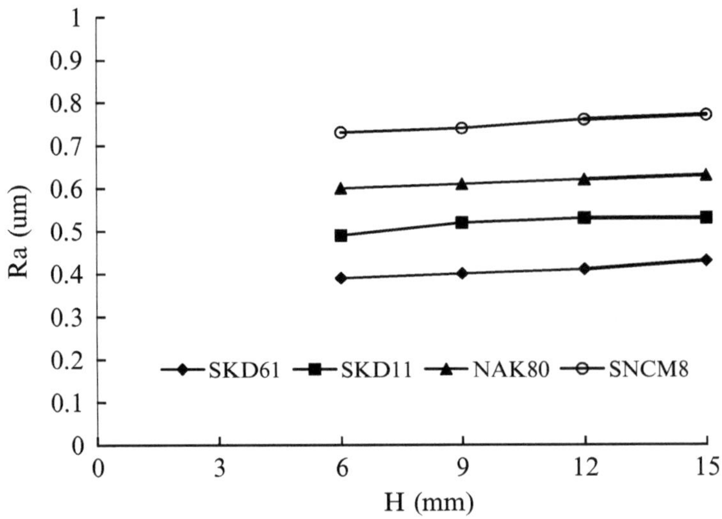

Fig. 3.132 Electropolishing at different height of partial ring Er (0 rpm, 4 L/min, continuous DC, 10 A, 2 mm/min, $t = 3$ mm, $\lambda = 0.5$ mm, $\theta' = 40°$, $R = 6$ mm, workpiece 600 rpm)

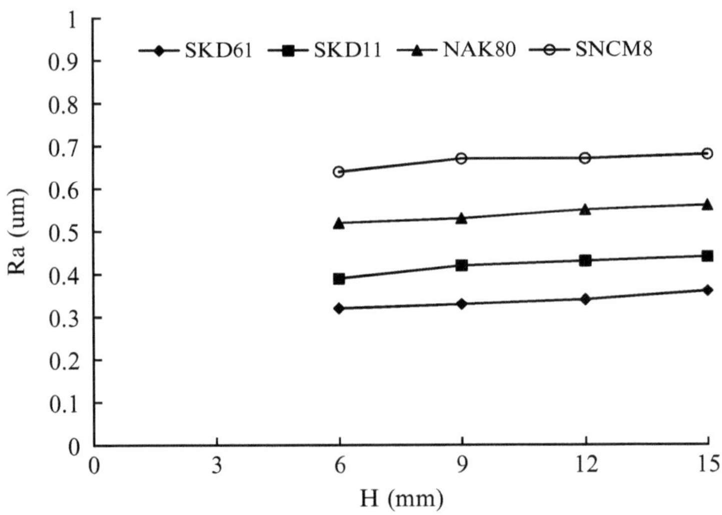

Fig. 3.133 Electropolishing at different height of partial ring Fr (0 rpm, 4 L/min, continuous DC, 10A, 2 mm/min, $d = 3$ mm, $L = 4$ mm, workpiece 600 rpm)

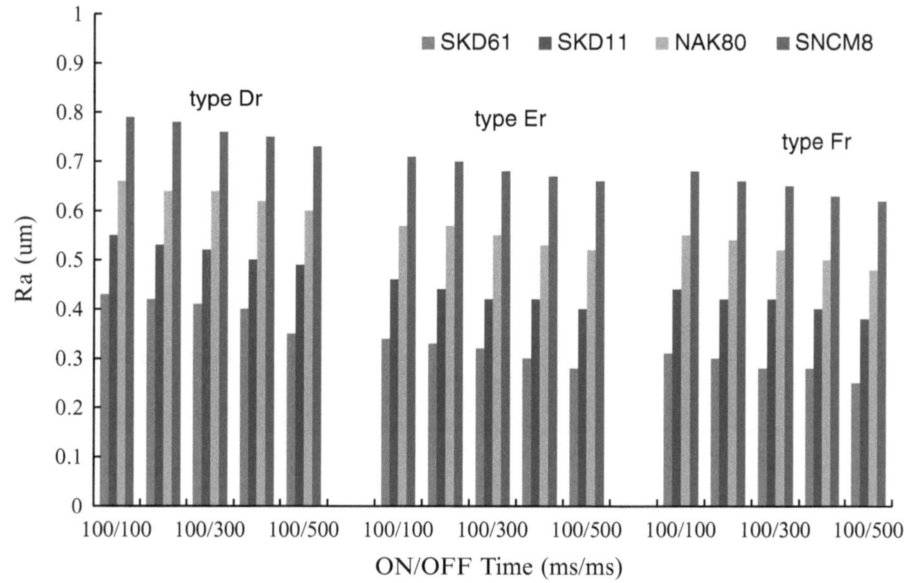

Fig. 3.134 Electropolishing with pulse direct current of different electrodes (SKD61, 0 rpm, 4 L/min, continuous DC, 10A, 2 mm/min, workpiece 600 rpm)

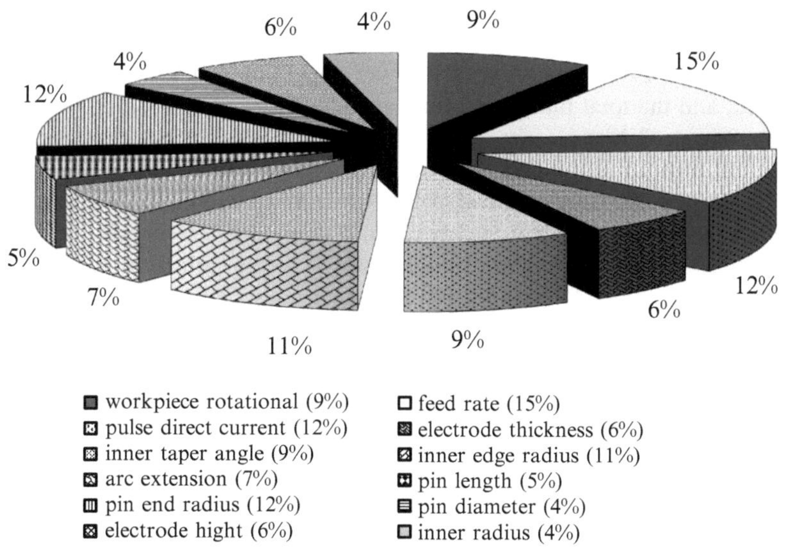

Fig. 3.135 The contribution pie of surface roughness improvement of partial ring (SKD61, 0 rpm, 4 L/min, 10 A, 2 mm/min, workpiece 600 rpm)

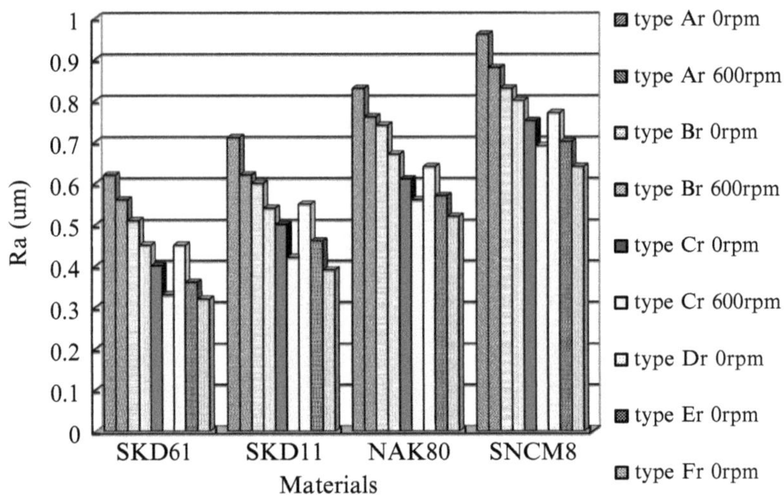

Fig. 3.136 Electropolishing for various materials (4 L/min, continuous DC, 10 A, 2 mm/min, workpiece 600 rpm)

rotation) and Fr are the best design. SKD61 achieves the fastest feed in electropolishing, when the amount of diameter reduction of other materials remains the same, as shown in Fig. 3.137. Figure 3.138 again confirms the effect of the optimal current rating in use of all electrodes. Pulse current has been found advantageous in electropolishing considering the discharge of the electrolytic product (see Figs. 3.127 and 3.134). However, a power supply with pulse function is more expensive, and the total machining time and cost will increase with the prolonged off-time. Figure 3.139 clearly shows the application of pulse current (100 ms/500 ms) to type Cr or Fr can only improve the surface roughness by 0.07 μm, while the design change of electrode from Ar to Cr or Dr to Fr produces much more significant improvement. Thus the use of pulse current is a secondary choice.

The improvement of surface roughness by electropolishing using different electrodes is compared in Fig. 3.140. The electrodes Cr and Fr perform the best among all features. In summary, the average contribution of surface finish improvement obtained through the workpiece rotation (9%), the electrode rotation (10%), using pulse direct current (12%), the electrode feed rate (15%), and the electrode design (54%) is shown in Fig. 3.141. The design of tool geometry leads to 54% of the improvement, compared to 46% from the influence of the process parameters, such as pulse direct current and motion control of the anode workpiece and the cathode electrode. For the purpose of practical use, the following two sets of conditions are suggested. First, one uses electrode Cr and 4 L/min, continuous DC, 10 A, 2 mm/min, $r' = 0.5$ mm, $d = 3$ mm, $L = 3$ mm, workpiece and electrode 600 rpm, one pin on electrode. Or, one uses electrode Fr nonrotational and 4 L/min, continuous DC, 10 A, 2 mm/min, $r' = 0.5$ mm, $d = 3$ mm, $L = 3$ mm, $H = 6$ mm, workpiece 600 rpm, one pin on electrode.

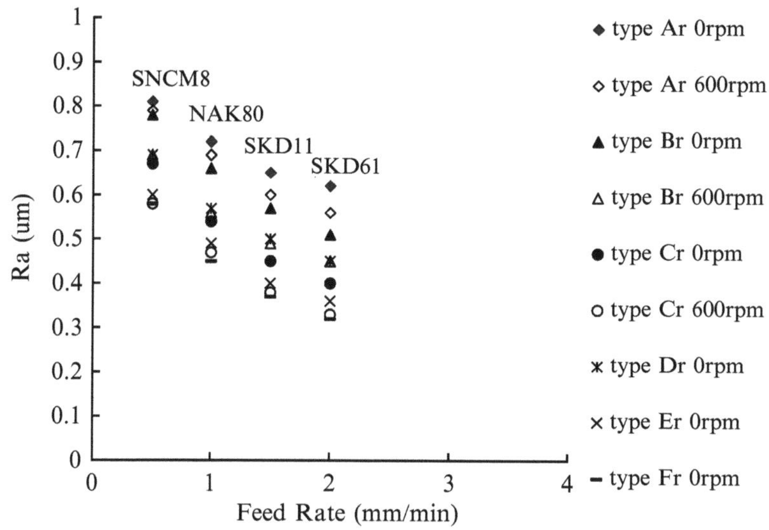

Fig. 3.137 Effects of workpiece materials on polishing feed rate (4 L/min, continuous DC, 10 A, workpiece 600 rpm)

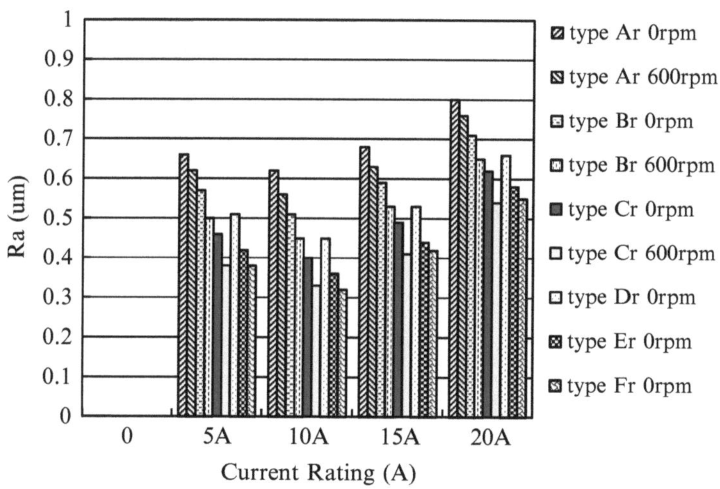

Fig. 3.138 Electropolishing at different current rating and electrode (SKD61, 4 L/min, continuous DC, 2 mm/min, workpiece 600 rpm)

3.11.3.4 Disc-Form Electrodes

Figure 3.142 shows the appropriate workpiece rotation (600–1,000 rpm) is advantageous for polishing. The polishing of SKD61 is the best, followed by SKD11, NAK80, and SNCM8. Figure 3.143 shows when the electrode type Ad faces the

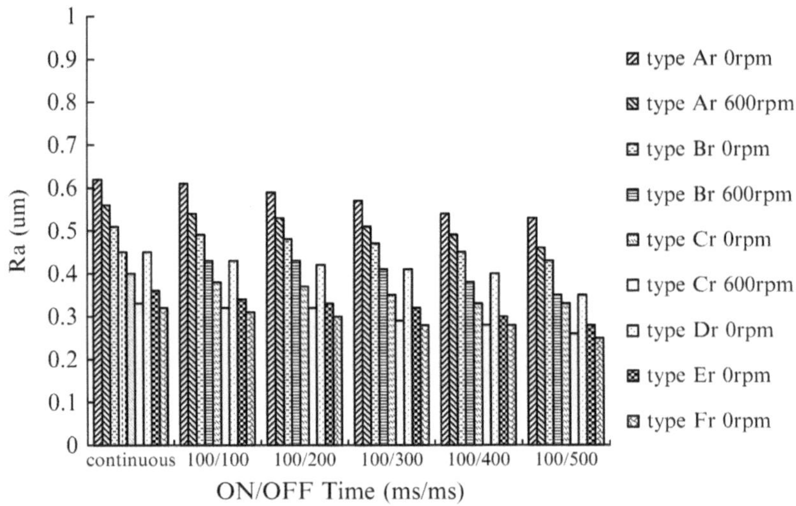

Fig. 3.139 Electrobrightening with different types of electrode using continuous and pulse direct current (4 L/min, continuous DC, 10 A, workpiece 600 rpm)

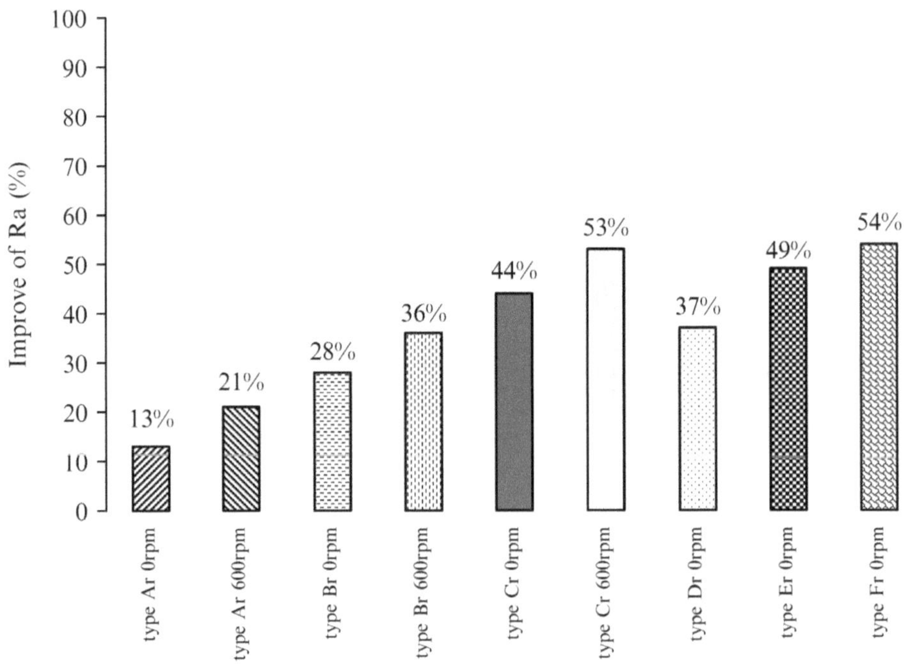

Fig. 3.140 The surface roughness improvement by use of different electrodes (4 L/min, continuous DC, 10 A, workpiece 600 rpm)

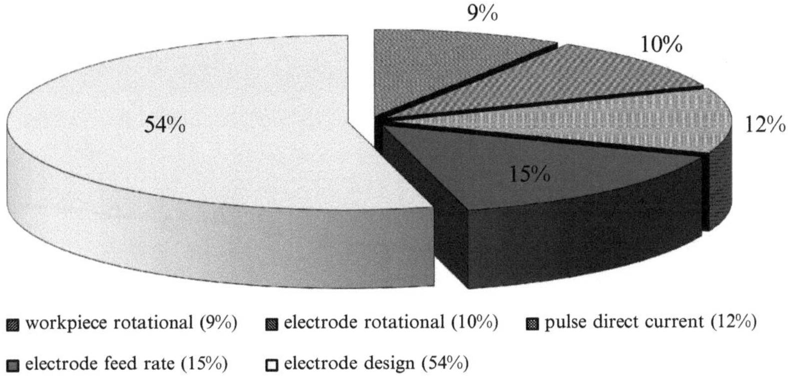

Fig. 3.141 The contribution pie of surface roughness improvement (SKD61, 4 L/min, 10 A, workpiece, and electrode 600 rpm)

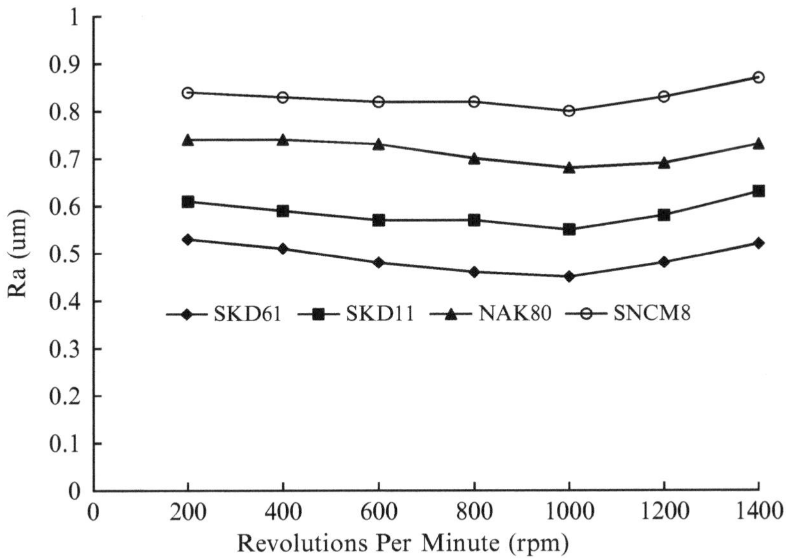

Fig. 3.142 Electropolishing at different rotational speed of workpiece (Type Ad, 0 rpm, 4 L/min, continuous DC, 10 A, 2 mm/min, t = 3 mm, μ = 40°, R = 0.5 mm)

opposite orientation, the polishing results are reversed. Figures 3.144, 3.145, and 3.146 show that good polishing can be achieved by adequate combination of electrode feed rate and current rating. An optimal effective current density in the working gap is achieved. Too low or too high current density causes insufficient polishing or overcutting, respectively. Figure 3.145 suggests the feed rate range of different workpiece materials is SKD61 1.5–2.5 mm/min, SKD11 1.0–2.0 mm/min, NAK80 0.5–1.5 mm/min, and SNCM8 0.5–1.0 mm/min. Figure 3.146 illustrates

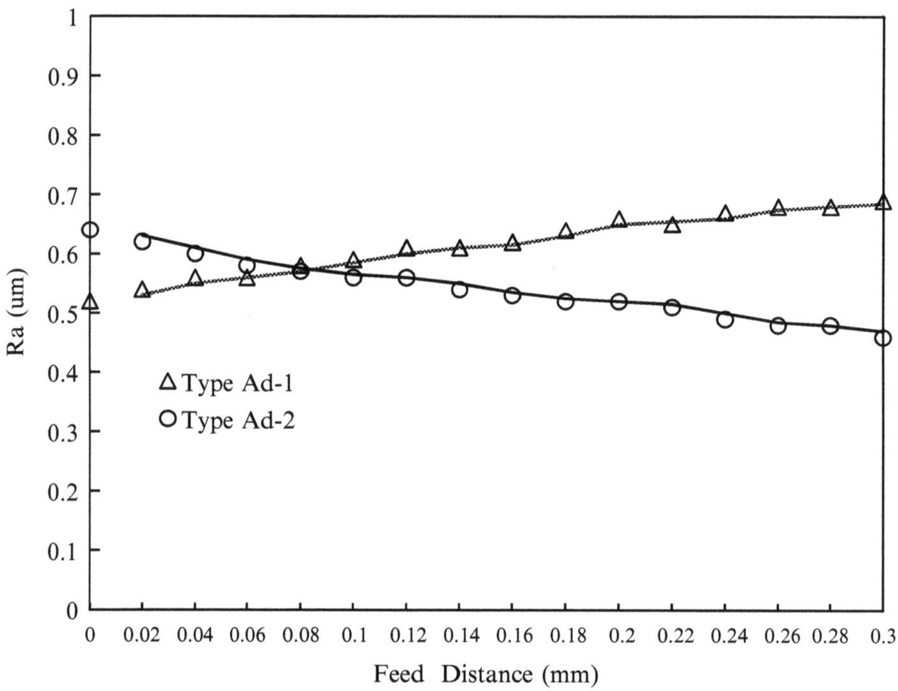

Fig. 3.143 Electropolishing compared with electrode orientation (Type Ad, 0 rpm, 4 L/min, continuous DC, 10 A, 2 mm/min, $t = 3$ mm, $\mu = 40°$, $R = 0.5$ mm, workpiece 600 rpm)

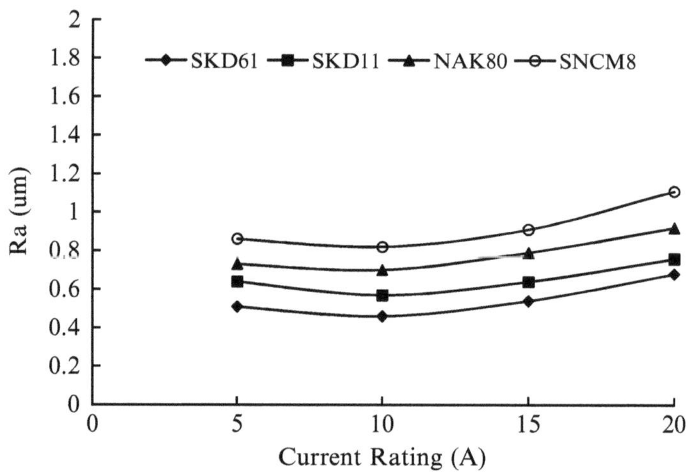

Fig. 3.144 Electropolishing at different current rating (Type Ad, 0 rpm, 4 L/min, continuous DC, 2 mm/min, $t = 3$ mm, $\mu = 40°$, $R = 0.5$ mm, workpiece 600 rpm)

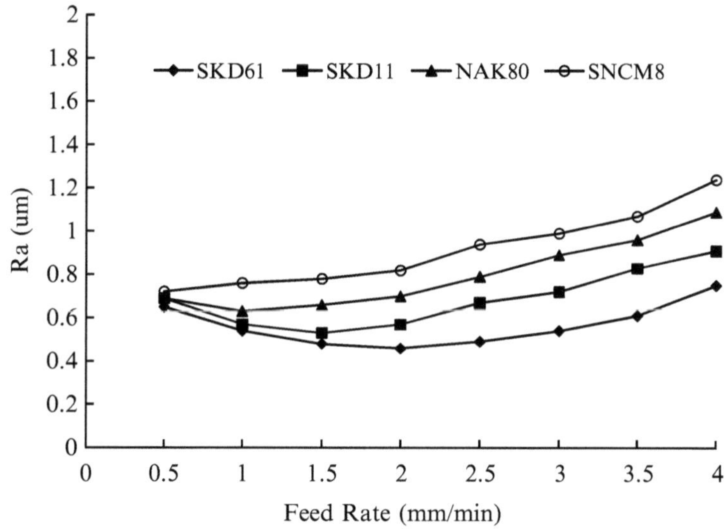

Fig. 3.145 Electropolishing at different feed rate of electrode (Type Ad, 0 rpm, 4 L/min, continuous DC, 10 A, t =3 mm, $\mu = 40°$, $R = 0.5$ mm, workpiece 600 rpm)

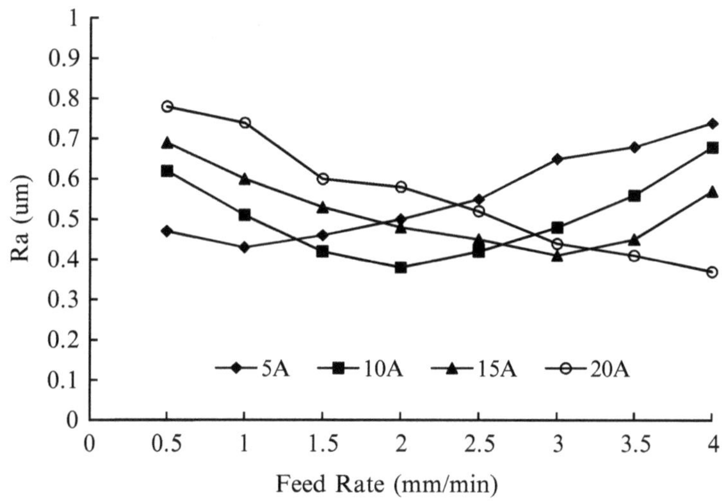

Fig. 3.146 Electrodepolishing at different feed rate of electrode (SKD61, Type Ad, 0 rpm, 4 L/min, continuous DC, $t = 3$ mm, $\mu = 40°$, $R = 0.5$ mm, workpiece 600 rpm)

good polishing of SKD61 can be achieved at adequate current rating and electrode feed rate, such as 5 A and 1 mm/min, 10 A and 2 mm/min, 15 A and 3 mm/min, and 20 A and 4 mm/min. Figure 3.147 shows the slower feed rate is, the larger is the reduction in diameter, since the time for material removal becomes longer. Figure 3.148 shows high rotational speed of electrode type A produces better

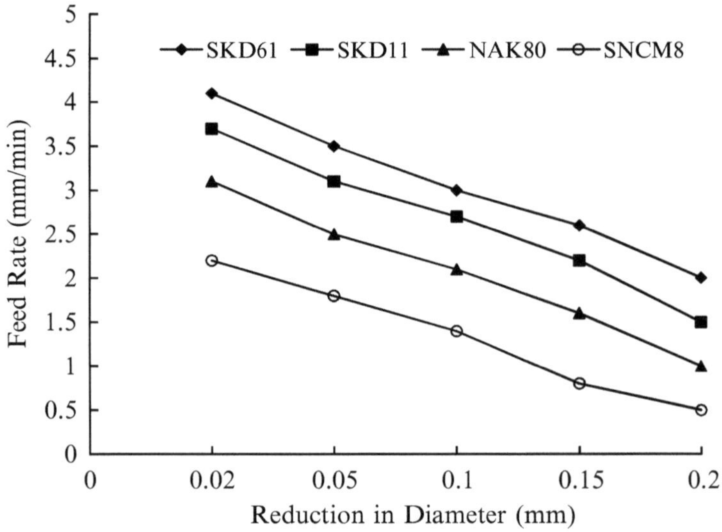

Fig. 3.147 Correlation between reduction in diameter and feed rate (SKD61, Type Ad, 0 rpm, 4 L/min, continuous DC, 10 A, 2 mm/min, $t = 3$ mm, $\mu = 40°$, $R = 0.5$ mm, workpiece 600 rpm)

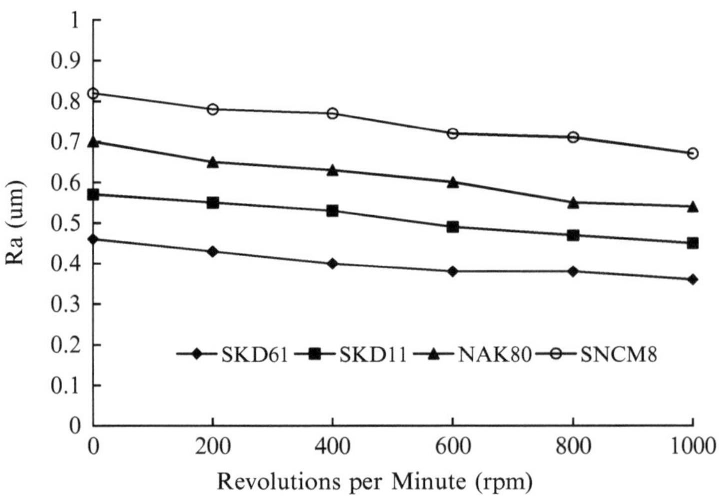

Fig. 3.148 Electropolishing at different rotational speed of electrode (Type Ad, 4 L/min, continuous DC, 10 A, 2 mm/min, $t = 3$ mm, $\mu = 40°$, $R = 0.5$ mm, workpiece 600 rpm)

polishing, since the rotational energy provides better discharge mobility. The adequate workpiece rotational speed (600–1,000 rpm) associated with higher electrode rotation produces better polishing, as shown in Fig. 3.149. The optimum is determined by the relative size between the electrode and workpiece. Figure 3.150 suggests the polishing using reduced disc electrode thickness, where the current

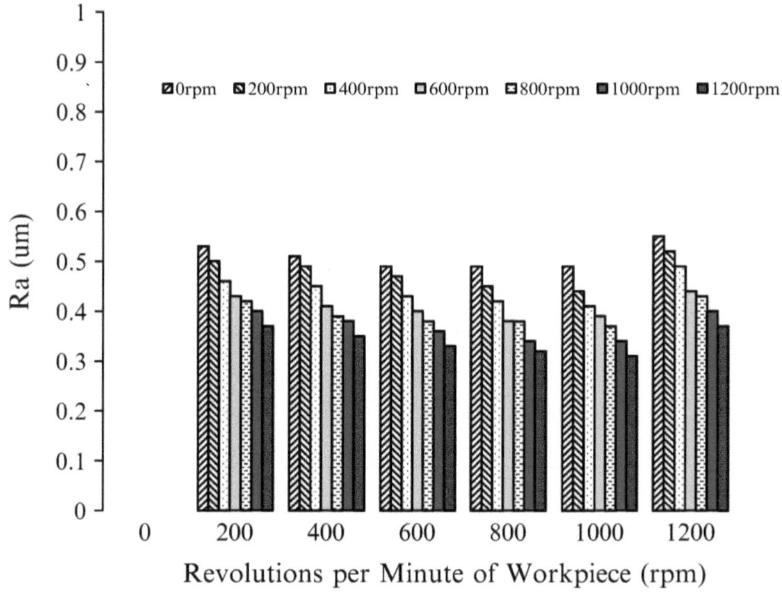

Fig. 3.149 Electropolishing at different rotational speed of electrode and workpiece (Type Ad, 4 L/min, continuous DC, 10 A, 2 mm/min, $t = 3$ mm, $\mu = 40°$, $R = 0.5$ mm) ([71], reprinted with permission)

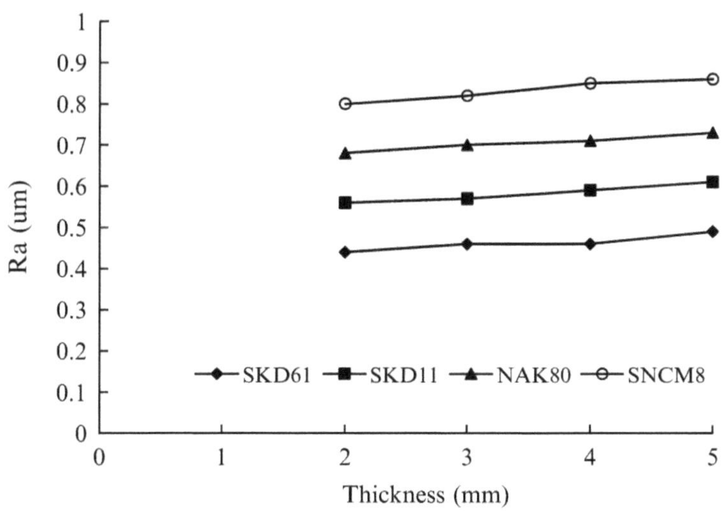

Fig. 3.150 Electropolishing at different electrode thickness (Type Ad, 0 rpm, 4 L/min, continuous DC, 10 A, 2 mm/min, $t = 3$ mm, $\mu = 40°$, $R = 0.5$ mm, workpiece 600 rpm)

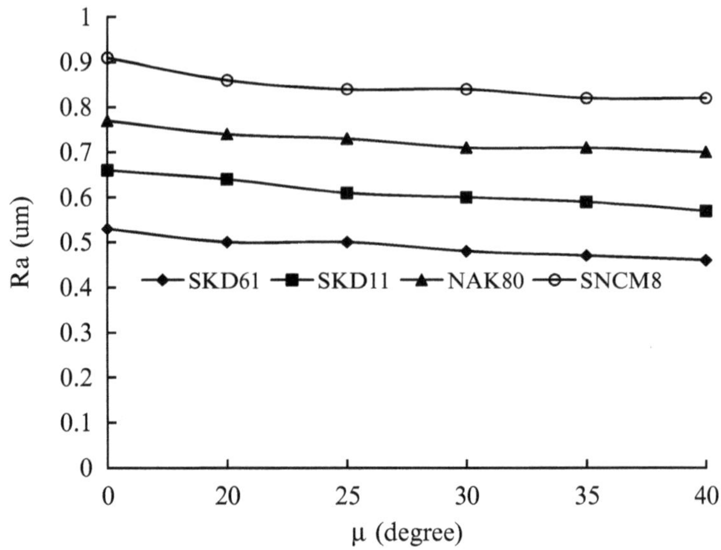

Fig. 3.151 Electropolishing at different taper angle (Type Ad, 0 rpm, 4 L/min, continuous DC, 10 A, 2 mm/min, $t = 3$ mm, $R = 0.5$ mm, workpiece 600 rpm)

density is increased and the secondary overcutting is reduced, the taper of the polished external cylindrical surface is controlled below 0.15° when the electrode thickness uses 3 mm. Figure 3.151 shows that the larger the taper angle is, the more effective is the polishing. Either reduced electrode thickness or increased taper angle releases the secondary overcutting. The adequate end radius (about 0.5–0.75 mm) should be taken into consideration, as shown in Fig. 3.152. If the end radius is too small (<0.5 mm), the current density becomes too large causing violent electrolytic reaction and nonuniformity of the surface finish. On the other hand, the large radius results in insufficient polishing energy and rough finish. It is worth noticing that small end radius associated with fast feed rate will keep the current density at the appropriate level; therefore better polishing can be obtained, as shown in Fig. 3.153. Figures 3.154 and 3.155 show that the large flute back rake angle and side rake angle slightly improve the polishing qualify for electrode type Bd and type Dd, since the flute incline can help the electropolishing products be discharged out of the fluted space, and the deep flute depth also provides mild improvement of the polishing (see Fig. 3.156). Figure 3.157 compares the polishing effect of different flute width of type Bd and type Dd. The ratio is defined as the flute width to the whole peripheral length. The results show that the wide flute is advantageous, since it provides sufficient space and time for dregs discharge, and reduces the process interference with the electropolishing products. Figure 3.158 further illustrates that the use of flute in the electrode type Bd with continuous direct current is comparable with the electrode type Ad with pulse direct current. The polishing effect of the latter is slightly (about 5%) better than the former, but the

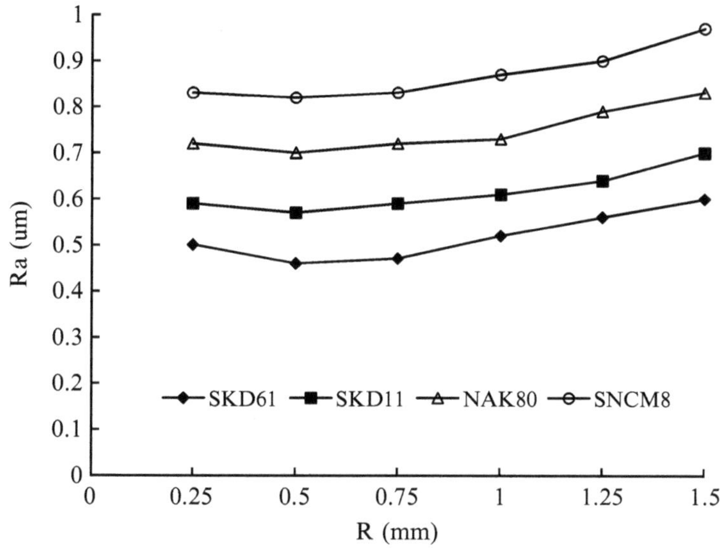

Fig. 3.152 Electropolishing at different disc end radius (Type Ad, 0 rpm, 4 L/min, continuous DC, 10 A, 2 mm/min, t = 3 mm, μ = 40°, workpiece 600 rpm)

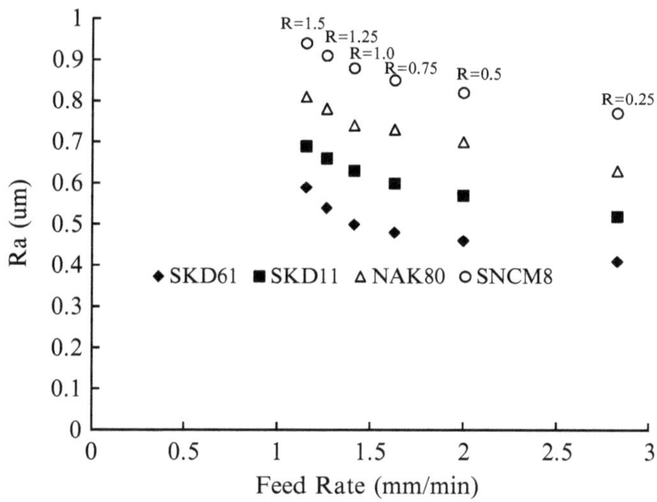

Fig. 3.153 Electropolishing at different edge radius and feed rate (Type Ad, 0 rpm, 4 L/min, continuous DC, 10 A, workpiece 600 rpm)

cost of the power supply is considerably higher and the process cycle time is increased. A good electrode design saves the production cost, as shown in this example. Figure 3.159 illustrates that adequate feed rate (1.5–2.5 mm/min) and longer pulse-off time or wider discharge flute are desired for fine finish.

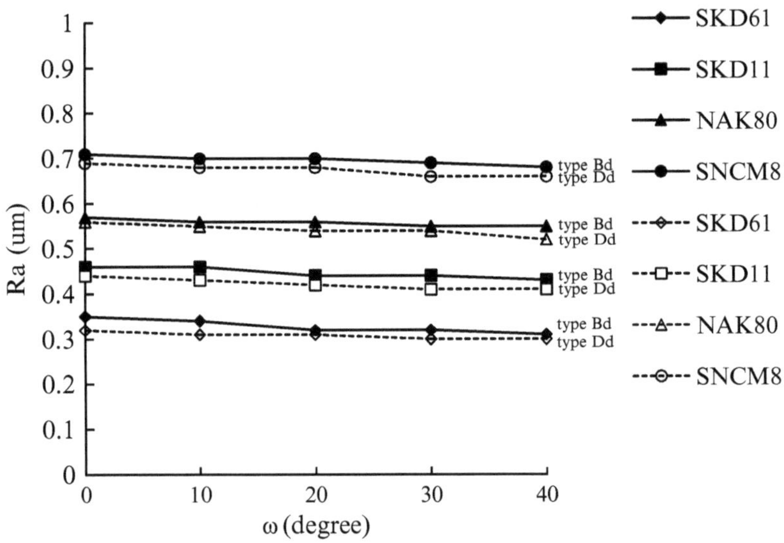

Fig. 3.154 Electropolishing with different flute back rake angle of electrode (Type B, 600 rpm, 4 L/min, continuous DC, 10 A, 2 mm/min, $t = 3$ mm, $h = 3$, μ, $\Psi = 40°$, $R = 0.5$ mm, workpiece 600 rpm)

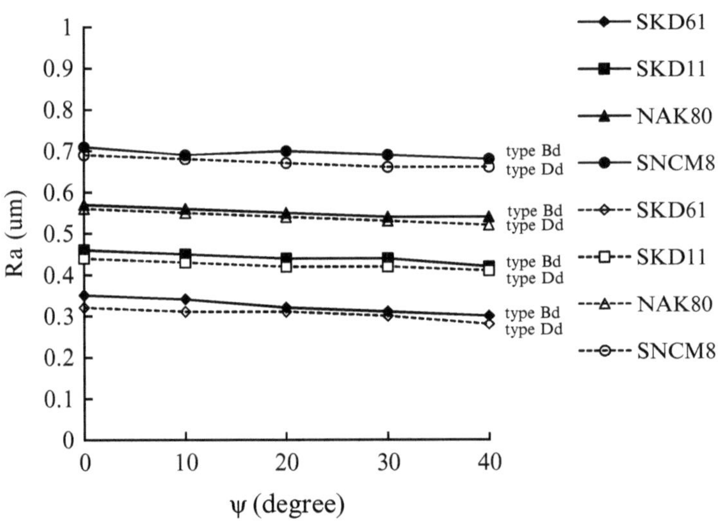

Fig. 3.155 Electropolishing with different flute side rake angle of electrode (Type Bd, 600 rpm, 4 L/min, continuous DC, 10 A, 2 mm/min, $t = 3$ mm, $h = 3$, μ and $\omega = 40°$, $R = 0.5$ mm, workpiece 600 rpm)

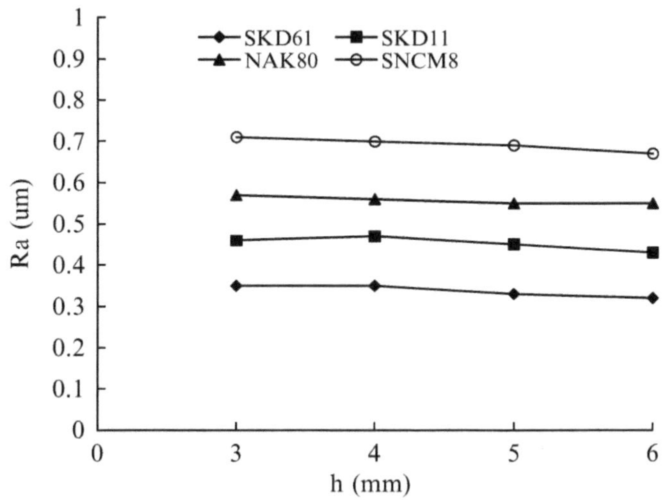

Fig. 3.156 Electropolishing at different electrode flute depth (Type Bd, 600 rpm, 4 L/min, continuous DC, 10 A, 2 mm/min, $t = 3$ mm, $\theta = 40°$, $R = 0.5$ mm, workpiece 600 rpm)

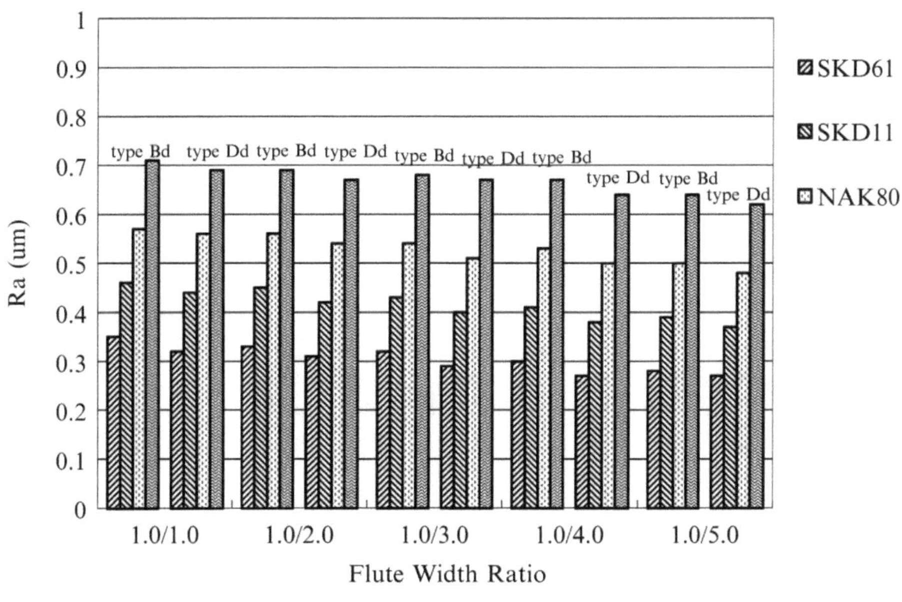

Fig. 3.157 Electropolishing with different flute width ratio of type Bd and type Dd (600 rpm, 4 L/min, continuous DC, 10 A, 2 mm/min, $t = 3$ mm, $h = 3$ mm, μ, ω, $\Psi = 40°$, R =0.5 mm, workpiece 600 rpm)

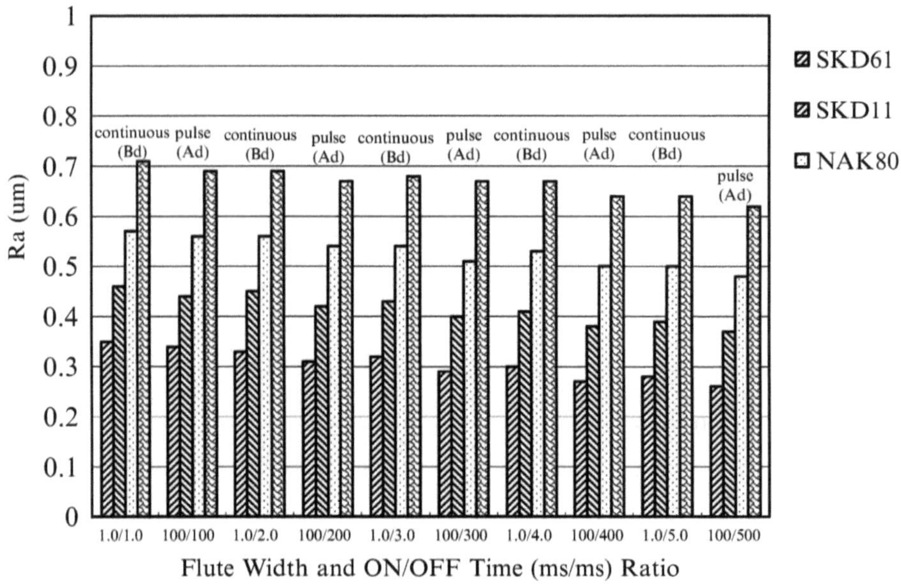

Fig. 3.158 Electropolishing with pulse direct current of type Ad and continuous direct current of type Bd (600 rpm, 4 L/min, 10 A, 2 mm/min, $t = 3$ mm, h = 3 mm, μ,ω, $\Psi = 40°$, $R = 0.5$ mm, workpiece 600 rpm)

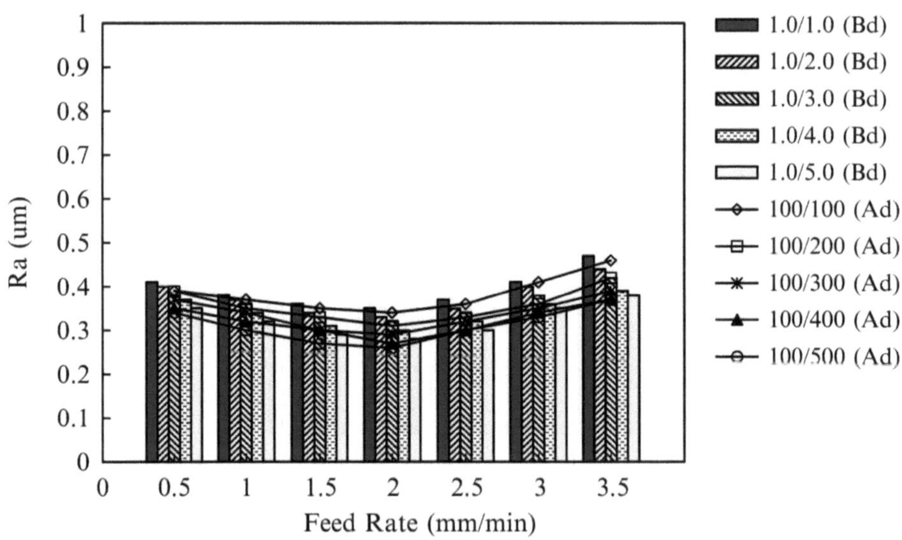

Fig. 3.159 Electropolishing using two techniques at different feed rate (SKD61, Type Ad and Bd, 600 rpm, 4 L/min, 10 A, $t = 3$ mm, $\theta = 40°$, $R = 0.5$ mm, workpiece 600 rpm)

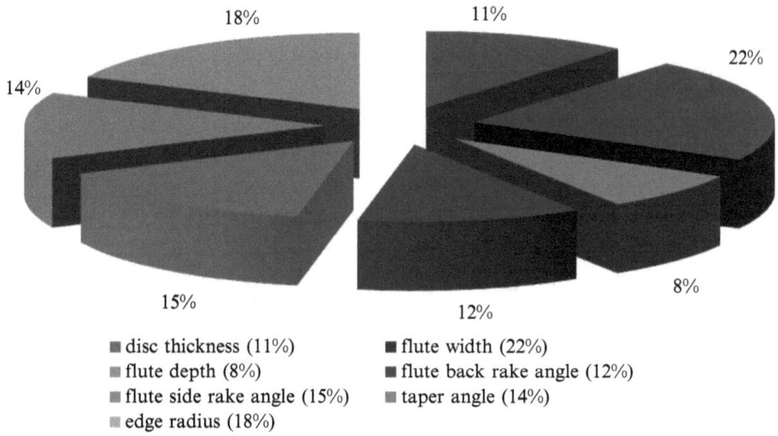

Fig. 3.160 The contribution pie of surface roughness improvement of electrode B (SKD61, 600 rpm, 4 L/min, continuous DC, 10 A, 2 mm/min, $t = 3$ mm, $h = 3$ mm, $\theta = 40°$, $R = 0.5$ mm, workpiece 600 rpm)

Figure 3.160 shows the contribution to surface finish improvement in use of electrode type Bd obtained through the disc thickness (11%), the flute width (22%), the flute depth (8%), the flute back rake angle (12%), the flute side rake angle (15%), the taper angle (14%), and the edge radius (18%). The flute width and edge radius contributes the most to the polishing.

As to the electrode type Ed, there are short pins attached on the disc. The discharge space for the electrolytic products is much more increased. Figures 3.161 and 3.162 show that long pin and appropriate pin end radius (about 0.5 mm) produce better polishing. Long pins can increase discharge space and reduce secondary overcutting. Whatever the pin end radius is too large or too small will produce insufficient or excessive current density, respectively, similar to the effects of the edge radius of electrode Bd. The polishing effect improves when the number of pins is decreased, as shown in Fig. 3.163. There is an average of 6–7% improvement for the number of pins decreased by one. The contribution to surface finishing in use of the electrode type Ed is shown in Fig. 3.164. Among the pin number (28%), the pin length (25%), and the pin end radius (47%), the pin end radius contributes the most. Figure 3.165 shows the polishing results of different type electrodes at the same feed rate. The electrode type Ed with pins attached on disc edge is the best among the five electrodes. Figure 3.166 shows the polishing results of different type electrodes at different current rating. The current rating of 10 A produces the best polishing. Based on the results obtained from Fig. 3.98, different materials are tested at their adequate feed rates, and the performance of various electrode design is clearly demonstrated, as shown in Fig. 3.167.

In summary, the improvement of surface roughness using different electrodes is shown in Fig. 3.168. The electrode type Ed with 600 rpm performs the best among all features.

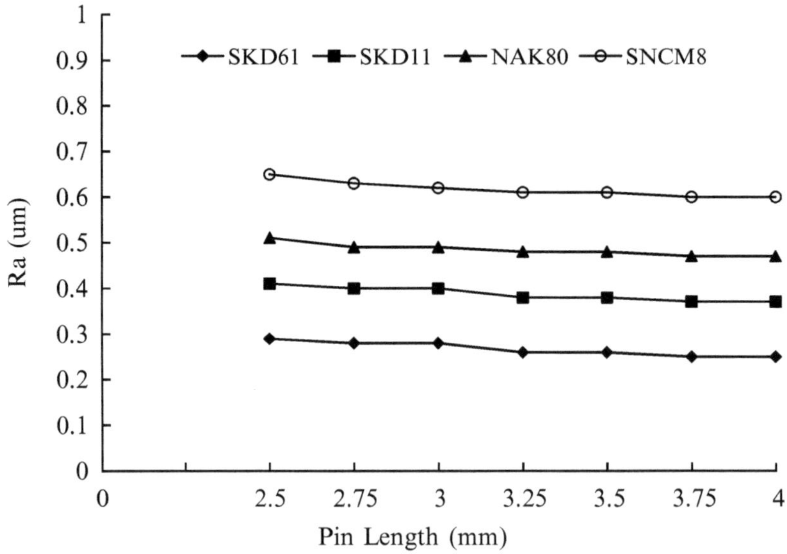

Fig. 3.161 Electropolishing at different pin length (type Ed, 600 rpm, four pins, 4 L/min, continuous DC, 10 A, 2 mm/min, $t = 3$ mm, $\theta = 40°$, $R = 0.5$ mm, workpiece 600 rpm)

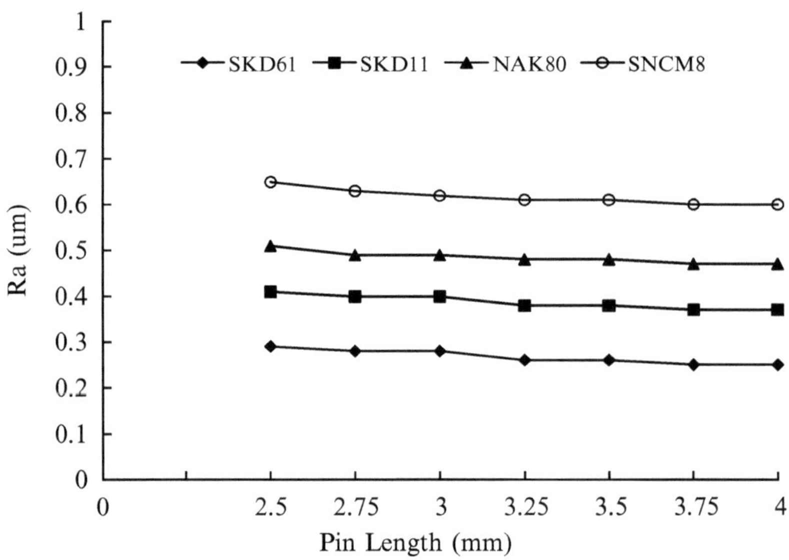

Fig. 3.162 Electropolishing at different pin end radius (type Ed, 600 rpm, four pins, 4 L/min, continuous DC, 10 A, 2 mm/min, $t = 3$ mm, $\theta = 40°$, workpiece 600 rpm)

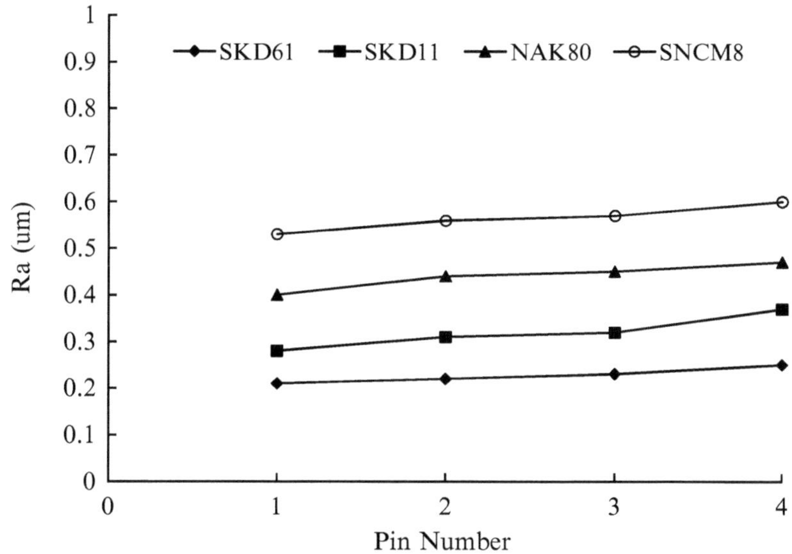

Fig. 3.163 Electropolishing at different number of pins (Type Ed, 600 rpm, 4 L/min, continuous DC, 10 A, 2 mm/min, t = 3 mm, θ = 40°, R = 0.5 mm, workpiece 600 rpm)

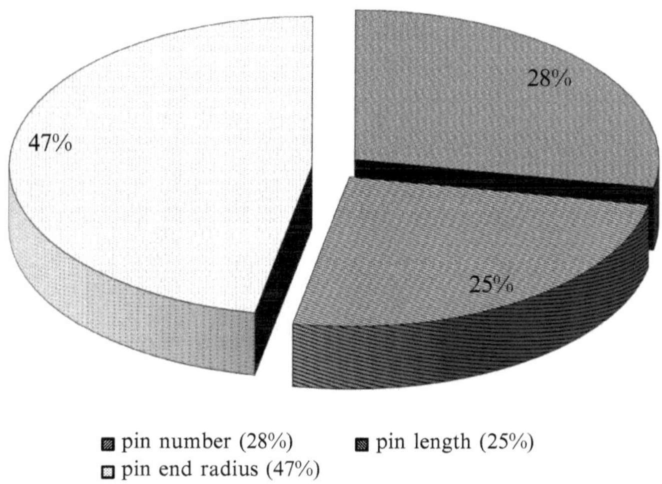

Fig. 3.164 The contribution pie of surface roughness improvement of type Ed (SKD61, 600 rpm, 4 L/min, continuous DC, 10A, t = 3 mm, 2 mm/min, workpiece 600 rpm)

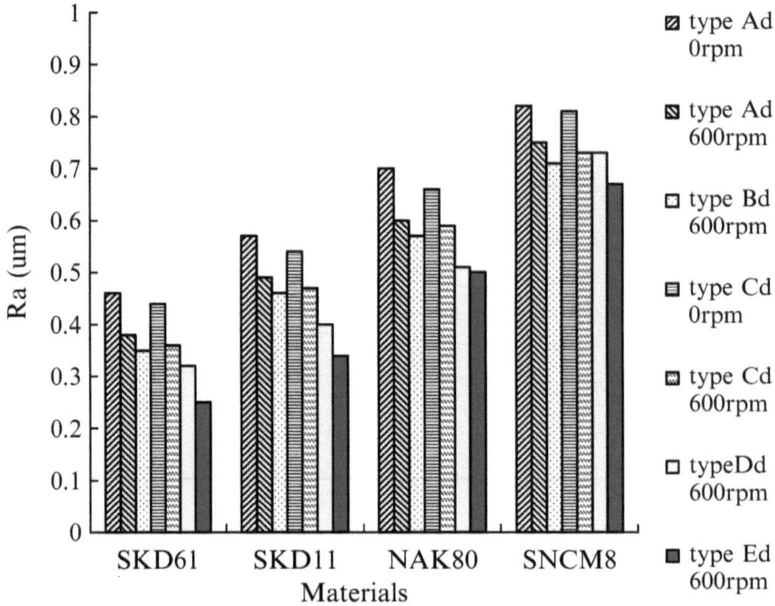

Fig. 3.165 Electropolishing for various materials (4 L/min, continuous DC, 10 A, 2 mm/min, $t = 3$ mm, $R = 0.5$ mm, workpiece 600 rpm)

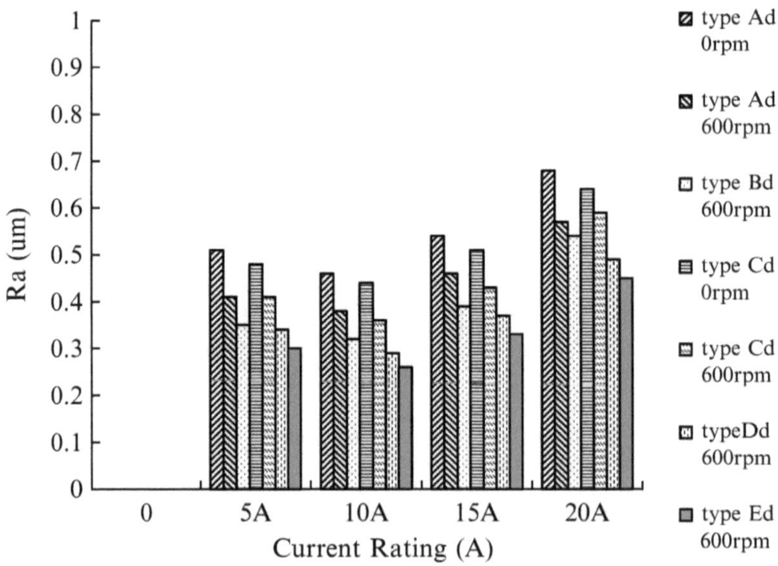

Fig. 3.166 Electropolishing at different current rating and electrode (SKD61, 4 L/min, continuous DC, 2 mm/min, $t = 3$ mm, $R = 0.5$ mm, workpiece 600 rpm)

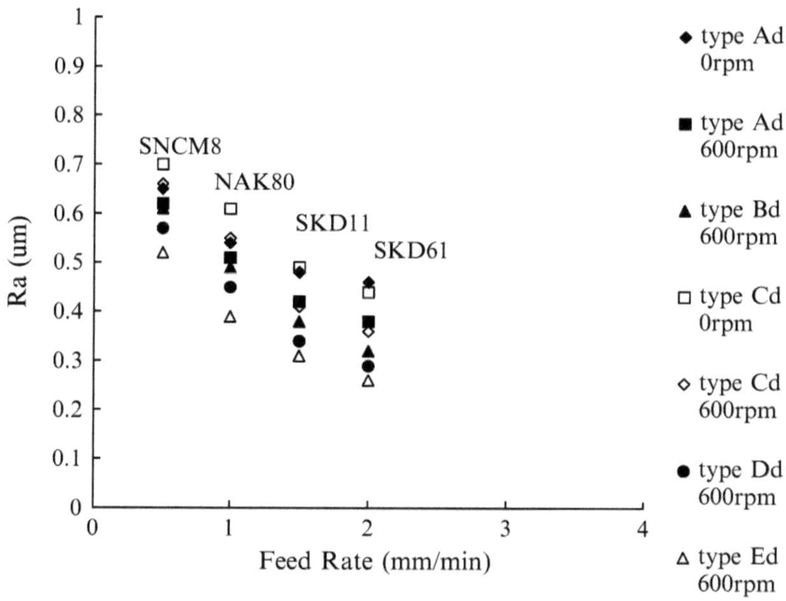

Fig. 3.167 Electropolishing with different types electrode at the same amount of material removal (4 L/min, continuous DC, 10 A, $R = 0.5$ mm, $t = 3$ mm, workpiece 600 rpm)

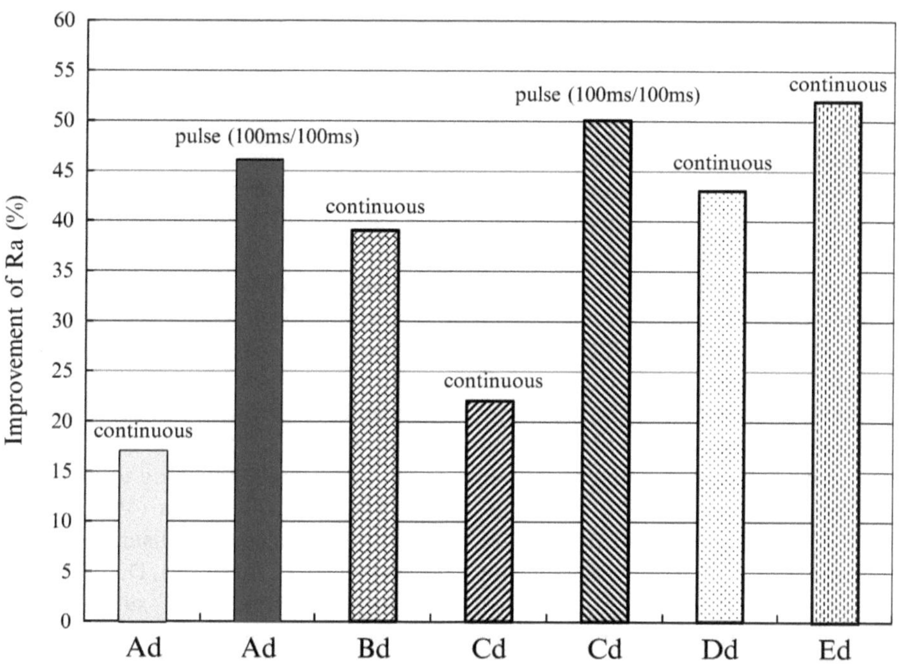

Fig. 3.168 The surface roughness improvement by use of different electrode (600 rpm, 4 L/min, 10 A, 2 mm/min, $t = 3$ mm, $R = 0.5$ mm, workpiece 600 rpm)

3.11.4 Ultrasonic-Aided Electropolishing

Electropolishing (EP), pulse-electropolishing (PEP), ultrasonic-aided electropolishing (UEP), and ultrasonic-aided pulse-electropolishing (UPEP) are compared. Two groups (mate and feeding) of six electrodes are used for comparison. The use of ultrasonics helps to elevate the polishing quality (see Fig. 3.169); higher frequency and greater power produce better polishing. Figure 3.170 shows the appropriate rotation (400–800 rpm) of electrode is advantageous for electropolishing (EP), while the ultrasonic-aided electropolishing (UEP) requires lower rotational speed to achieve the same polishing. Namely the ultrasonics is more beneficial to the higher rotational speed. Among the six electrodes, the best polishing comes from the electrode with large discharge space, namely the electrode Fm with round pin of off-center performs the best (see Fig. 3.171). The comparison of improvement of surface roughness in electropolishing (EP) using different electrodes is shown in Fig. 3.172. The best result in electropolishing is obtained by the electrode type Fm (70%), followed by Em (59%), Dm (51%), Cm (38%), Bm (25%), and Am (17%). In case of ultrasonic-aided pulse-electropolishing, it elevates the polishing effect in average of 24%. The improvement ratios are Fm (72%), Em (61%), Dm (53%), Cm (40%), Bm (28%), and Am (19%). The improvement of ultrasonic-aided pulse-electropolishing only reaches 3% compared to the ultrasonic-aided electropolishing. The average contribution of surface finish improvement obtained through the electrode rotation (12%), the pulse (16%), the ultrasonic (21%), and the electrode geometry (51%) is shown in Fig. 3.173. As mentioned above, the electrode type Fm with ultrasonic-aided pulse-electropolishing can achieve the best polishing effect. Since the resting period (pulse-off time) for dregs discharge takes more working time, one should use ultrasonic-aided electropolishing to save the polishing time.

The second part of the experiments undergoes polishing by the electrodes of three groups, i.e., ring-form electrodes, arrow-head electrodes, and disc-form electrodes. Both the workpiece and the electrode rotate. Figure 3.174 shows that the best polishing comes from the electrode Fu and Eu (disc-form), followed by Du and Cu (arrow-head), and Bu and Au (ring-form). The disc-form electrodes provide the widest open space than other electrodes, thus produce the best polishing. The arrow-head electrodes has discharge flute of double slopes, which is beneficial to dregs discharge. The ring-form electrodes appear in a form of non-open space, which can provide an electropolishing process for workpiece without rotation (such as drawing, form rolling, or extrusion). Although each group of the electrodes has different polishing effect, its application varies. Figure 3.175 shows the improvement ratios of surface roughness by ultrasonic-aided electropolishing using different electrodes are Fu (58%), Eu (56%), Du (54%), Cu (49%), Bu (4%), and Au (27%). The ultrasonic-aided pulse-electropolishing elevates the polishing effect in average at to 22%. The improvement ratios are Fu (60%), Eu (60%), Du (56%), Cu (53%), Bu (44%), and Au (31%). The average contribution of surface finish improvement obtained through pulse (14%), ultrasonics (20%), electrode rotation (13%), and electrode geometry (53%) is shown in Fig. 3.176. One should use

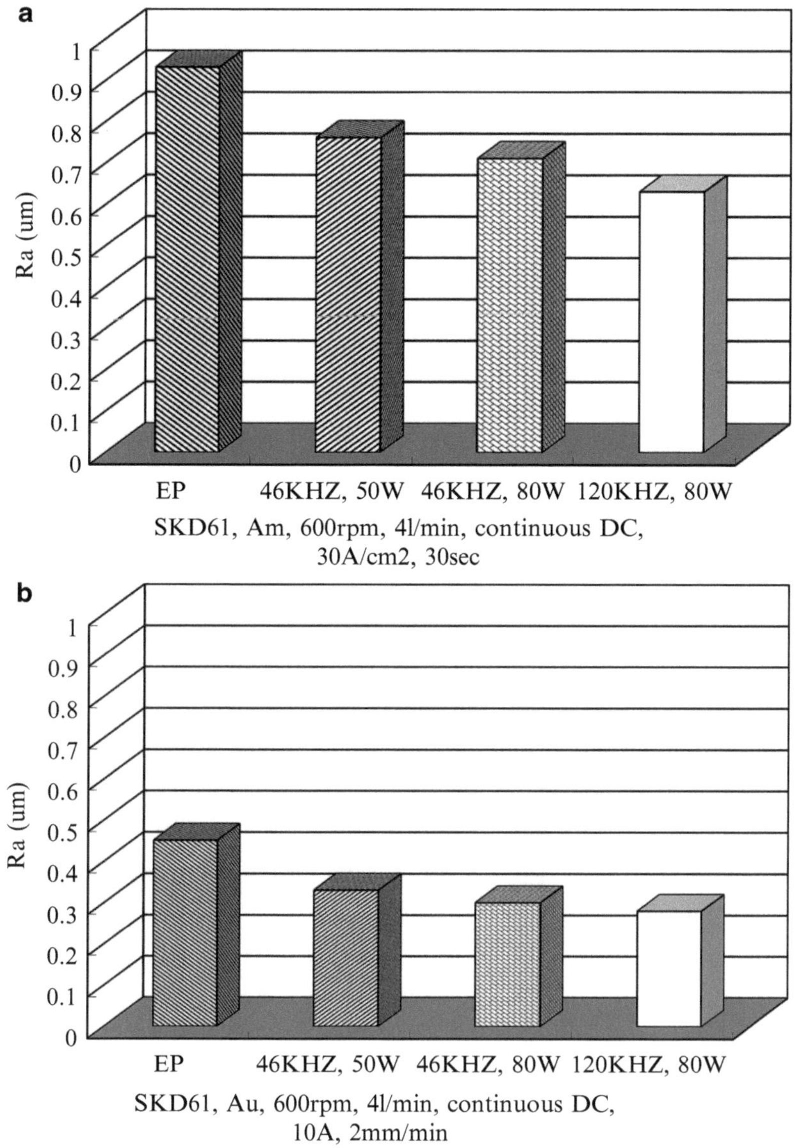

Fig. 3.169 Effects of frequency and power of ultrasonic in electropolishing ([72, 73], reprinted with permission)

ultrasonic or pulse to help electropolishing, while the use of pulse combined with the ultrasonic achieves limited effect and extends the polishing time. The use of ultrasonic not only improves the polishing but also saves the polishing cost.

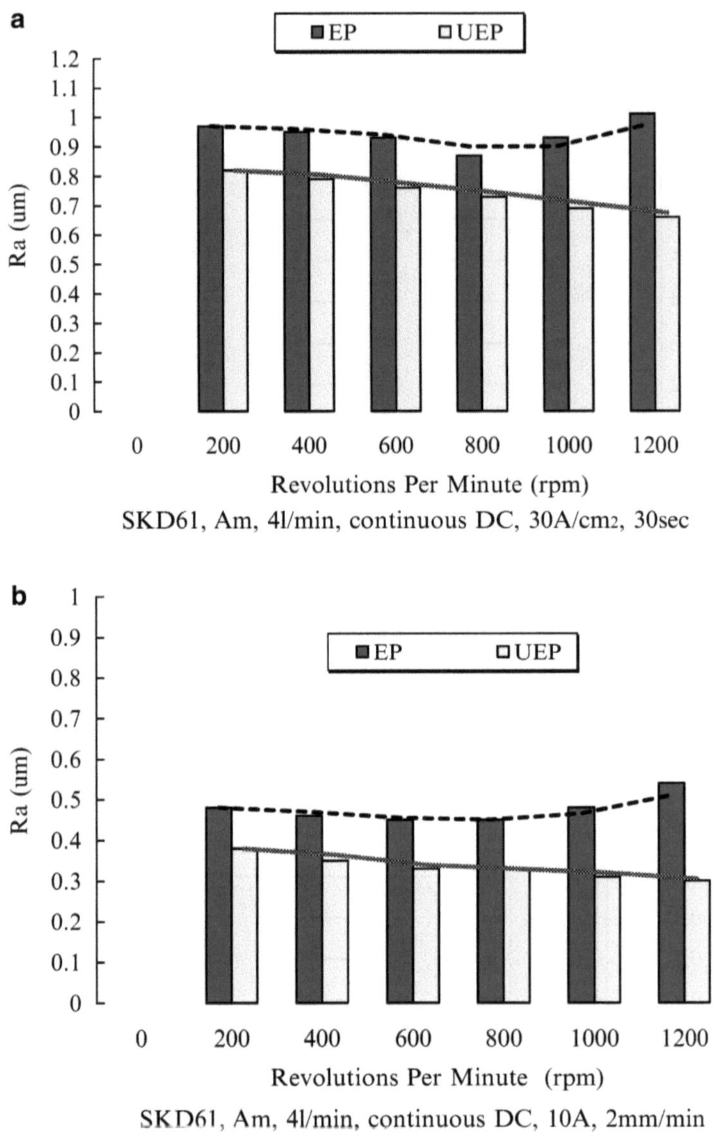

Fig. 3.170 Effects of rotational speed of electrode ([72, 73], reprinted with permission)

3.11.5 Electropolishing of Spherical Surface

Figure 3.177 shows that an adequate gap width between the electrode and spherical surface produces a better finish. The proper gap width can not only facilitate discharging electrolytic products but also adequately provide the prompt and smooth flow of the electrolyte, which further assists in discharging electrolytic products.

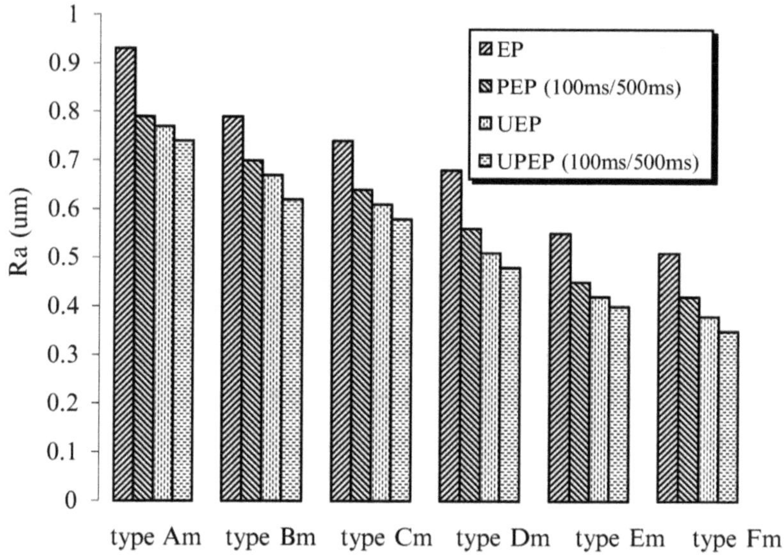

Fig. 3.171 Comparison of electropolishing with different mate-electrodes (SKD61, 600 rpm, 4 L/min, 30 A/cm^2, 30 s)

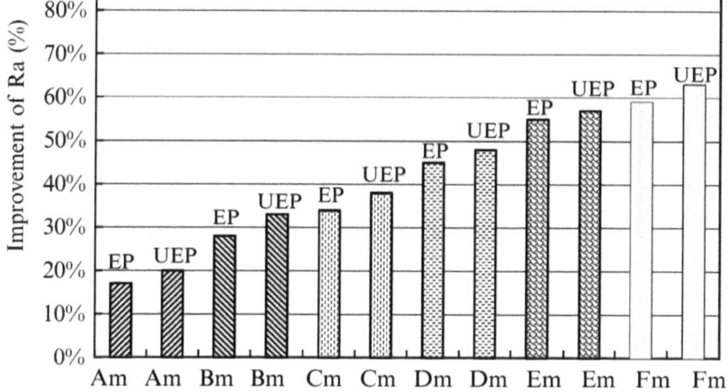

Fig. 3.172 The surface roughness improvement of ultrasonic-aided electropolishing for different mate-electrodes (SKD61, 600 rpm, 4 L/min, continuous DC, 30 A/cm^2, 30 s) ([73], reprinted with permission)

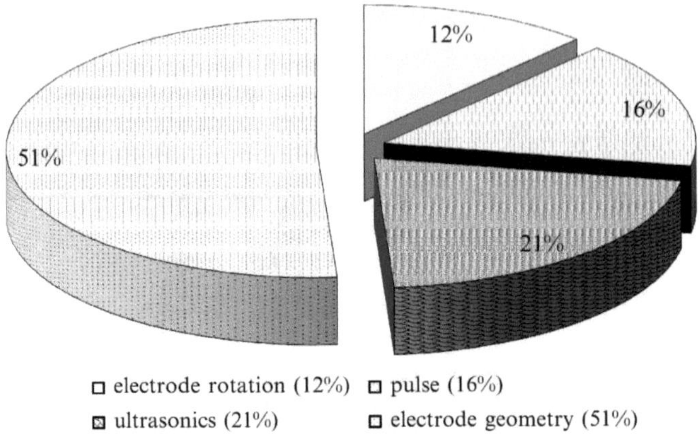

Fig. 3.173 The contribution pie of surface finish improvement using mate-electrodes (SKD61, 600 rpm, 4 L/min, DC, 30 A/cm^2, 30 s) ([73], reprinted with permission)

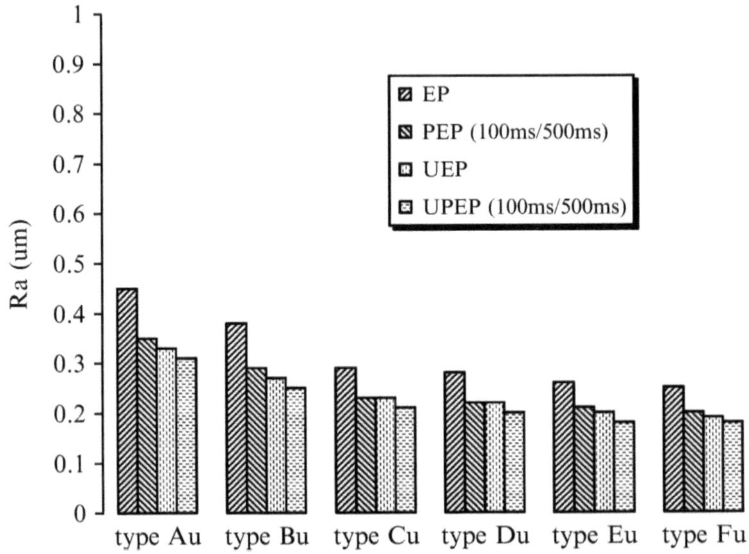

Fig. 3.174 Comparison of electropolishing with feeding electrodes (SKD61, 600 rpm, 4 L/min, 10 A, 2 mm/min) ([73], reprinted with permission)

One can see that the polishing effect is better when the gap width is between 0.3 and 0.4 mm. A small gap width makes the discharge of electrolytic depositions from the gap difficult, and the finish effect is reduced. A large gap width limits the effect of electropolishing (EP). Thus, a gap width of 0.4 mm is more effective in the current experiment. Figure 3.178 shows that the optimal polishing time and

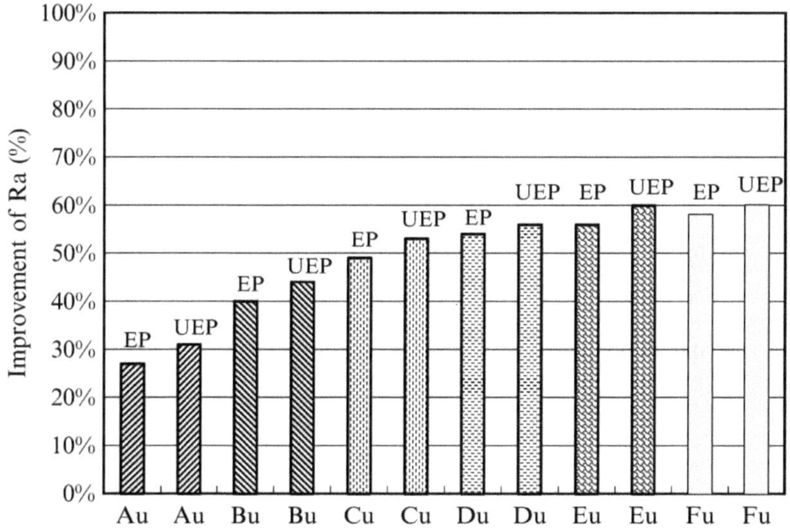

Fig. 3.175 The surface roughness improvement of ultrasonic-aided electropolishing for different feeding electrodes (SKD61, 600 rpm, 4 L/min, continuous DC, 10 A, 2 mm/min) ([73], reprinted with permission)

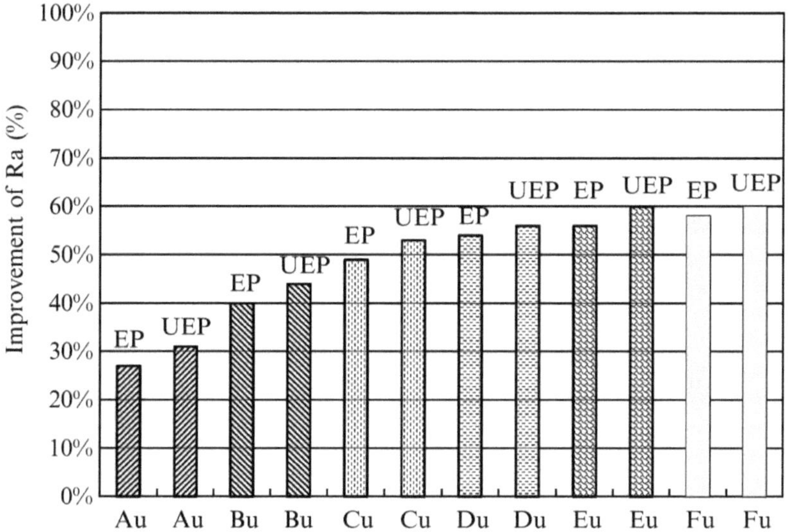

Fig. 3.176 The contribution pie of surface finish improvement using feeding electrodes (SKD61, 600 rpm, 4 L/min, DC, 2 mm/min) ([73], reprinted with permission)

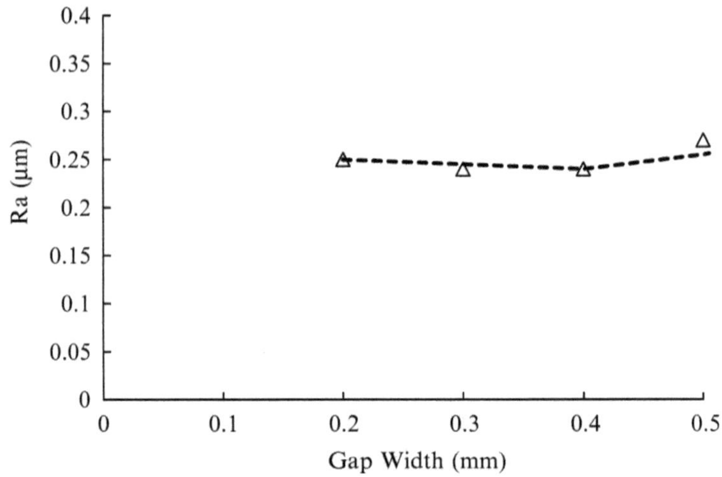

Fig. 3.177 Electropolishing at different gap width between electrode and workpiece (NAK80, $NaNO_3$, 20 %wt, 15 L/min, 60 A/cm^2, 23 s, continuous DC)

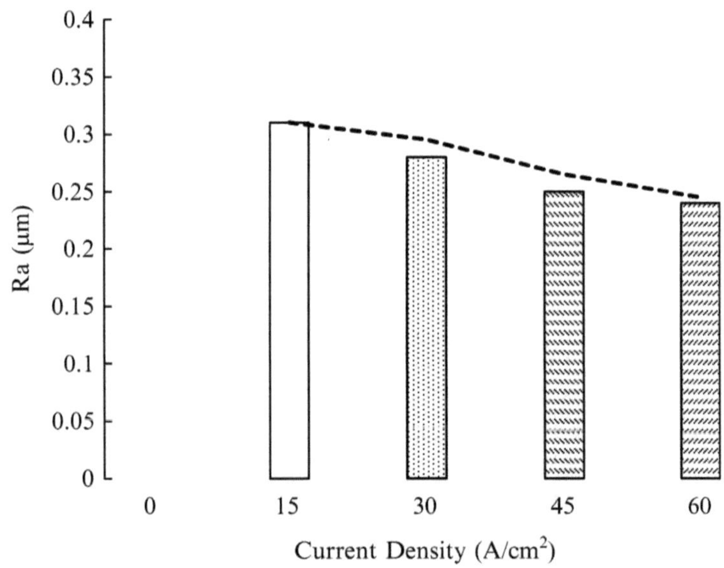

Fig. 3.178 Effects of current density of electropolishing (NAK80, $NaNO_3$, 20 %wt, 15 L/min, 23 s, continuous DC)

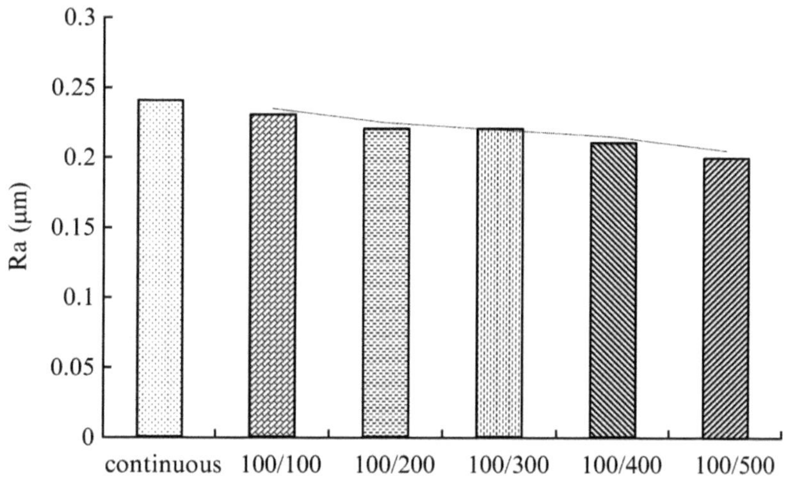

Fig. 3.179 Electropolishing through continuous and pulsed direct current (NAK80, $NaNO_3$, 20 % wt, 15 L/min, 60 A/cm^2, continuous DC)

current density are correlated. The higher the current density (60 A/cm^2), the shorter the time (23 s) needed for a better finish to be achieved by EP process. The machining costs were inversely proportional to the polishing time. The results indicate that large current density required the shortest time, and its finishing effect was the best. Figure 3.178 also illustrates that ultrasonics indeed helps the electropolishing, since dreg discharge by ultrasonics is very effective. The use of a lower current density can thus be replaced by a higher current density aided by ultrasonics to promote finish efficiency.

Figure 3.179 shows the effects of the pulsed direct current. A longer off-time is slightly more advantageous because the discharge of dregs and cuttings of rotational ultrasonic-aided pulse-electropolishing (UREP) during the off-time is more complete. The effects of the ultrasonic-aided facilitate a prompt discharging of the electrolytic products during the off-time. However, the machining time of the pulsed direct current is longer and the cost is raised. Figure 3.180 illustrates the effects of the diameter of the wire electrode. A small diameter provides more open space for dregs discharge and produces a better finish effect in the current study. Figure 3.181 compares the use of the rotation of the spherical workpiece. A higher rotational speed of the spherical workpiece produces a better polishing. A high rotational speed brings kinetic energy of the electrolyte for dreg discharge and produces a better effect for the electropolishing process. The electrolytic products can be promptly discharged depending on the cooperation of the auxiliary effect from the ultrasonic-aided and the higher rotational speed of the spherical workpiece. Figure 3.182 shows the evaluation of the finish effect of four process features. Both the workpiece rotation and the pulsed current can further improve the electropolishing. The working table rotation constructs a rotational effect and a

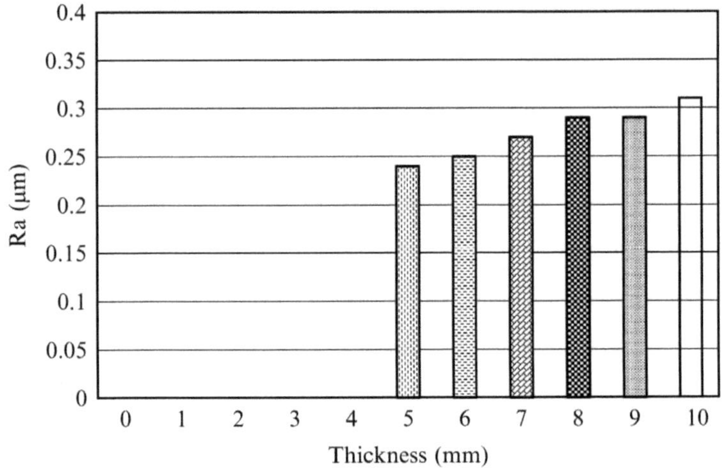

Fig. 3.180 Electropolishing at different electrode thickness (NAK80, $NaNO_3$ of 20 %wt, 15 L/min, 60 A/cm^2, 23 s, continuous DC, 120 kHz/80 W)

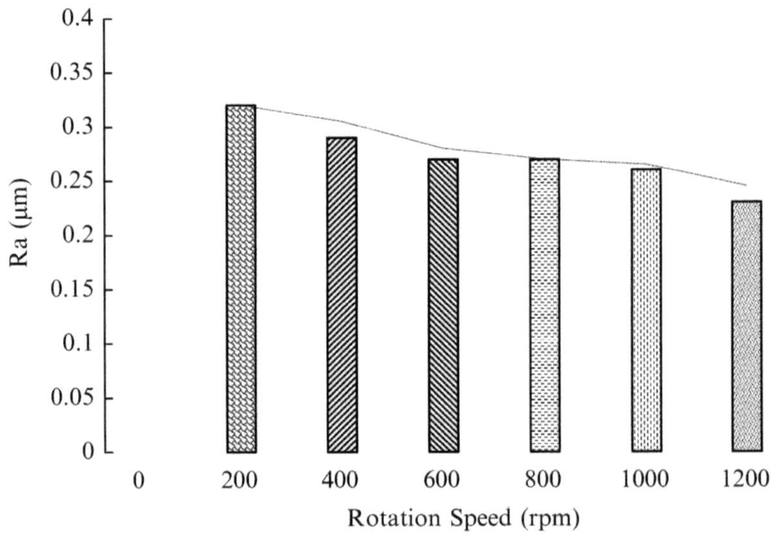

Fig. 3.181 Rotational effects of electropolishing (NAK80, $NaNO_3$, 20 %wt, 15 L/min, 60 A/cm^2, 23 s, continuous DC, 120 kHz/80 W)

continuity dregs discharge effects through the ultrasonic-aided; thus the polishing performs better, and also possesses significant economic advantage. The finish time while using the workpiece rotation and the ultrasonic assistance will not be prolonged as with the pulsed current. In fact, the off-period is often several times as long as the on-period. Owing to the effective discharge of electrolytic dregs and

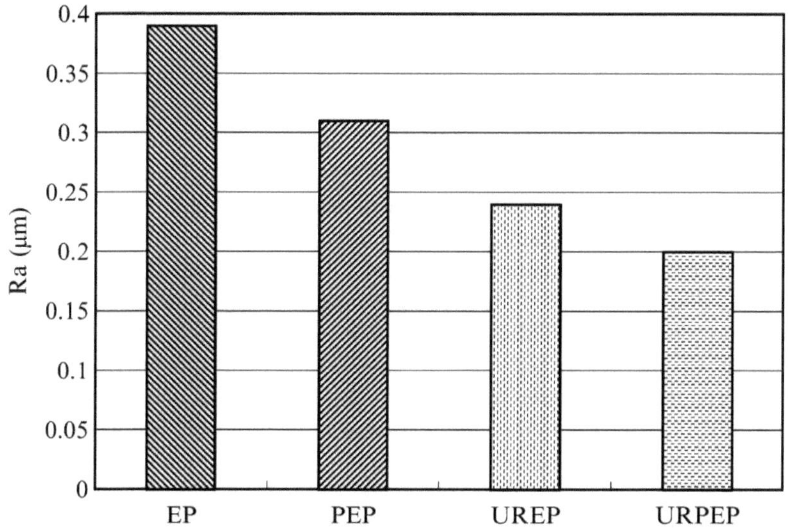

Fig. 3.182 Effects of evaluation of the finishing effect of four process (NAK80, $NaNO_3$, 20 %wt, 15 L/min, 60 A/cm^2, 100 ms/500 ms, 120 kHz/80 W)

cost savings, ultrasonic-aided electropolishing with workpiece rotation (UREP) is the recommended process feature rather than the ultrasonic-aided pulsed electropolishing with workpiece rotation (UPEP).

3.12 Conclusions

The electrochemical boring erosion profile for a flat-end moving electrode is discussed in Sect. 3.1. The iteration integral shows the formation of the hole over time and with a changing electrode gap. Once the hole is bored through, the prediction is modified to convert the virtual material removal underneath the bottom plane a real removal on the sidewall. The electrical current is unchanged before and after boring; hence the same amount of removed material that occurs with the redistributed electric charges at the machined surface is applied in the analysis. The proposed model is useful in the design of an electrochemical boring task to form a small hole or opening, with or without specific wall taper. The model agrees fairly well with experiment for both stationary and moving electrodes. The machining of complex shapes, such as spherical, cylindrical, and flat surfaces, is predictable by the model, provided there are no undercuts, or no irregular electrolyte flow caused by abrupt changes in the surface profile.

The optimized electrodes set a new standard in electropolishing. They provide the advantages of using economical equipment, providing a low-cost process, and allow controllable material removal compared with conventional electrochemical machining (ECM); use of these electrodes requires no complex workpiece pre-polishing as soakage-electropolishing does. The process provided a rapid improvement in surface roughness. Among commonly used die materials, SKD61 shows the best electropolishing surface finish because it has the greatest current efficiency, followed by SKD11, NAK80, and SNCM8. Use of an inserted electrode with a discharge flute significantly improves the surface finish, and a helix flute performs better than a straight flute. In addition, the presence of a water hole in the electrode is effective. The use of a pulsed direct current slightly improves the polishing effect, although it raises the machining costs due to prolonged cycle times. The borer-cutter electrode is a better design than the presence of a cycle lip on the leading edge of the electrode cylinder. The helix flute electrode effectively guides the transportation of the discharged dregs, while the borer-cutter electrode provides the greatest discharge space. Inserted electrodes are suitable for small or medium size holes, while feeding electrodes are suitable for large or deep holes. The working time for the former is short provided the power supply is sufficient, while the latter requires longer polishing times. The obtained surface roughness's from either electropolishing after drilling or electrobrightening after reaming are similar. Though the polishing time of the electropolishing is greater, the total process cycle time can be shorter, since no reaming is required.

A turning tool with greater cutting and clearance angles, and smaller cross section provides larger discharge space and produces a smoother external cylindrical surface. The end clearance angle and the end cutting edge angle are the most significant factors in tool design, while the side cutting edge angle plays a negligible role in electropolishing. A smaller nose radius increases current density and provides increased feed rates, resulting in better polishing. Adequate workpiece revolution speed is advantageous for polishing. The most convenient application is the use of the electropolishing turning tool as an electrode, while a conventional turning tool requires slight modification for optimal electropolishing. An arrowhead electrode with small end radius, small taper angle, and large fluted negative rake and side rake angles provide a large discharge space for swarf. A small end radius also provides a high current density and fast feed rate. The polishing effect of a double-fluted cone arrowhead electrode surpasses the single-flute design by increasing discharge space and flushing mobility during electropolishing. The same principle applies to the flute opening of the flat arrowhead design. A large flute rake angle and deep fluted thin electrode are also advantageous for polishing. The rotating arrowhead electrode with discharge flute is a novel design that outperforms the turning tool electrode.

The thin ring-form electrode with large inner taper angle has a large volume for swarf discharge, and thus improves polishing quality. An adequately sharp inner edge rounding produces high current density, which provides both a fast feed rate and effective polishing. The short, narrow partial ring with large inner radius provides a large discharge space, which again is advantageous for polishing quality. The design change from an inner taper to inner pins for both full- and partial-ring electrodes performs well. The full-ring electrode is suitable for workpieces with various cross sections, and is applicable to machining the external profile.

Thin disc-form electrodes with large taper angle feature a larger discharge space, and thus produce a smoother polished surface. A smaller edge radius produces larger current densities and so enables a greater feed rate. A discharge fluted electrode with large reverse flute rake angle, side rake angle, wider flute, and deeper flute depth performs well. Direct current pulse electropolishing is slightly more effective than using a fluted electrode with continuous direct current, but machining time and cost increase. A disc electrode with small end radius pins attached on its edge is the best design. The use of pulsed direct current has a less significant effect on electropolishing than an optimal electrode does. An equivalent pulse effect arises from the discharge flute of the disc-form electrode, while the disc-form electrode with pins provides the largest discharge space.

Ultrasonic-aided electropolishing provides ultrasonic energy to the electrolyte, and is beneficial to swarf discharge for electropolishing or pulse-electropolishing. Pulse-electropolishing and ultrasonic-aided electropolishing can both elevate polishing quality. A long pulse-off time, high frequency, and large ultrasonic power are optimal for good polishing. For a given polishing time, ultrasonically aided polishing has a slightly greater effect on polishing quality than pulsing has. The effect of ultrasonic pulse-electropolishing is limited compared to that without pulsing; however, the cost using ultrasonics is low. In summary, the electrode design has the most significant impact on polishing quality compared to other processing parameters, and an adequate electrode design with optimized parameters and secondary equipment can provide a satisfactory improvement to the finished surface.

References

1. Fortana MG (1986) Corrosion engineering. McGraw-Hill, New York
2. Jain VK, Pandey PC (1980) Precis Eng 2:195–206
3. Rajurkr KP et al (1992) J Mater Proc Technol 35:83–91
4. Riggo OL, ad Locke CE (1981) Anodic protection. Plenum Press, New York
5. McGeough JA (1974) Principle of electrochemical machining. Chapman & Hall, London
6. Bhattacharyya B, Doloi B, Sridhar PS (2001) J Mater Proc Technol 113:301–305
7. Kunieda M, Yoshida M, Yoshida H (1993) ASME 64:693–699
8. Datta M, Shenoy RV, Romankiw LT (1993) ASME 64:675–692
9. Masuzawa T, Tonshoff HK (1997) Annals of the CIRP 46/2:621–628
10. Hardisty H, Mileham AR, Shirvani H (1995) Proc Instn Mech Engrs 210:109–118
11. Reddy MS, Jain VK, Lai GK (1988) J Eng Ind 110:111–118
12. Kőnig W, Humbs HH (1977) Annals of the CIRP 26/1:83–86
13. Narayanan OH, Hinduja S, Noble CF (1980) Int J Mach Tool Des Res 26:323–338
14. De Silva AKM, Altena HSJ, McGeough JA (2000) Annals of the CIRP 49/1:151–155
15. Misra VN, Khangaonkar PR, Dokras VM (1971) J Sci Ind Res 30:342–348
16. Hopenfeld J, Cole RR (1966) ASME J Eng Ind 88(4):455–461
17. McGeough JA (1974) Principles of electrochemical machining. Chapman and Hall, London
18. McGeough JA (1988) Advanced methods of machining. Chapman and Hall, London
19. De Silva A, Mcgeough JA (1986) Proc Instn Mech Engrs 200(B4):237–246
20. Jain VK, Nanda VN (1986) Precis Eng 8(1):27–33

21. Hopenfeld J, Cole RR (1969) ASME J Eng Ind 91(3):755–763
22. Hoare JP, LaBoda MA (1969) Electrochem Sci 16(2):199–203.
23. Phillips RE (1986) Carbide and Tool Journal 18(6):12–14
24. Wilson J (1971) Practice and theory of electrochemical Machining. John Wiley, New York, pp 79–161
25. Dietz H, Gunther KG, Otto K (1973) Annals of the CIRP 22(1):61–62
26. Risko DG (1992) Society of manufacturing engineers, technical papers, pp 192–225
27. Rasch FO et al (1978) Anneals of the CIRP 27(2):561–563
28. Kashcheev VD, Merkulova NS, Davydov AD (1966) Applied Electrical Phenomena 5:207–214
29. Rybalko AV, Dikusar AI (1995) ISEM-XI. pp 491–504
30. Datta M, Landolt D (1983) J Appl Electrochem 13:795–802
31. Fadaie-Tehrani A, Atkinson J (1995) ISEM-XI, pp 543–552
32. Chin D-T, Mao K-W (1974) J Appl Electrochem 4:155–161
33. Noto K, Okudairira H, Kawafune K (1973) Annals of the CIRP 22:63–66
34. Radhakrishnan V, Krishnaiah Chetty OV, Achyutha BT (1980) Wear 68(1):1–6
35. LaBada MA, Mc Millan LM (1967) Electro Technol 5(7–8):340–345
36. Mileham AR, Harrey SJ, Stout KJ (1986) Wear 109:207–214
37. Datta M, Landolt D (1981) Elector Acta 26(7):899–907
38. Bejar MA, Gutierrez F (1993) J Mat Proc Technol 37:691–699
39. Bannard J (1977) J Appl Electrochem 7:267–270
40. Fadaie-Tehrani A, Atkinson J (1995) ISEM-XI, pp 543–552
41. LaBada MA, Mc Millan LM (1967) Electro Technol 5(7–8):340–345
42. Shen WM (1995) The study of polishing of electric discharge-machined mold with ECM, M. Sc. Thesis, National Yunlin Institute of Technology, Taiwan
43. Rajurkar KR (1995) Annals of the CIRP 44:177–180
44. Chuchro M, Ruszaj A, Zybura-Skrabalak M (1995) The influence of electrochemical disollution process conditions on machined surface geometry. The Institute of Metal Cutting, Cracow, Poland, pp 521–531
45. Acharya BG, Jain VK, Batra JL (1986) Precis Eng 8(2):88–96
46. Louter SP, Cook NH (1973) J. Eng. for Industry 95(4):992–996
47. Masuzawa T (1987) Annals of the CIRP 36:123–126
48. Budzynski AF (1986) ISEM8. Moscow
49. SaKai S, Masuzawa T, Itou S (1988) JSEME 22(43):18–28
50. Rozenberg LD, Kazemtsev VF, Makrov LO, Yakhimovich DF (1964) Ultrasonic cutting. Consultants Bureau, New York
51. Opitz H, Heitmann H, Becker-Barbrock V (1967) Annals of the CIRP 15:177–180
52. Wood RW (1927) Philosophical Magazine. 7 Sept 1927. pp 417–436
53. Shaw MC (1956) Microtechnic. 10(6):257–265;88–96
54. Dikushin VI, Barke VN (1958) Stanki i Instr 5:1058–1066
55. Komaraiah M, Manan MA, Narasimha Reddy P, Victor S (1988) Precis Eng 10(2):59–65
56. Komaraiah M, Narasimha Reddy P (1991) Int J Prod Res 29(11):2177–2187
57. Komaraiah M, Narasimha Reddy P (1993) Int J Mach Tools Manufact 33(3):495–505
58. Khairy ABE (1990) Wear 137:187–198
59. Wang B, Fang Z (1996) J Atmos Sci 53:2786–2802
60. Gilmore R (1991) Seventh international Conference on computer-aided Production Engineering. vol 28(12), pp 139–148
61. Reddy MS et al (1988) J Eng Ind 110:111–118
62. Hocheng H, Hsu CC (1995) J Mater Process Technol 48:255–266
63. AZCUE JM (1995) Intern J Environ Anal Chem 62:137–145
64. Watmough DJ (1993) Ultrasonic 32(4):315–317
65. Institute of Advanced Manufacturing Sciences INC (1980) Machining data handbook. 3rd edn. vol 2(18) p 11

66. Hocheng H, Pa PS (1999) Electropolishing and electrobrightening of holes using different feeding electrodes. J Mater Process Technol 89–90:440
67. Hocheng H, Pa PS (2003) Effective form design of electrode in electrochemical smoothing of holes. Int J Adv Manuf Technol 21(12):995
68. Hocheng H, Pa PS (2002) The application of a turning tool as the electrode in electropolishing. J Mater Process Technol 120(1):6
69. Hocheng H, Pa PS (2004) Design of arrow-head electrode in electropolishing of cylindrical part, research trends. Int J Mater Prod Technol 20(4):312
70. Hocheng H, Pa PS (2000) Ring-form electrode in electropolishing of external cylindrical surface. Int J Elec Mach 5:7
71. Hocheng H, Pa PS (2003) Electropolishing of cylindrical workpiece of tool materials using disc-form electrodes. J Mater Process Technol 142(1):203
72. Hocheng H, Pa PS (2003) Continuous secondary ultrasonic electropolishing of an SKD61 cylindrical part. Int J Adv Manuf Technol 21(4):238
73. Hocheng H, Pa PS (2001) Electrode form design and ultrasonic aid in electropolishing of holes. JSEME: Int J Elec Mach 5:7
74. Hocheng H, Gao PS, Lin SC (2005) Generation of erosion profile of through hole in electrochemical boring using stepwise moving electrode. Int J Manuf Technol Manag 7(2–4):268–286

Chapter 4
Chemical Mechanical Polishing

H.Y. Tsai, H. Hocheng, and Y.L. Huang

Abstract Chemical mechanical planarization (CMP) has emerged as an indispensable processing technique for planarization in submicron multilevel VLSI. An analytic model of the material removal rate is proposed for CMP. The effects of applied pressure and polishing velocity are derived by considering the chemical reaction as well as the mechanical bear-and-shear processes. The material removal rate is less linearly correlated to the pressure and relative velocity than that predicted by the frequently cited empirical Preston equation [1]. The effects of CMP kinematic variables on wafer nonuniformity are also investigated. The significance of velocity uniformity is demonstrated by both analysis and experiment. For the endpoint detection, an accurate in situ monitoring method can significantly improve both yield and throughput. A model for CMP polishing pad temperature that is capable of predicting the CMP endpoint in situ is established, based on the total consumed kinematic energy. The process endpoint is detectable by application of the proposed regression method to the measured temperature rise. In addition, the chapter develops an endpoint monitoring method that uses acoustic emissions that occur during CMP. The method considers differences in friction characteristics between the polishing pad and the copper metal overlay. For the flow of slurry between wafer and pad, this study provides a visualized characterization of the amount and distribution of the fluid film between wafer and pad. Digital photographs taken through the transparent carrier and dyed fluid are used to analyze the fluid film.

Keywords Nonuniformity • Kinematic analysis • Endpoint detection • Pad temperature • Acoustic emissions • Flow of slurry • Fluid film • Dielectric film • Copper • Bare silicon

H.Y. Tsai (✉) • H. Hocheng • Y.L. Huang
Department of Power Mechanical Engineering, National Tsing Hua University,
No. 101, Sec. 2, Kuang-Fu Rd, Hsinchu city 30013, Taiwan ROC
e-mail: hytsai@pme.nthu.edu.tw

H. Hocheng and H.-Y. Tsai (eds.), *Advanced Analysis of Nontraditional Machining*,
DOI 10.1007/978-1-4614-4054-3_4,

4.1 Introduction

Historically chemical mechanical planarization (CMP) has been used to polish optical flat surfaces. As the wiring density in high-performance chips increases and the device size scales down to submicrons (as shown in Table 4.1), planarization technology becomes indispensable during both device fabrication and formation of multilevel interconnects. CMP has emerged recently as a processing technique for higher degree of planarization in submicron VLSI, and is widely accepted for planarizing interlevel dielectrics and selective removal of aluminum, tungsten, copper, and titanium overburden following metal filling of studs and interconnects.

A schematic diagram of a CMP is shown in Fig. 4.1 [2] (view from top and side). The wafer is held face-down by a rotating carrier pressed against the polishing pad, which is rotating as well. A down force is applied onto the wafer. The polishing colloidal slurry, which consists of chemical reagents, is dispensed onto the pad surface. The slurry is distributed across the pad by the centrifugal force of the rotating pad to wet the polishing pad and the wafer. The chemical reaction in CMP softens the dielectric or hardens the metal surface material of the wafer. The treated material is then removed mechanically by fine abrasives. In balance between the chemical and mechanical actions, the process removes material continuously.

Although quite a few papers discuss the experimental results and the effects of the process parameters, the material removal mechanism is addressed less. Some authors consider the material removal mechanism at the direct contact between the wafer and the pad, while the others consider there is a thin fluid film between the wafer and the pad.

Warnock presented the dependence of the polish rate on the wafer shape, though it is completely a phenomenological model [3]. Yu et al. presented a physical CMP model that includes the effects of polishing pad roughness and dynamic interaction between pad and wafer. Two new feature-scale-polishing mechanisms based on

Table 4.1 Roadmap by Semiconductor Industry Association

Year	2001	2002	2003	2004	2005	2006	2007
Minimum feature size (nm)	130	115	100	90	80	70	65
Maximum substrate diameter (mm)	200	300	300	300	300	300	300
DRAM							
Numbers of metal levels	3	4	4	4	4	5	5
Contacts (nm)	165	140	130	110	100	90	80
Memory Size (mega bits)	512			2,000			6,000
MPU							
Numbers of metal levels	7	7–8	8	8	8–9	9	9
Contacts (nm)	165	140	122	100	90	80	75
Microprocessor speed (MHz)	1,684			5,173			11,511
Depth of focus (μm)	0.7	0.6	0.6	0.5	0.5	0.4	0.4

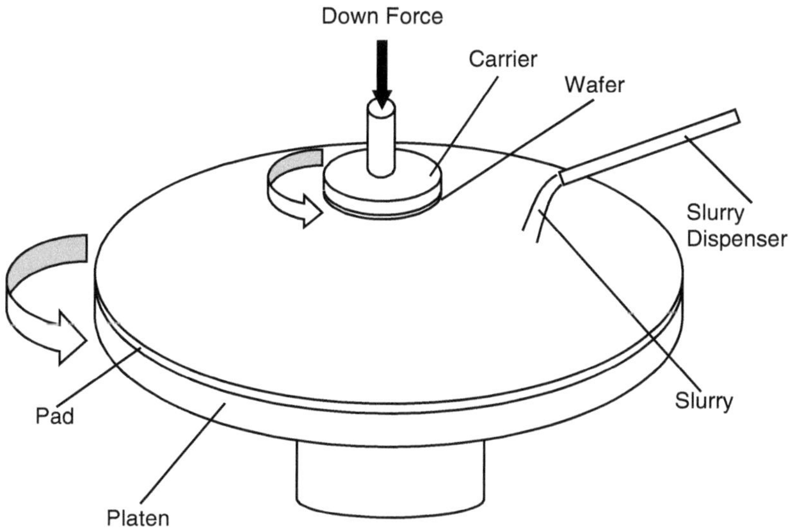

Fig. 4.1 Schematic diagrams of a CMP equipment [2], reprinted with permission

asperity theory are proposed and investigated experimentally. However, it is not clear whether or how the asperity affects the global quality of planarization [4]. Warnock and Yu et al. both derive the reduction rate of the step height assuming the direct contact between pad and wafer without considering the role of abrasive flow in between.

Liu assumes that the abrasive cuts the wafer under the loaded pressure and the relative velocity between the wafer and the pad. The model is capable of delineating the role of the mechanical properties of the slurry particles and the surface film to be polished. Liu assumes that the deformed volume can be decomposed of three regions, namely the regions of elastic deformation, plastic deformation, and microcutting. He also assumes the deformation volume during contact to be approximately equal to the volume of microcutting and calculates the removal rate by the Hertzian elastic equation. The results were also compared with the wear model of Cook [5], which is derived from the glass polishing process [6]. Cook and Liu both believe that the removal of wafer is primarily due to the direct contact between abrasives and wafer, and the Hertzian Contact is adopted. Cook's results did not reveal the detail of material removal rate, and Liu did not characterize the role of flow.

Runnels presented two CMP models in 1994. One is the tribology analysis of CMP, and the other is the feature-scale fluid-based erosion model for CMP. Runnels considers that the knowledge of the stress distribution on the wafer surface is required in order to explain the variation of material removal rate on a wafer during CMP. This study analyzes the fluid film between the wafer and pad, and demonstrates that hydroplaning is possible for standard CMP processes. The importance of wafer curvature, slurry viscosity, and rotation speed on the thickness of the fluid film is also demonstrated. Although the analysis is novel, the model can only estimate the thickness of the film between pad and wafer, and Runnels admitted that the assumption of the spherical wafer needs modification [7]. The other physical-based feature-scale erosion model is critical to the understanding of CMP, because it establishes the link between the chemical effects at the particle scale and the process conditions at the wafer scale utilizing a hydrodynamic slurry layer. Given some assumptions, the fluid flow is modeled by the steady-state two-dimensional Navier–Stokes equations for incompressible Newtonian flow [8]. Runnels later implemented Warnock's phenomenological erosion model. The improved development allows for closer fits to the experimental data and uniform discretization of the wafer feature [9]. Although Runnels takes into account that it is the slurry flow erodes the pattern of wafer, his model does not account for the effects of the abrasives in the slurry.

Tseng and Wang combined Runnels' and Cook's work without physical interpretation. A model of removal rate was obtained to account the dependence of the removal rate on the down pressure and the rotation speed during CMP process based on the theory of elasticity [10].

Although CMP has been widely practiced in semiconductor manufacturing for a few years, the mechanism of the material removal remains to be explored. The macroscopic empirical Preston equation obtained from glass polishing is often considered the major reference. A model based on individual abrasive material removal in a flow field is proposed. It considers the abrasives remove the material by a bear-and-shear process, while the slurry flow between the wafer and the pad plays a role in transporting the chemical means. Preston's equation can be properly interpreted and used with modification. The analysis and the independent experimental results are discussed in the following.

In addition, the role of the uniform polishing velocity in the planarization was highlighted [11, 12]. Though it is of great importance in the process recipe in CMP practice, the mechanism of influence of the polishing velocity on nonuniformity and the analytic correlation between them remain unexplored. Based on the optimal selection of the speed of platen and carrier for CMP has been analytically derived [13], the kinematic analysis is used to predict the nonuniformity in CMP affected by various kinematic variables. Experimental results are also obtained and discussed.

In the metal CMP process, the previously deposited metal layer is removed, leaving the metal line and the metal in the contact vias or plugs. An evenly polished surface satisfies two requirements, namely to ensure the good metal step coverage and to provide a field within the lithography depth of focus, that contact vias and metal wires can be well patterned [14]. The accumulated variation in film thickness will impair the process. CMP has become a key process for the Cu interconnects, low-κ materials, and dual damascene process.

As more wafers are polished by CMP, the pad starts to degrade and affects its planarizing ability. In order to ensure the wafers are properly planarized and to control the constant yield of CMP, the online monitoring and endpoint detection techniques are needed.

Table 4.2 Thermal monitoring for CMP

US Patent No.	Date	Assignee
5,196,353	Mar 23, 1993	Micron Technology
5,597,442	Jan 28, 1997	Taiwan Semiconductor Manufacturing Co. Ltd.
5,643,050	Jul 1, 1997	Industrial Technology Research Institute
5,647,952	Jul 15, 1997	Industrial Technology Research Institute
5,722,875	Mar 3, 1998	Tokyo Electron, IPEC-Planar
5,834,377	Nov 10, 1998	Industrial Technology Research Institute
5,891,352	Apr 6, 1999	Luxtron
6,007,408	Dec 28, 1999	Micron Technology
6,077,452	Jun 20, 2000	Luxtron
6,077,783	Jun 20, 2000	LSI Logic
6,110,752	Aug 29, 2000	Luxtron
6,150,271	Nov 21, 2000	Lucent Technology

Pad temperature has been used to monitor metal CMP process [15–25]. Table 4.2 lists patents using thermal methods to monitor a CMP process. In these studies, raw signal of pad temperature is recorded directly and is thus easily affected by environment and slurry flow. Most CMP machines are equipped with pad temperature sensor also to monitor stable chemical reaction during polishing. By monitoring pad temperature change, the polishing of different layers of materials can be detected. A physical model is proposed to explain the pad temperature distribution, which agrees with the polishing results, and the endpoint can be detected. In addition, an endpoint monitoring method using acoustic emissions during the chemical mechanical polishing is developed since the coefficient of friction between the pad and dielectric layer is distinguishably lower than the one between the pad and the copper overlay.

While the effects of other process parameters on material removal rate and nonuniformity have been presented [13, 26, 27], the role of slurry flow is not fully understood. Basically, the abundant and uniformly distributed slurry supply between pad and wafer is desired. However, how the process recipe affects the slurry distribution needs to be investigated. An adequate slurry film between polishing pad and wafer ensures the successful CMP. This challenge becomes more severe as the wafer size increases to 300 mm and beyond. The analytical solution for the flow field is of limited feasibility due to the ill undefined boundary conditions. The current study presents the experimental results of flow visualization that reveal the influence of the process parameters on the amount as well as distribution of the fluid film between pad and wafer.

Runnels and Eyman described the slurry flow assuming the wafer to be a surface of large radius of curvature. This model was able to predict the thickness of fluid film [7]. The technique of dual laser-induced fluorescence was employed to observe the flow field. The flow was affected by the surface condition of pad, while the platen speed exerted secondary influence. The rinsing location and pad surface showed effects on the uniformity of fluid distribution [28]. A two-dimensional simulation model was established to calculate the velocity and pressure of flow field. The flow underneath wafer was assumed uniform. Back flow was predicted at wafer front and acceleration would occur at wafer tail [29]. The independent experimental results of pressure distribution of flow field showed inconsistency with the simulation [30]. To explore the effects of process parameters on the fluid film, the current study includes more extensive process parameters than the previous investigation, such as pad groove design, carrier speed, flow rate, and wafer size. An effective while less complicated experimental method is used for the characterization of the amount and distribution of the fluid film between wafer and pad.

4.2 Material Removal Rate of Dielectric Films and Bare Silicon Wafers

4.2.1 Bear-and-Shear Process Model

4.2.1.1 Chemical and Mechanical Action

First consider the chemical reaction of silicon dioxide,

$$\mathrm{SiO_2} + 2\mathrm{H_2O} \rightarrow \mathrm{Si(OH)_4}. \tag{4.1}$$

There exists a rate constant in a chemical reaction. The rate constant often increases rapidly with the increasing temperature. A rough rule valid for typical reactions in solution is that it doubles or triples for each increase of 10°C in temperature. Arrhenius pointed out that the constant k fits the expression [31]

$$k = C\mathrm{e}^{-E_\mathrm{a}/RT} \tag{4.2}$$

where R is the gas constant, E_a is the Arrhenius activation energy and C is the pre-exponential factor (or frequency factor). The unit of C is the same as k. The unit of E_a is the same as RT, namely energy per mole (kcal/mole or kJ/mole). C and E_a characterize the reaction.

As mentioned above, the reaction rate is proportional to the rate constant at the constant molecular concentration of water in CMP. Therefore, the material removal rate in view of the chemical action is proportional to the rate constant. Once the top surface is removed mechanically, the silicon dioxide film exposed to the slurry will keep on reacting chemically. These reactions are in sequence, that is, the chemical reaction first and then the mechanical removal repeatedly. Besides, the following facts exist. The dissolution rate of glass into water increases with compressive stress [32]; therefore the hydrate reaction rate of silica is accelerated by the pressure of the slurry particle against the surface layer of wafer. The compressive stress decreases the activation energy, thus increases the rate constant and the chemical reaction rate. If the chemical reaction were the limiting step, the material removal rate would not increase with the solid content in the slurry [33]. Hence the mechanical removal is considered the bottle neck in the material removal process chain and is worthy of analysis. In other words, the material removal is proportional to the mechanical removal rate.

Once the products ($Si(OH)_4$) are produced by the chemical reaction, the slurry particles carried by the slurry flow remove the products. The polishing by mechanical means in CMP is further considered as follows.

1. The slurry film between pad and wafer supports the down pressure, and the carrier (i.e., wafer) holds a nonparallel but slightly inclined configuration relative to pad
2. The slurry particles carrying kinetic energy approach and remove the molecules of the wafer surface material

The major aspects in the process are then described in the following.

4.2.1.2 Thrust Bearing Analogy

The thrust load of a rotary machinery is often counterbalanced by self-acting or hydrodynamic bearings shown in Fig. 4.2 [35]. A thrust plate attached to, or forming part of, the rotating shaft is separated from the sector-shaped bearing pads by a lubricant film. The load-carrying capacity of the bearing arises entirely from the pressures generated by the geometry of the thrust plate over the bearing pads.

The simplest form of the pivoted-pad bearing provides straight-line motion only and consists of a flat surface sliding over a pivoted pad as shown in Fig. 4.3 [34]. If the pad is in equilibrium under a given set of operating conditions, any change in these conditions (such as load, speed, or viscosity) will alter the pressure distribution and thus momentarily shift the center of pressure. This event will create a moment that causes the pad to change its inclination and the shoulder height (S_h). A pivoted-pad slider bearing is thus supported at a single point so that the angle of the inclination becomes a variable and has much better stability than a fixed inclined slider under varying conditions of operation.

This behavior is similar to the relationship between the gimbal mechanism of the carrier and the pad during CMP, except that the positions of the gimbal mechanism of the carrier and the pad are reversed. The location of the shoe pivot can be found from the equilibrium of moments acting on the shoe about the point. For all practical purposes, the force due to friction in the pivot is ignored. The minimum fluid film thickness in designing a pivoted-pad thrust bearing is found as

$$h_0 \propto \left(\frac{\mu V}{F}\right)^{1/2} \tag{4.3}$$

where μ is the viscosity of the film, V is the velocity, and F is the applied normal load. Although the exponential values other than 1/2 have been proposed for various applications, Bhushan and Runnels considered it 1/2 in the CMP process [7, 33].

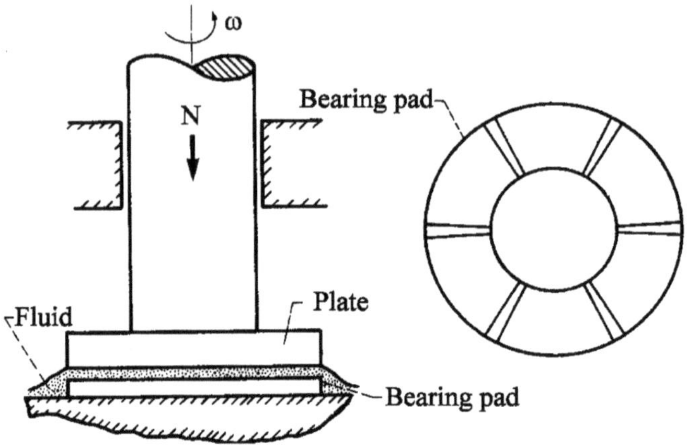

Fig. 4.2 Thrust bearing [34], reprinted with permission

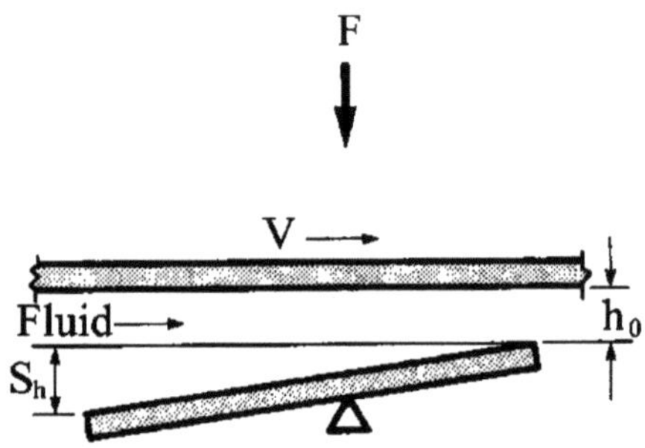

Fig. 4.3 Simple pivoted-pad bearing [34], reprinted with permission

4.2.1.3 Couette Flows in Turbulence

In Fig. 4.4 [34], two infinite plates are 2 h apart, and the upper plate moves at speed V relative to the lower. The fluid pressure p is assumed constant. The upper plate is held at temperature T_1 and the lower plate at T_0. These boundary conditions are independent of coordinates x or z (called infinite plate); hence it follows that $u = u(y)$ and $T = T(y)$. The reduced basic equations are

Continuity Equation:

$$\frac{\partial u}{\partial x} = 0 \tag{4.4}$$

Momentum Equation:

$$\frac{\mathrm{d}^2 u}{\mathrm{d}y^2} = 0 \tag{4.5}$$

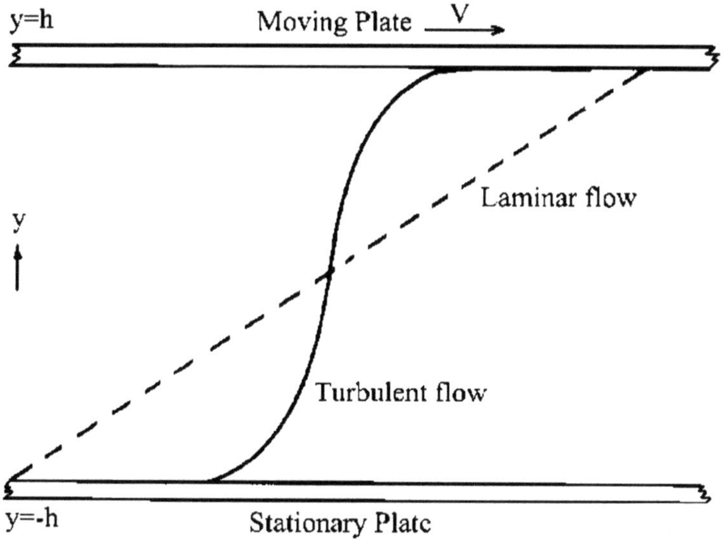

Fig. 4.4 Schematic of couette flow between two plates [34], reprinted with permission

Energy Equation:

$$k\frac{d^2T}{dy^2} + \mu\left(\frac{du}{dy}\right)^2 = 0 \quad (4.6)$$

where the continuity merely verifies the assumption that $u = u(y)$. Equation (4.5) can be integrated twice to obtain

$$u = C_1 y + C_2. \quad (4.7)$$

Using the no-slip boundary conditions, that is, $u(-h) = 0$ and $u(+h) = V$, one obtains $C_1 = V/2$ h and $C_2 = V/2$. Thus the velocity distribution is

$$u = \frac{V}{2}\left(1 + \frac{y}{h}\right). \quad (4.8)$$

This is the exact steady-flow solution of the Navier–Stokes equation, called laminar flow, and has a smooth-streamline character. It is well known that all laminar flows become unstable beyond a finite value of a critical parameter (Reynolds Number). A different type of flow then ensues on an entirely new fluctuating flow regime, called turbulent.

The pattern changes into a randomly fluctuating flow, sketched by Reichardt in1956 as shown in Fig. 4.3, varies slightly with Reynolds number and increases the wall shear by two orders of magnitude [36]. Therefore, the relationship between velocity and the position from the middle of the gap to the upper wall is expressed as

$$u = ay^n \quad (4.9)$$

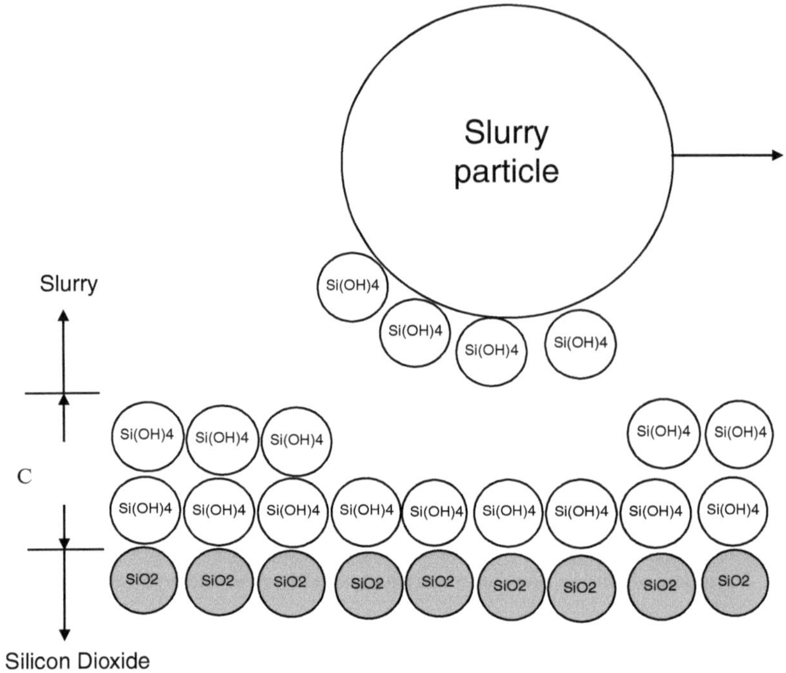

Fig. 4.5 Schematic of the chemical products carried by the slurry particle [34], reprinted with permission

where n is a number larger than 100. The no-slip condition is $u = V/2$ at $y =$ h, or $a = \frac{Vh^{-n}}{2}$; hence the wall shear can be determined

$$\tau = \mu \frac{\partial u}{\partial y} = \mu n a h^{n-1} = \frac{\mu n V}{2h}. \tag{4.10}$$

Equation (4.10) tells the wall shear becomes two orders of magnitude larger. The wall shear stress produced by Couette flows will be applied to the derivation of the relative polishing velocity in CMP.

4.2.1.4 Energy Balance

When a slurry particle binds to the molecule $Si(OH)_4$ of the wafer surface and removes the molecule, the shear force must be larger than the binding force of the molecule, and larger energy than the surface energy of the generated new surface must be transmitted.

Consider a slurry particle pulling the molecules off wafer surface, as shown in Fig. 4.5 [34]; the shear stress of slurry flow is assumed uniformly distributed around the particle. In order to remove one molecule from surface, the following condition must be satisfied [37],

$$\tau\left(\frac{\pi D_{\mathrm{p}}^{2}}{4}\right)D_{\mathrm{m}} \geq 2\gamma A_{\mathrm{m}}N \tag{4.11}$$

where γ is the surface energy between the surface and sub-layer molecule, D_{p} is the diameter of the slurry particle, D_{m} and A_{m} are the diameter and the cross-section area of the $Si(OH)_4$ molecule, respectively, and N is the number of molecules to be removed.

In this case, N is considered quite large because the ratio of the the diameter of the molecule to the diameter of the abrasive is about 1/100. The left hand side of (4.11) is an approximation of the energy transmitted to wafer surface to move the particle for a distance D_{m}, where it is defined that the molecule will be sheared off once the moving distance is larger than D_{m}. The right hand side of (4.11) is the surface energy of the newly generated surface, when one molecule is moved. (4.11) can be rearranged

$$N = \frac{\tau D_{\mathrm{p}}^{2}}{2\gamma D_{\mathrm{m}}}. \tag{4.12}$$

Equation (4.12) indicates that the number (N) of removed $Si(OH)_4$ is proportional to the wall shear stress. The role of the slurry particles in the film between the pad and the wafer is to shear off the chemical products; meanwhile the $Si(OH)_4$ is continuously produced.

4.2.1.5 Material Removal Rate

The volumetric material removal rate ($\dot{M}$) is determined by the volume (∇) of removed molecules of the chemical products (4.12) and the encounter frequency (f) of the slurry particles on the removed surface,

$$\dot{M} \propto N \cdot \nabla \cdot f = \frac{\tau D_{\mathrm{p}}^{2}}{2\gamma D_{\mathrm{m}}} \cdot \frac{\pi D_{\mathrm{m}}^{3}}{6} \cdot f \tag{4.13}$$

where f is a function of the solid content and the characteristics of the turbulent flow.

One recalls (4.10) that the wall shear is proportional to the reciprocal of the film thickness. Substituting (4.3) into (4.10), one obtains

$$\tau = \frac{\mu n V}{2h} \propto \mu n V\left(\frac{\mu V}{F}\right)^{-1/2} \tag{4.14}$$

Since the force F is equal to PA, (4.14) can be reduced to

$$\tau \propto n(\mu VPA)^{1/2} \tag{4.15}$$

where A is the area of the wafer.Combining (4.13) and (4.15), one obtains the volumetric material removal rate

$$\overset{\bullet}{M} \propto f \cdot \left(\frac{nD_{\mathrm{p}}^2 D_{\mathrm{m}}^2}{12\pi\gamma}\right)(\mu A)^{1/2}(PV)^{1/2}. \tag{4.16}$$

The removal rate in thickness per unit time is obtained by

$$\overset{\bullet}{H} = \frac{\overset{\bullet}{M}}{A}. \tag{4.17}$$

It shows the removal rate increases with the increasing polishing speed and pressure to the power of 1/2. The experimental results are discussed in the following.

4.2.2 Experimental Findings

The experiments are carried out on both silicon dielectric film and bare silicon wafer by two polishers, IPEC372M and Peter-WaltersAC1400, respectively.

4.2.2.1 IPEC 372M Polisher

The parameters of the material removal experiments are listed in Table 4.3. The experimental results of polishing silicon dioxide are shown in Fig. 4.6 [34]. The material removal rate increases with the increasing velocity by the power of 0.65, slightly larger than the predicted 0.5. The reason lies in that a small portion of the wafer is polished in direct contact mode as a result of not ideally conformed pad (particularly at wafer edge), where the material removal rate is linearly proportional to the polishing velocity [38]. The empirical Preston equation based on glass polishing also implies this part of process. In fact, the total material removal rate consists of the contribution from the direct contact mode as well as the noncontact mode, that is,

$$\begin{aligned} \overset{\bullet}{H} &= \alpha K_1 V^{0.5} + \beta K_2 V^{1.0} \\ 0.5em &= K_3 V^{0.65} \end{aligned} \tag{4.18}$$

Table 4.3 Parameters of experiment of silicon dioxide

Polisher	IPEC 372M Polisher
Pad type	Rodel IC1400
Slurry type	Cabot SS25 (1:1 dilute by DI Water) pH 10.8
Slurry flow rate (ml/min)	150
Down pressure (psi)	5, 8, 10
Back pressure (psi)	3
Platen speed (rpm)[a]	10, 20 , 30, 40, 50, 60, 70, 80
Polishing time (min)	1

[a]Carrier speed is equal to platen speed

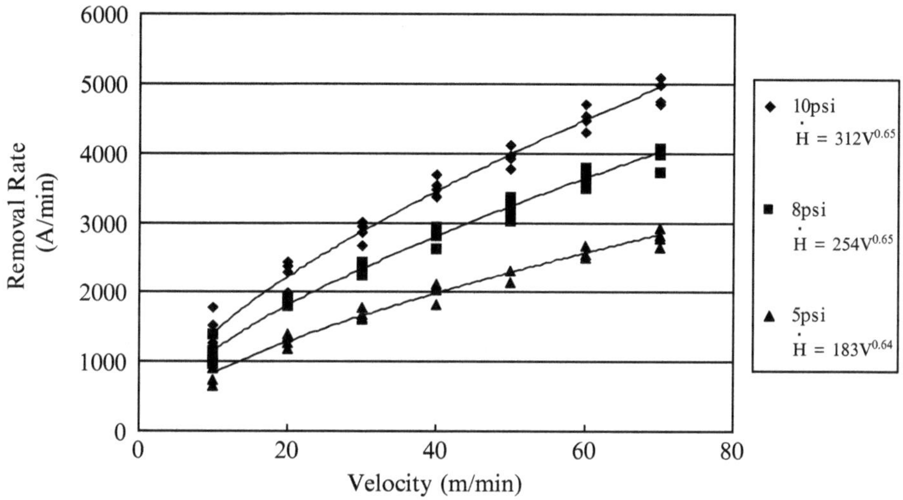

Fig. 4.6 Relationship between removal rate of SiO2 and polishing velocity [34], reprinted with permission

where α and β represent the fraction of the noncontact and the contact area between wafer and pad during CMP, respectively, and $\alpha + \beta = 1$. K_1, K_2, and K_3 are the proportional constants between the removal rate and velocity.

As shown by (4.3), the minimum film thickness increases with the velocity; thus the fraction of the contact area (β) decreases with the increasing velocity. The effect of the polishing velocity in the current experiment is indicated by its exponent of 0.65, indeed between 0.5 and 1.0, close to 0.5.

4.2.2.2 Peter-Wolters AC 1400-P Polisher

The parameters of the material removal experiments are listed in Table 4.4. The experimental results of bare silicon wafer polishing are shown in Fig. 4.7 [34].

Table 4.4 Parameters of experiment of bare silicon

Polisher	Peter-Wolters AC1400-P Polisher
Pad type	Rodel-Nitta Suba 600
Slurry type	Rodel-Nitta Nalco 2350 (1:20 dilute by DI water) pH 10.3
Slurry flow rate (ml/min)	750
Down pressure (psi)	1.42
Sun gear speed (rpm)	15
Ring gear speed (rpm)	8, 4, 0.1, −4, −8
Upper/lower pad speed (rpm)	−5.7/−20, −5.7/−13, −5.7/−5.7, −5.7/1, −5.7/8, −16.7/−16.7, −11.2/−11.2, −22.2/−22.2, −0.18/−0.18
Polishing time (min)	10

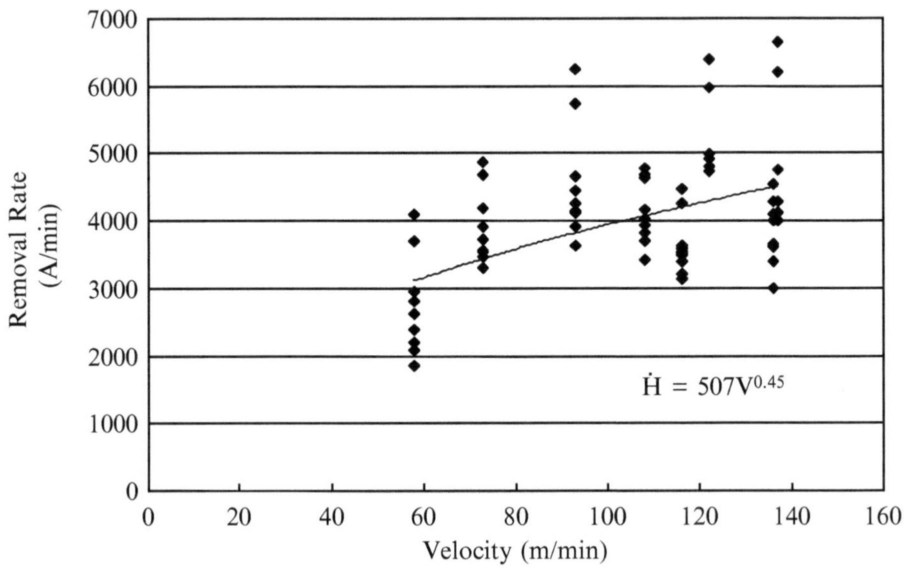

Fig. 4.7 Relationship between removal rate of bare silicon wafer and polishing velocity [34], reprinted with permission

The material removal rate increases with the increasing relative velocity by the power of 0.45, slightly lower than the predicted 0.5. This is attributed to the fact that the wafer is purposefully not held tightly within the carrier of the polishing machine. Designed by the machine builder to avoid the danger of stress-induced cross-wafer cracking, there is a clearance between wafer and inner rim of carrier. The wafer is thus allowed to rotate slightly within the carrier due to the fluid drag; therefore the wafer velocity is slightly lower than the carrier velocity. The actual relative velocity between wafer and pad is less than the calculated velocity assuming full engagement. As a result, the effect of the polishing velocity contributes less

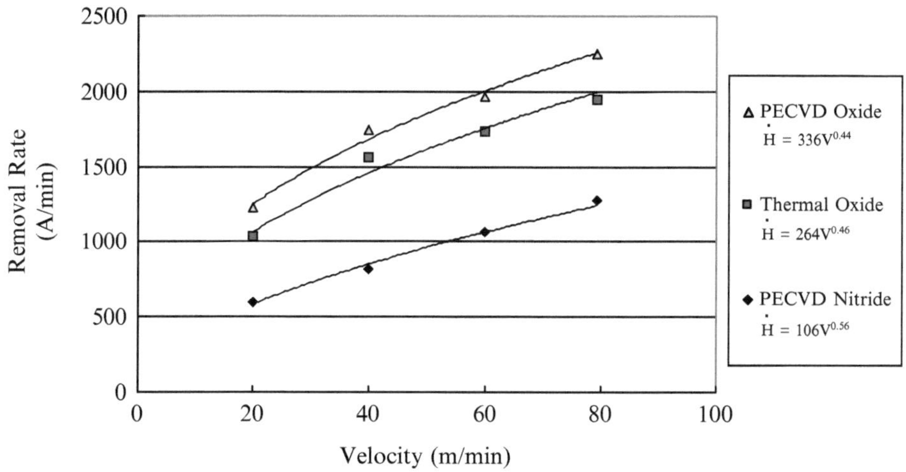

Fig. 4.8 Relationship between removal rate and polishing velocity in sun's experiment [34], reprinted with permission

to the material removal rate than expected. The exponent of the velocity is slightly less than 0.5 in use of this type of machine for polishing bare silicon wafer.

4.2.2.3 Case Findings

Case 1

Sun et al. published the experimental results of the removal rate of dielectrics as a function of the platen rotation speed [39, 40]. It was redrawn in Fig. 4.8 [34]. The material removal rate increases with the increasing velocity by the power between 0.44 and 0.56 for different dielectric thin films. These results exactly fall into the prediction in the current model of removal rate.

Case 2

Tseng and Wang proposed a model to predict the removal rate increasing with the velocity by the power of 0.5 on thermal dioxide film [10, 41]. Their experimental results were redrawn in Fig. 4.9 [34]. One notices that the polishing speed is in fact proportional to the carrier speed [13]. The power law of the velocity lies between 0.42 and 0.54, mostly 0.50. These experimental results agree well with the current model.

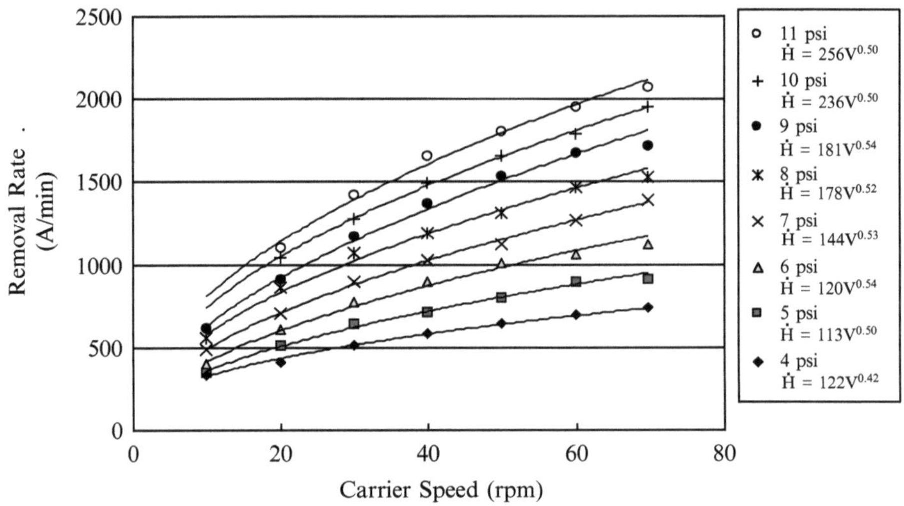

Fig. 4.9 Relationship between removal rate and carrier speed in Tseng's experiment [34], reprinted with permission

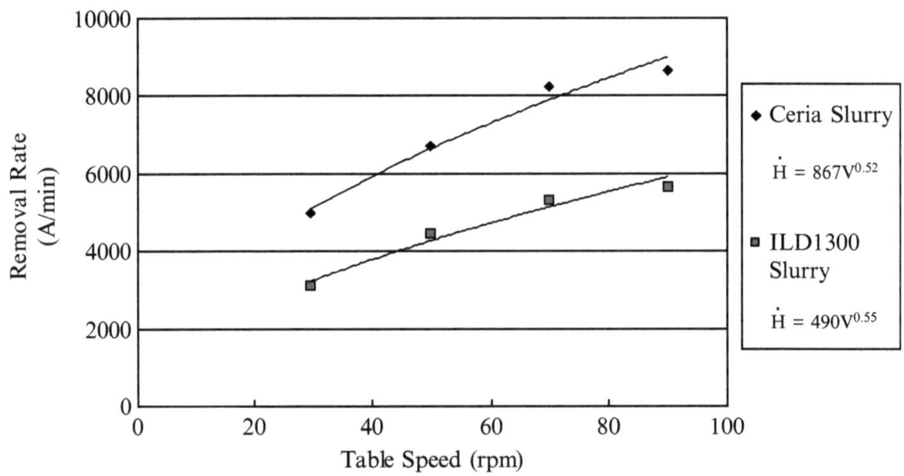

Fig. 4.10 Relationship between removal rate and table speed in Evans' experiment [34], reprinted with permission

Case 3

Evans et al. published the experimental results of the removal rate of dielectrics as a function of the table speed in 1998 [42]. It was redrawn in Fig. 4.10 [34]. One also notices the polishing speed is proportional to the table speed. The oxide removal

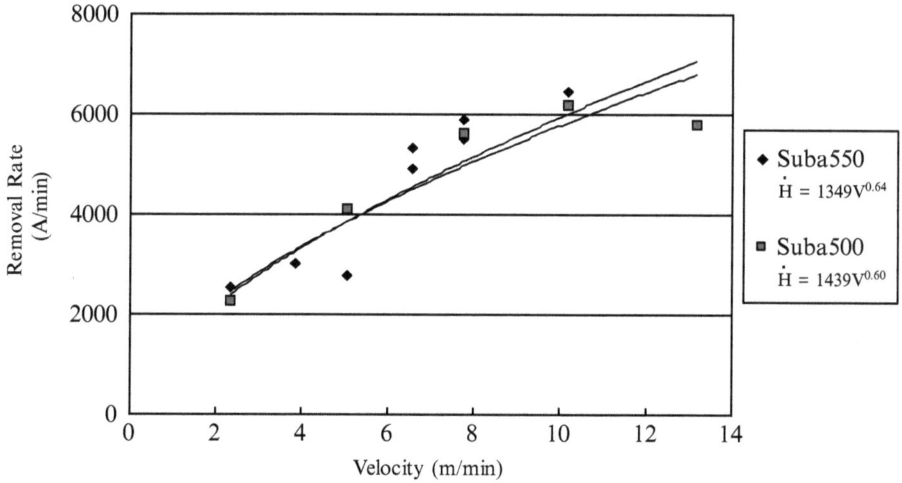

Fig. 4.11 Relationship between removal rate and polishing velocity in Steigerwald's experiment [34], reprinted with permission

rate increases with the increasing velocity by the power of 0.52 and 0.55 for ceria slurry and ILD1300 slurry, respectively. These results follow the prediction in the current study.

Case 4

In the case of metal CMP, Steigerwald et al. showed the experimental results of the copper polish rate versus the polishing velocity using Strassbaugh 6CU polisher, Suba500 pad, and Suba550 pad. The results are redrawn in Fig. 4.11 [40]. The removal rate increases with the increasing polishing velocity by the power of 0.64 and 0.60 for Suba550 and Suba500, respectively. The results are considered a positive reference to the prediction of the current model.

Extended Support

The proposed model also predicts that the removal rate increases with the encounter frequency *f*, which is determined by the solid particle content in the slurry. Bhushan's experimental results prove that the removal rate increases with the increasing solid content [33]. Huang's work concluded when the abrasives and the wafer surface are attracted to each other, the removal rate will be higher where the attraction is stronger [43]. These effects can be well explained by the current model. The slurry particles and the wafer surface being attracted to each other will increase the encounter frequency and the removal rate as a result.

4.3 Effects of Kinematic Variables on Nonuniformity

4.3.1 Kinematic Analysis

4.3.1.1 Polishing Speed and Nonuniformity

Figure 4.12 [44] illustrates the kinematic parameters of CMP. The relative velocity of any point A on the wafer with respect to the pad ($\vec{V}$) is shown as

$$\vec{V} = \vec{V}_c + \vec{u} - \vec{V}_p = \vec{r}_w \times \vec{\omega}_c + \vec{u} - \vec{r}_p \times \vec{\omega}_p \tag{4.19}$$

where $\vec{V}_c$ is the carrier velocity, $\vec{V}_p$ is the pad velocity at that point, and $\vec{u}$ is the carrier translation speed. The magnitude of the velocity is

$$V = \left|\vec{V}\right| = \left|\vec{r}_w \times \vec{\omega}_c - \vec{r}_p \times \vec{\omega}_p + \vec{u}\right|. \tag{4.20}$$

The nonuniform polishing velocity between the wafer and the pad at any point is thus determined by kinematic analysis. The maximum variation of the polishing speed is

$$\begin{aligned} |V_{max} - V_{min}| &= \left|[R \times \omega_c + (e - R) \times \omega_p] - [(e + R) \times \omega_p - R \times \omega_c]\right| \\ &= 2R \times \left|(\omega_p - \omega_c)\right| \end{aligned}. \tag{4.21}$$

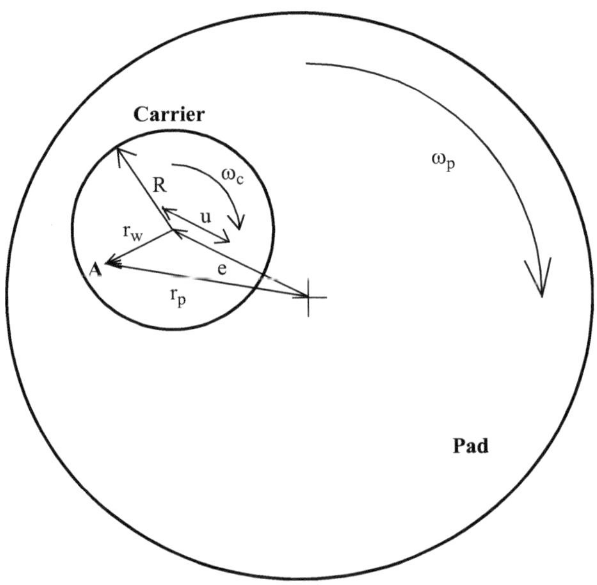

Fig. 4.12 Parameters in the kinematic analysis [44], reprinted with permission

The magnitude of relative speed affects the material removal rate, while its direction does not. From the existing proposed material removal rate models, one recognizes that the material removal rate is proportional to the relative velocity between the pad and the wafer surface by the power of 1 or 0.5 [1, 8, 10, 38]. Whatever the power of the relative velocity can be, the amount of material being polished is proportional to the integral of the relative velocity. Based on the Preston equation, the integral of the velocity along a period of time (T) indicates the amount of material being polished (M),

$$M = \int_0^T \dot{M}\, \mathrm{d}t = K \cdot P \int_0^T V\, \mathrm{d}t = K \cdot P \cdot \tilde{V} \tag{4.22}$$

where K is the proportionality constant, P is the applied pressure, and $\tilde{V}$ is the integral of velocity V. Equation (4.22) can derive the nonuniform material removal resulted from the nonuniform velocity. For this purpose, the nonuniformity in material removal across the wafer can be predicted by

$$\text{Nonuniformity}\,(\%) = \frac{\tilde{V}_{\text{std}}}{\tilde{V}_{\text{avg}}} \times 100\% \tag{4.23}$$

where $\tilde{V}_{\text{std}}$ is the standard deviation of the velocity integral on wafer, and $\tilde{V}_{\text{avg}}$ is the average velocity integral.

Examples of the calculated polishing speed from (4.20) are shown in Fig. 4.13 [44], where the diameter of wafer (D) is 150 mm, and e, the distance between the centers of pad and wafer, is 160 mm, and u is zero. Figure 4.13 shows the platen speed majorly determines the level of the polishing speed, while the carrier speed controls the cycle period. The amplitude of the variation of polishing speed is dependent on the difference of the rotational speed between the platen and the carrier.

4.3.1.2 Effects of Kinematic Variables

Platen Speed ω_{p}

Since r_{p} is much larger than r_{w}, and u is quite small compared with V_{c} and V_{p}, ω_{p} is the dominant of the polishing speed (V). In fact, the orientation of ω_{p} and ω_{c} is set identical; otherwise the variation of the relative speed at a given point will be too large for practice. One notices from Fig. 4.13 that the polishing speed increases with ω_{p}. Figure 4.14 [44] summarizes 16 kinematic conditions during CMP. It tells that the nonuniformity increases with ω_{p} when $\omega_{\text{p}} > \omega_{\text{c}}$, decreases with ω_{p} when $\omega_{\text{p}} < \omega_{\text{c}}$, and the sensitivity of variation increases with ω_{c}.

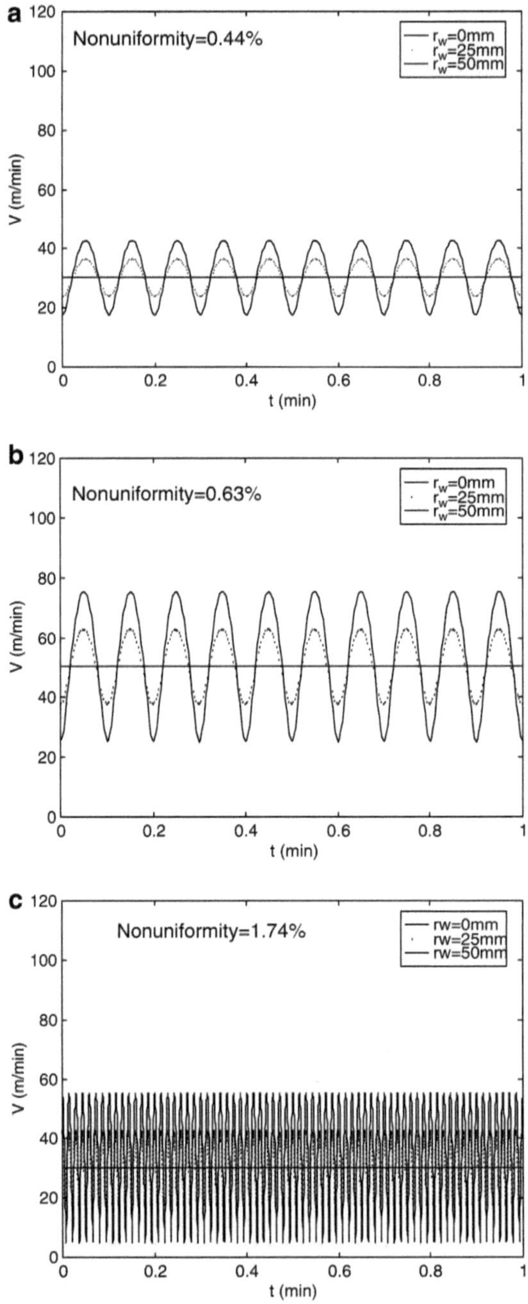

Fig. 4.13 Polishing speed and calculated nonuniformity ($D = 150$ mm, $u = 0, e = 160$ mm) [44], reprinted with permission. (**a**) $\omega_p = 30$ rpm, $\omega_c = 10$ rpm, (**b**) $\omega_p = 50$ rpm, $\omega_c = 10$ rpm, (**c**) $\omega_p = 30$ rpm, ω_c=70 rpm, (**d**) ω_p=50 rpm, ω_c=70 rpm, (**e**) $\omega_p = 30$ rpm, $\omega_c = -10$ rpm

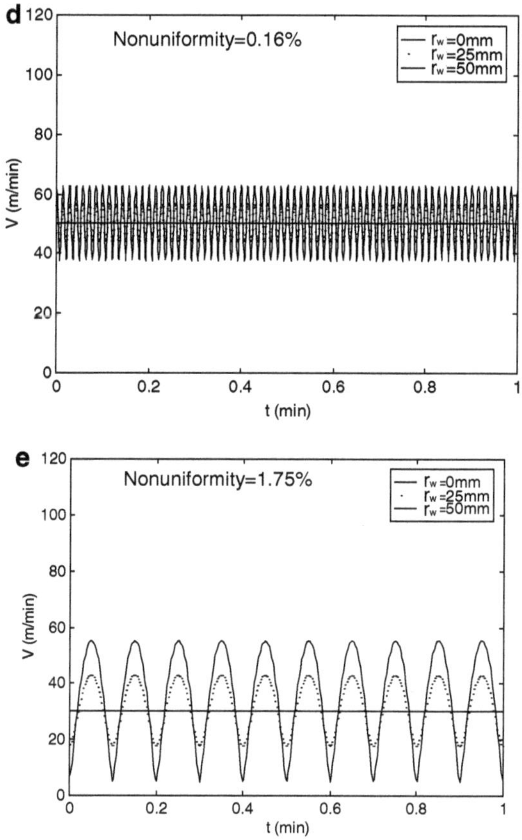

Fig. 4.13 (continued)

Carrier Speed ω_c

ω_c mainly affects the resulted cycle frequency of the polishing speed, the frequency increases with increasing ω_c. Its effects on the nonuniformity is revealed in Sec. Platen Speed. ω_c should have an appropriate value adapted to ω_p, which is usually obtained from the processing experience.

Figure 4.14 predicts that the lowest nonuniformity is zero when $\vec{\omega}_p$ and $\vec{\omega}_c$ are equal. Though this single result is acknowledged by shop floor and researcher, the complete kinematic map is established in the present work. The ratio of ω_c to ω_p can be used to describe the relationship between the rotational speeds and the nonuniformity. Several representative cases are shown in Fig. 4.14 and Table 4.5. In the case of $-1 \leq \frac{\omega_c - \omega_p}{\omega_p} \leq 1$, nonuniformity remains less than 1%, and grows worse to greater than 1% otherwise. The results suggest the process window for a uniform CMP in consideration of the kinematic uniformity as following,

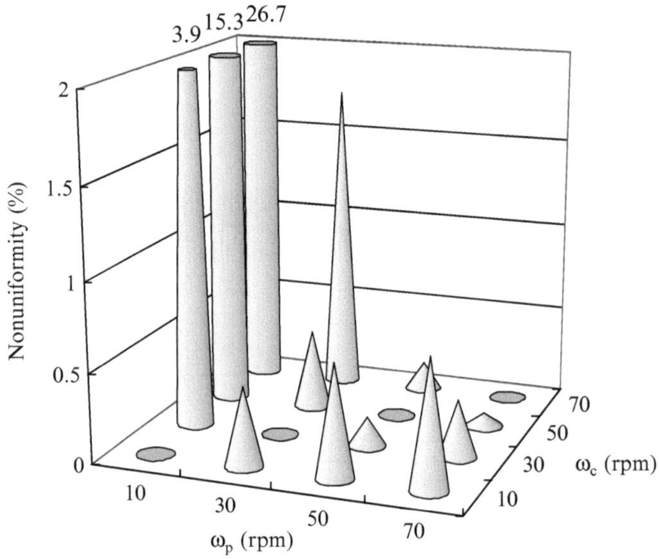

Fig. 4.14 Theoretical relationship between the rotational speeds and nonuniformity (D = 150 mm, $u = 0$, e = 160 mm) [44], reprinted with permission

$$\eta = \left| \frac{\omega_c - \omega_p}{\omega_p} \right| \leq 1 \quad (4.24)$$

where η is called the index of speed window.

This proposed rule implies the rotational speeds of carrier and platen should be close to each other, and the high platen speed is advantageous kinematically. The former is more feasible and will be examined experimentally in the following section. The latter, however, is limited by the healthy slurry distribution and the chemical interaction between particles and wafer. Too large rotational speed of platen cannot provide sufficient time for the both conditions vital for the successful planarization. Figure 4.15 [44] shows the relationship between the nonuniformity and the speed window index η for carrier eccentricity of 160 mm. It shows the nonuniformity increases with increasing deviation between carrier speed and platen speed. However, the effects are milder at the range shown by (4.24), and the nonuniformity becomes absolutely unacceptable when η goes beyond 2.

Carrier Eccentricity *e*

Figure 4.16 [44] and Table 4.5 show the nonuniformity is slightly improved when the carrier translates at large eccentricity relative to the pad center, worsened at small eccentricity. The polishing speed increases gradually as the carrier translates outward on the pad, where the pad velocity increases. The influence is similar to

Table 4.5 Calculated nonuniformity

			Nonuniformity (%)								
			$u = 0$, $e = 160$ mm			$u = 1$ mm/s, $e = 160$–220 mm			$u = 1$ mm/s, $e = 230$–290 mm		
ω_p (rpm)	ω_c (rpm)	η	$D = 150$ mm	$D = 200$ mm	$D = 300$ mm	$D = 150$ mm	$D = 200$ mm	$D = 300$ mm	$D = 150$ mm	$D = 200$ mm	$D = 300$ mm
10	10	0.00	0.00	0.00	0.00	0.00	0.00	0.00	0.00	0.00	0.00
10	30	2.00	3.86	6.65	15.29	8.25	14.25	32.71	4.43	7.68	17.36
10	50	4.00	15.29	23.19	32.78	32.61	54.42	87.74	17.36	30.14	63.25
10	70	6.00	26.66	32.60	38.59	65.33	86.98	109.0	39.26	62.38	93.59
30	10	0.67	0.44	0.76	1.74	0.94	1.63	3.72	0.50	0.87	1.99
30	30	0.00	0.00	0.00	0.00	0.00	0.00	0.00	0.00	0.00	0.00
30	50	0.67	0.44	0.76	1.74	0.94	1.63	3.72	0.50	0.87	1.99
30	70	1.33	1.74	3.00	6.78	3.72	6.46	14.63	1.99	3.46	7.87
50	10	0.80	0.63	1.09	2.49	1.35	2.35	5.35	0.72	1.25	2.86
50	30	0.40	0.16	0.27	0.63	0.34	0.59	1.35	0.18	0.31	0.72
50	50	0.00	0.00	0.00	0.00	0.00	0.00	0.00	0.00	0.00	0.00
50	70	0.40	0.16	0.27	0.63	0.34	0.59	1.35	0.18	0.31	0.72
70	10	0.86	0.72	1.25	2.85	1.55	2.69	6.14	0.83	1.44	3.28
70	30	0.57	0.32	0.56	1.28	0.69	1.20	2.75	0.37	0.64	1.47
70	50	0.29	0.08	0.14	0.32	0.17	0.30	0.69	0.19	0.16	0.37
70	70	0.00	0.00	0.00	0.00	0.00	0.00	0.00	0.00	0.00	0.00

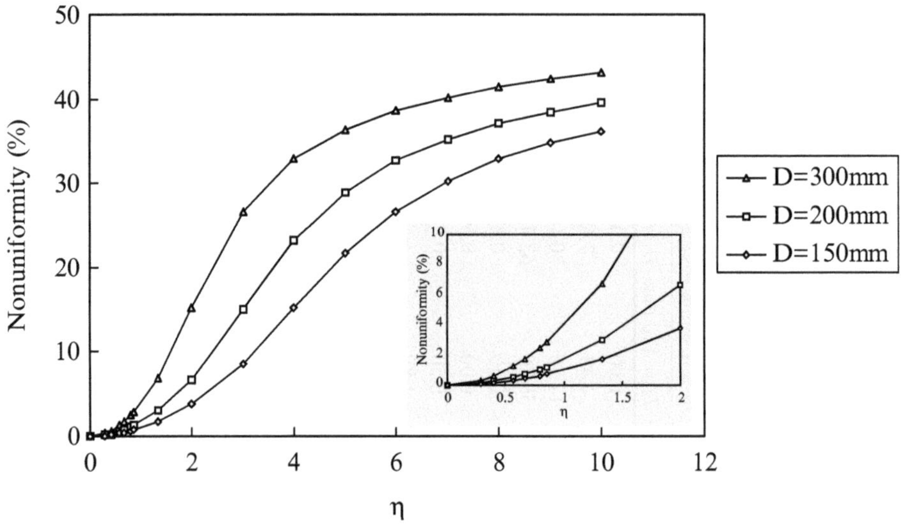

Fig. 4.15 Relationship between nonuniformity and the speed window index ($u = 0, e = 160$ mm) [44], reprinted with permission

that discussed in Sec. Platen Speed. The relationship between the nonuniformity and the eccentricity e at varying speed window index (η) is shown in Fig. 4.17 [44]; the nonuniformity increases with increasing eccentricity. The far rim of pad provides large polishing speed relative to carrier speed, which is advantageous for the kinematic uniformity, as discussed in Sec. Platen Speed and Carrier Speed.

Carrier Translation Speed u

The effect of carrier translation on the nonuniformity is not significant, as shown in Table 4.5 and Fig. 4.18 [44]. The results of $u = 0$ occur at $e = 160$ mm. The translation speed plays no major role in planarization from the kinematic point of view. The current application of carrier translation in CMP, however, is primarily attributed to the maximum utilization of the pad area.

Wafer Size D

Table 4.5 and Figure 4.19 [44] show the nonuniformity drastically increases with the wafer size (D). The currently discussed kinematic effects possess increasing significance as the challenge of larger wafer size and lower nonuniformity becomes more rigorous. The process window of the kinematic conditions will be significantly narrowed when large wafer is planarized.

4.3.1.3 Absolute Uniformity of Velocity

When $\vec{\omega}_p$ equals to $\vec{\omega}_c$, (4.20) becomes

$$V = \left|(\vec{r}_w - \vec{r}_p) \times \vec{\omega}_c + \vec{u}\right| = \left|\vec{u} - \vec{e} \times \vec{\omega}_c\right|. \tag{4.25}$$

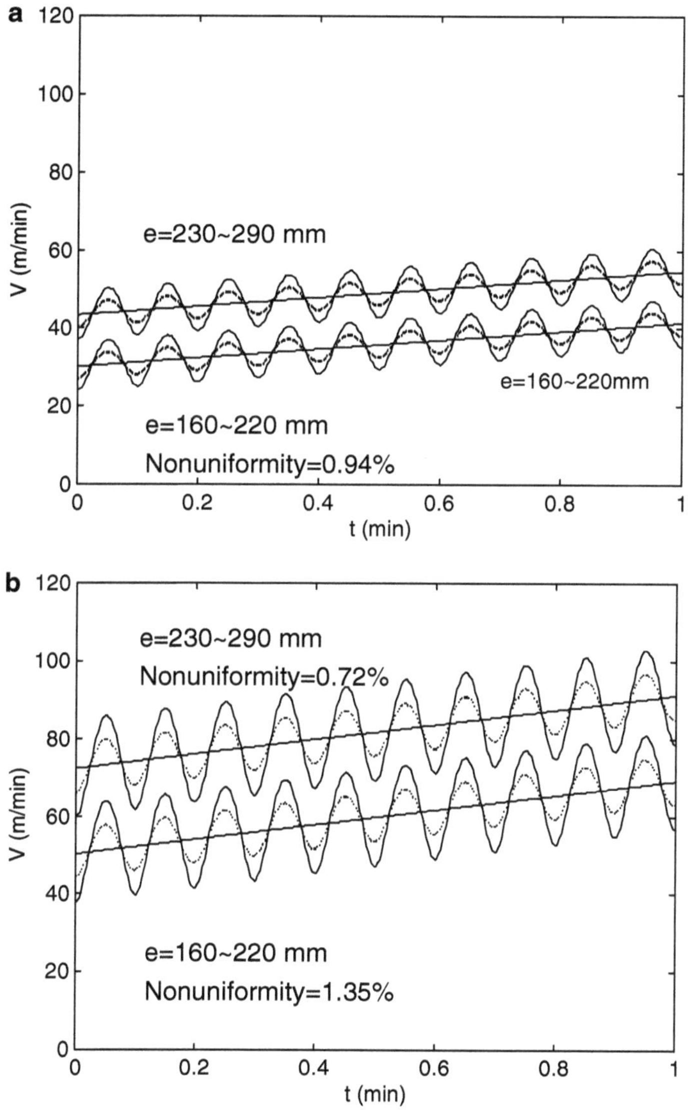

Fig. 4.16 Polishing velocity associated with the carrier translation ($u = 1$ mm/s) [44], reprinted with permission. (**a**) $\omega_p = 30$ rpm, $\omega_c = 10$ rpm, (**b**) $\omega_p = 50$ rpm, $\omega_c = 10$ rpm, (**c**) $\omega_p = 30$ rpm, $\omega_c = 70$ rpm, (**d**) $\omega_p = 50$ rpm, $\omega_c = 70$ rpm

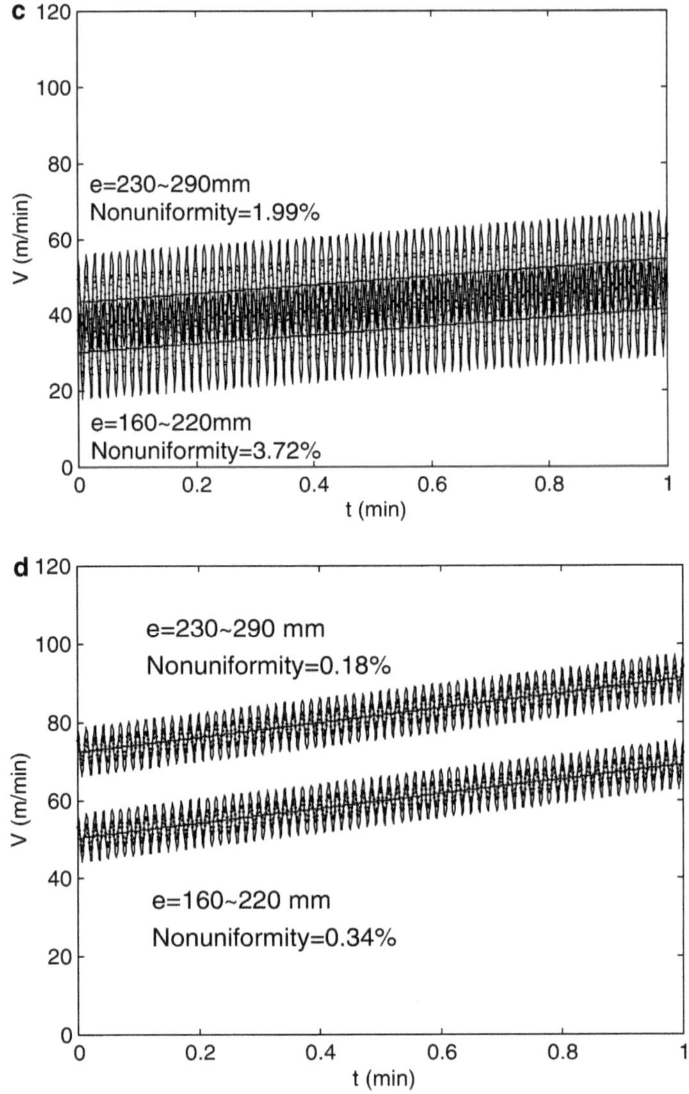

Fig. 4.16 (continued)

Since the values of e, ω_c and u are preselected, the polishing velocity will be kept constant, that is, will not vary with the rotation of both platen and wafer. While such an observation has been reported [24, 25], the significance of the derived results are not fully explored. In this case, the relative velocity is constant at any point across the wafer surface and is called the *absolute velocity uniformity*. The platen and carrier rotate so that any point on the wafer surface can be polished uniformly, that

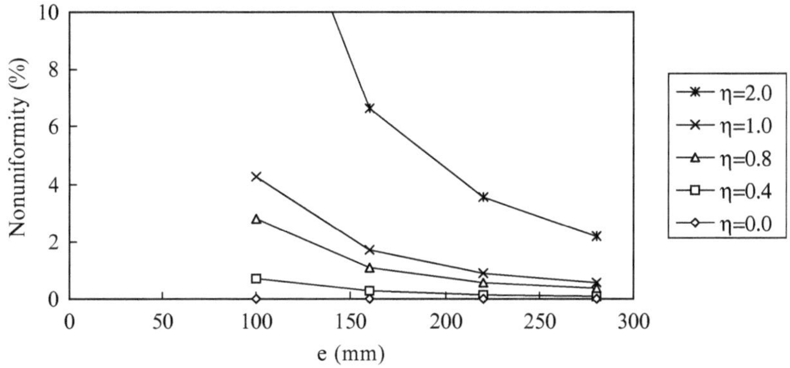

Fig. 4.17 Relationship between nonuniformity and eccentricity ($D = 200$ mm, $u = 0$) [44], reprinted with permission

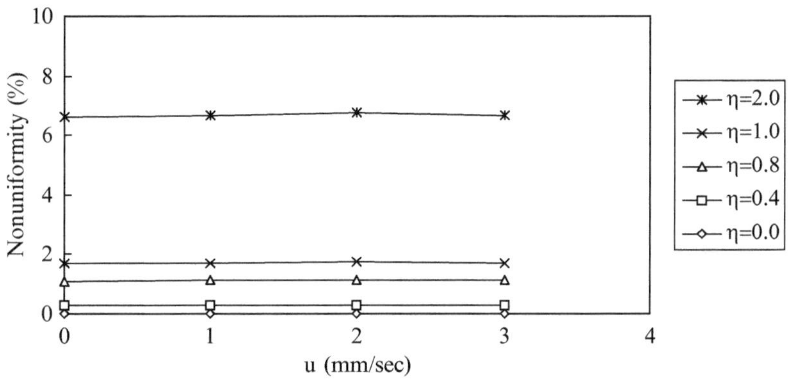

Fig. 4.18 Relationship between nonuniformity and translation speed ($D = 200$ mm, $e = 130$–190 mm) [44], reprinted with permission

is, the amount of material removal remains constant based on Preston equation. In this case, the nonuniformity is kinematically zero due to the constant velocity at any point on the wafer surface. In a complete rotational period of the wafer, the polishing direction finishes a cycle of 360°, as shown in Fig. 4.20a [44].

When the eccentricity becomes zero, the relative speed is linearly proportional to r_w, as shown in Fig. 4.20b [44]. The velocity field on the wafer is called concentric velocity uniformity. When the eccentricity is infinite, the polishing speed approaches to $r_p \times \omega_p$, which is the case of a linear abrasive-belt polishing in one direction, as shown in Fig. 4.20c [44]. The velocity pattern is called linear velocity uniformity. The case of absolute velocity uniformity is experimentally investigated in the following.

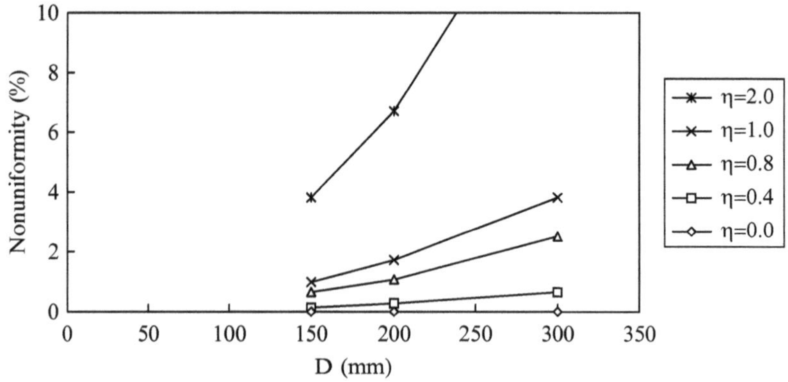

Fig. 4.19 Relationship between nonuniformity and wafer size ($u = 0$, $e = 160$ mm) [44], reprinted with permission

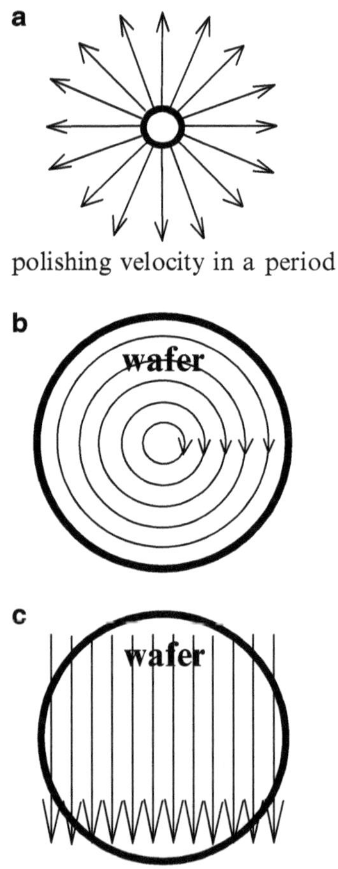

Fig. 4.20 Special cases of uniform velocity [44], reprinted with permission. (**a**) Absolute velocity uniformity ($\vec{\omega}_p = \vec{\omega}_c$), (**b**) concentric velocity uniformity ($e = 0$), (**c**) linear velocity uniformity ($e \to \infty$)

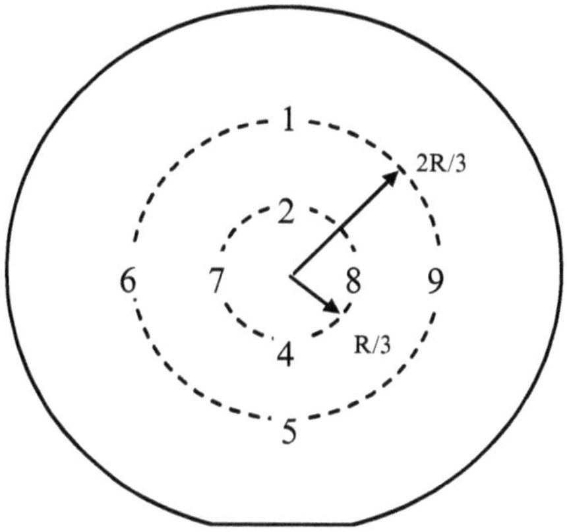

Fig. 4.21 The locations of measurement on the wafer [44], reprinted with permission

Table 4.6 Experimental parameters

Down force (psi)	10, 8, 5
Back pressure (psi)	3
Platen speed (rpm)	10, 30, 50 ,70
Carrier speed (rpm)	10, 30, 50, 70
Carrier translation (mm/s)	1

4.3.2 Experimental Findings

All the test samples in the present study were grown on P-type (100) 150 mm silicon wafers. The films of silicon dioxide were produced by wet oxidation, in which the silicon was exposed to the oxidizing ambient H_2, O_2 at 980°C. The experiments were carried out on a Westech 372M polisher using IC 1400 pad and CABOT SS25 slurry. The slurry consists of fumed silica particles suspended in alkali solution. After each polishing, the pad was conditioned. The thickness of the dielectric films was measured by Nanometrics 2100XP at nine different spots across the wafer, as shown in Fig. 4.21 [44]. Table 4.6 shows the experimental parameters. The polish rate is defined as the removal rate averaged over the nine locations.

Figure 4.20a tells the zero nonuniformity occurs when the platen speed equals to the carrier speed in consideration of the kinematic uniformity. As discussed in Sec. Carrier Speed, good nonuniformity is expected when ω_c and ω_p are close to each other. Figure 4.22 [44] shows the experimental relationship between the rotational speeds and the nonuniformity at different down pressures. One identifies that low nonuniformity occurs along the diagonal of $\omega_p = \omega_c$ in most cases. High values of nonuniformity are located at large ω_c and low ω_p, as predicted in Fig. 4.15. The experimental kinematic conditions for low nonuniformity at various down pressure are summarized in Fig. 4.23 [44]. One notices that whatever down pressure is, low nonuniformity occurs when the carrier speed is close to the platen speed. For example, the lowest nonuniformity at platen speed 50 rpm lies at the carrier speed of 50 rpm. This result agrees with the kinematic analysis.

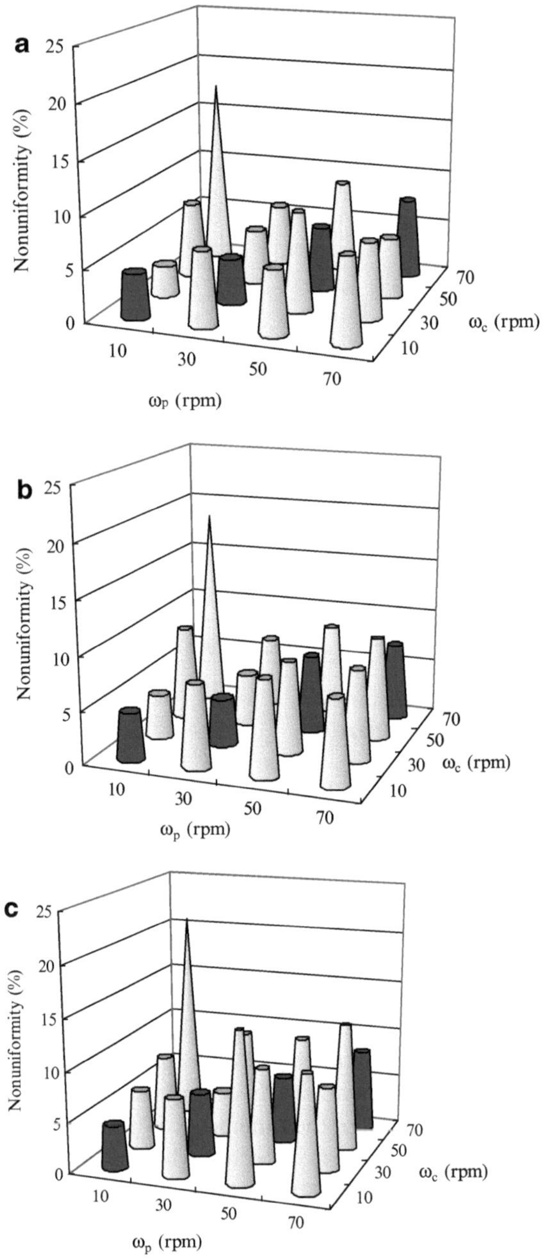

Fig. 4.22 The experimental relationship between the rotational speeds and nonuniformity [44], reprinted with permission. (**a**) Down pressure 10 psi, (**b**) down pressure 8 psi, (**c**) down pressure 5 psi

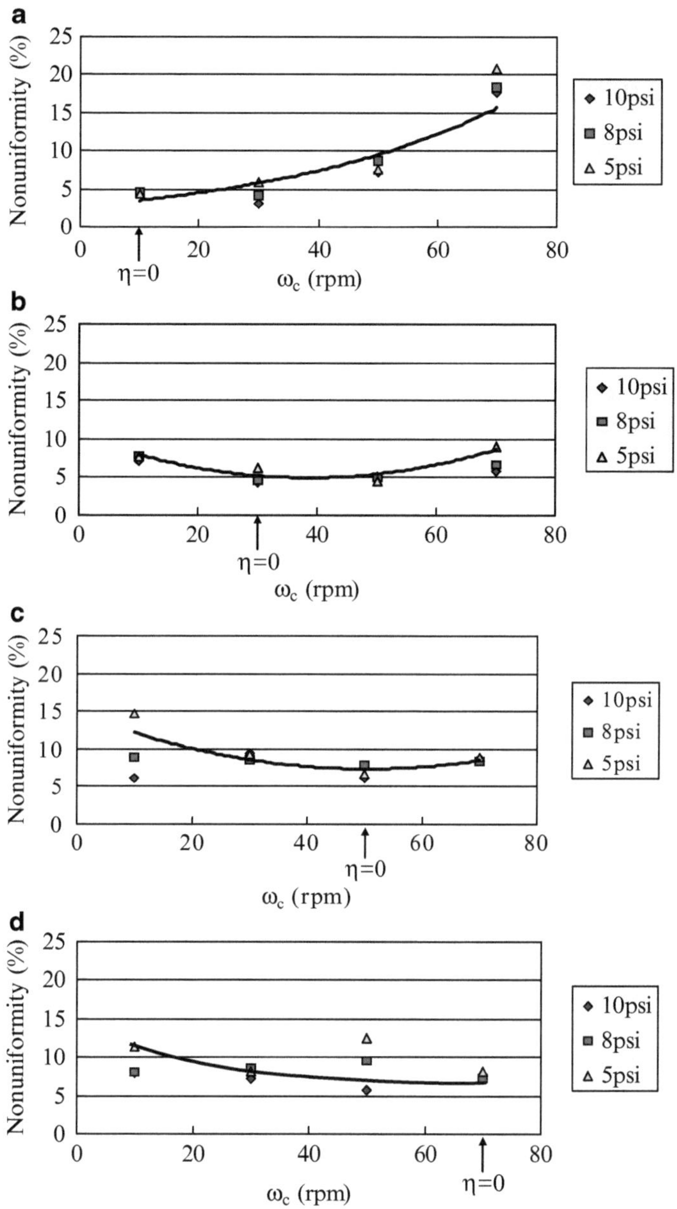

Fig. 4.23 The experimental kinematic conditions for low nonuniformity [44], reprinted with permission. (**a**) $\omega_p = 10$ rpm, (**b**) $\omega_p = 30$ rpm, (**c**) $\omega_p = 50$ rpm, (**d**) $\omega_p = 70$ rpm

4.4 In Situ Endpoint Detection in Polishing of Copper Overlay

4.4.1 By Pad Temperature

4.4.1.1 Principle of Thermal Endpoint Detection

Kinematics Analysis in CMP

During a metal CMP process, it is desired to remove the excess metal layer on dielectric film. At the beginning of the process, the area of the metal film A_{metal} being polished by the pad equals to the wafer surface area A_{wafer}. When the area of the metal film is reduced to that of metal via, the endpoint is reached. Besides, the polish arm exerts a downward pressure P onto the wafer, and the coefficients of friction μ_{metal} and $\mu_{\text{dielectric}}$ exist between the pad and the metal layer and the dielectric layer, respectively. The resulting frictional forces are

$$F_{\text{f,metal}} = \mu_{\text{metal}} \times A_{\text{metal}} \times P \tag{4.26}$$

$$F_{\text{f,dielectric}} = \mu_{\text{dielectric}} \times (A_{\text{wafer}} - A_{\text{metal}}) \times P \tag{4.27}$$

where $F_{f,\text{metal}}$ is the frictional force between the pad and the metal film, and $F_{f,\text{dielectric}}$ is the frictional force between the pad and the dielectric film. The energy consumed by the two layers during polishing is

$$U_{\text{metal}} = F_{\text{f,metal}} \times V_{\text{metal}} \times t \tag{4.28}$$

$$U_{\text{dielectric}} = F_{\text{f,dielectric}} \times V_{\text{dielectric}} \times t \tag{4.29}$$

where V_{metal} and $V_{\text{dielectric}}$ denote the relative speed between the pad and the metal layer and the dielectric layer, respectively, and t represents the polishing time. Thus the total consumed energy is

$$U_{\text{total}} = U_{\text{metal}} + U_{\text{dielectric}} \tag{4.30}$$

Substituting (4.26) to (4.29) into (4.30),

$$\begin{aligned} U_{\text{total}} &= F_{\text{f,metal}} \times V_{\text{metal}} \times t + F_{\text{f,dielectric}} \times V_{\text{dielectric}} \times t \\ &= \mu_{\text{metal}} \times A_{\text{metal}} \times P \times V_{\text{metal}} \times t + \mu_{\text{dielectric}}(A_{\text{wafer}} - A_{\text{metal}}) \\ &\quad \times P \times V_{\text{dielectric}} \times t \\ &= P \times t \times [\mu_{\text{metal}} \times A_{\text{metal}} \times V_{\text{metal}} \\ &\quad + \mu_{\text{dielectric}} \times (A_{\text{wafer}} - A_{\text{metal}}) \times V_{\text{dielectric}}] \\ &= P \times t \times A_{\text{wafer}} \times [\mu_{\text{metal}} \times A_p \times V_{\text{metal}} \\ &\quad + \mu_{\text{dielectric}} \times (1 - A_p) \times V_{\text{dielectric}}] \end{aligned} \tag{4.31}$$

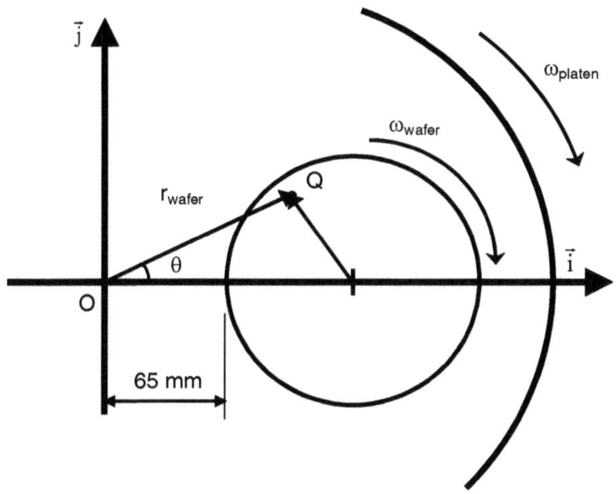

Fig. 4.24 Definition of kinematic conditions [45], reprinted with permission

where the partial area $A_p = A_{metal}/A_{wafer}$ is the ratio of the area of metal layer to wafer area and ranges from 1 to 0.

One can see that a proportionality constant exists between the energy consumption and the polishing speed (4.28 and 4.29). The consumed energy eventually dissipates as heat causes a rise in pad temperature. The relative polishing speed V_{metal} and $V_{dielectric}$ of any point on the pad with respect to the wafer is shown as

$$
\begin{aligned}
V &= \left|\vec{V}_{wafer} + \vec{V}_{oscillation} - \vec{V}_{platen}\right| \\
&= \left|\vec{\omega}_{wafer} \times \vec{r}_{wafer} + \vec{V}_{oscillation} - \vec{\omega}_{platen} \times \vec{r}_{platen}\right| \\
&\left\| \begin{vmatrix} \vec{i} & \vec{j} & \vec{k} \\ 0 & 0 & \omega_{wafer} \\ r_{wafer}\cos\theta - X_{wafer} & r_{wafer}\sin\theta & 0 \end{vmatrix} + V_{oscillation}\vec{i} \right. \\
&\left. + \begin{vmatrix} \vec{i} & \vec{j} & \vec{k} \\ 0 & 0 & \omega_{platen} \\ r_{wafer}\cos\theta & r_{wafer}\sin\theta & 0 \end{vmatrix} \right\| \\
&\times \left| [V_{oscillation} + r_{wafer}\sin\theta \times (\omega_{platen} - \omega_{wafer})]\vec{i} \right. \\
&\left. + L[r_{wafer} \times \cos\theta \times (\omega_{wafer} - \omega_{platen}) - X_{wafer} \times \omega_{wafer}]\vec{j} \right|
\end{aligned} \tag{4.32}
$$

where $\vec{V}_{wafer}$ denotes the wafer velocity at the point, $\vec{V}_{platen}$ represents the platen velocity at that point, and $\vec{V}_{oscillation}$ is the carrier oscillation speed. Figure 4.24 [45] presents the definition of the parameters, where for a point Q

$$\theta = \omega_{platen} \times t. \tag{4.33}$$

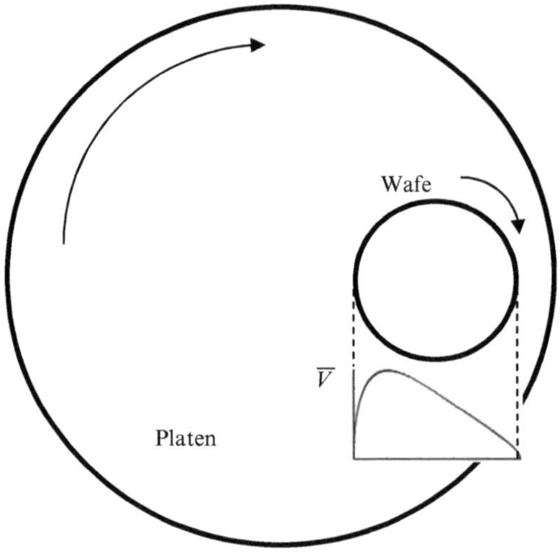

Fig. 4.25 Distribution of speed integral on pad during CMP [45], reprinted with permission

Substituting Eq. (8) into Eq. (7),

$$V = |[V_{\text{oscillation}} + r_{\text{wafer}} \sin(\omega_{\text{platen}} \cdot t) \times (\omega_{\text{platen}} - \omega_{\text{wafer}})]\vec{i} + [r_{\text{wafer}} \times \cos(\omega_{\text{platen}} \cdot t) \times (\omega_{\text{wafer}} - \omega_{\text{platen}}) - X_{\text{wafer}} \times \omega_{\text{wafer}}]\vec{j}\,|. \quad (4.34)$$

In CMP process, both the pad and the wafer rotate; hence the polishing speed is varying across the pad, from the moment the pad touches the wafer to the moment the pad loses contact with the wafer. In other words, the total consumed kinematic energy during the contact period can be calculated by the substitution of (4.34) into (4.31) and by the integral of the polishing speed over that period,

$$\begin{aligned} U_{\text{total}} &= PA_{\text{wafer}} \int_{t_1}^{t_2} [\mu_{\text{metal}} A_{\text{p}} V_{\text{metal}}(t) + \mu_{\text{dielectric}}(1 - A_{\text{p}})\, V_{\text{dielectric}}(t)]\text{d}t \\ &= P\,A_{\text{wafer}}[\mu_{\text{metal}} A_{\text{p}} \int_{t_1}^{t_2} V_{\text{metal}}(t)\, \text{d}t + \mu_{\text{dielectric}}(1 - A_{\text{p}}) \int_{t_1}^{t_2} V_{\text{dielectric}}(t)\text{d}t] \\ &= P\, A_{\text{wafer}}[\mu_{\text{metal}} A_{\text{p}} \bar{V}_{\text{metal}} + \mu_{\text{dielectric}}(1 - A_{\text{p}})\bar{V}_{\text{dielectric}}] \end{aligned} \quad (4.35)$$

where $\bar{V}_{\text{metal}}$ and $\bar{V}_{\text{dielectric}}$ are called the speed integrals. In (4.35), A_{p} remains constant during the period from t_1 to t_2. t_1 is the time when the pad touches the wafer, and t_2 is the time when the pad leaves the wafer.

Figure 4.25 [45] shows the speed integral along the diameter of wafer in radial direction on pad while polishing the copper film on wafer surface by (4.35) as an example. $\bar{V}_{\text{metal}}$ and $\bar{V}_{\text{dielectric}}$ are the integrals of the velocity over the entire contact period for a point on the pad. Thus they are constant at the same radius on the pad under a preset kinematic polishing condition. The speed integral on each point of

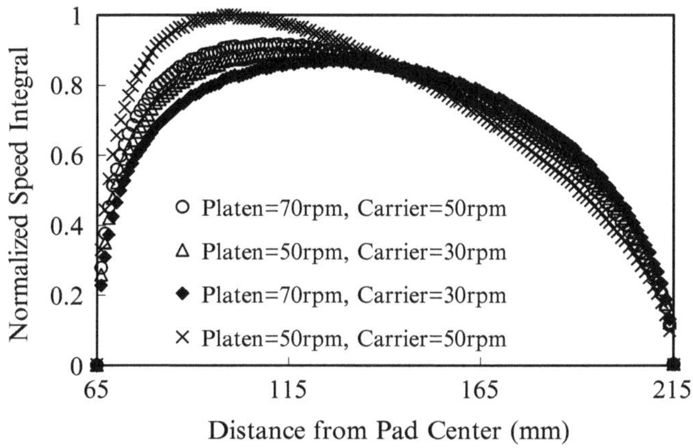

Fig. 4.26 Speed integral at various polishing conditions [45], reprinted with permission

pad associated with the down pressure and the coefficient of friction between wafer and pad determine the amount of energy consumed during the period of the kinematics process, as seen in (4.35). Since this energy eventually turns to heat, there exists a correlation between the consumed energy and the temperature rise

$$\begin{aligned} T &= K\ U_{\text{total}} \\ &= K\ P A_{\text{wafer}}\left[\mu_{\text{metal}} A_p \bar{V}_{\text{metal}} + \mu_{\text{dielectric}}(1 - A_{\text{p}})\bar{V}_{\text{dielectric}}\right]. \end{aligned} \tag{4.36}$$

where K is a proportionality constant found by substitution of the coefficient of friction between copper film and pad (found in reference [42], μ_{metal}) and that between oxide film and pad (found in reference [43], $\mu_{\text{dielectric}}$) into (4.36). Secondly, the measured ΔT, the preset P and areas, and the calculated $\bar{V}_{\text{metal}}$ and $\bar{V}_{\text{dielectric}}$, all acquired from the experiment, were substituted into (4.36). K was then calculated. It varies with the thermal properties of pad material, slurry, and the environment. Equation (4.36) shows the speed integral indeed can be used as an indicator of the pad temperature in CMP. Figure 4.26 [45] shows the speed integral of different experimental kinematic conditions.

In the above analysis, the following assumptions were made. (1) There was minor heat conduction through pad, wafer, slurry, and atmosphere. (2) The amount of heat produced from the exothermic chemical reaction was insignificant compared to the heat generated by mechanical friction, since the amount of polished copper material is very small. (3) The wafer was in full contact with pad during process. (4) The pressure exerted by the carrier was constant through the process. (5) The rotational rates of both platen and carrier were constant through the process.

Endpoint Monitoring

In (4.36), one can see that the pad temperature rise varies with down force, coefficients of friction, speed integral, and the area of metal layer and dielectric layer. During the

Table 4.7 Kinematic parameter design for copper wafers

Wafer No.	Platen speed (rpm)	Carrier speed (rpm)
1	70	50
2	50	30
3	70	30
4	50	50

process, the down force, the kinematic parameter, and the coefficients of friction are constant, while the area of the metal layer (or the dielectric layer) varies. At the beginning of the process, the pad polishes the metal layer only. Thus, a part of the metal layer is removed as transition, until the endpoint of the CMP process, when the pad polishes the dielectric layer. The portion of metal via is relatively small. By analyzing the behavior of temperature rise, one can tell that the slope of the temperature rise keeps constant at the beginning when metal layer only is being polished. The slope varies when metal layer and dielectric layer are both being polished in the transition. When the slope shows constant behavior again, which means that only dielectric layer is being polished, the endpoint is identified.

4.4.1.2 Experimental Findings

The IPEC 372M polisher of single-head and dual-platen wafer polisher with a 6-inch wafer carrier is used. The experimental consumables include a RODEL IC-1400K-grooved pad attached to the primary platen, with a RODEL Politex pad on the final platen. The backing film on the carrier assembly is a RODEL DF-200 carrier film. Pad conditioning is performed with a nylon brush after each polishing. The slurry of pH value 3–4 with Al_2O_3 abrasive with particle size of 0.1–0.3 μm was used.

The wafers used in this study were 6-inch with 1 μm of thermal oxide as the dielectric layer. The wafer surface was sputtered with a very thin layer of gold to improve the adhesion between the copper layer and the oxide layer. Then the wafers were electroplated with 1,500 Å copper film. Nine points on the wafers were measured by a four-point probe for actual copper film thickness.

The down pressure and the backside pressure in this study were set constant as 4.0 psi and 0.2 psi, respectively. Table 4.7 lists the platen speeds and the carrier speeds in this study. The slurry flow rate was set as 100 ml/min. The carrier did not oscillate during the process. The thermal image camera AGEMA Thermovision® 900 System was used to measure the pad temperature during CMP processes, as depicted in Fig. 4.27 [45]. This image system has the capability of capturing two-dimensional thermal responses on the polishing pad with a temperature resolution of 0.1 K and 272 × 136 pixels in each image. The continuous recording of the thermal data sequence is also performed. The measured thermal image through the transparent media is calibrated every time before experiment, thus the coherence of the measured area is secured.

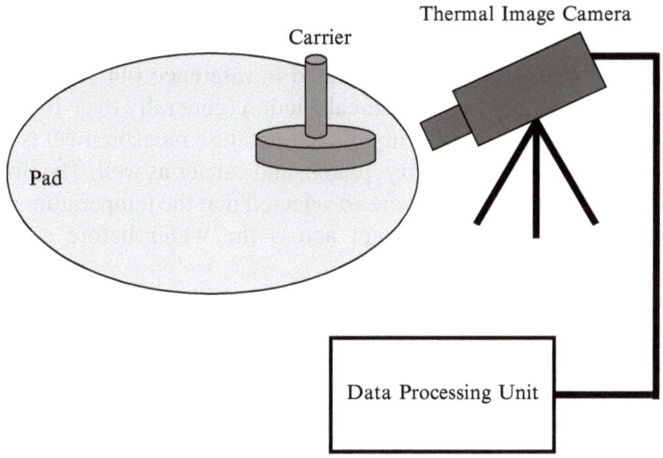

Fig. 4.27 Experimental setup [45], reprinted with permission

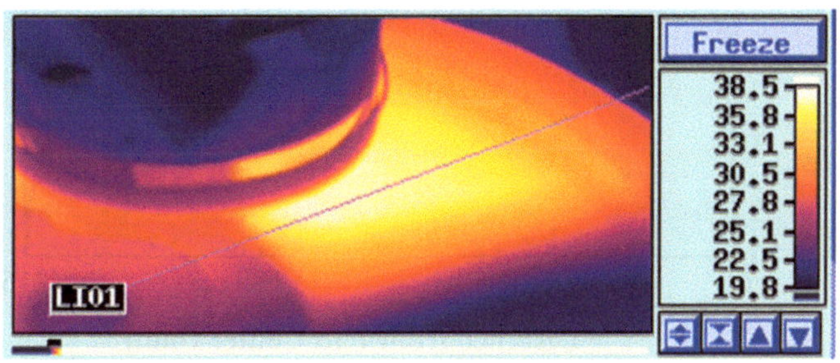

Fig. 4.28 Thermal image on pad [45], reprinted with permission

Pad Temperature Measurement and Analysis

The thermal images obtained from the experiments display the temperature of the working area. The measurement path was targeted at the part of the pad that the wafer just passes through and extends from the pad center to the outskirt of the pad, as the LI01 shown in Fig. 4.28 [45]. One can see that the center region of the pad, which does not polish the wafer, is cooler than the polishing area. Another cool region is at the edge of the pad, which is cooled by the environment. The warm area is where the major polishing takes place.

The temperature of the center part of the pad is used as the reference temperature, and the differences between the measured temperature along the scanning path and the reference temperature are used to verify the analysis of (4.36). The metal oxide molecules do require energy while being removed from the bulk. But the current experiment discusses a macroscopic temperature variation over the entire polishing pad rather than the microscopic energy dissipation. Thus the microstructural breakage was not considered in this chapter. The amount of heat generated by the exothermic reaction was not measured in this study. However, the thickness of the removed copper layer is so thin, that the amount of heat produced by the reaction (maximum a few watts, 0.89 W as indicated in reference [44]) is negligible compared to the heat generated by mechanical friction (generally over 100 W, 180 W in current case). Another concern during the temperature measurement is that the heat can dissipate through open air, slurry, platen, and carrier as well. The timing and the locations of measurement on pad were so selected that the temperature was detected while these points on pad just swept across the wafer before significant heat dissipation could take place.

Figure 4.29 [45] shows the comparisons between kinematics analysis and experimental results of the temperature rise. The temperature measurement shows minor

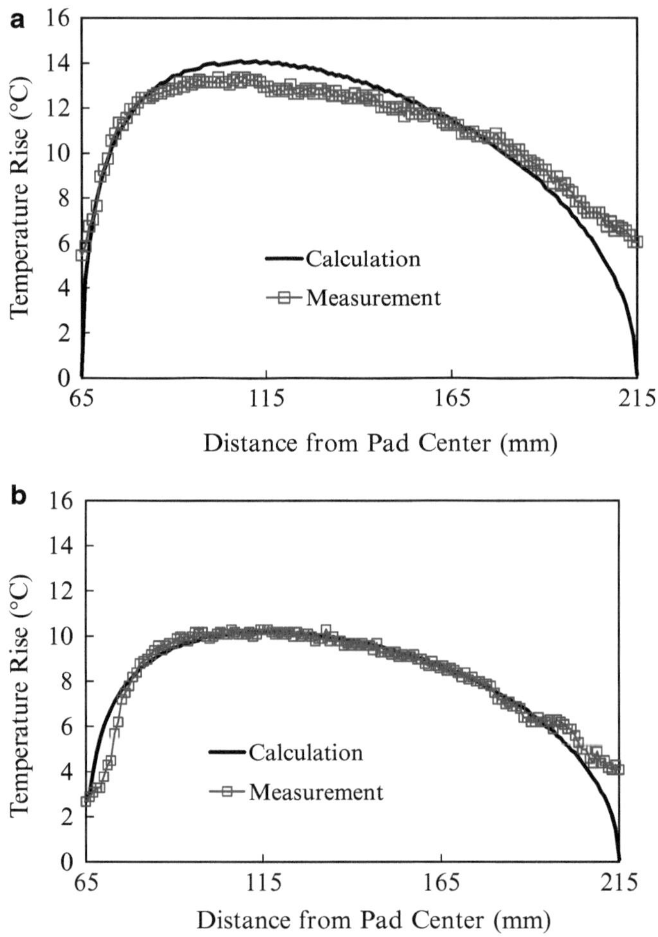

Fig. 4.29 Comparison of calculations and measured temperature rise on pad [45], reprinted with permission. (**a**) Platen speed = 70 rpm and carrier speed = 50 rpm, (**b**) platen speed = 50 rpm and carrier speed = 30 rpm, (**c**) platen speed = 70 rpm and carrier speed = 30 rpm, (**d**) platen speed = 50 rpm and carrier speed = 50 rpm

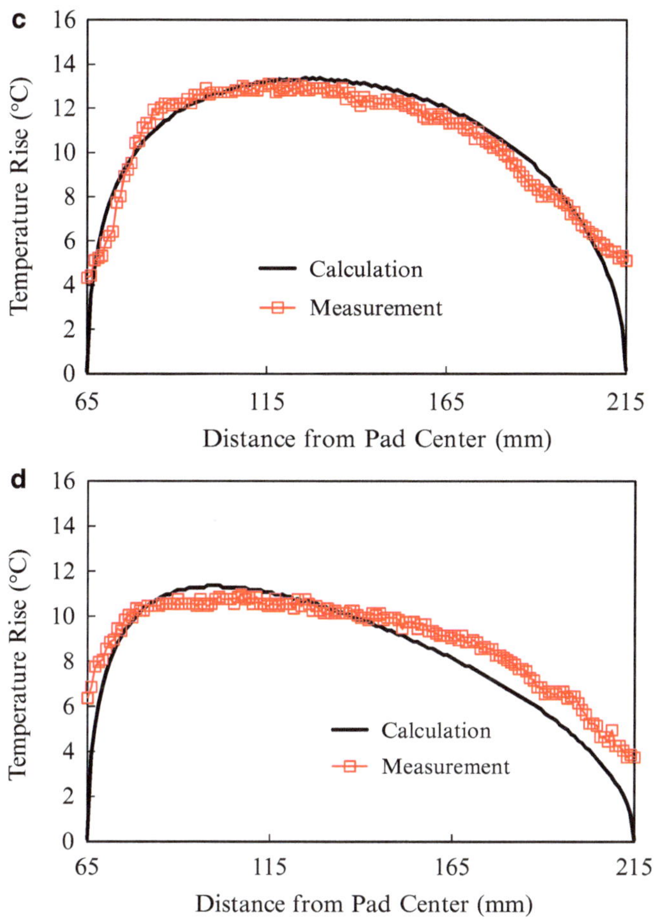

Fig. 4.29 (continued)

scattering due to the local disturbance of the slurry flow when the thermal images were taken. Figure 4.30 [45] shows that the proposed kinematics analysis is able to describe the temperature rise across the pad.

In (4.36), the coefficient of friction between the metal film and the pad and the coefficient of friction between the dielectric film and the pad are essential data. The coefficient of friction between copper film and pad was measured 0.7 [46], and the coefficient of friction between oxide film and pad was reported 0.35 [47], smaller than the former. With substitution of the coefficients of friction into (4.36), the speed integral under different copper polishing conditions can be obtained and the constant K is found 0.003. The uncertainty in the measurements will induce a linear shift of the calculation as shown in (4.36). However, the monitoring strategy is based on the pattern transition of the temperature rise (generated by the intrinsic difference in the coefficients of friction when polishing different materials) rather than the

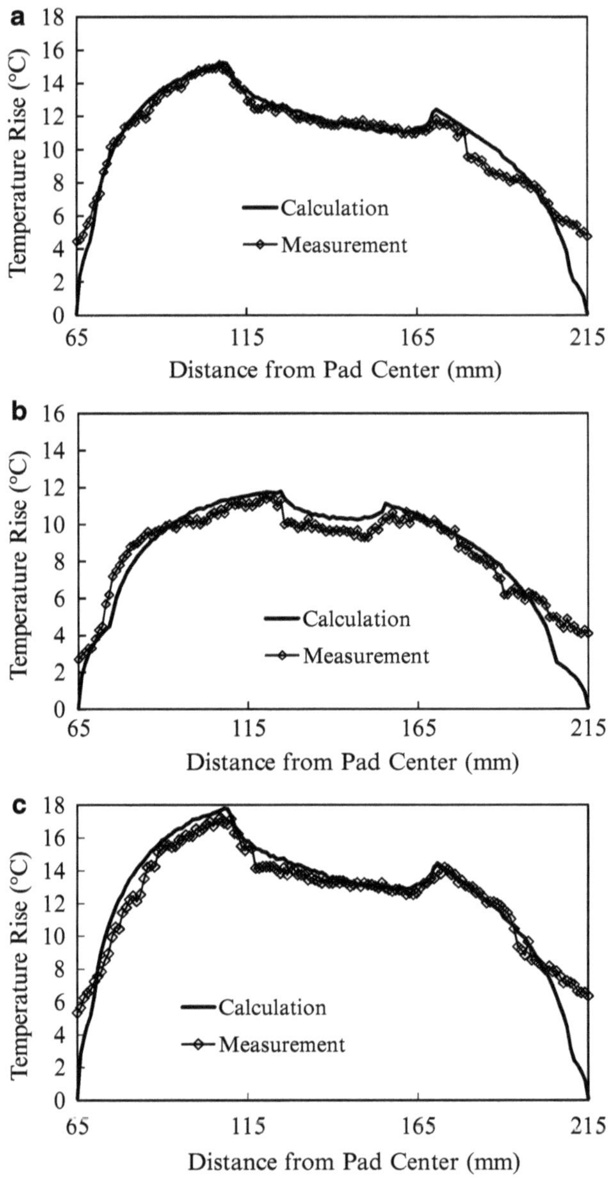

Fig. 4.30 Comparison of calculated and measured temperature rise with metal layer partially removed [45], reprinted with permission. (**a**) Platen speed = 70 rpm and carrier speed = 50 rpm, 37th second, (**b**) platen speed = 50 rpm and carrier speed = 30 rpm, 32nd second, (**c**) platen speed = 70 rpm, carrier speed = 30 rpm, 25th second, (**d**) platen speed = 50 rpm, carrier speed = 50 rpm, 43rd second

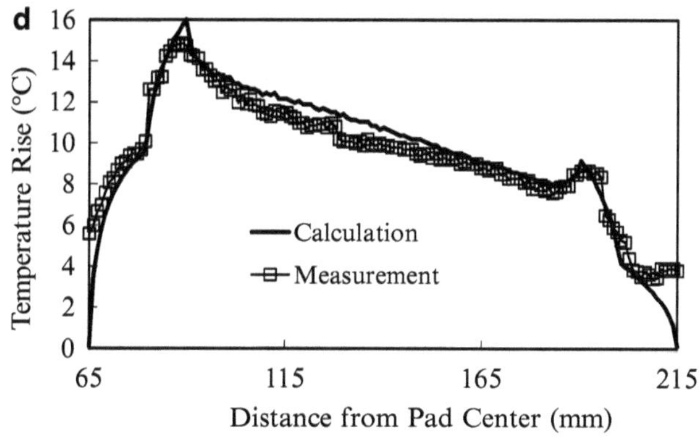

Fig. 4.30 (continued)

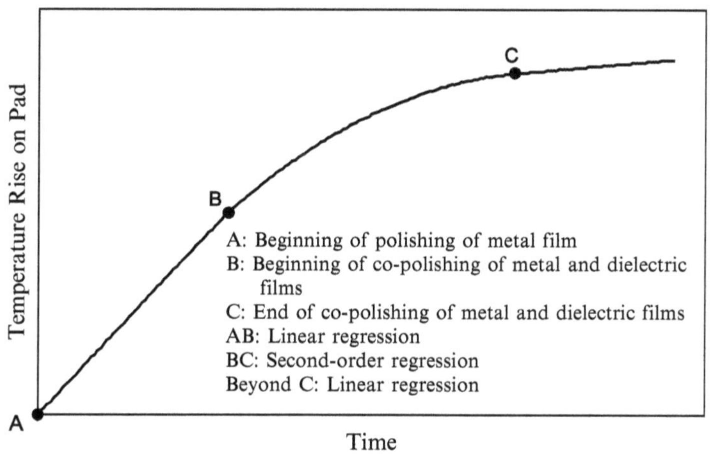

Fig. 4.31 Schematic of monitoring strategy [45], reprinted with permission

absolute level of the temperature rise, as elaborated in the next section. The shift of the calculation remains coherent with the pattern of the experimental temperature rise and does not affect the adopted online monitoring technique.

When the pad polishes both metal and dielectric simultaneously, the temperature distribution will no longer be smooth (as shown in Fig. 4.30); Fig. 4.31 [45] shows the pad temperature distribution in this circumstance. The dip shape in the middle part of the curve indicates the partial removal of copper film in that region on the wafer. The exposed dielectric has lower coefficient of friction than copper; thus the polishing energy and the resulted temperature rise are reduced. The measured temperature rise and the calculated value show fair agreement in the current study.

A significant research on the dynamic system of the thermal behavior in CMP has been recently published [48]. The comparison between their results and the current measurements is made as follows. (1) The reference stated that for every degree increase in the temperature from the incoming slurry to outgoing slurry, the rate of heat flow for copper is 17.4 W. The temperature rise measured in the current paper lies around 12°; namely it indicates the input heat flow rate would be around 200 W based on this reference, which is indeed close to the frictional power input from the spindle (180 W) in the current study. (2) The measurement of temperature rise shown in Fig. 4.26 of [48] ranges from 10° to 18° (Fig. 4.32 in [48] shows about 14°), which is consistent to our results (around 12°). The difference can be attributed to the machine specifications, location of measurement, and pad/slurry properties.

Based on the above comparisons, the current study of the thermal behavior is found consistent with other references.

Endpoint Detection

As discussed in Sect. 4.3.1.1, the temperature rise is a function of both the coefficients of friction between the pad and the metal film as well as that between the pad and the dielectric film. The analysis of the temperature rise distribution tells the progress of metal film removal. In order to monitor CMP process in situ and more efficiently, several points of the temperature on the scanning path are recorded. In the beginning phase, the pad polishes the metal layer only; thus the temperature rises steadily with time. When part of the metal film is removed in the transition phase, the pad polishes both metal and dielectric film. Since the coefficients of friction for dielectric film is less than copper, the temperature rises less rapidly than in the previous phase and varies with the relative amount of metal film left on wafer surface. When the metal film on wafer surface is fully removed in the end phase, the temperature rises steadily and milder as compared to polishing metal film in the beginning period. The second order curve can happen when a very thick Cu film is polished. However, in most cases of CMP processes (including the currently investigated case), the Cu film is so thin that the temperature increases linearly until the underlying oxide shows.

Based on this thermal behavior, a monitoring strategy can be developed in use of the regression technique. When monitoring, the pad temperature is constantly measured. The least-square regression of first and second order is continuously carried out for all existing temperature measurement. The errors of two regressions are compared to the other continuously. It is true that in the beginning period, the linear regression indeed shows less error than the second-order regression until the transition phase when both metal and dielectric films are polished and the behavior of temperature rise changes. After the error of linear regression is found larger than the second-order regression for a period of five continuous measurement, the beginning of transition is identified. Similarly, after the error of the linear regression is found again less than the second-order regression for a period of five continuous measurement, the end of transition is identified. Namely, the metal film is fully removed, or the endpoint is reached. The error of all regression is less than 5%. Figure 4.31 shows the monitoring strategy schematically. Figures 4.32 and 4.33 [45] show the history of temperature rise on pad at the location of radial distance 110 mm and 140 mm from O in Fig. 4.24, respectively, at different polishing conditions.

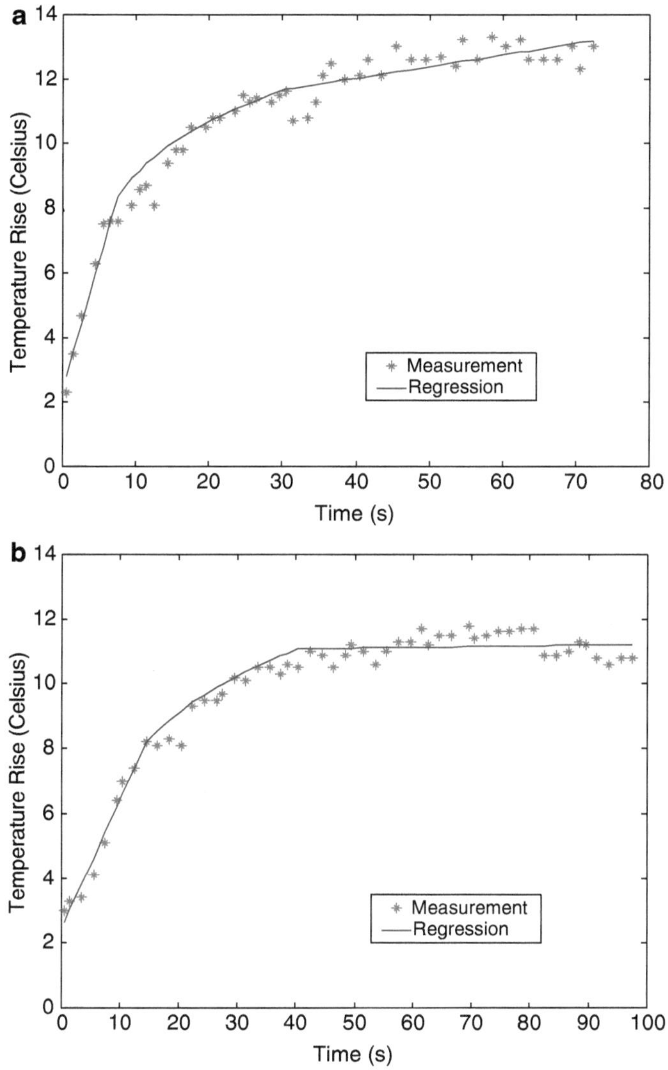

Fig. 4.32 Monitoring of temperature rise on pad passing wafer at radial distance at 110 mm [45], reprinted with permission. (**a**) Platen speed = 70 rpm, carrier speed = 50 rpm, (**b**) platen speed = 50 rpm, carrier speed = 30 rpm, (**c**) platen speed = 70 rpm, carrier speed = 30 rpm, (**d**) platen speed = 50 rpm, carrier speed = 50 rpm

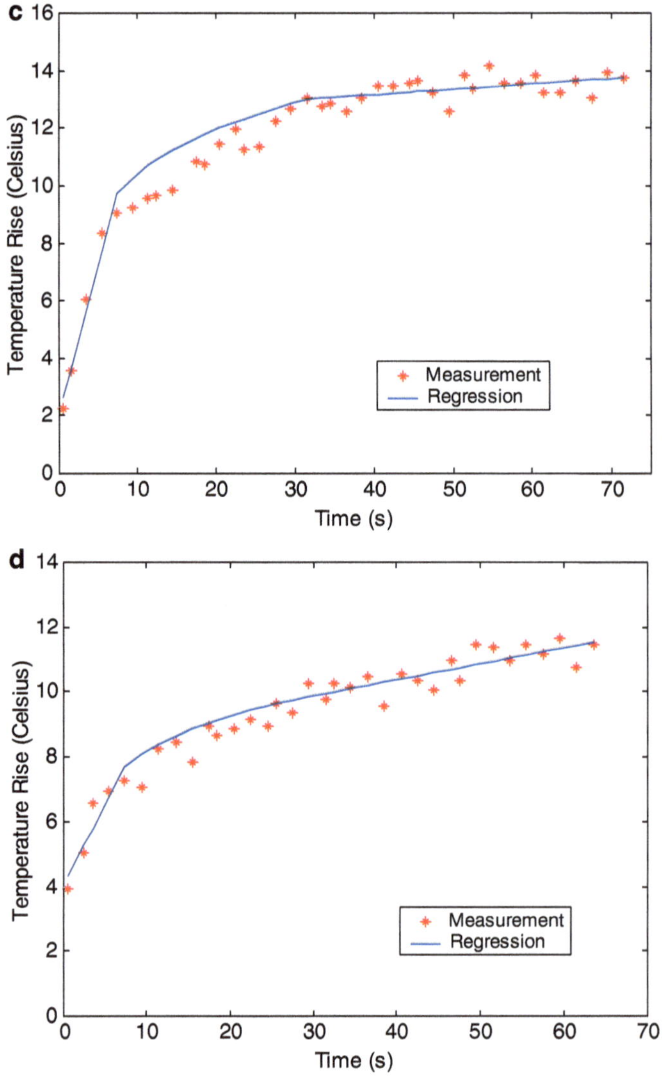

Fig. 4.32 (continued)

The temperature rises at these points were found to follow the pattern described above. At different location, the time of reading the endpoint is shown different. Monitoring of more locations on pad should and have been carried out, so that the actual endpoint of the process is identified as the latest one.

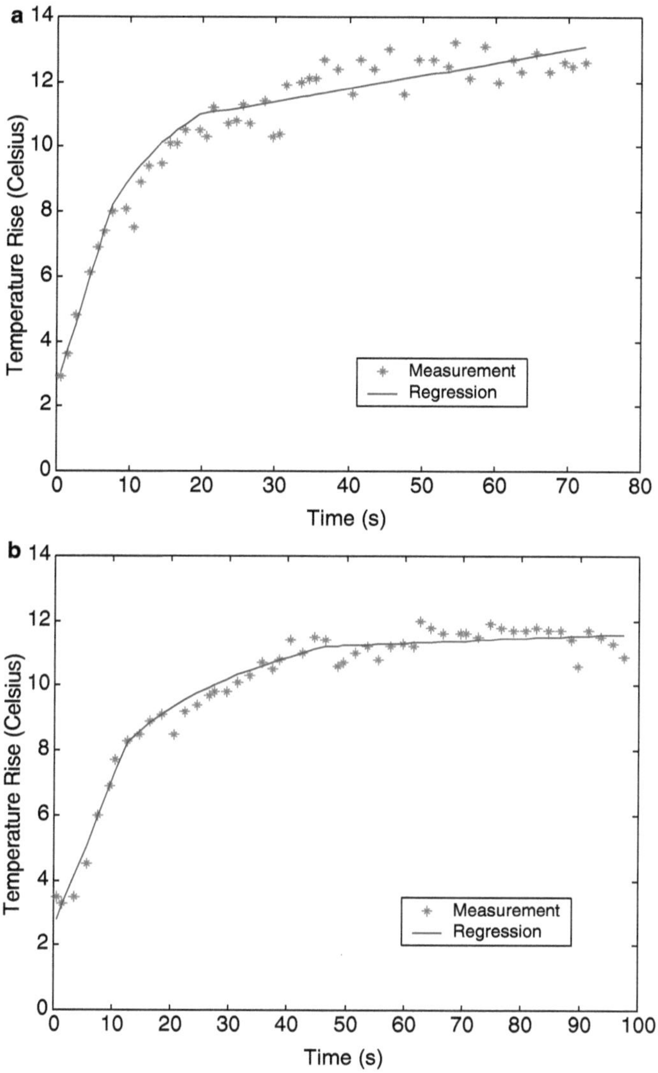

Fig. 4.33 Monitoring of temperature rise on pad passing wafer at radial distance at 140 mm [45], reprinted with permission. (**a**) Platen speed = 70 rpm, carrier speed = 50 rpm, (**b**) platen speed = 50 rpm, carrier speed = 30 rpm, (**c**) platen speed = 70 rpm, carrier speed = 30 rpm, (**d**) platen speed = 50 rpm, carrier speed = 50 rpm

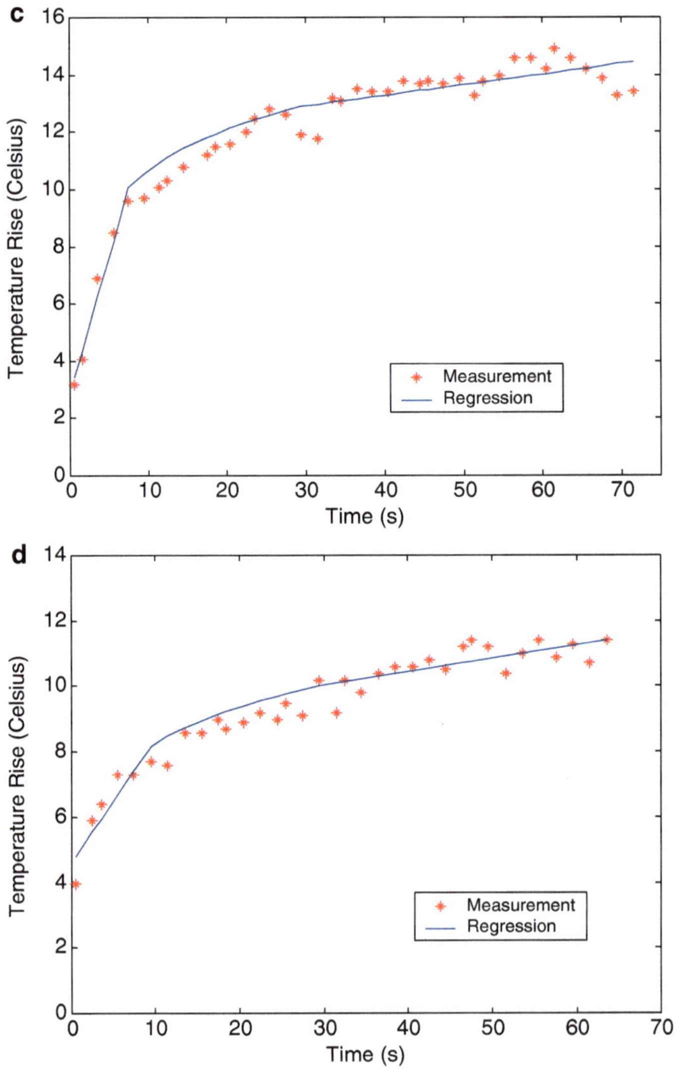

Fig. 4.33 (continued)

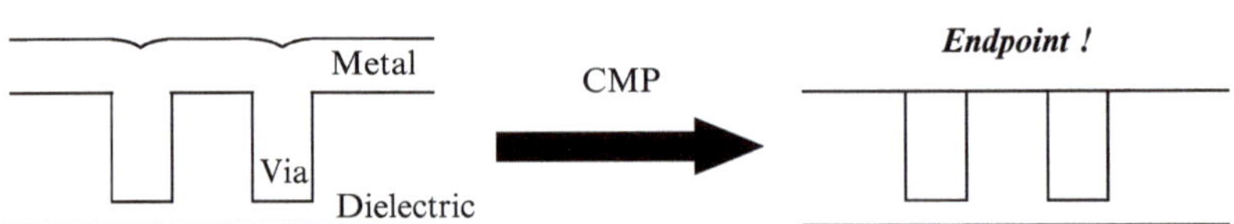

Fig. 4.34 Schematic of endpoint in metal CMP [2], reprinted with permission

4.4.2 By Acoustic Emissions

4.4.2.1 Principles of Endpoint Detection

Since the wafer surface is in full contact with the polishing pad, detecting its endpoint is a challenge. There are patented methods to answer the need for in situ monitoring and endpoint detection. The reported working principles can be categorized into seven, including optical, electrical, acoustical/dynamic, thermal, frictional, chemical/electrochemical methods, and others [49]. The optical method is among the most often employed methods in monitoring CMP processes. For a patterned wafer, however, the existence of the metal lines often interferes with the optical measurement. Besides, the measurement at only one point on wafer often does not provide the correct information across the whole wafer. The electrical methods mostly use the embedded electrodes in the pad or carrier to measure the changes in the electrical conductivity, capacity, impedance, or resistance during CMP processes to detect the process endpoint. These installations require a modified carrier and platen to accommodate the electrodes and a complex wiring route, which seriously limit the acceptance in practice. As to the acoustic emissions (AE), the merit is their inherent high frequency feature can keep the acquired signals away from the low frequency noise. The temperature on pad has been used to monitor metal CMP process. Quite a few CMP machines are equipped with a platen temperature controller or a pad temperature-measuring module to ensure stable chemical reaction during polishing. It can perform the temperature measurement at a distance from the pad without attaching sensors with the entangled wires. The thermal method is limited for metal CMP due to the low sensitivity of the temperature change of the oxide layer on wafer. The frictional methods measure the change of the torque, the motor current, or the motor voltage applied to the carrier or the platen to monitor the endpoint of removal of the metal layer. However, the acquired signals often contain considerable noise, or the signal is too weak, that limits its applications. The chemical methods mainly measure the particle size, the presence of certain vapor, the electrochemical potential, zeta potential or conductivity to detect the endpoint. The disadvantage is the probes have to contact the moving wafer. The response time is also a concern.

Many of these methods require the rearrangement of the machine structure and wires; some can only be implemented on certain types of machines. Besides, each monitoring method has its advantage and disadvantage, a multi-sensor system compensating the individual drawbacks is considered a good approach to the accurate in situ endpoint detection. In fact, some CMP machines now are equipped with optical as well as thermal or frictional sensors. Hence the exploration of various monitoring techniques is essential for the advancement toward accurate CMP endpoint detection. The acoustic emissions as monitoring tool for CMP endpoint are further explored in this study.

The acoustic emission has been used for tool condition monitoring since the late 1970s [50–53]. In metal cutting, the common AE sources are attributed to the plastic deformation in the shear zone and the tool/chip interface, rubbing of the tool on the tool/chip interface and the machined workpiece surface, chip breakage and entanglement, chipping and breakage of cutting tool [54, 55]. AE can also be used to investigate the abrasive process [56]. The frequency spectrum and the signal magnitude are obtained from the acoustic sensors attached to the polishing equipment for further analysis [49, 57–63]. Tang et al. used AE sensors to monitor a dielectric CMP process. The RMS signals were found closely related to the average material removal rate. The microscratching during CMP can be also detected [58, 59]. The results are encouraging for the current research on the metal CMP process, which uses different acoustic emissions from polishing the metal and dielectric layers for endpoint detection.

One notices the coefficient of friction between the pad and dielectric layer, 0.35, is distinguishably lower than that between the pad and the copper overlay, 0.7 [46, 47]. Secondly, the metal layer on the wafer surface is scratched by the abrasive particles in the slurry giving out acoustic emission signals. When the upper copper layer is completely removed, namely the process endpoint, as shown in Fig. 4.34 [2], the slurry particles start to scratch the dielectric material beneath, which will send out different acoustic emission signals. Thus the emitted acoustic signals during CMP of metal overlay can be used to differentiate the borderline.

Using the above-mentioned characteristic difference in the coefficients of friction between the pad and copper and dielectric films, one can monitor the temperature on polishing pad to detect the transition from polishing the copper layer to the dielectric layer. The kinematic power is consumed for polishing work and eventually becomes measurable thermal emissions. A correlation between the kinematic parameters and the pad temperature with the associated monitoring technique has been proposed [64, 65]. The first derivative of the thermal emissions keeps constant at the beginning phase when the metal overlay is being polished. It decreases in the transition when metal layer and dielectric layer are both being polished simultaneously across the wafer, and becomes constant again when the dielectric layer is being polished eventually. The polishing endpoint is thereby identified.

The purpose of the current study is to explore an in situ endpoint detection method using the acoustic emissions with reference to the alternative technique using thermal emissions. The result of this study is expected to be feasible for different CMP machines without significant machine modification.

4.4.2.2 Experimental Findings

The CMP machine in the experiment is an IPEC 372M wafer polisher carrying 150 mm wafer with 1 μm of oxide dielectric layer and 1500 Å copper overlay film. The polishing consumables include a grooved pad, the slurry abrasive of 0.1 M HNO_3 solution, and Al_2O_3 particles of 0.1–0.3 μm in size and weight percentage of

Table 4.8 Polishing process conditions

	1	2	3	4
Down pressure	4.0 psi			
Backside pressure	0.2 psi			
Slurry flow rate	100 ml/min			
Platen speed	70 rpm	50 rpm	70 rpm	50 rpm
Carrier speed	50 rpm	30 rpm	30 rpm	50 rpm

3–5%. The down pressure and the backside pressure were set 4.0 psi and 0.2 psi, respectively. The slurry flow rate was 100 ml/min. The polishing conditions are shown in Table 4.8.

The AE sensor is required to be placed as close to the acoustic source, that is, the polished wafer, as possible. In the current study, a PAC wide-band AE sensor S9208 is placed at the backside of the carrier in a hole drilled on the plastic cover of the carrier assembly; thus the AE sensor can be mounted on the metal part of the carrier. A slip ring is specially designed and put in the carrier assembly so that the tongues inside the slip ring transfer the AE signal from the inner ring on rotating carrier to the stationary outer ring. The inner and outer rings are kept connected all the time. The inherent high-frequency feature of AE signals can separate with the low-frequency noise. The AE noise caused by the friction of the slip rings is about 50–80 KHz, which is well below the higher-than-200 KHz signals produced in polishing wafer. The O-ring between the plastic cover and the metal part of the carrier can further isolate the undesired noise. The AE signals acquired by the sensor are transferred to the 49 dB preamplifier with a built-in band pass filter of 95 KHz to 1 MHz. The sampling rate is five mega samples per second. A threshold level was set according to an online pretest to cancel the background noise and to serve as a reference for the waveform properties. Two preliminary calibration tests verify that the AE signals are sent from the wafer surface and immediately correctly received by the AE sensor at the backside of the carrier. The acoustic emissions do not transfer via the pad and slurry.

In the experiment of monitoring using the thermal emissions, the AGEMA Thermovision® 900 System is installed by the wafer polisher to measure the pad temperature during CMP processes. This image system captures the two-dimensional thermal emissions on the polishing pad with the temperature resolution of 0.1 K and 272 × 136 pixels in each image. The measured thermal image is calibrated before each experiment to ensure the coherence of the measurement [65].

The energy index and root-mean-square (RMS) value of the acoustic emissions are used. They are calculated as in the following.

$$\text{Energy Index} = \frac{1}{R}\int_0^{\infty} V^2(t)dt \tag{4.37}$$

$$\text{RMS} = \sqrt{\frac{1}{\Delta t}\int_0^{\Delta t} V^2(t)dt} \tag{4.38}$$

where t is the integration time, R is the electrical resistance of the measuring circuit, and $V(t)$ is the level of measured acoustic signal. The calculated AE energy is proportional to the true AE energy in the event [66]. It is later normalized in the experimental analysis.

Figure 4.35 [2] shows the energy index versus the polishing time. The metal film is first softened by the chemical in the slurry, followed by the removal of copper by the slurry abrasives sending out significant elastic waves. It took a few seconds to pick up the preset down force and backside pressure at the beginning. When the metal film is fully removed, the chemical in the slurry does not react with the exposed dielectric oxide layer. The particles in the slurry scratching the surface remove little amount of the hard oxide material, and the AE signals decrease correspondingly to a lower level. At this moment the process endpoint is called. The polished wafer is unloaded and examined for the removal of the copper overlay.

As indicated in Fig. 4.35, the energy index can identify the endpoint at varying polishing conditions of rotational rates of platen and carrier. The four condition sets well represent the polishing recipe in practice. The time of the endpoint is determined by these conditions. The correlation between the processing parameters and the time of endpoint involves both mechanical and chemical aspects and remains an empirical observation and practice. In general, slower polishing speed (at lower rotational rates of platen and carrier) needs longer time due to lower removal rate, based on the widely accepted empirical Preston equation [1]. Hence the case (d) of platen speed $=$ 50 rpm and carrier speed $=$ 30 rpm shows the latest endpoint. The numerical normalization of the signal values was performed by dividing the difference between each measurement and the minimum measured value in the four experimental conditions with the difference between the maximum and the minimum of the measured values. The RMS of the AE signal shows the same results, as shown in Fig. 4.36 [2]. For the patterned wafer, the overlay copper film on the dielectric layer will be removed left with the copper filled in vias. The AE signals can nevertheless tell the endpoint in time because they are insensitive to the quite small area of the exposed copper on dielectric layer (less than 1% of the whole dielectric area).

Polishing the metal film sends out more acoustic emissions than the dielectric film; in Fig. 4.35 one further notices the ratio is about 2. The number can also be found corresponds to the ratio of the coefficients of friction between the polishing pad and these films, 0.7 and 0.35, respectively.

The results are comparable to the existing references using the alternative thermal monitoring technique under the same process conditions, as shown in Table 4.9 [65]. The slight delay of 0–15% in use of the thermal method is attributed to the less agile response of the heat transfer in the system (including the slurry and polymer pad) than the acoustic wave propagation, which suggests the advantage for using the current explored method. It demonstrates the endpoint in copper CMP can be monitored in situ by acoustic emissions for the deep-submicron integrated circuits manufacturing.

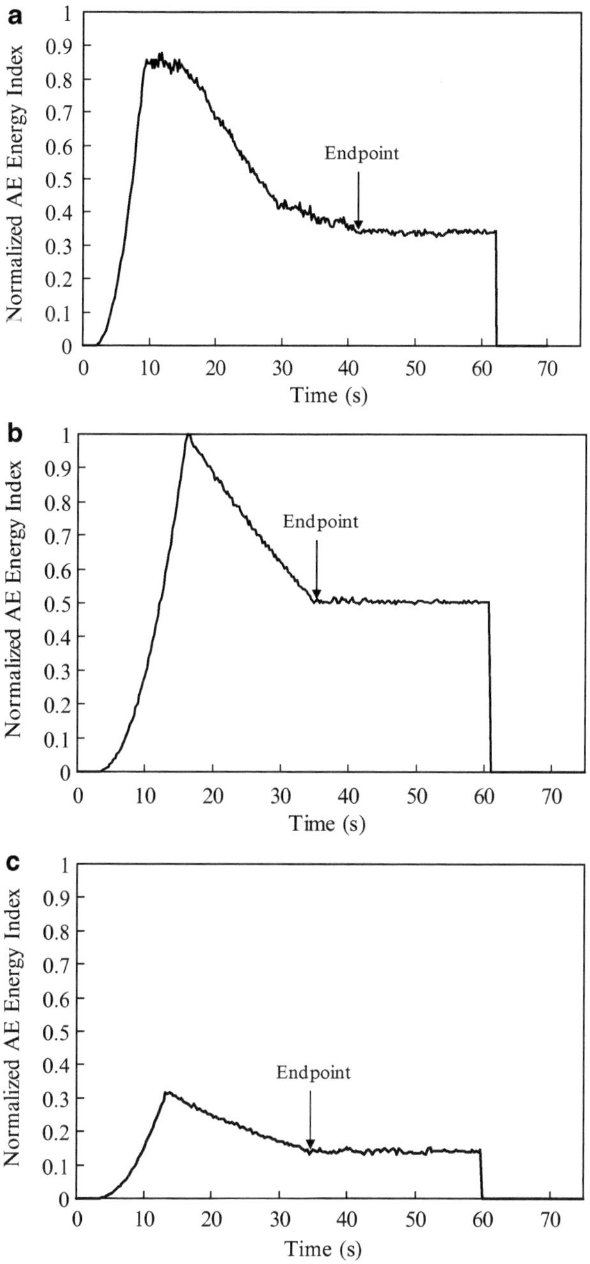

Fig. 4.35 Energy index of acoustic emissions during copper CMP [2], reprinted with permission. (**a**) Platen speed = 70 rpm, carrier speed = 50 rpm, (**b**) platen speed = 70 rpm, carrier speed = 30 rpm, (**c**) platen speed = 50 rpm, carrier speed = 50 rpm, (**d**) platen speed = 50 rpm, carrier speed = 30 rpm

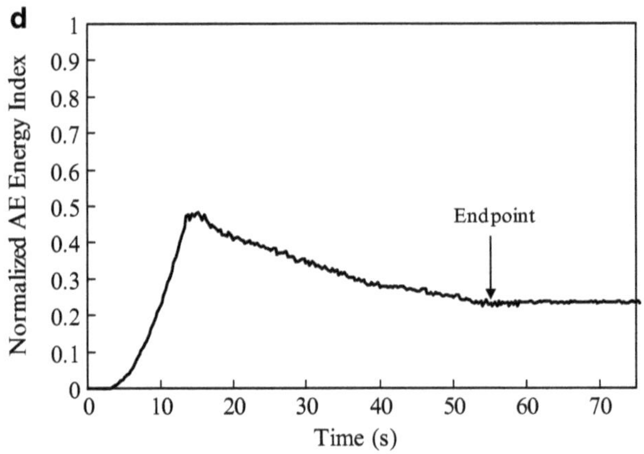

Fig. 4.35 (continued)

4.5 Flow Field

The systematic view of experimental setup is shown in Fig. 4.37(a) [67]. The single-head CMP machine has a transparent carrier body to visualize the flow underneath. Glass wafers of 200 and 300 mm in diameter are used in lieu of silicon and the red dye is added to enable the observation. The concentration of red dye used in the current experiments is 1.5 g per 1000 g water, while the saturation dissolution concentration is 325 g per 1000 g water at the temperature of 25°C. A CCD camera and image processing board are mounted to capture the flow and transfer into 352 × 240 pixel image at the interval of 10 s during the entire course of 100 s. The linearity of camera is proved before use. The transparent carrier is shown in Fig. 4.37b [67], and Fig. 4.37c [67] illustrates the obtained picture. To study the effects of various pad morphology, three common Rodel pads are tested—(1) the perforated ICI1000-SUBAIV, (2) the concentric K grooved ICI1400-SUBAIV, and (3) the K-XY grooved ICI1000 with larger groove cross section than (2). The pad was carefully cleaned before every single experiment. Besides, the picture of pad before every single experiment was taken as a reference. Then every captured image in the experiment was calibrated by this reference image to eliminate the influence of any possible sticking red dye as well as other disturbances, such as lighting variation and pad reflection. The pad grooves would not accumulate the red dye because the red dye is fully dissolved and well suspended in the slurry without precipitation. One also notes that each pad was used for only 48 times, which is much less than its design life. The confidence intervals of 95% of the measured data are shown in the figures by the error bars. Quite a few CMP operating parameters are included to investigate their effects on the slurry film, namely the carrier speed, platen speed, rinsing location, flow rate, and wafer size. Table 4.10 shows the range of experimental parameters.

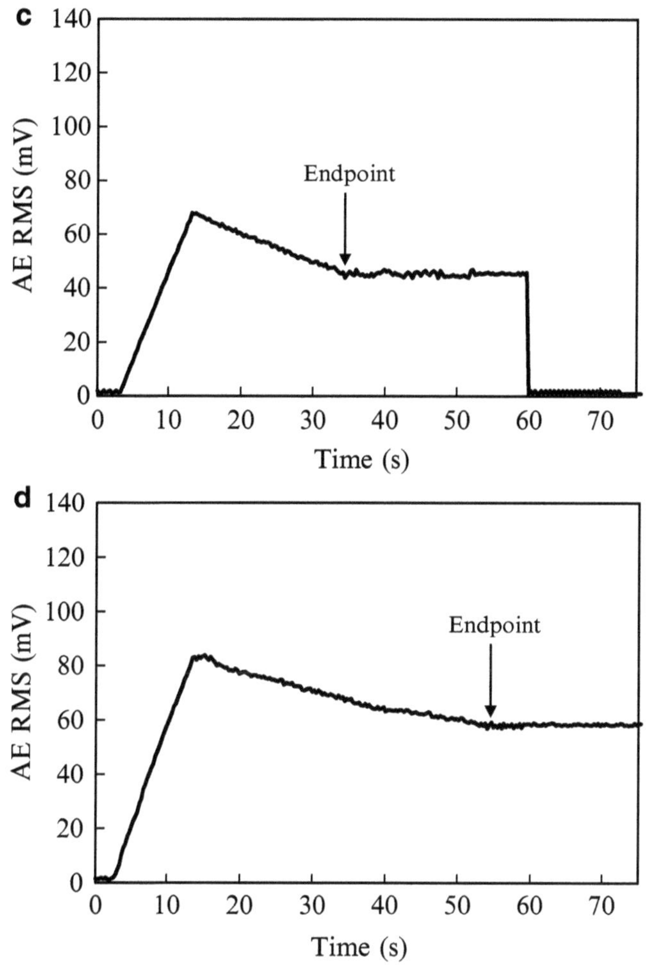

Fig. 4.36 RMS of acoustic emissions during copper CMP [2], reprinted with permission. (**a**) Platen speed = 70 rpm, carrier speed = 50 rpm, (**b**) platen speed = 70 rpm, carrier speed = 30 rpm, (**c**) platen speed = 50 rpm, carrier speed = 50 rpm, (**d**) platen speed = 50 rpm, carrier speed = 30 rpm

Based on the acquired digital image, the mean gray value (MGV) and its nonuniformity (NU) are analyzed. The former indicates the amount of flow between pad and wafer. The latter defined by the standard deviation divided by MGV will reveal the uniformity distribution of the flow film. Both quantities provide visualized characterization of the fluid between pad and wafer. The elapsed time to reach the steady state of these quantities is also considered a significant information obtained from this experiment.

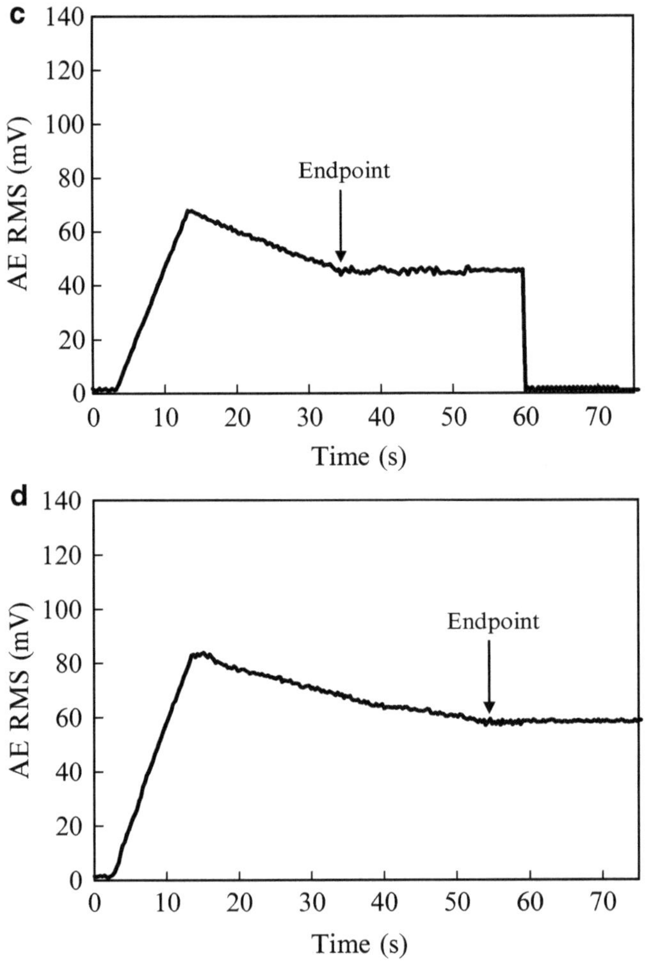

Fig. 4.36 (continued)

Table 4.9 Endpoint time determined by thermal and AE methods

	Platen = 70 rpm, Carrier = 50 rpm	Platen = 70 rpm, Carrier = 30 rpm	Platen = 50 rpm, Carrier = 50 rpm	Platen = 50 rpm, Carrier = 30 rpm
Thermal method	40 s	42 s	41 s	62 s
AE method	41 s	35 s	34 s	55 s

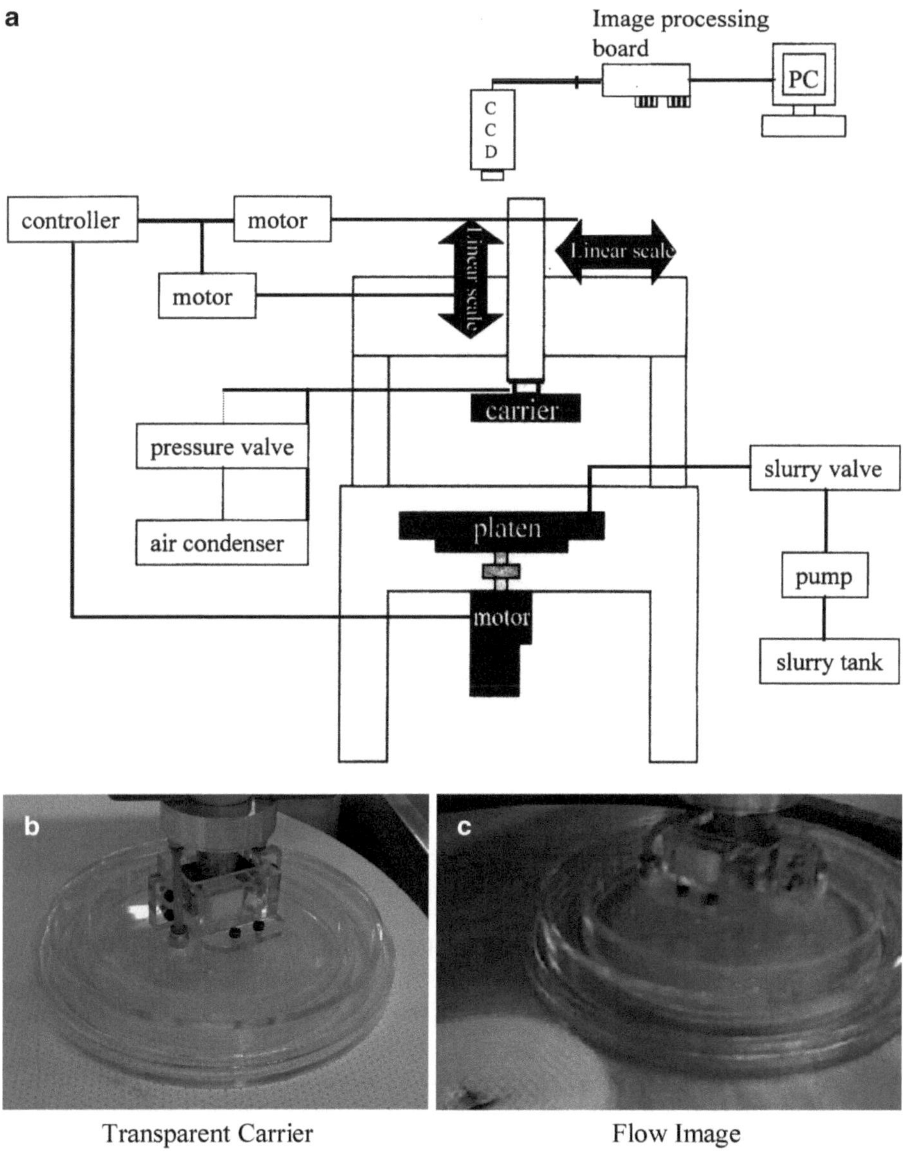

Fig. 4.37 Experimental setup [67], reprinted with permission. (**a**) Systematic view, (**b**) transparent carrier, (**c**) flow image (90 s, 60 rpm, 60 rpm, 150 ml/min, 25 %, 200 mm, IC1000 SUBAIV)

4.5.1 Elapsed Time to Steady State

The steady state of the slurry film between wafer and pad is reached when both the quantity and distribution of the fluid film do not change with time. As the experiment started, the red color intensity and its distribution change drastically, while

Table 4.10 Experimental parameters

Platen speed (rpm)	30/60
Carrier speed (rpm)	30/60
Flow rate (ml/min)	150/300
Rinsing location (% of pad radius from platen center)	25/50/75
Wafer size (mm)	200/300
Pad (Rodel)	IC1000-SUBAIV/IC1400-SUBAIV/IC1000

Table 4.11 Effects of process parameters on elapsed time to steady state

Process parameters		Elapsed time to steady state (sec)
Platen speed (rpm)	30	59
	60	57
Carrier speed (rpm)	30	61
	60	55
Flow rate (ml/min)	150	60
	300	55
Rinsing location (% of pad radius)	25	52
	50	58
	75	64
Down force (psi)	1	59
	2	64
Wafer size (mm)	200	61
	300	51
Pad (Rodel)	IC1000/SUBAIV	51
	IC1400/SUBAIV	53
	IC1000	64

they became stable as time elapsed. The elapsed time to steady state is defined by the time when the change rate of both items (namely MGV and NU) is found below 10%. Bearing in mind that only a couple of minutes elapse for each CMP process in industry, the following results provide valuable information of how agile the process goes steady. The effects of processing parameters on elapsed time to steady state are summarized in Table 4.11. Table 4.11 shows the steady state of the fluid film between pad and wafer is reached around 50 s after start. Large platen speed slightly shortens the transition period by spreading fluid more effectively, and Table 4.11 shows similar effects of the carrier speed as well. Doubling the supply of slurry decreases the fluid development time by 10%, while adding fluid at location farther from the platen center takes longer time to reach the steady state, as also illustrated in Table 4.11. The same table demonstrates large down force will narrow the gap between pad and wafer; thus the fluid encounters more flow resistance and needs more time to distribute. At the same carrier speed, wafer with larger diameter provides higher velocity for better fluid blending, although the rinsing fluid per unit wafer area would decrease. One more thing showed in

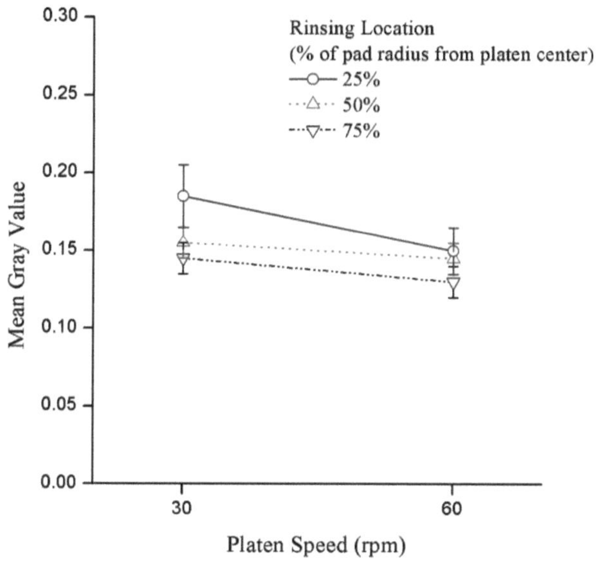

Fig. 4.38 Effect of platen speed on mean gray value [67], reprinted with permission

Table 4.11 is that the shorter developing time for 300 mm wafer to reach the steady state. Besides, Table 4.11 shows the perforated pad is the most advantageous for steady-state evolution, followed by K and K-XY grooved pads. More and large grooves will disturb the isotropic fluid distribution process by guiding the fluid along particular directional grooves.

4.5.2 Mean Gray Value

MGV indicates the amount of fluid contained between pad and wafer. Increasing the platen speed drives fluid toward the platen rim instead of staying with the wafer owing to the increased centrifugal force. On the other hand, rinsing toward the platen center demonstrates more pronounced influence to reduce this negative effect, as shown in Fig. 4.38 [67]. Increasing the carrier speed causes less significant influence at predetermined platen speed and rinsing location, as Fig. 4.39 [67] shows. These two figures also strongly indicate that rinsing at central region on platen helps the fluid fill the gap between pad and wafer, or the fluid tends to be driven outward, leaving the pad incompletely rinsed. Figure 4.40 [67] shows increasing the wafer from 200 to 300 mm will significantly reduce the amount of slurry filled between pad and wafer. A remedy is to considerably increase the flow rate. Figure 4.41 [67] summarizes the experimental results of the pad influence. The

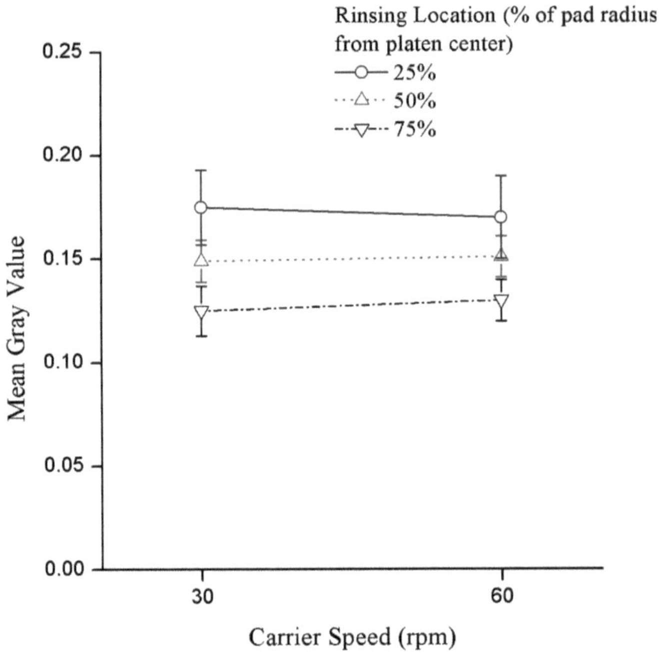

Fig. 4.39 Effect of carrier speed on mean gray value [67], reprinted with permission

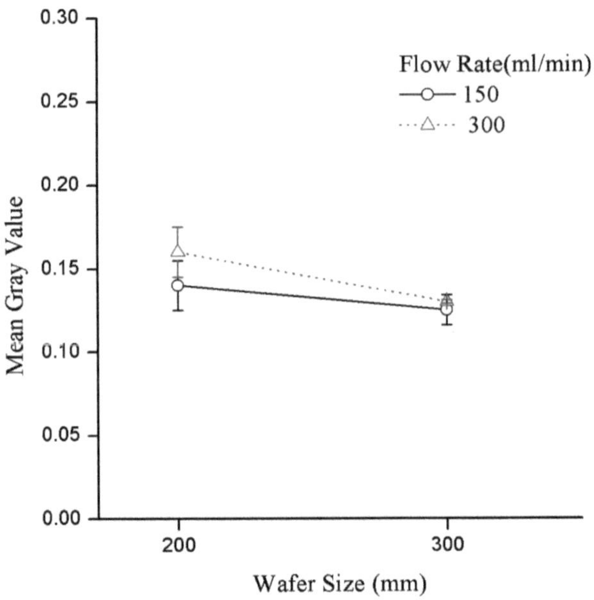

Fig. 4.40 Effect of wafer size on mean gray value [67], reprinted with permission

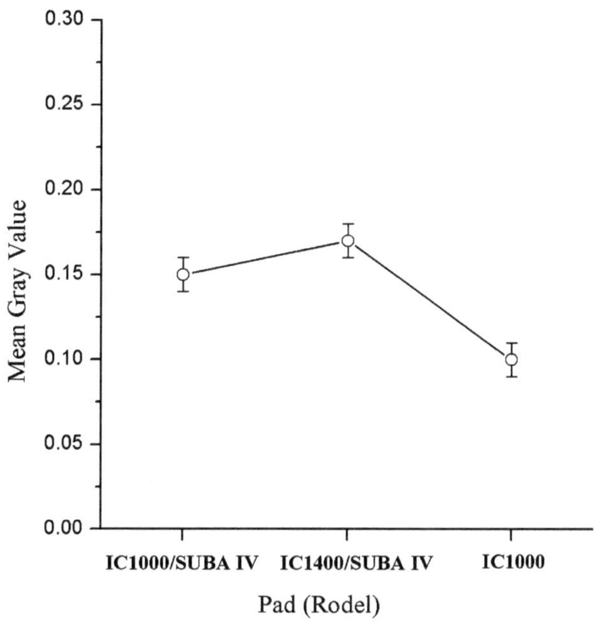

Fig. 4.41 Effect of pad on mean gray value [67], reprinted with permission

Table 4.12 Summary of experimental results of fluid film between pad and wafer

	Rapid steady state	Fluid amount	Fluid uniformity	Recommendation
Increasing platen speed	⇡	↓	—/↓ (rinsing close to pad center)	Careful
Increasing carrier speed	⇡	↑	↑	Yes
Increasing flow rate	↑	↑	—	Yes
Rinsing toward center	↑	↑	⇡	Yes
Increasing wafer size	↑	↓	↑ (at low carrier speed)	Careful
More grooves on pad	↓	↓	⇣	No

↑: more advantageous, ⇡: milder advantageous, —: no effect, ↓: more disadvantageous, —/↓: milder disadvantageous

perforated and K-grooved pads are beneficial in carrying more fluid film, while X-Y grooves reduce the amount of working fluid by possibly holding the majority of fluid in the grooved passaes.

The effects of processing parameters on the fluid amount between pad and wafer are summarized in Table 4.12.

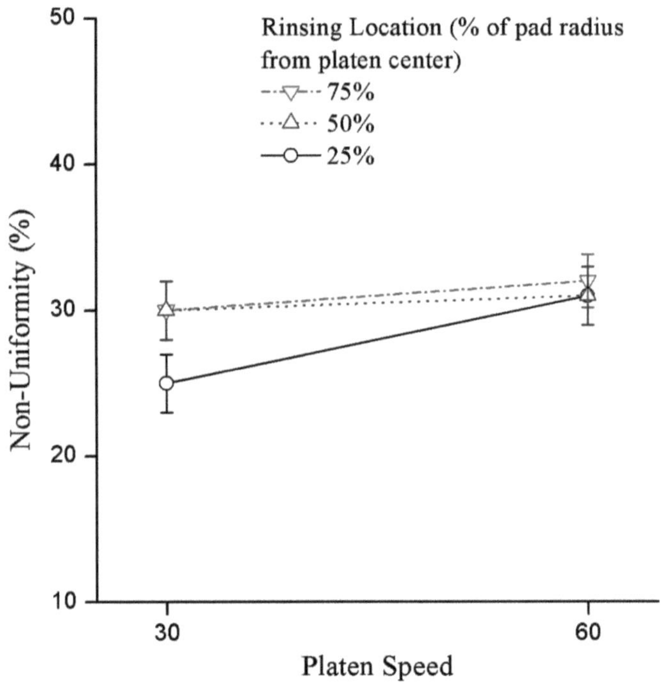

Fig. 4.42 Effect of platen speed on nonuniformity [67], reprinted with permission

4.5.3 Nonuniformity

Since the red color intensity (gray value) directly indicates the amount of slurry, the nonuniformity of red color intensity shows the nonuniformity of slurry. It is calculated by

$$\text{Nonuniformity} = \frac{\text{Standard deviation of gray value}}{\text{Mean gray value}} \times 100\ \%. \tag{4.39}$$

It is an indicator of the distribution of the slurry across the wafer. Larger number means worse uniformity. Figure 4.42 [67] shows the platen speed and the rinsing location play limited role in the fluid uniformity. However, the advantage of rinsing close to the pad center (1/4 radius) associated with low platen speed is very clear. When rinsing near center, high platen speed causes more chaotic fluid flow by large centrifugal force. In this case, although the steady state is reached fast by high speed (see Table 4.11), both the amount and uniformity of fluid are sacrificed. On the other hand, the carrier speed is able to help distribute the fluid uniformly across the wafer, as Fig. 4.43 [67] demonstrates. Rotation about the wafer itself instead of the platen is advantageous for uniformity. The flow rate does not affect the uniformity conclusively. However, more fluid supply

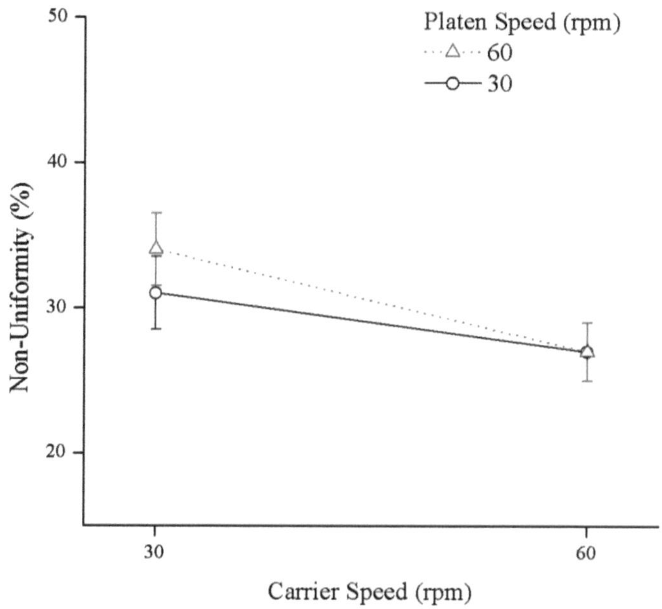

Fig. 4.43 Effect of carrier speed on nonuniformity

leads to more scattered distribution features. The results are shown in Fig. 4.44 [67]. Increasing the wafer size to 300 mm worsens slurry distribution to certain level regardless the carrier speed is high or low; namely the negative effect of wafer size is exclusively dominant, as revealed in Fig. 4.45 [67]. The advantage of using high carrier speed (as shown in Fig. 4.43) vanishes when the large 300 mm wafer is processed. Figure 4.46 [67] shows the experimental results of the pad design. More pad grooves give more scattered and undesired results of uniformity. The grooves act as the dominant fluid passage in defined directions. Compared to the isotropic perforation on pad, the groove design is considered disadvantageous for the uniform distribution of fluid.

The effects of processing parameters on the fluid uniformity between pad and wafer are summarized in Table 4.12.

4.6 Conclusions

The clearance between the wafer surface and pad is defined by the minimum fluid film thickness, as determined by the parameters, including load, velocity, and viscosity, in a manner similar to the design of a thrust bearing. The abrasives present in the slurry flow mechanically remove the chemical treatment products from the wafer surface. According to the bear-and-shear process model, the material removal rate increases with increasing polishing velocity by a power of approximately 1/2.

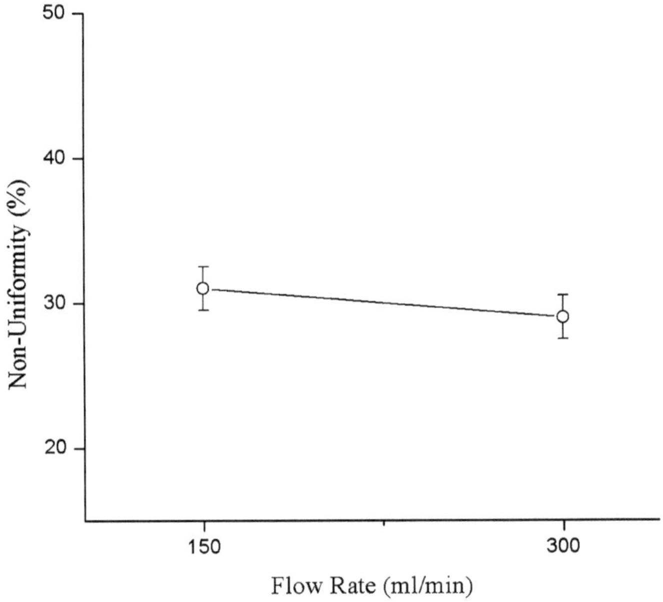

Fig. 4.44 Effect of flow rate on nonuniformity [67], reprinted with permission

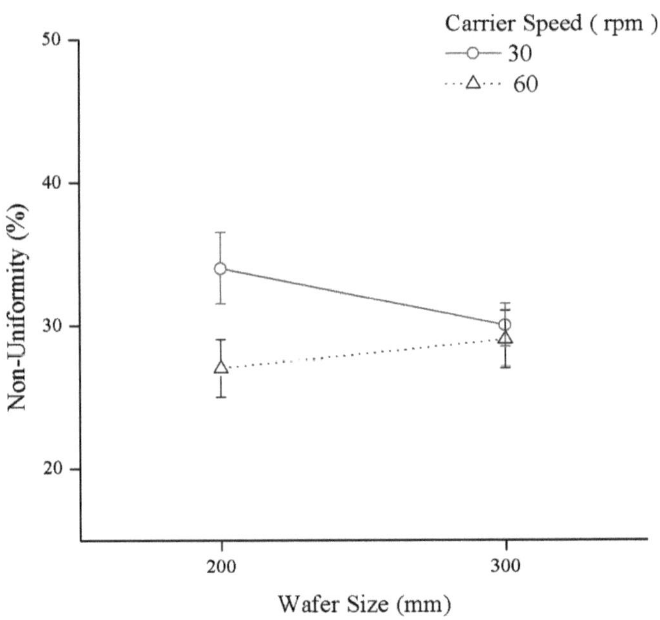

Fig. 4.45 Effect of wafer size on nonuniformity

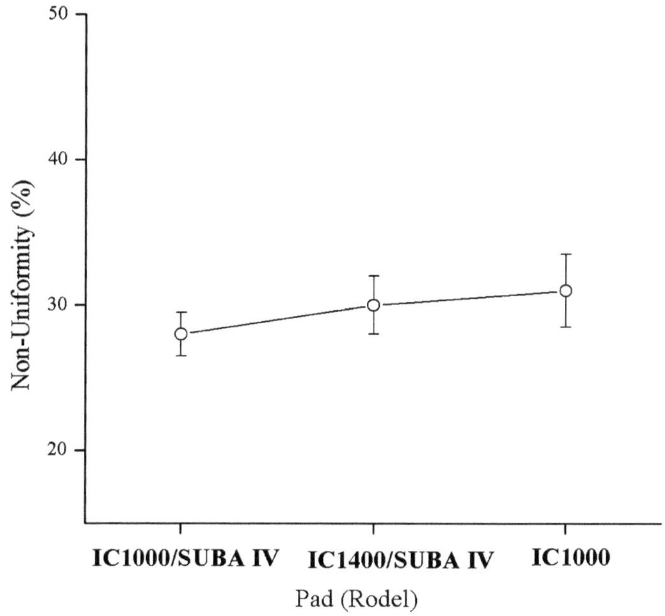

Fig. 4.46 Effect of pad on nonuniformity [67], reprinted with permission

This correlation was verified by the polishing experiments of other authors and independent researchers. Modification of the widely cited empirical Preston equation, based on glass polishing, can provide an analysis for extended dielectrics CMP applications. This paper summarizes representative works, and serves as a guide for practical prediction of material removal rates during CMP.

The results of a kinematic analysis for nonuniformity in CMP suggest that nonuniformity is determined chiefly by the ratio of carrier speed and platen speed, and by wafer size. The speeds should be kept as close to each other as possible for good uniformity in wafer planarization. The speed window index, η, describes the kinematic effect on nonuniformity. Carrier translation has a limited effect on uniformity, and the carrier's eccentricity contributes to reduce the nonuniformity. While non-kinematic concerns, such as the selection of consumables, are a part of the planarization process, our proposed model contributes to the overall process design, and will gain increasing significance as the challenge of larger wafer sizes and the need for reduction in nonuniformity becomes more rigorous.

In a metal CMP process, the pad works to remove an overlaid metal layer. Frictional forces and the rotations of the pad, metal layer, and the dielectric layer absorb energy and cause pad temperature to rise. Since both the pad and the wafer rotate, each location on the pad experiences a varying relative speed during polishing from the moment that the point of the pad touches the wafer, to the moment the point of the pad loses contact with the wafer. The total absorbed kinematic energy during this contact period can be calculated from the integral of the relative polishing speed over the contact time. However, the frictional forces between the pad, metal layer, and the dielectric layer are different due to the various coefficients of friction. We established a temperature rise model based on the integral of the relative polishing speed and the frictional force; we then verified this model by monitoring the pad temperature in situ.

After the overlaid metal layer is partially removed, the pad begins to polish the lower dielectric layer. Since the coefficient of friction between the pad and dielectric is significantly lower than that between the pad and the metal layer, pad temperature increases more slowly than it does during metal layer polishing. In applying the proposed regression method, the process endpoint is identifiable from the change in rate of temperature rise.

CMP has become an indispensable process in deep submicron semiconductor manufacture. In a metal CMP process, the endpoint is reached when the overlay metal layer is removed. An effective in situ endpoint monitoring method is needed to improve the yield rate and throughput of CMP. Acoustic emissions that occur during polishing the metal and dielectric layers are different due to the layers inherent differences in tribology. The coefficient of friction between the polishing pad and the dielectrics is lower than that between the pad and the metal. The acoustic emissions decrease in magnitude when the metal film is removed. Either the measured RMS signal or the energy index can indicate the process endpoint effectively. The experimental results and the existing alternative endpoint monitoring method, using pad temperature, support the proposed method

Preliminary experimental results of the influence of CMP processing on the amount and uniformity of the fluid film between pad and wafer are provided. The fluid film is visualized through the use of transparent carrier and dyed fluid. The digitally analyzed picture, captured by CCD, correlates the amount and uniformity of the fluid between pad and wafer varying with some of the CMP process parameters. Abundant and uniform slurry distribution between pad and wafer is desired for fast material removal and good uniformity across the wafer. The preliminary observation of the effects of the process parameters is summarized in Table 4.12. A large platen speed drives the slurry to the outer rim of the pad. The speed is positive only when slurry is dispensed at the pad's central region. A high carrier speed helps distribute the slurry uniformly. A greater slurry supply is more advantageous for both the amount and the uniformity of slurry, though this will increase the costs of consumables. The slurry should be rinsed at the pad center to improve supply at the polishing site. Increasing the wafer size unavoidably presents a greater challenge to provide an adequate polishing slurry supply across the wafer. One approach to solve this is to increase the carrier speed in conjunction with rinsing at the pad center. X–Y grooves on the pad do not necessarily deliver the assumed performance improvement above that of perforated or K-grooved pads.

References

1. Preston FW (1927) J Soc Glass Tech 11:214
2. Hocheng H, Huang YL (2007) Trans Semiconductor Manufact 20(3):306
3. Warnock J (1991) J Electrochem Soc 138(8):2398
4. Yu TK, Yu CC, Orlowski M(1993) A statistical polishing pad model for chemical mechanical polish, IEDM Technical Digest, p. 865
5. Cook LM (1990) J Non-Cryst Solids 120:152
6. Liu CW, Dai BT, Yeh CF (1996) J Electrochem Soc 143(2):716
7. Runnels SR, Eyman LM (1994) J Electrochem Soc 141(6):1698
8. Runnels SR (1994) J Electrochem Soc 141(7):1900
9. Runnels SR, Olavson T (1995) J Electrochem Soc 142(6):2032
10. Tseng WT, Wang YL (1997) J Electrochem Soc 144(2):L15
11. Patrick WJ, Guthrie WL, Standley CL, Schiable PM (1991) J Electrochem Soc 138(6):1778
12. Cooke F, Brown N, Prochnow E (1976) Opt Eng 15(5):407
13. Hocheng H, Tsai H.Y.,Chen L.J. (1997) A kinematic analysis of CMP based on velocity model, in: Proc. 2nd Int. CMP-. MIC, p.277
14. Chang CY, Sze SM (1996) ULSI technology. McGraw-Hill, New York
15. Chen L-J, Huang Y-L, Lin Z-H and Chiou H-W (1998) Pad thermal image end-pointing for CMP process, Proc. 3rd Int'l CMP-MIC, p 28
16. Lin Z-H, Chiou H-W, Shih S.-Y., Kuo L-H, Chen L-J and Hsia C(1999) Study of tungsten CMP endpoint window, Proc. 4th Int'l CMP-MIC, p65
17. Sicurani E, Fayolle M, Gobil Y, Morand Y and Tardif F (1996) W-CMP consumables evaluation-electrical results and end-point detection, Proc. the Advanced metallization and interconnect systems for ULSI applications, 561
18. Fayolle M, Sicurani E, Morand Y (1997) Microelectronic Eng 37(38):347
19. Wang YL, Liu C, Feng MS, Tseng WT (1998) Mat Chem Phys 52(1):17
20. Springer G (1999) Dependence of wafer carrier motor current and polish pad surface temperature signal on CMP consumable conditions and Ti/TiN liner deposition parameters for tungsten CMP endpoint detection, Proc. 4th Int'l Chemical-Mechanical Planarization for ULSI Multilevel Interconnection Conference (CMP-MIC), The Institute for Microelectronics InterConnection, p. 45
21. Stein DJ, Hetherington DL (1999) SPIE 3743:112
22. Sue L, Lutzen J, Gonzales S, Wertsching F, Golzarian R (1999) MRS Proc 566:109
23. Beckage PJ, Lukner R, Cho W, Edwards K, Jester M, Shaw S (1999) SPIE 3882:118
24. Sugimoto F, Arimoto Y, Ito T (1995) Jap J Appl Phys, Part 1-Regular Papers 34(12A):6314
25. Litvak HE, Tzeng HM (1996) Semicon Int'l 19(8):259
26. Nanz G, Camilletti LE (1995) IEEE Trans Semiconductor Manufact 8:4
27. Steigerwald JM, Murarka SP, Gutmann RJ (1997) Chemical mechanical planarization of microelectronic materials. Wiley, New York
28. Coppeta J, Rogers C, Philipossian A, Kaufman F (1997) Mat Res Soc Symp Proc 447:95
29. D.P.Y. Bramono, L.M. Racz, in: CMP-MIC Conference, Numerical flow visualization of slurry in a chemical mechanical planarization process, 1998, p. 185
30. Levert J, Shan L, Danyluk S(1998) Pressure Distribution at Silicon-Polishing Pad interface, Proc. Silicon Machining, 1998 Spring Topical Meeting, 96–100
31. Levine IN (1983) Physical chemistry. McGraw-Hill, New York, p 522
32. Nogami M, Tomozawa M (1984) J Am Ceram Soc 67(2):151
33. Bhushan M, Rouse R, Lukens JE (1995) J Electrochem Soc 142(11):3845
34. Hocheng H, Tsai HY, Su YT (2001) J Electrochem Soc 148(10):G581
35. Hamrock BJ (1994) Fundamentals of fluid film lubrication. McGraw-Hill, New York, p 170
36. White FM (1991) Viscous fluid flow. McGraw-Hill, New York, p 113
37. Su YT, Hung TC, Cheng YY (1995) Wear 188:77
38. Hocheng H, Tsai HY (1998) ASME Int'l Mech Eng Cong and Expo 8:33

39. Sun SC, Yeh FL, Tien HZ (1994) Mat Res Soc Symp Proc 337:139
40. Steigerwald JM, Murarka SP, Gutmann RJ (1997) Chemical mechanical planarization of microelectronic materials. John Wiley & Sons, New York, p 154
41. Chen D-Z, Lee B-S (1999) J Electrochem Soc 146(2):744
42. Evans DR, Ulrich BD, Oliver MR (1998) Chemical–mechanical planarization for ULSI multilevel interconnection. In: Wade TE (ed) VMIC Proceedings Series, Santa Clara, USA 347, 1998
43. Huang CW (1998) Master Thesis, Nat'l Chiao Tung Univ., Taiwan
44. Hocheng H, Tsai HY, Tsai MS (2000) Int J Mach Tool Manufact 40(11):1651
45. Hocheng H, Huang YL (2004) Trans Semiconductor Manufact 17(2):180
46. Liang H, Xu GH (2002) Scripta Materialia 46:343
47. Sikder AK, Giglio F, Wood J, Kumar A, Anthony M (2001) J Electron Mater 30(12):1520
48. White D, Melvin J, Boning D (2003) J Electrochem Soc 150(4):G271
49. Hocheng H, Tsai HY, Huang YL (2001) Key Engineering Materials. Trans Tech, Switzerland, p 1
50. Iwata K, Moriwaki T (1977) Annals of the CIRP 26:21
51. Dornfield DA (1984) Acoustic emission monitoring and analysis in manufacturing. The American Society of Mechanical Engineers, New York
52. Dornfeld DA, Cai HG (1984) ASME Trans, J Eng for Ind 106:28
53. Inasaki I (1991) Annals of the CIRP 40:359
54. Dornfeld DA, Kannatey-Asibu E (1980) Int'l J Mech Sci 22(285)
55. Kannatey-Asibu E, Dornfeld DA (1981) ASME Trans, J of Eng for Ind 103:330
56. Chang YP, Hashimura M, Dornfeld DA (1996) Annals of the CIRP 45:331
57. Fukuroda A, Nakamura K, Arimoto Y, (1995) In situ CMP monitoring technique for multi-layer interconnection, Technical Digest Int'l Electron Devices Meeting, Proc. the 1995 Int'l Electron Devices Meeting, p. 469
58. Tang JS, Unger C, Moon Y, Dornfeld D (1997) Proc Mater Res Soc Symp 476:115
59. Tang JS, Dornfeld D, Pangrle SK, Dangca A (1998) J Electron Mater 27(10):1099
60. Hetherington DL, Stein DJ (1999) The Int'l Soc Opt Eng 3884:24
61. Kojima T, Miyajima M, Akaboshi F, Yogo T, Ishimoto S, Okuda A (2000) IEEE Trans Semiconductor Manufact 13(3):293
62. Colgan M, Morath C, Tas G, Grief M (2001) Solid State Tech 44(2):67
63. Sampson RK (2002) Use of Acoustic Spectral Analysis for Monitoring/Control of CMP Process, US Patent 6424137
64. Hocheng H. and Huang Y-L(2001) Preliminary Study of Endpoint Detection for Chemical Mechanical Planarization Process Using Acoustic Emission, Proc. 18th Int'l VLSI Multilevel Interconnection Conf., p. 107
65. Huang YL (2002) Ph.D. Dissertation, National Tsing Hua University, Taiwan
66. Miller RK, Mclentire P (1987) Nondestructive testing handbook, volume 5, acoustic emission testing. American Society for Nondestructive Testing, USA
67. Hocheng H, Cheng CY (2002) Trans Semiconductor Manufact 15(1):45

Chapter 5
Ultrasonic Machining

K.L. Kuo, H. Hocheng, and C.C. Hsu

Abstract Ultrasonic machining (USM), using shaped tools, high-frequency mechanical motion, and abrasive slurry is effective for materials of extreme hardness or brittleness. Unlike other nonconventional machining methods such as laser beam and electrical discharge machining, USM does not thermally damage the workpiece. This is important for the longevity of materials in service. However, the tool experiences wear, which causes a reduction machining efficiency. Composite materials offer advantages in structural applications because of their high specific strength and directional properties. In many applications, composites are cured in their final shape; however, machining is necessary at both the prepreg and product stages. In traditional drilling, delamination and splintering at the edges of holes often occur due to anisotropy and the lamination of composite materials. USM is suitable for such materials for its mode of material removal, which utilizes small individual abrasives. In this chapter, we discuss on-line tool-wear monitoring during USM, the effect of abrasive and drilling parameters on material removal rate, hole clearance, edge quality, tool wear, and surface roughness of composites for application in the manufacturing industry. This chapter shows that USM can provide greater profits than other nonconventional machining processes offer.

K.L. Kuo (✉)
Department of Vehicle Engineering, National Taipei University of Technology, Taipei, Taiwan, ROC
e-mail: klkuo@ntut.edu.tw

H. Hocheng • C.C. Hsu
Department of Power Mechanical Engineering, National Tsing Hua University, Hsinchu, Taiwan, ROC
e-mail: hocheng@pme.nthu.edu.tw

H. Hocheng and H.-Y. Tsai (eds.), *Advanced Analysis of Nontraditional Machining*,
DOI 10.1007/978-1-4614-4054-3_5,

Keywords Tool wear • Ultrasonic machining • Resonance frequency • Monitoring • Drilling • Composite material • Material removal rate • Oscillation • Abrasive • Delamination • Vibration amplitude • Surface roughness

5.1 Introduction

For machining hard and brittle materials, ultrasonic machining (USM) is among the most effective methods, regardless the work material is electrically conductive or not. In general, USM is relatively suitable for metals, ceramics, and composites [1]. Wood first noted the application of high-frequency (about 70 kHz) sound waves for machining in 1927 [2]. The first patent of USM was issued in 1945 [3]. Rozenberg et al. review the fundamental study of ultrasonic cutting process and the machine tool. The book was published over 30 years ago and provided valuable knowledge of USM [4]. In regard to material removal mechanism of USM, Shaw considers the direct hammering to be the major cause for material removal [5]. Rozenburg and Kremer et al. propose that the material removal also involved the erosion on the work surface due to cavitation effect of the abrasive slurry [4, 6]. The abrasive particles are excited by the tool vibrating at ultrasonic frequency, e.g., 20 kHz, and amplitude of tens of micrometers. A continuous flow of the abrasive slurry flushes away the chips removed in the process. The grain size of the abrasive shows great effect on surface roughness and cut accuracy [7], and the ratio of material hardness to the modulus of elasticity tends to influence considerably the material removal rates, surface roughness, and cut accuracy. USM has been variously termed ultrasonic drilling; ultrasonic abrasive machining; ultrasonic cutting; ultrasonic dimensional machining, and slurry drilling [8]. The rotary tool was found superior to conventional static tool in USM [9].

The mechanical properties of work material influence the USM. Dam et al. show that tough materials give low removal rate, high tool wear, and fair surface roughness [10]. Costa et al. studied zirconia and white cast iron abrasive-wear under two-body and three-body configurations [11]. The results show that the performance of zirconia is favorable only for low-severity wear conditions and for intermediate and high degrees of transformability. While the difficult-to-cut material is effectively machined, the tool of ultrasonic machine also wears. The tool unavoidably experiences wear by the highly abrasive particles and machining chips. Anantha Ramu et al. discuss several tools and work materials with respect to the penetration rate and tool wear in ultrasonic drilling [12]. The tool wear is affected by the vibration amplitude of tool and the static load applied upon tool. Large vibration amplitude increases the kinetic energy of the abrasives, which wears the tool tip severely. Large static load, on the other hand, depresses the free vibration of the abrasives, hence the tool wears slowly [13]. The reduction of tool length will cause the change of vibrating frequency eventually leading to reduction of USM efficiency. An on-line technique to monitor the tool wear is desired. The monitoring of tool wear during USM is not found in the existing literature. The present study offers an on-line technique to detect the wear length of tool.

Advanced fiber-reinforced plastics are being increasingly used due to their high specific strength and stiffness, thermal resistance, and damping capacity. Ceramic composites provide tailored properties by changing the fiber reinforcing manner and designing proper matrix microstructure. During the past decade, a number of articles have been reported on the investigation of mechanical properties, fabrications, modeling, and applications [14–16]. The C/SiC composite material possesses superior mechanical properties and high oxidation resistance leading to an important role in high-temperature applications, e.g., high-temperature oven, space plane, etc. [17]. Drilling is among the most frequently used machining methods needed for parts production in industrial applications. Since C/SiC composite material is hard and brittle, ultrasonic drilling is considered suitable for the material [18]. Ultrasonic drilling effectively reduces the concerns often encountered in drilling of composite materials, such as delamination, fiber pull-out, microcracking, and burrs. Besides, it induces no thermal, chemical, electrical, or metallurgical threat to the workpiece.

In this chapter, the monitoring of the tool wear and the processing for ultrasonic drilling of composite material are investigated and discussed. In comparison to other feasible machining processes, USM possesses advantages concerning machinability and cost.

5.2 On-Line Tool-Wear Monitoring

5.2.1 Resonance Frequency of Tool

The profile of the horn-shaped tool bar used in the USM can be categorized into step, catenoidal, exponential, and conical [12]. Other profiles are the combinations of these basic shapes. Horns of special profiles, e.g., Gaussian horn and Fourier horn, are hard to produce, and the magnification rate of the amplitude of these special profiles is not superior to the basic shapes. Hence they are rarely used. The exponential horn is defined by the exponential variation of the cross-section of the horn. Figure 5.1 shows three kinds of the exponential horn, including the outer exponential profile, the wedge-shaped exponential profile, and the inner exponential profile of the horn. The resonance conditions vary with the profiles of the horn. The horn used in the current study is shown in Fig. 5.1a and a tool for cutting or polishing is welded on it.

The relationship between the ultrasonic velocity (c), the wavelength (λ), and the frequency (f) is as follows

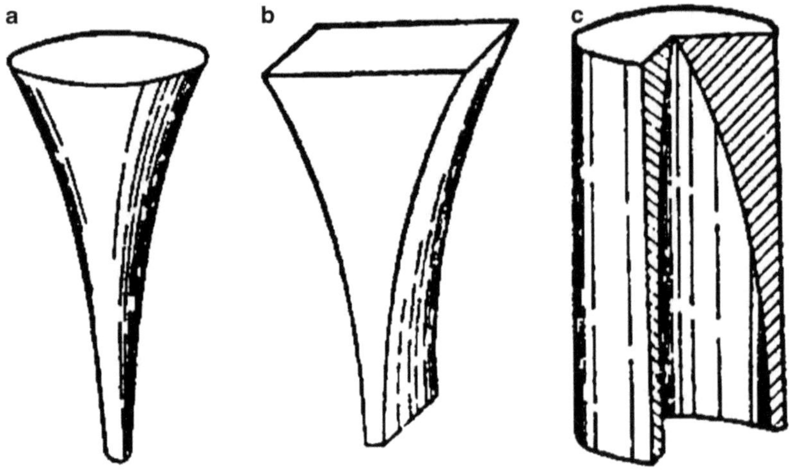

Fig. 5.1 Exponential horn. (**a**) Outer exponential. (**b**) Wedge-shaped exponential. (**c**) Inner exponential [2], reprinted with permission

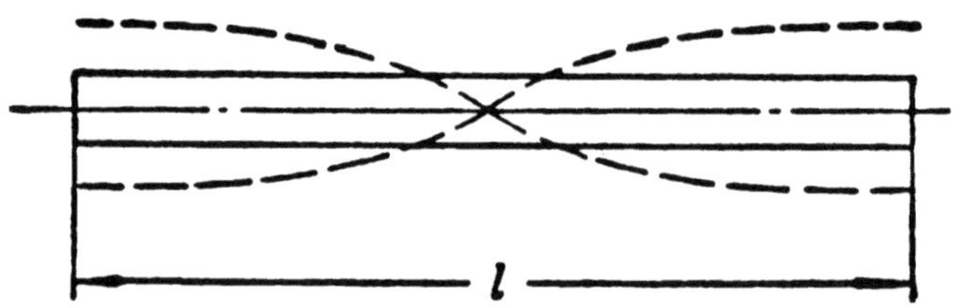

Fig. 5.2 Half-wave length resonance mode of straight cylinder [2], reprinted with permission

$$c = \lambda f \tag{5.1}$$

The design of the ultrasonic horn is to make the length (l) equal to the half wavelength traveled in the horn for the best magnification. An example of straight cylinder is shown in Fig. 5.2. If the resonance frequency is f, then

$$l = \frac{c}{2f} \tag{5.2}$$

Figure 5.3 shows the various cross-sectional of the horn cylinder. The cross-sectional area of the horn is

$$S_x = S_1 e^{\gamma x} \tag{5.3}$$

where γ is tape constant of the exponential horn. The general longitudinal vibration is written as

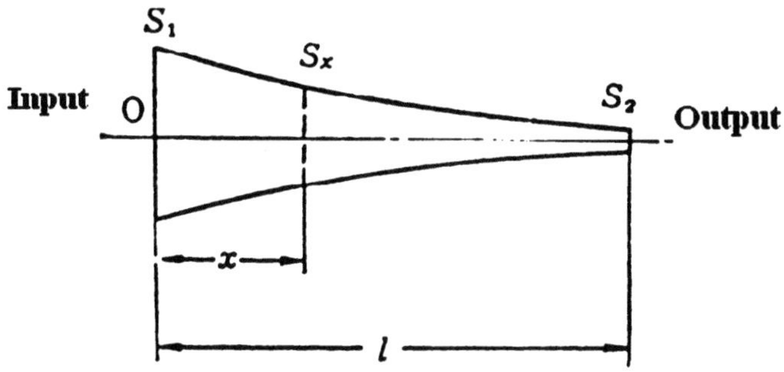

Fig. 5.3 Outer exponential cylinder of horn [2], reprinted with permission

$$\rho S \frac{\partial^2 u}{\partial t^2} = E \frac{\partial}{\partial x}\left(S \frac{\partial u}{\partial x}\right) \tag{5.4}$$

where E is the Young's modulus, ρ is the density, u is the displacement, and t is the time. The boundary conditions are

$$\frac{\partial u}{\partial x} = 0 \text{ at } x = 0 \text{ and } x = l \tag{5.5}$$

The resonance condition of the exponential horn is obtained [12].

$$l_n = \frac{nc}{2f_n}\sqrt{1 + \tan^2\phi_n} \tag{5.6}$$

where

$$\tan\phi_n = \frac{1}{2\pi n}\log_e\left(\frac{S_2}{S_1}\right) = \frac{\gamma l_n}{2\pi n} \tag{5.7}$$

and n denotes the nth resonance mode, $n = 1, 2, 3, \ldots..$

The best condition of the horn during USM occurs at the first resonance mode (namely the half wavelength resonance, $n = 1$ as mentioned above). One obtains the following relationships

$$l = \frac{c}{2f}\sqrt{1 + \tan^2\phi} = \frac{c}{4\pi f}\sqrt{4\pi^2 + \gamma^2 l^2} \tag{5.8}$$

$$l = \frac{c}{4\pi f}\sqrt{4\pi^2 + [\log_e(S_2/S_1)]^2} \tag{5.9}$$

where

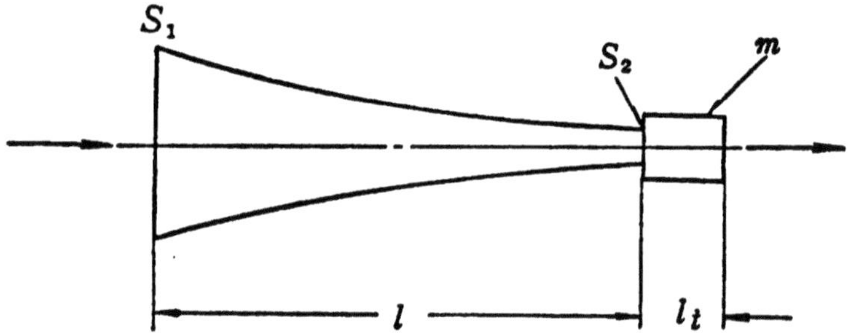

Fig. 5.4 Exponential horn with a tool at the front end [1], reprinted with permission

$$\tan\phi = \frac{1}{2\pi}\log_e(S_2/S_1) = \frac{\gamma l}{2\pi} \tag{5.10}$$

There is a tool attached to the front end of the horn, as shown in Fig. 5.4. The tool has an uniform cross-section with length l_t and mass m. Its equivalent mass m_t for analysis is expressed by

$$m_t = m\frac{\tan(\alpha_t l_t)}{\alpha_t l_t} \tag{5.11}$$

One notices the coefficient $\frac{\tan(\alpha_t l_t)}{\alpha_t l_t}$ is generally larger than 1, hence the equivalent m_t is larger than m.

When the first resonance mode (half wavelength resonance) occurs

$$\log_e(S_2/S_1) = 2\tan\phi\left\{\tan^{-1}\left(\frac{-\cos\phi}{\sin\phi + \frac{S_2\rho c}{\omega m_t}}\right) + \pi\right\} \tag{5.12}$$

where ϕ can be calculated, as S_1 and S_2 are provided. The correlation between the resonance frequency and the length l at resonance is given by [12].

$$l = \frac{c}{2f}\sqrt{1+\tan^2\phi} \tag{5.13}$$

Equation (5.13) provides the feasible application of the resonance frequency for monitoring of the progressive tool wear in association with reduction of tool length. The experimental results are shown in the next section.

5.2.2 Experimental Results and Discussions

Figure 5.5 shows the USM system in the present work. A transducer is used to generate the required ultrasonic vibration at ultrasonic frequency and small amplitude. A horn is connected to the transducer to magnify the vibration. A tool made of copper is fixed at the bottom of the horn to transmit the vibration energy to the machining area. During the process, the vibrating tool energizes the abrasive slurry to

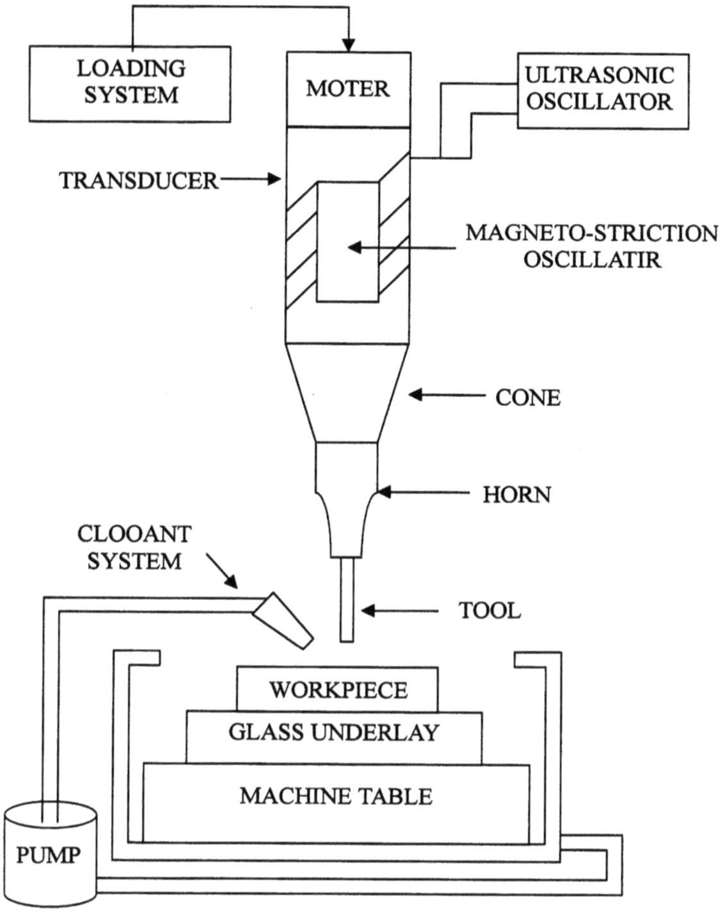

Fig. 5.5 Schematic diagram of an ultrasonic machining system

carry out the material removal of the work. Motion up to three degrees of freedom (linear motion in two directions and a rotation at the base) is controlled simultaneously by the PC-based controller. The static load applied to the tool is determined by the weight difference between the counterweights and the transducer and the horn. The weight of the transducer and the horn acts on one end of the lever assembly via an arm and a wheel, while the counterweights are placed on the other end of the lever assembly. By changing the counterweights, one can adjust the static load of the tool.

Based on the theory described above, a set up of the tool-wear monitoring system using the frequency feedback is developed, as shown in Fig. 5.6. When the tool wears, the total mass of the horn and the tool decreases, and the resonance frequency of the transducer changes. The resonance frequency is fed to the generator, and the generator will adjust itself to a new driving frequency coping with the frequency of the transducer. Using the frequency signal from the generator, one can

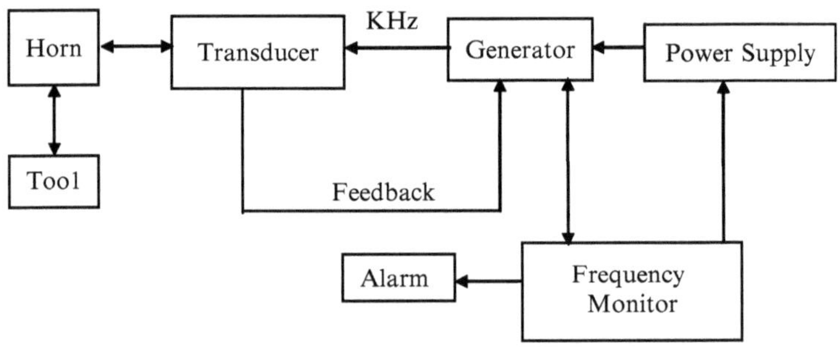

Fig. 5.6 Schematic diagram of the tool-wear monitoring system

monitor the tool wear continuously. In case of serious tool wear beyond a certain limit or tool breakage, the power supply of the machining system can be shut down to stop the machining process. An alarm can be set up in parallel.

In the current study, the length of the exponential horn is 100 mm, and the diameters of both ends are 20 and 8 mm, respectively. The constant C is 5.2×10^5 cm/s. The data of the brazed copper tool is as follows. The diameter is 3 mm, the tool length is 3.3 cm, the density is 8.9 g/cm^3, and the sound velocity of the copper tool is 3.7×10^5 cm/s. Its mass is 2.076 g.

The equivalent mass m_t is obtained by (5.11). Substituting m_t into (5.12) and (5.13), one obtains the theoretical value of the resonance frequency 26,852 Hz. The equivalent mass m_t changes as the tool wears. Hence the resonance frequency associated with the progressive tool wear can be calculated.

Two sets of different horn and initial tool length are investigated in the current study. Table 5.1 shows the theoretical and experimental frequencies varying with the tool wear. The resonance frequency increases with the reduced length of the worn tool, as also shown in Fig. 5.7. The nearly constant deviation of the experimental value from the theoretical value is around 5%, which is a reasonable result of the lack of consideration of the damping, the joint backlash and the mass of the attached machine structure existing in the machining system.

As shown in Table 5.1, the experimental resonance frequency changes from 25.528 to 25.794 kHz when the tool length wears from the initial length of 33.1 to 28.5 mm. It tells the tool shortening of 4.6 mm by increase of frequency of 266 Hz. An approximately linear relation exists at the early stage of tool wear, hence 1 Hz difference of frequency implies the 17.3 μm wear loss of the tool length. Figure 5.7 tells the tool wear affects the resonance frequency more significantly at the initial stage than later. The monitoring of the tool wear is therefore more sensitive at beginning, when the USM is operated at the designed tool length for maximum efficiency. However, the alarm frequency of the tool wear can be set at any value based on need.

Figure 5.8 shows the monitoring resolution of tool wear in the current USM system using the resonance frequency. Although the resolution is not a constant in such a machine tool, an excellent resolution down to less than 20 μm is obtained by the variation of 1 Hz in frequency monitoring. The resolution maintains around 50 μm per Hz, even when the tool wears 50% of its original length. This resolution

Table 5.1 Tool length and resonance frequency

Tool length (mm)	Resonance frequency (KHz)		
	Experiment	Theory	Error (%)
(a) Tool Set A (mass of horn 126.045 g, diameters 20 and 8 mm, initial tool length 33.1 mm)			
33.1	25.528	26.852	4.93
28.5	25.794	27.204	5.18
24.8	25.923	27.36	5.25
20.1	26.042	27.486	5.25
15.1	26.124	27.573	5.26
10.1	26.157	27.608	5.26
5.2	26.188	27.632	5.23
(b) Tool Set B (mass of horn 126.192 g, diameters 20 and 8 mm, initial tool length 36.1 mm)			
36.1	25.299	26.324	3.89
30.6	25.655	26.886	4.58
25.1	25.906	27.211	4.80
20.1	26.009	27.336	4.85
15.2	26.088	27.422	4.86
10.2	26.144	27.457	4.78
6.2	26.188	27.481	4.71

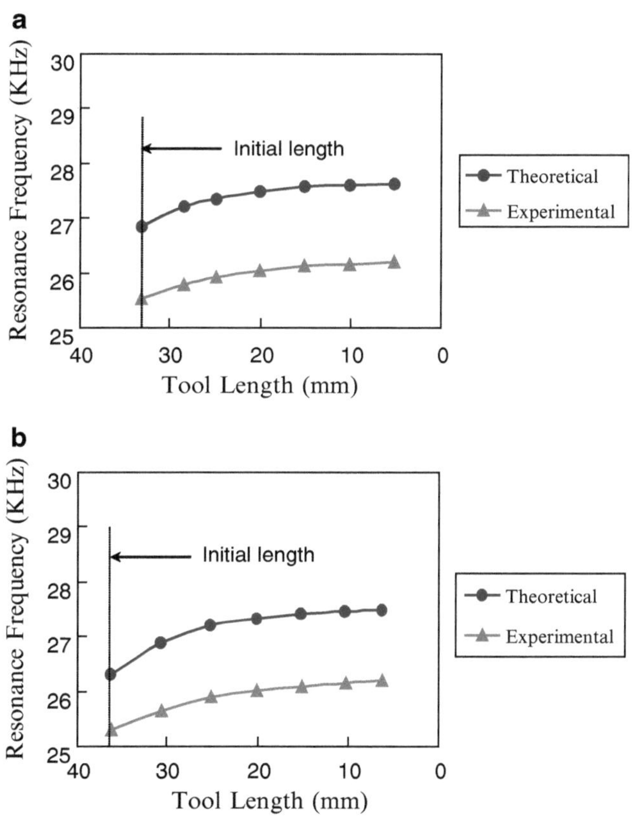

Fig. 5.7 Relationship between tool length and resonance frequency. (**a**) Tool Set A. (**b**) Tool Set B

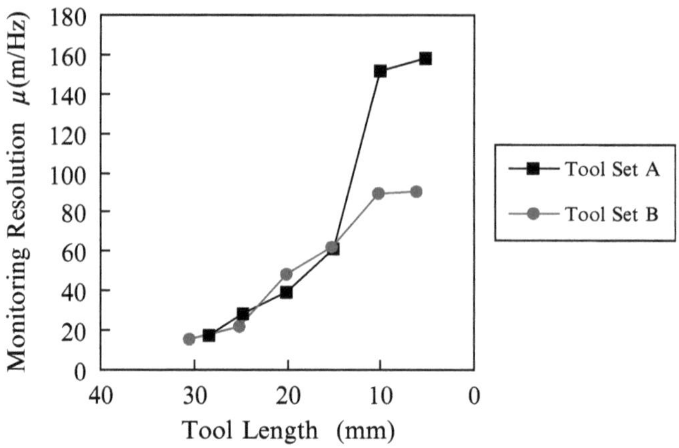

Fig. 5.8 Monitoring resolution of tool wear using resonance frequency

is sufficient for industry. In the practice of tool-wear monitoring, a master curve correlating the resonance frequency and the tool length must be drawn in advance. This curve can be obtained by the theoretical analysis shown in Sect. 5.2.1, and 5–10% modification to this result. The reasonable extent of modification can be conveniently determined by the comparison of the measured frequency with the theoretical value before the machining starts, when the tool is new.

5.3 Drilling of Carbon/Epoxy and Carbon/Peek Composite Materials

5.3.1 Experiment and Materials

The experiments were conducted on an 800 W ultrasonic drilling machine with tuned frequency of 20 kHz [19]. The experimental setup is shown in Fig. 5.5. Boron carbide or silicon carbide abrasives were mixed with water at various volume concentrations. The abrasive slurry is circulated and supplied to cutting tool. The mild steel tool is a circular tube of 10 mm outside diameter (OD) and 2.1 mm thickness brazed on the toolholder. The drilling conditions are listed in Table 5.2.

The materials are the thermoset-based Carbon/Epoxy and thermoplastics-based Carbon/PEEK. Sixteen layers of prepreg were laminated and cured.

1. *Carbon/Epoxy*

    (a) 123°C at atmospheric pressure for 60 min, followed by
    (b) 183°C at 1.05 MPa for 120 min, followed by
    (c) Air cooling to room temperature

Table 5.2 Drilling conditions

Abrasives [mesh number]	S_iC/#150, #220, #400, #600 B_4C/#220
Volume concentration [%]	S_iC/13.4%, 18.6% B_4C/14.7%, 18.7%, 22.8%, 25.6%
Electric current [ampere]	0.4, 0.5, 0.6, 0.8
Feed rate [mm/min]	0.3, 0.4, 0.5, 0.6

2. *Carbon/PEEK*

   (a) 390°C at 1.4 MPa for 20 min, followed by
   (b) 390°C at 5.5 MPa for 20 min, followed by
   (c) Air cooling to room temperature

The laminate thickness is 2.0 and 230 mm by 210 mm in area. The fiber volume content is 60%. 2^4 factorial design of Taguchi method was employed to reduce the number of experimental cycles.

The material removal mechanism is often complex to construct a sound theoretical model. Experiments involving a number of parameters are time-consuming and are liable to incomprehensible results. Dimensional analysis provides a preliminary insight at reduced experimental expense. For the average surface roughness (Ra, £gm) and the material removal rate (V', mm^3/s), the dominant parameters are tile tool feed rate (f, mm/min), abrasive size (S, £gm), abrasive density (£1, g/cm^3) and hardness (H, Nt/m^2), input energy of tool oscillation (W, watt), and abrasive flow rate (A, cm^3/s). Being nondimensionalized by PI theorem, the relationship can be obtained as:

$$\mathrm{Ra/S} = \mathrm{f(A/fS^2}, \rho \mathrm{S^2f^3/W, Hf\ S^2/W)} \tag{5.14}$$

and

$$\mathrm{V'/fS^2} = \mathrm{f(A/fS^2}, \rho \mathrm{S^2f^3/W, Hf\ S^2/W)} \tag{5.15}$$

The experimental results are then incorporated. The relationship between the nondimensionalized surface roughness and material removal rate and machining parameters can be shown as:

$$\frac{Ra}{S} = 0.19\left(\frac{A}{fS^2}\right)^{0.13}\left(\frac{\rho S^2 f^3}{W}\right)^{0.21}\left(\frac{HfS^2}{W}\right)^{-0.43} \tag{5.16}$$

$$\frac{\dot{V}}{fS^2} = 3.6\left(\frac{A}{fS^2}\right)^{0.33}\left(\frac{\rho S^2 f^3}{W}\right)^{0.37}\left(\frac{HfS^2}{W}\right)^{-0.74} \tag{5.17}$$

Further discussions of the effects of individual machining parameters are summarized in the following section.

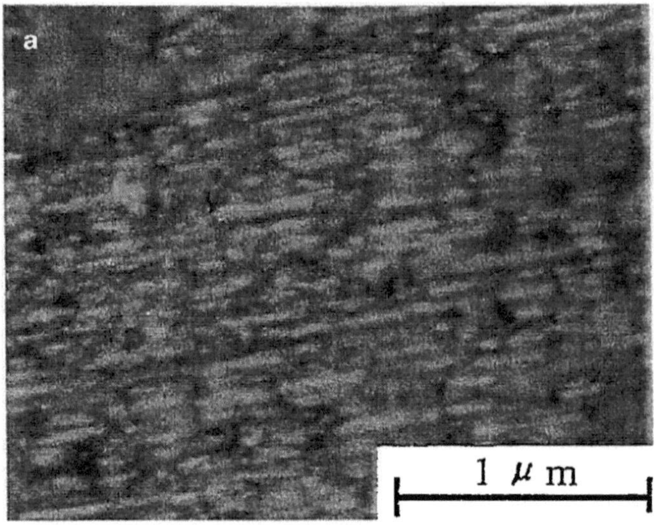

#220 B_4C c = 18.7%, Carbon/Epoxy i = 0.5amp, f = 0.4mm/min

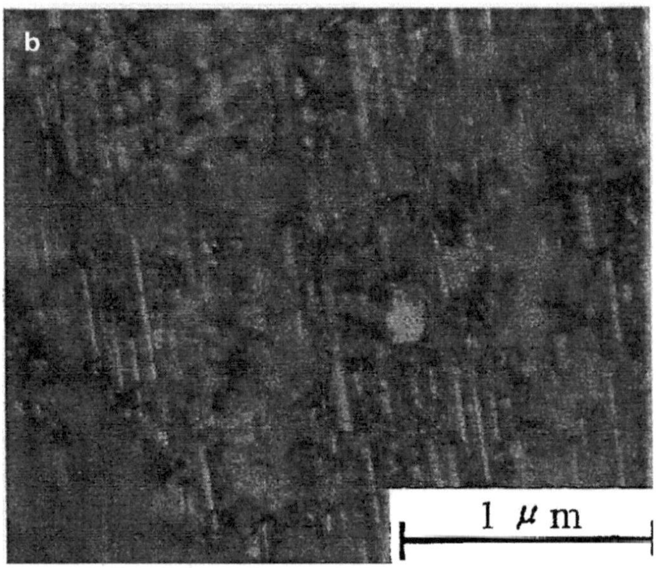

#220 B_4C c = 25.6%, Carbon/PEEK i = 0.8amp, f = 0.3mm/min

Fig. 5.9 Optical microscope photograph of machined surface under cutting tool [19], reprinted with permission. (**a**) #220 B_4C $c = 18.7$ %, carbon/epoxy $i = 0.5$ amp, $f = 0.4$ mm/min; (**b**) #220 B_4C $c = 25.6$ %, carbon/PEEK $i = 0.8$ amp, $f = 0.3$ mm/min. c: abrasive volume concentration; i: input current; f: tool feed

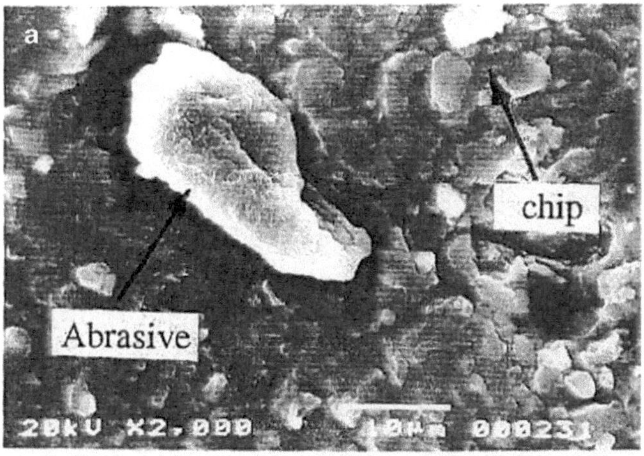

#220 B_4C c = 18.7%, Carbon/PEEK i = 0.8amp, f = 0.4mm/min

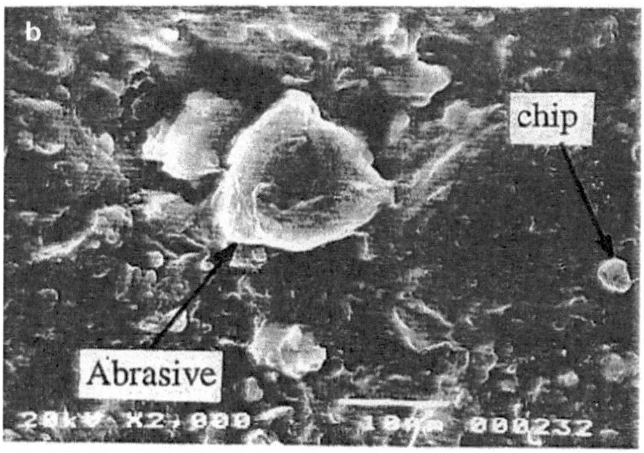

#220 B_4C c = 22.8%, Carbon/Epoxy i = 0.6amp, f = 0.6mm/min

Fig. 5.10 SEM photograph of abrasive and chip [19], reprinted with permission. (**a**) #220 B_4C $c = 18.7\%$, carbon/PEEK $i = 0.8$ amp, $f = 0.4$ mm/min; (**b**) #220 B_4C $c = 22.8\%$, carbon/epoxy $i = 0.6$ amp, $f = 0.6$ mm/min. c: abrasive volume concentration; i: input current; f: tool feed

5.3.2 Drilling Results and Discussions

The machined surface is shown in Fig. 5.9. There are craters on the workpiece caused by abrasive particle hammering or impact. The USM consists of continual material removal in small craters. Figure 5.10 illustrates abrasive particles and chips after machining. The size of the chip is about one-tenth of the abrasive particle, or several micrometers. The chips are produced by the hammering or free impacting of abrasives. The SEM photographs (Figs. 5.11 and 5.12) of the machined surface show plastic deformation of matrix and brittle fracture of fibers.

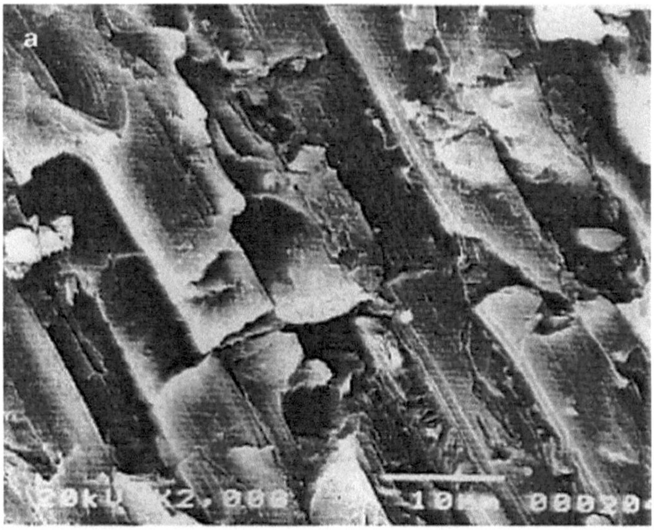

#220 B_4C c = 18.7%, Carbon/Epoxy i = 0.5amp, f = 0.4mm/min

#220 B_4C c = 25.6%, Carbon/PEEK i = 0.5amp, f = 0.5mm/min

Fig. 5.11 SEM photograph of fiber failure [19], reprinted with permission. (**a**) #220 B_4C $c = 18.7\%$, carbon/epoxy $i = 0.5$ amp, $f = 0.4$ mm/min; (**b**) #220 B_4C $c = 25.6\%$, carbon/PEEK $i = 0.5$ amp, $f = 0.5$ mm/min. c: abrasive volume concentration; i: input current; f: tool feed

No significant difference exists between carbon/epoxy and carbon/PEEK in the tested range of cutting conditions.

The surface roughness values obtained in USM are plotted in Figs. 5.13, 5.14, 5.15, and 5.16. The major parameters are the grain size, the input current of

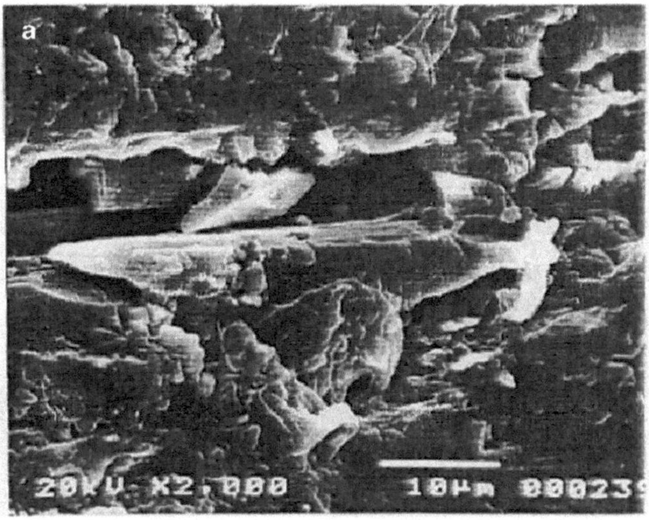

#220 B_4C c = 25.6%, Carbon/PEEK i = 0.4amp, f = 0.6mm/min

#220 B_4C c = 18.7%, Carbon/Epoxy i = 0.8amp, f = 0.6mm/min

Fig. 5.12 SEM photograph of plastic deformation of matrix [19], reprinted with permission. (**a**) #220 B_4C $c = 25.6\%$, carbon/PEEK $i = 0.4$ amp, $f = 0.6$ mm/min; (**b**) #220 B_4C $c = 18.7\%$, carbon/epoxy $i = 0.8$ amp, $f = 0.6$ mm/min. c: abrasive volume concentration; i: input current; f: tool feed

ultrasonic oscillations, and the concentration of abrasives. The feed rate of tool has negligible effect on surface roughness. The increasing roughness value with the grain size and the energy of ultrasonic oscillations is explained by the produced larger craters on the workpiece. Larger amount of material removal by each

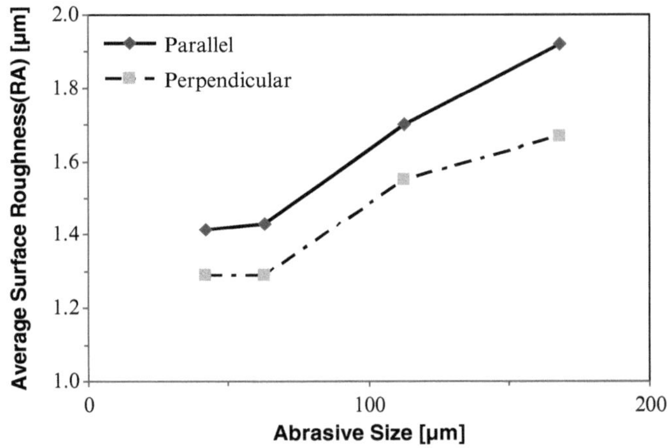

Fig. 5.13 Influence of grain size on surface roughness

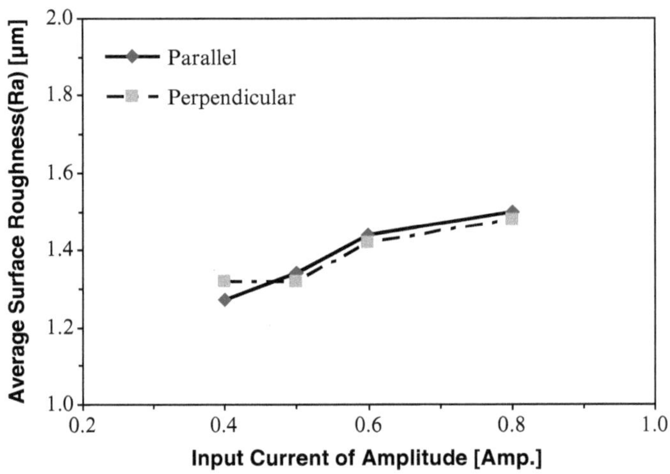

Fig. 5.14 Influence of input current on surface roughness

abrasive results in rougher surface. The higher the concentration of abrasives is, the more frequent the abrasives scrape on the workpiece, hence the machined surface is slightly rougher. The feed rate of tool determines the rate of generation of the new surface only, it has no effect on the number and the size of produced craters on the machined surface.

The optical microscope photograph (Fig. 5.17) and the above figures show uniform surface roughness parallel and perpendicular to fibers. Table 5.3 shows the measured values. Besides, no significant difference exists between thermoset and thermoplastic composite material. These facts are explained by the micro-

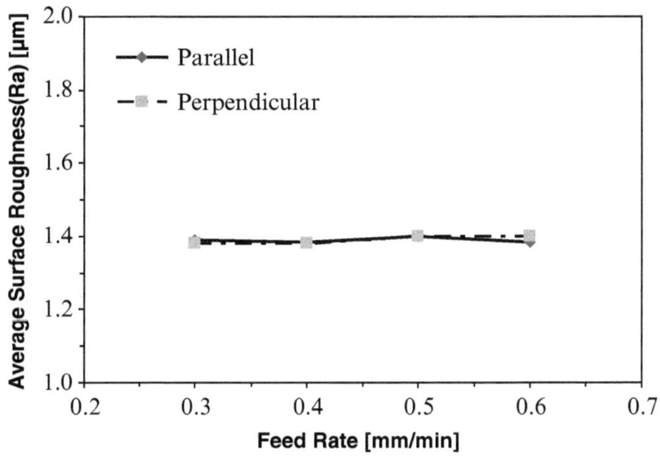

Fig. 5.15 Influence of tool feed on surface roughness

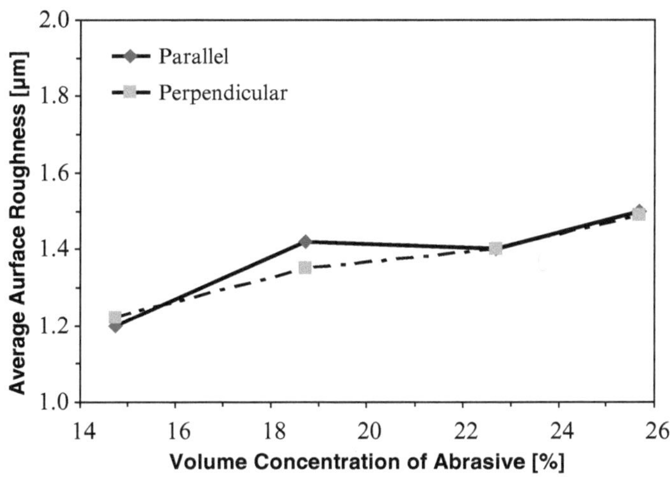

Fig. 5.16 Influence of volume concentration of abrasives on surface roughness

material removal by abrasives, which involves little macro matrix deformation, thus the produced surface is independent of the fiber direction or the matrix thermoplasticity. The same discussions apply to the negligible difference in material removal rate in Table 5.3.

The optical projection photographs (Figs. 5.18 and 5.19) show clean hole edge, no delamination occurs. At high energy input, the thermoplastics-based carbon/PEEK presents minor fuzz at exit due to the relatively viscous deformation of

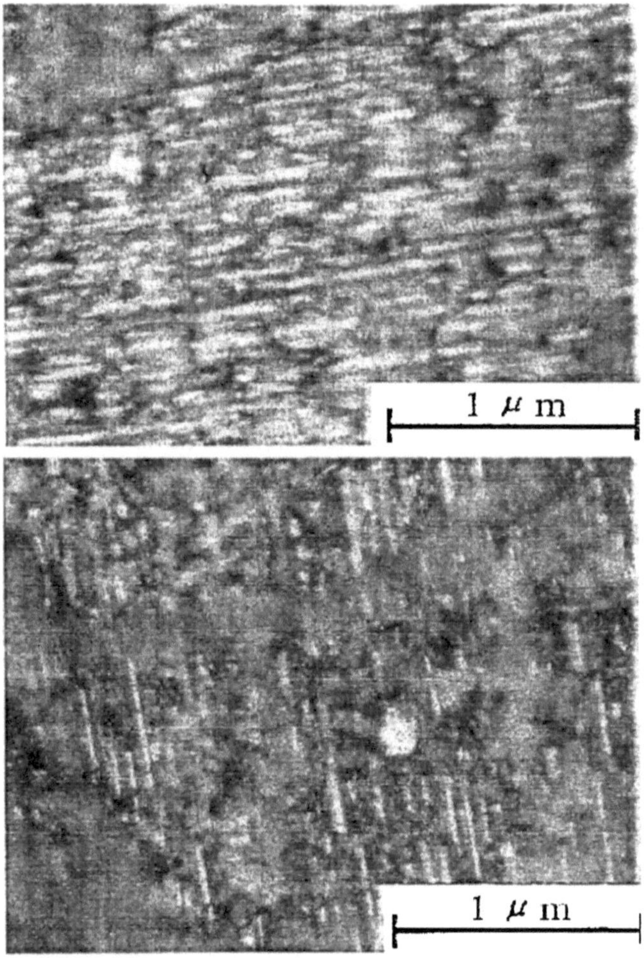

#220 B_4C c = 18.7%, Carbon/Epoxy i = 0.5amp, f = 0.4mm/min

Fig. 5.17 Optical microscope photograph of machined hole wall [19], reprinted with permission. #220 B_4C $c = 18.7\%$, carbon/epoxy $i = 0.5$ amp, $f = 0.4$ mm/min. c: abrasive volume concentration; i: input current; f: tool feed

Table 5.3 Material removal rate and surface roughness in ultrasonic cutting of C/epoxy and C/PEEK

		Surface roughness (μm)	
	Removal rate	//Fibers	⊥Fibers
Carbon/epoxy	0.249	1.36	1.34
Carbon/PEEK	0.245	1.39	1.42

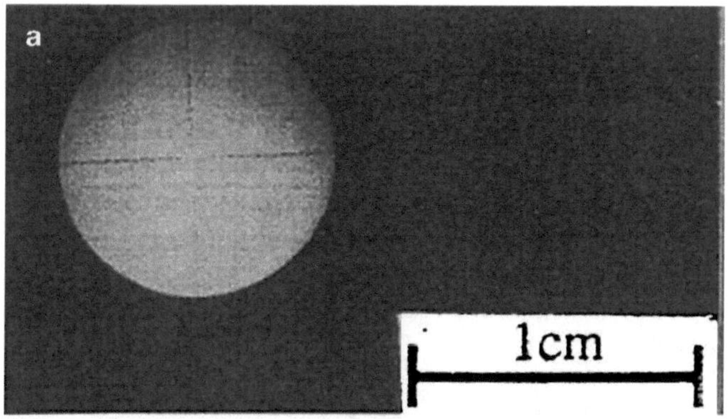

#220 B_4C c = 22.8% i = 0.6amp, f = 0.6mm/min

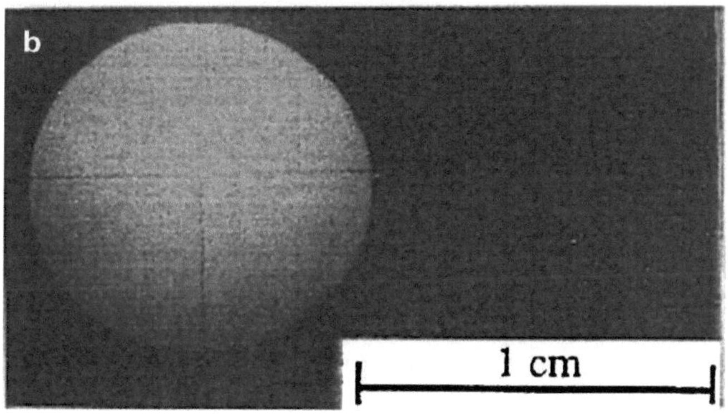

#220 B_4C c = 25.6% i = 0.8amp, f = 0.3mm/min

Fig. 5.18 Optical projection of hole exit of carbon/epoxy [19], reprinted with permission. (**a**) #220 B_4C $c = 22.8$ % $i = 0.6$ amp, $f = 0.6$ mm/min; (**b**) #220 B_4C $c = 25.6$ % $i = 0.8$ amp, $f = 0.3$ mm/min. c: abrasive volume concentration; i: input current; f: tool feed

matrix and decreased cooling by slurry toward exit. With the aid of liquid nitrogen, the viscosity of thermoplastics is reduced, thus the exit edge is free of fuzz. Figure 5.20 illustrates the SEM photograph of the hole exit. The thick thermoplastic recast covering fibers and the built up edge at exit can be seen in (a), while (b) shows less thermoplastic flow of file matrix with neat hole wall and recognizable fibers are seen.

The diameter clearance between the produced hole and the cutting tool is enlarged with the grain size and the energy of ultrasonic oscillations, as shown in Figs. 5.21 and 5.22. The feed rate of tool or the volume concentration of abrasives does not affect the hole clearance significantly.

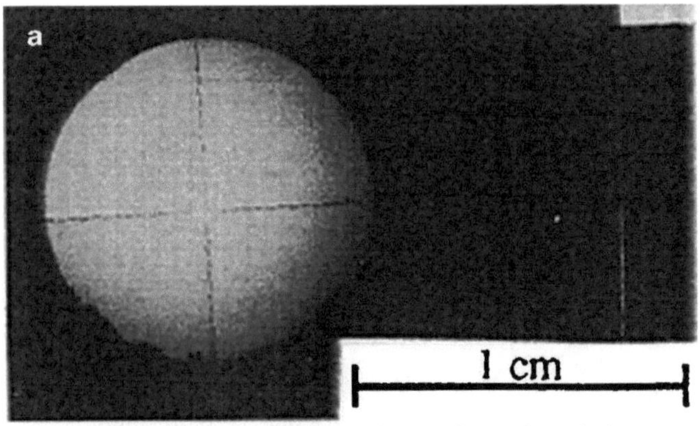

#220 B_4C c = 25.6% i = 0.4amp, f = 0.6mm/min

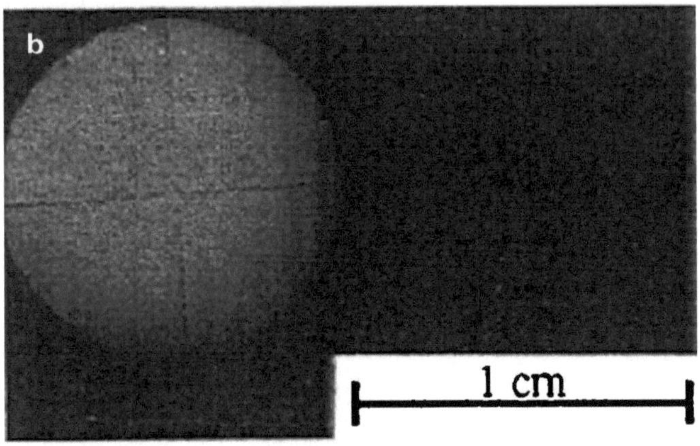

#220 B_4C c = 14.7% i = 0.8amp, f = 0.6mm/min

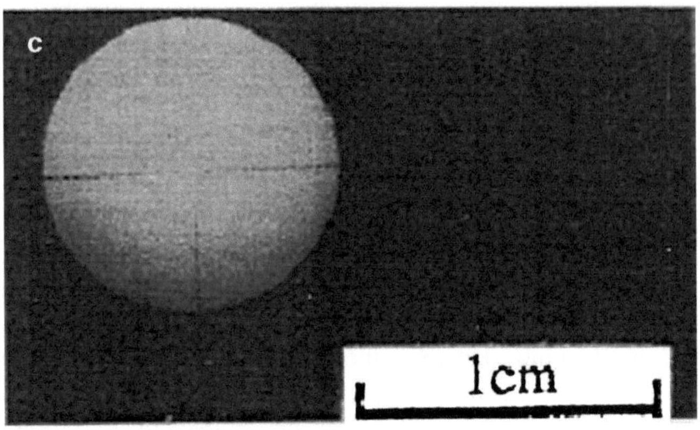

#220 B_4C c = 14.7% w / Liq. N2 i = 0.8amp, f = 0.6mm/min

Fig. 5.19 Optical projection of hole exit of carbon/PEEK [19], reprinted with permission. (**a**) #220 B_4C $c = 25.6\%$ $i = 0.4$ amp, $f = 0.6$ mm/min; (**b**) #220 B_4C $c = 14.7\%$ $i = 0.8$ amp, $f = 0.6$ mm/min; (**b**) #220 B_4C $c = 14.7$ % w/Liq. N_2 $i = 0.8$ amp, $f = 0.6$ mm/min. c: abrasive volume concentration; i: input current; f: tool feed

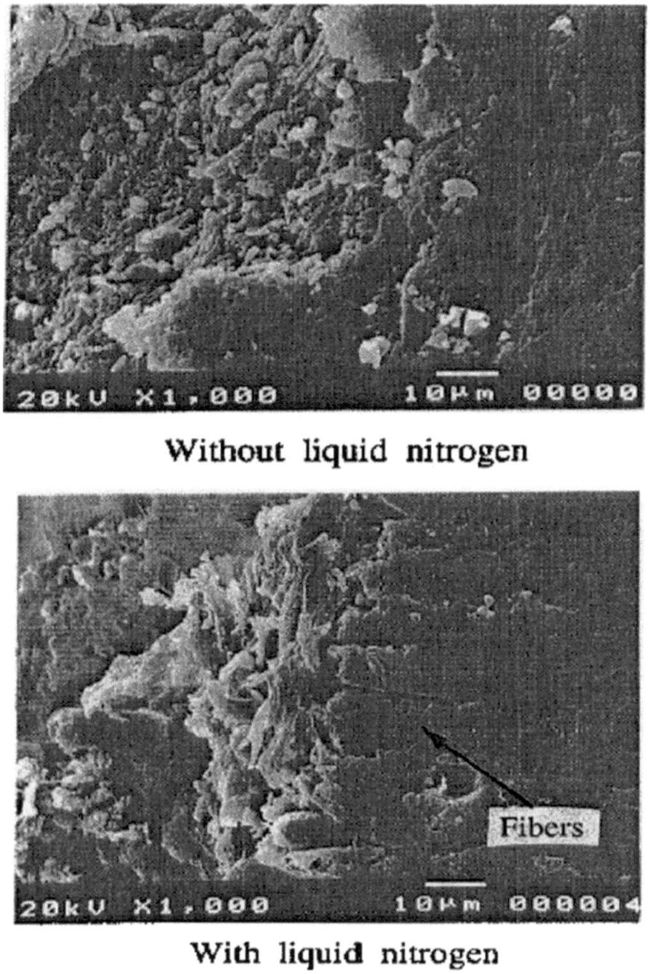

Fig. 5.20 SEM photograph of machining with and without liquid nitrogen [19], reprinted with permission. #220 B_4C $c = 14.7\%$, carbon/PEEK $i = 0.8$ amp, $f = 0.6$ mm/min

5.3.3 Analysis of Drilling Economy

Three nonconventional machining processes are compared, i.e., abrasive water-jet, laser, and ultrasonic. The main cost factors are the investment, economic life, salvage value, annual revenue, and expenditures. For the current workpiece, water-jet or laser cuts in average 15 times faster than USM. However, ultrasonic machine can readily perform multi-hole production for equal production rate. The following analysis presents the results of single-hole and multi-hole ultrasonic cutting.

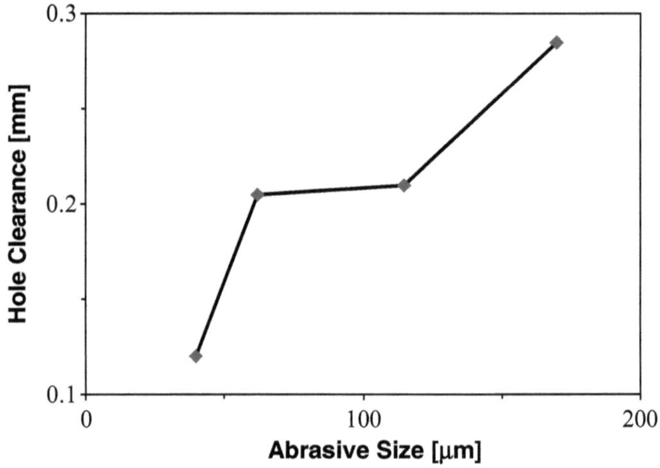

Fig. 5.21 Effect of abrasive size on hole clearance [19], reprinted with permission

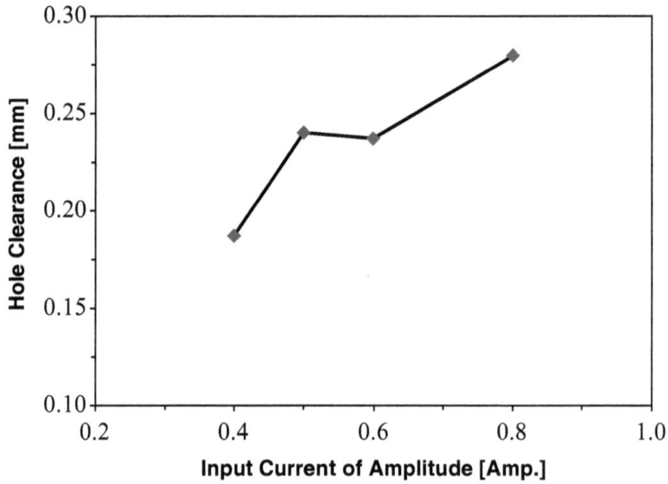

Fig. 5.22 Effect of input current on hole clearance [19], reprinted with permission

The annual and investment expenditures are shown in Tables 5.4 and 5.5. The analysis is based on 8 h/day and 240 days/year. All values are in US Dollars. Using the annual worth method or present worth method (see Tables 5.6 and 5.7), the USM is less economical than the other two processes in the case of single-hole production, due to its much slower material removal rate. With multi-hole production, however, USM prevails over the other two processes.

Table 5.4 Annual expenditure analysis [unit: US Dollar]

Process	Item of expenditure	Quantity of consumption	Annual expenditure	Total cost
Laser machining	Gas	164 m^3/month	19,872	36,832
	Electricity	10 kW/h	1,536	
	Maintenance	1,600/year	1,600	
Abrasive water-jet machining	Nozzle set	One set/100 h	6,912	36,112
	Garnet abrasives	436 kg/day	23,040	
	Electricity	3.75 kW/h	5,760	
	Maintenance	400/year	400	
Ultrasonic machining	S_iC abrasives	4 kg/day	7,680	12,488
	Electricity	3 kW/h	4,608	
	Maintenance	200/year	200	

Table 5.5 Investment analysis [unit: US Dollar]

	Ultrasonic machining				
	Single-hole/ machine	Five holes/machine	Fifteen holes/machine	Laser machining	Abrasive water-jet machining
Investment	180,000	36,000	12,000	160,000	100,000
Economic life	Six years			Nine years	Nine years
Salvage value	1,800	360	120	1,600	1,000
Annual expenditure	187,320	37,464	12,488	36,832	36,112

The lowest rate of return: 15%

Table 5.6 Profit analysis by Annual Worth Method [unit: US Dollar]

	Ultrasonic machining				
	Single-hole/ machine	Five holes/ machine	Fifteen holes/ machine	Laser machining	Abrasive water-jet machining
Capital recovery = investment* (A/P, i %, N)-Salvage*(A/F, i %, N)	47,358	9,472	3,157	33,436	20,897
Annual expenditure	187,320	37,464	12,488	36,832	36,112
Annual cost	234,678	46,936	15,645	70,268	57,009

i%: lowest rate of return; N: period; F: future worth; A: uniform payment; P: present worth

5.4 Drilling of Carbon/SiC Composite Materials

5.4.1 Experiment

The woven carbon cloth was adopted as reinforcement, and the matrix was obtained from the pyrolysis of polysilazane [20]. Slurry infiltration and multi-impregnation were applied to densify the composites. In order to further enhance the density of

Table 5.7 Profit analysis by Present Worth Method [unit: US Dollar]

	Ultrasonic machining				
	Single-hole/ machine	Five holes/ machine	Fifteen holes/ machine	Laser machining	Abrasive water-jet machining
Investment	180,000	36,000	12,000	160,000	100,000
First reset (investment-salvage) *(P/F, i %, N)	77,036	15,407	51,358	45,033	28,145
Second reset (investment-salvage) *(P/F, i %, N)	30,278	6,056	2,019	0	0
Annual expenditure (annual expenditure) *(P/A, i %, N)	1,147,897	229,579	76,526	225,706	221,294
Salvage after 18 years-salvage* (P/F, i %, N)	−145	−29	−9.7	−129	−81
Present worth-cost	1,435,065	287,013	95,671	430,610	349,359

i %: lowest rate of return; N: period; F: future worth; A: uniform payment; P: present worth

the composites, heat treatment over 1800°C was used. The fabrication processing of C/SiC composite material is shown as follows.

(I) Preparation of Raw Materials
PAN carbon cloth #3085 supplied by Porcher Industries in France, matrix precursor of polysilazane NCP-200 supplied by Nichimen Corporation in Japan, and 95% β-SiC nanosize powder (<30 nm) supplied by MTI corporation in USA were adopted as raw materials. The nanosize powder mixing with toluene for part of samples was placed in the ultrasonic machine for 20 min and solid polysilazane NCP-200 was dissolved in toluene aided by ultrasonic vibration. The carbon cloths were then dipped into the mixtures of dispersed powder and matrix precursor for 10 min. The impregnated carbon cloths were dried in an oven at 115°C for 7 h.

(II) Hot Press
Twelve layers of dried prepreg were laminated in a stainless mold, the pressure from 150 to 1,000 psi and temperature up to 300°C were applied gradually. The laminates were held at 300°C for 2 h and then cooled to room temperature.

(III) Pyrolysis
The carbon/ceramic precursor composites were placed in a tube furnace at 1,100°C under N_2 atmosphere for 1 h to convert the polysilazane NCP-200 into ceramics. Five cycles of impregnations and pyrolysis were performed for densification of the composites, this type of specimens is denoted as Specimen 1. In order to improve the crystalline of the composites, higher temperature at 1,600°C was adopted for the densification after the first and the fifth precursor impregnations. Heat treatment at 1,800°C was performed to examine the change of crystallinity and variation of mechanical property. The specimens treated at 1,600°C and 1,800°C were denoted as Specimens 2 and 3, respectively. The composites with and without adding of nanosize powder are categorized as N and Y, respectively.

Table 5.8 Drilling conditions

Abrasives and mesh size	SiC/#150, #220, #400
	B_4C, Al_2O_3/#220
Abrasives volume concentration (%)	SiC/10, 25, 32
	B_4C, Al_2O_3/10
Electric current (A)	0.9, 1.1, 1.3
Loading pressure (MPa)	0.4, 1.1, 2.1
Workpiece and thickness (mm)	N1/2.35 ± 0.5

Six composites, N1, N2, N3, Y1, Y2, and Y3, are obtained in total. The experiments of ultrasonic drilling were conducted on an 800 W ultrasonic drilling machine with tuned frequency of 20 KHz. The experimental setup is shown in Fig. 5.5. The B_4C, SiC, or Al_2O_3 abrasives were mixed with water at various volume concentration. The abrasive slurry is circulated and supplied to the gap between workpiece and cutting tool. The stainless steel tool is a circular tube of 4.51 mm OD and 3.80 mm ID brazed on the toolholder. The drilling conditions are listed in Table 5.8.

The density of workpiece was measured by the water immersion method. The open porosity was measured by a boiling-water penetration method described in ASTM C-20. The flexural strength was measured by three-point bending method with a span-to-depth ratio of 30/(2.35 ± 0.5) and a crosshead speed of 0.125 mm/min. All tests were performed on an Instron test machine following the procedures of ASTMD-790. The morphology and microstructure of the carbon/ceramic composites were analyzed by scanning electron microscopy (SEM) and X-ray diffraction (XRD). The material removal rate was calculated by measuring the machining time and the volume of removed material. The hole clearance was measured by profile projector. The edge quality was evaluated by hole-edge photograph. The tool wear was measured by vernier.

5.4.2 *Results and Discussions*

5.4.2.1 Ultrasonic Drilling

Since N1 (heat treatment temperature 1,100°C) demonstrates the highest flexural strength among the other composites, it is selected for investigating the feasibility of ultrasonic drilling of the C/SiC composite.

(I) Material Removal Rate

Figure 5.23 shows the material removal rate is proportional to the static load applied to the tool. The abrasives gain higher impact force to remove material. The effects of grain size on material removal rate are shown in Fig. 5.24. It shows that there is an optimal size where the material removal rate is maximum. The experimental work of Fukumote shows the optimal value depends upon the amplitude of the tool oscillation [21]. Coarser abrasives cause more extensive fracture

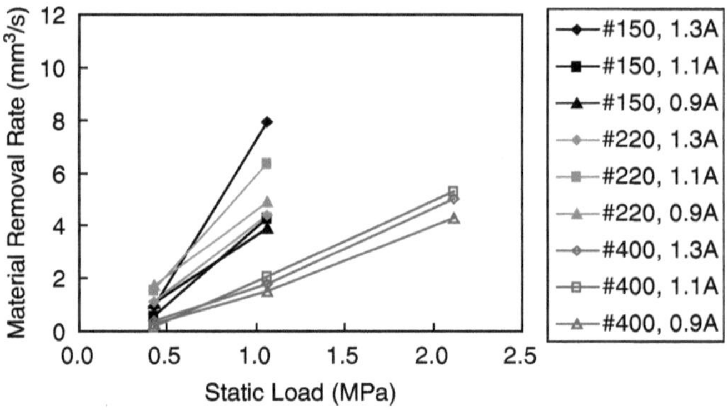

Fig. 5.23 Influence of static load on material removal rate (SiC, 10%) [20], reprinted with permission

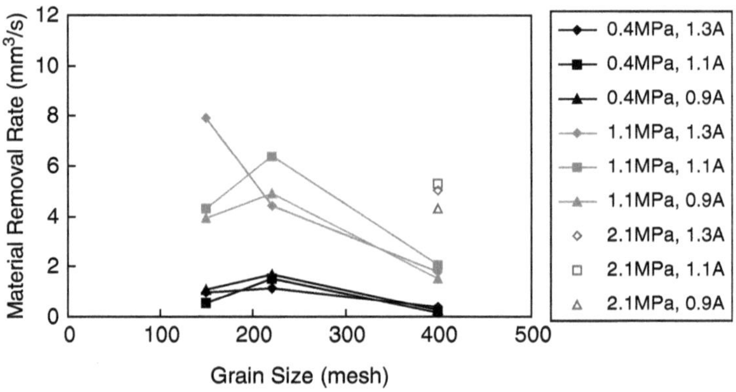

Fig. 5.24 Influence of grain size on material removal rate (SiC, 10%) [20], reprinted with permission

association of material during the hammering of abrasives, thus the material removal rate increases. However, increasing the size of the abrasives at the same level of slurry concentration means reduction of the real number of abrasives. The material removal rate then decreases. The vibration energy of the abrasives affects the optimum of the abrasive size, as mentioned in the above reference. The effects of various grains on material removal rate are shown in Fig. 5.25. The material removal rate shows the maximum value when B_4C grain is adopted, because the hardness of B_4C is the highest, followed by SiC and Al_2O_3.

The effects of vibration amplitude(determined by electric current) on material removal rate are also shown in Figs. 5.23, 5.24, and 5.25. They show that the material removal rate increases with an increase in the static load. As the vibration amplitude increases up to a certain value (electric current = 1.1 A), the material removal rate increases due to the kinetic energy of abrasives. A further increase of vibration amplitude beyond this value results in a reduction in material removal

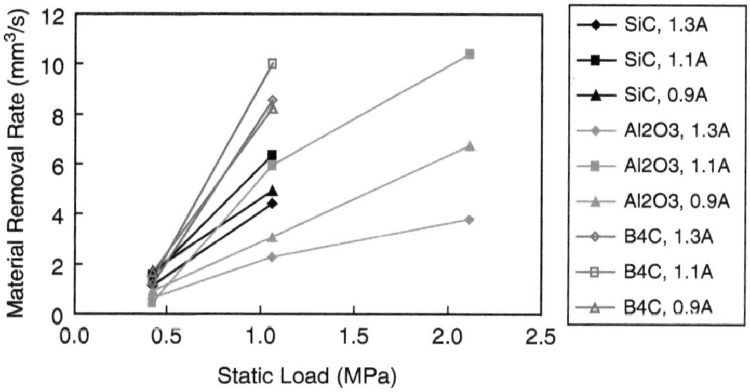

Fig. 5.25 Influence of various grains on material removal rate (#220, 10 %) [20], reprinted with permission

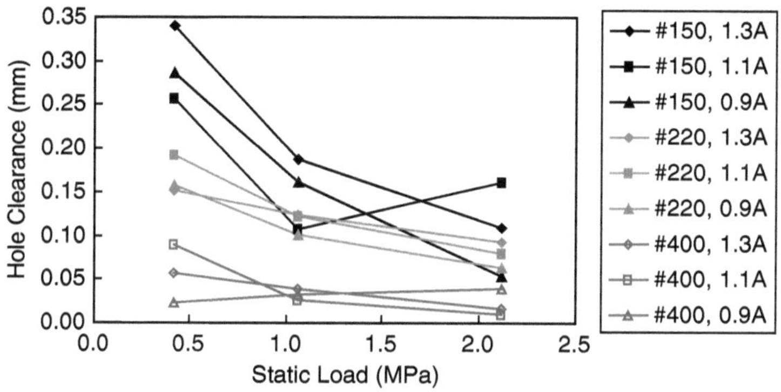

Fig. 5.26 Influence of static load on hole clearance (SiC, 10 %) [20], reprinted with permission

rate. The reduction in material removal rate was attributed to the rapid increase in the interference between abrasives motion, or "an excessive increase in alternate grains and a weakening of the bond" [22].

(II) Hole Clearance

The static load and vibration amplitude influence on the hole clearance, as shown in Fig. 5.26. They show the hole clearance decreases with an increase in the static load, since the machining time decreases and the transverse vibration of the tool is depressed with an increase in the static load. As the vibration amplitude increases, the hole clearance increases as a result of the enhanced vibration of the tool and the abrasives in the gap.

The effects of grain size on hole clearance are shown in Fig. 5.27. It shows that the diameter clearance between the drilled hole and the cutting tool is enlarged with the grain size, because the passage of the grains is cut large. Figure 5.28 shows that different grains do not affect the hole clearance significantly.

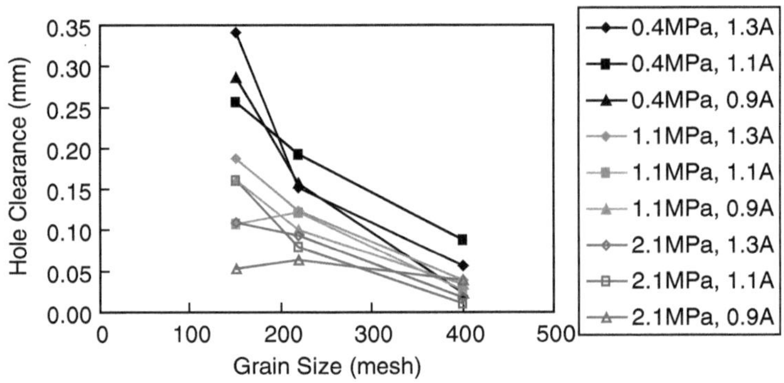

Fig. 5.27 Influence of grain size on hole clearance (SiC, 10 %) [20], reprinted with permission

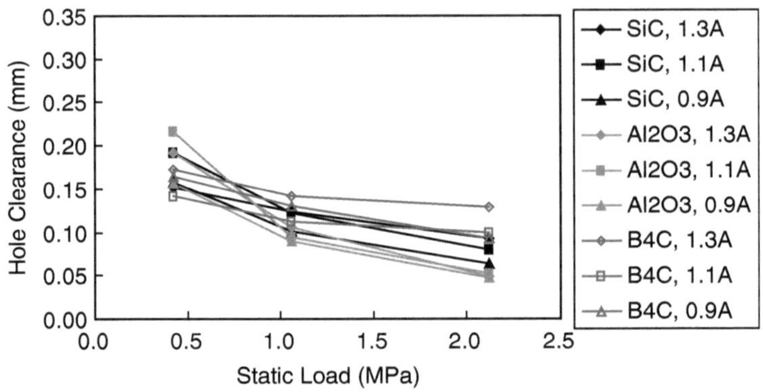

Fig. 5.28 Influence of various grains on hole clearance (#220, 10 %) [20], reprinted with permission

(III) Edge Quality

Influences of static load on edge quality are shown in Fig. 5.29. It shows the delamination and splintering at hole exit are more serious with an increase in the static load. When machining is progressing toward the hole exit, the uncut thickness of workpiece becomes insufficient to resist the push-out of material. Figure 5.30 shows the best edge quality when B_4C grain is adopted. B_4C showing the highest hardness cuts material in the neatest manner, followed by SiC and Al_2O_3. The edge quality at hole entrance is good, as shown in Figs. 5.29 and 5.30. They are free of fuzz or fiber pull-out, and unaffected by machining conditions.

(IV) Tool Wear

In the 81 tests, the tool weight reduces 0.2 g and the tool length is worn 3.15 mm in total. In other words, the tool wear is 38.9 μm per hole, which is a quite positive machinability index of the C/SiC composite material in ultrasonic drilling.

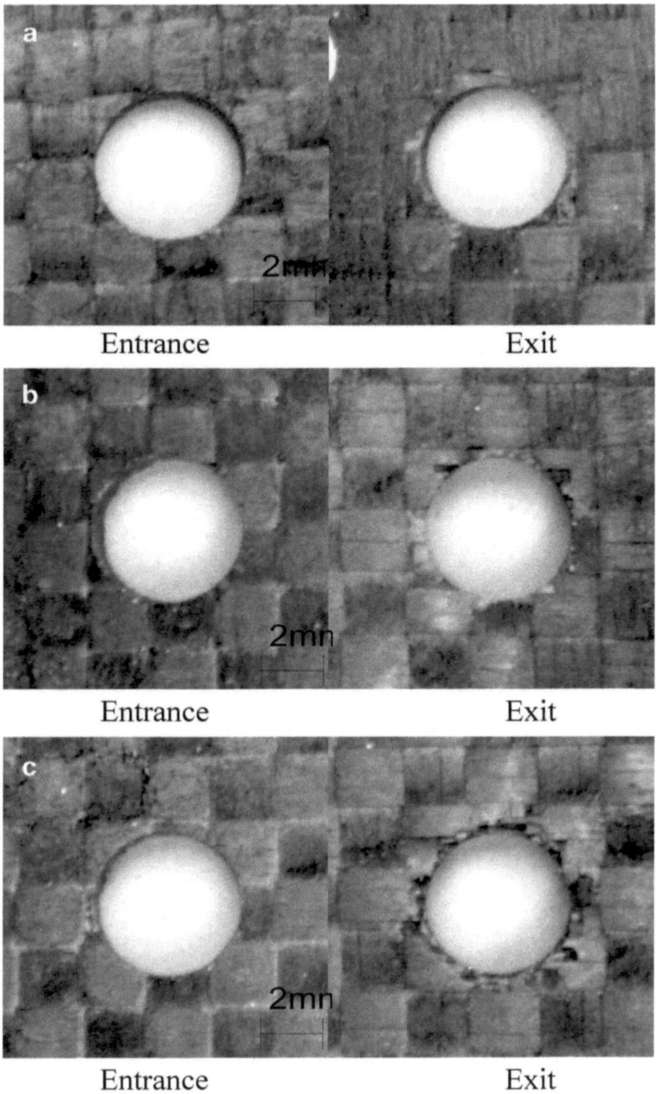

Fig. 5.29 Influence of static load on edge quality (SiC, #220, 10 %, 0.9 A) [20], reprinted with permission. (**a**) 0.4 MPa, (**b**) 1.1 MPa, (**c**) 2.1 MPa

5.4.2.2 Alternative Machining Processes

Water-jet, laser, and conventional drilling are other alternatives to the presented machining process. In consideration of machining quality, water-jet machining often produces delamination, while laser machining produces thermal stress and heat affected zone in material. The ultrasonic drilling rarely causes such a concern.

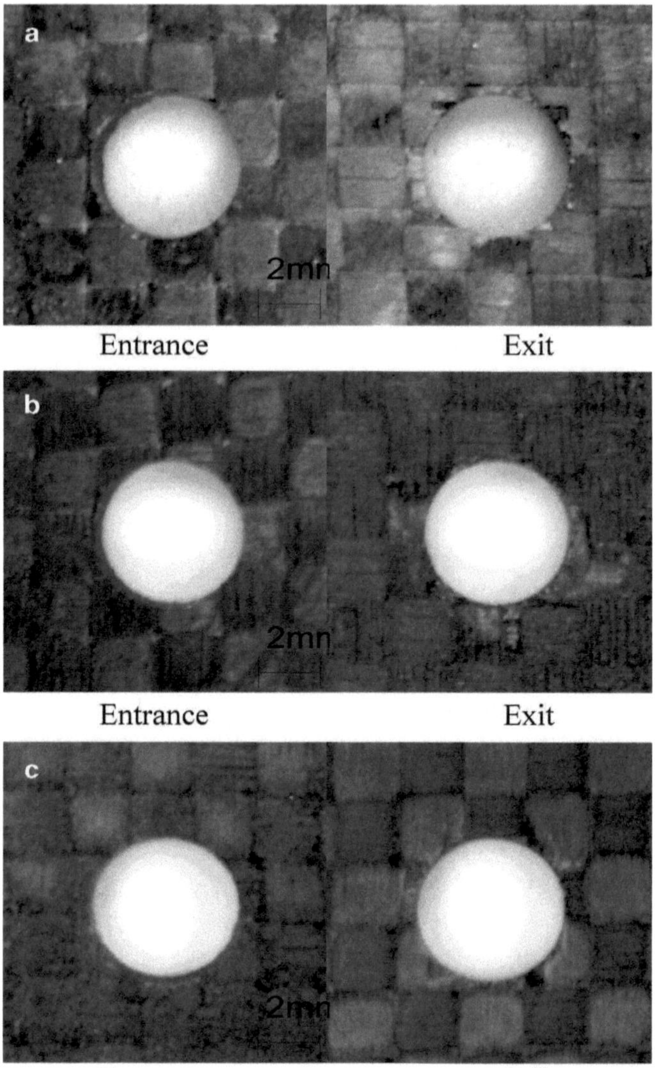

Fig. 5.30 Influence of various grains on edge quality (#220, 10 %, 0.9 A, 1.1 MPa) [20], reprinted with permission. (**a**) SiC, (**b**) Al_2O_3, (**c**) B_4C

The conventional drilling must be proceeded with high spindle speed to achieve good hole edge, as shown in Fig. 5.31. In consideration of machining cost, because ultrasonic drilling can provide simultaneous multi-hole drilling, it is more economical than the above-mentioned processes. Hence ultrasonic drilling will demonstrate strong competition to the conventional as well as the nonconventional machining processes for this composite material.

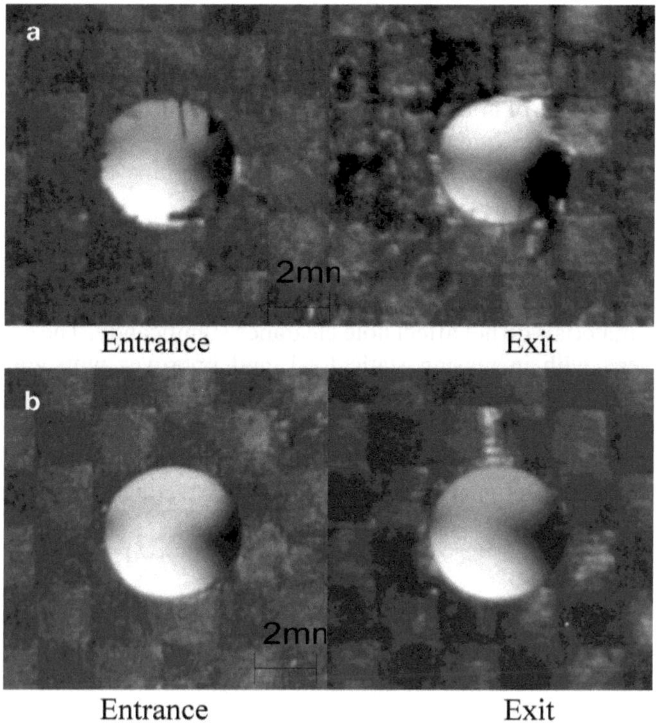

Fig. 5.31 Edge quality of conventional drilling (feed rate 0.12 mm/rev) [20], reprinted with permission. (**a**) 500 rpm, (**b**) 1,500 rpm

5.5 Conclusions

This chapter presented a literature review of ultrasonic machining used to monitor the tool wear correlated to the tool length and the resonance frequency, and to improve the quality of holes when drilling composites. Reduction in tool wear, and the resulting improvements in the quality of holes are achievable by selecting appropriate machining parameters. We come to the following conclusions from the extensive review of damage-free USM cutting efficiency.

1. A theoretical analysis conventionally applied to horn design reveals an effective correlation between tool length and the resonance frequency. An industrial USM system could readily adopt this method to enhance its function without major effort.
2. The main cutting mechanism of ultrasonic drilling is abrasive particle hammering or impacting of the workpiece to remove material in micro-craters. The obtained surface roughness increases with grain size, energy of the ultrasonic oscillations, and the concentration of abrasives and it is independent of the tool feed rate, fiber direction, or matrix thermo-plasticity.

3. The thermoplastics-based carbon/PEEK composite produces minor fuzz at exit, due to its thermo-viscosity. With the aid of liquid nitrogen, the produced exit edge is neat. The grain size and the energy of ultrasonic oscillations significantly affect the clearance of holes. Multi-hole production increases the productivity of ultrasonic drilling of composite materials.
4. In ultrasonic drilling of C/SiC composite materials, the material removal rate increases with static load and grain hardness.
5. Optimal grain size and vibration amplitude produce the maximum removal rate. Hole clearance increases with grain size and vibration amplitude, while it decreases with static load.
6. The grain materials do not affect hole clearance significantly. The edge quality at exit worsens with increasing static load, and improves with grain hardness, though regardless of grain hardness, the edge quality at the hole entrance remains good.
7. Tool wear is minor. Ultrasonic drilling is a reliable method among the conventional and nonconventional machining techniques available for C/SiC composite materials.

References

1. Hocheng H, Kuo KL (2002) On-line tool wear monitoring during ultrasonic machining using tool resonance frequency. J Mater Process Technol 123:80–84
2. Wood RW and Loomis (1927) Philosophical Magazine 7:417–436
3. Balamuth LA (1945) Method of abrading. British Patent 602801
4. Rozenberg LD, Kazantsev VF, Makarov LO, Yakhimovich DF (1964) Ultrasonic cutting. Consultants Bureau, New York, pp 80–85
5. Shaw MC (1956) Ultrasonic grinding. Microtechnic 10(6):257–265
6. Kremer D, Ghabrial SM, Moisan A (1981) The state of the art of ultrasonic machining. Ann CIRP 30:107–110
7. Komaraiah M, Manan MA, Reddy PN, Victor S (1988) Investigation of Surface Roughness and Accuracy in Ultrasonic Machining. Precis Eng 10(2):59–65
8. Thoe TB, Aspinwall DK, Wise MLH (1998) Review on ultrasonic machining. Int J Mach Tools Manuf 38(4):239–255
9. Komaraiah M, Manan, MA., Narasimha R "(1991) Rotary Ultrasonic Machining–A New Cutting Process and its Performance. Int J Prod Res 29(11):2177–2187
10. Dam H., Quist P., Schreiber M. (1995) Productivity, surface quality and tolerances in ultrasonic machining of ceramics. J Mater Process Technol 51:358–368
11. Costa HL, Pandolfelli VC, Mello JDB (1997) On the abrasive wear of zirconias. Wear 203:626–636
12. Anantha Ramu BL, Krishnamurthy R, Gokularathnam CV (1989) Machining performance of toughened zirconia ceramic and cold compact alumina ceramic in ultrasonic drilling. J Mech Work Technol 20:365–375
13. Hocheng H, Kuo KL, Lin JT (1999) Machinability of Zirconia Ceramics in Ultrasonic Drilling. Mater Manuf Process 14(5):713–724
14. Kuo WS, Chou TW (1995) Multiple Cracking of Unidirectional and Cross-PlyCeramic Matrix Composites. J Am Ceram Soc 78(3):745–755

15. Fedou R, Langlais F, Naslain R (1993) Model for the Isothermal Isobaric Chemical Vapor Infiltration (CVI) in a Straight Cylindrical Pore. 2. Application to the CVI of SiC. J Mater Syn Process 1(2):61
16. Tai NH, Chou TW, Ma CCM (1994) Effects of Deposition Mechanisms in the Modeling of Forced-Flow/Temperature-Gradient Chemical Vapor Infiltration. J Am Ceram Soc 77 (3):849–851
17. Chawla KK (1993) Ceramic matrix composites. Chapman & Hall, London
18. Komaraiah M, Manan MA, Reddy PN, Victor S (1988) Investigation of Surface Roughness and Accuracy in Ultrasonic Machining. Precis Eng 10(2):59–65
19. Hocheng H, Hsu CC (1995) Preliminary Study of Ultrasonic Drilling of Fiber-Reinforced Plastics. J Mater Process Technol 48:255–266
20. Hocheng H, Tai NH, Liu CS (2000) Assessment of ultrasonic drilling of C/SiC composite material. Compos Part A 31:133–142
21. Kainth GS, Nandy A, Singh K (1979) On the Mechanics of Material Removal in Ultrasonic Machining. Int J Mach Tool Des Res 19:33–41
22. Pei ZJ, Khanna N, Ferreira PM (1995) Rotary Ultrasonic Machining of Structural Ceramics—A Review. Ceram Eng Sci Proc 16(1):259–278

Chapter 6
Water Jet Machining

H. Hocheng, H.Y. Tsai, and K.R. Chang

Abstract Water jet drilling, in spite of its advantages of no tool wear and thermal damage, often creates delamination in composite laminate at bottom. An analytical approach to study the delamination during drilling by water jet piercing is presented. This model predicts an optimal water jet pressure for no delamination as a function of hole depth and material parameters.

Moreover, the kerf formation of a ceramic plate cut by an abrasive waterjet is discussed. The mechanism and the effectiveness of material removal are studied. The kerf is slightly tapered with wider entry due to decreased cutting energy with kerf depth. A high-power input per unit length produces a small taper but a wide slot.

Abrasive waterjet is adequate for machining of composite materials thanks to minimum thermal or mechanical stresses induced. The feasibility of milling of composite materials by abrasive waterjet is discussed. The basic mechanisms of chip formation, single-pass milling, double-pass milling followed by the repeatable surface generation by multiple-pass milling are studied. High volume removal rates as well as a neat surface are desired. Based on the results of single-pass milling tests, this chapter discusses the double-pass milling considering the effect of lateral feed increments. The study then extends to six-pass milling. The obtained surface roughness from the six-pass milling is expressed as a function of the width-to-depth ratio and the lateral increment. With the knowledge of the volume removal rate and the surface roughness as well as the effects of the major process parameters, one can proceed to design a milling operation by abrasive waterjet.

Keywords Composite laminates • Drilling • Mechanism of delamination • Micro-machining • Kerf • Milling • Removal rate • Width-to-depth ratio • Surface roughness

H. Hocheng (✉) • H.Y. Tsai • K.R. Chang
Department of Power Mechanical Engineering, National Tsing Hua University, Hsinchu, Taiwan, ROC
e-mail: hocheng@pme.nthu.edu.tw

H. Hocheng and H.-Y. Tsai (eds.), *Advanced Analysis of Nontraditional Machining*, DOI 10.1007/978-1-4614-4054-3_6, © Springer Science+Business Media New York 2013

6.1 Introduction

Composite materials are used increasingly in high-performance applications because of their superior-specific strength and stiffness. However, the macroscopically distinct multiphases of the material structure make such materials difficult to machine with conventional tools. While much work has been done in studies of both theoretical and experimental mechanics of composites, little has been done in the area of manufacturing processes, particularly in machining. Since the 1970s, machining data on composite materials have been reported. These works reveal the drastic wear of cutting tools and damage to the composite laminate in the form of chipping, cracking, and delamination [1, 2]. Particularly in drilling, delamination around drilled holes is widely observed. Some researchers discussed the effects of tool wear [3], the others reported the optimal processing variables. [4, 5]. Ho-Cheng and Dharan constructed the first model predicting the critical thrust force at the onset of delamination during drilling [6].

All these papers provide insight into the response of composite laminate to machining. However, machining of composites with conventional tools is associated with inherent problems such as tool wear and thermal damage. Another technique for machining composites is the water jet. Water jet cutting uses high-pressure water instead of a solid tool and hence eliminates the above-mentioned problems. Besides, minimal dust, high speed cutting, and multidirectional cutting capability are offered by this technique. By adding abrasives, one can improve the cutting performance to cut almost any material regardless of its hardness and strength.

As early as 1974, Mohaupt and Burns provided an equation of energy balance predicting the depth of cut of polymers as a function of nozzle diameter, nozzle pressure, and feed rate. Since the cutting theory includes a material property of energy absorbed per unit volume during cutting, which is unknown, it requires preliminary experiments to determine several coefficients in the equation [7]. Research water jet machining flourished in the 1980s [8–10]. Hashish published a modeling study of metal cutting with abrasive water jets [11, 12]. He used the existing erosion theories of cutting wear and deformation wear to calculate the depth of cut. He also ran a series of tests of water jet machining [12]. Recently, researchers presented valuable cutting data for composite materials using water jet [5, 13]. Currently water jet has been well recognized as a powerful tool in machining, particularly for advanced materials.

Since the material removal by water jet is driven by high water pressure, the workpiece is subject to a mechanical force and, in certain cases, damage is caused by this force. Koenig and coworkers reported damage at the tool exit due to the jet force acting perpendicular to the composite laminae. They investigated experimentally the effects of process parameters and thickness and composition of the material. The water pressed in between laminate layers during cutting is suggested as the main cause of damage [5]. Hashish found that the hole piercing of composite laminates by high-pressure water jet resulted in fracture, cracking, or delamination. He reasoned that the shock loading of water and hydrodynamic pressurization are responsible for the damage. Reducing the pressure or the jet size and/or supporting the material at the bottom can eliminate the undesired failure at exit [12].

Advanced ceramics have been applied increasingly in the optical, electronic, mechanical, and biological industries due to their inherent superior high temperature strength, hardness, wear resistance, and proven electrical, optical thermal, and magnetic properties [14]. Currently, the popular mechanical stock-removal methods available for shaping and finishing ceramic components are grinding, honing, lapping, and polishing. Since ceramics are extremely hard, diamonds are often chosen as cutting materials. Nevertheless, the combination of high hardness and strength causes considerable wear on tools, which leads to higher machining cost and less accurate dimensions. In addition, the low material removal rate results in longer machining time and high processing cost [15].

In order to extend the applications of ceramic materials, efficient machining methods must be explored to meet the demands of various ceramic components. Among these, several nontraditional machining methods, such as laser [16], electro-discharge [17], ultrasonic [18], and abrasive waterjet, have emerged as promising machining methods.

The high-pressure abrasive waterjet was first commercialized in 1983, and has been used in industry for the linear cutting and piercing of steel, cast iron, super alloys, glass, and composites [19]. Research into the machinability of ceramics by abrasive waterjet has been limited. Early reports show that this machining technology can be applied effectively for cutting advanced ceramics [20, 21]. The abrasive waterjet possesses some advantages over the traditional grinding process for reduced interfacial temperature, tool wear, and machining time. In 1987, Hunt et al. [22] conducted an experimental study to determine the surface finish characteristics of sintered aluminum oxides machined by an abrasive waterjet, a beam force transducer being used to quantify the surface finish and the cutting efficiency with three grades of alumina. It was found that the workpiece reaction force can be used as a process parameter for determining and controlling the quality of the machined surface, this force increasing with the traverse speed, and being related directly to the surface finish: the higher is this force, the rougher is the surface finish. In 1991, Hashish [23] discussed the general characteristics of surfaces machined by an abrasive waterjet in terms of surface texture and surface integrity, the latter including microstructural change, cracking, work-hardening, and heat-affected zones. The effects of operation parameters on aspects of surface texture such as waviness, kerf taper, burr height, and surface finish in thin sheet metals, were presented and discussed.

The abrasive waterjet possesses some advantages over the traditional milling process for reduced interfacial temperature, tool wear, and machining time. It especially surpasses in milling of FRPs for significant improvement of cut quality without burrs or delamination. The high-pressure abrasive waterjet was commercialized in 1983, and has been used in industry for linear cutting and piercing steel, cast iron, super alloys, glass, and composites [19]. A waterjet at transonic speed carrying abrasive particles provides an effective means for removal of difficult-to-machine material. In the machining process, no chemical, thermal, and electrical threat is induced to the workpiece. Dust is also eliminated. Research of milling of composite materials by abrasive waterjet is however limited. In 1989, an investigation of milling with abrasive waterjet was issued by Hashish [11]. Later he discussed the general characteristics of machined surfaces by abrasive waterjet in terms of surface texture and surface integrity [24].

In Sect. 6.2, the phenomenon of delamination damage produced during the water jet drilling of composite laminates is studied. The mechanism of separation of laminae is described by the system of a circular plate under central pin load. A model is proposed that relates the delamination of the laminate to water jet pressure and composite material properties. This approach is based on the theory of delamination in twist drilling [25]. Experimental results from other investigators are compared with this analysis.

The purpose of Sect. 6.3 is to explore the machinability of ceramics by an abrasive waterjet. Experimental results for slot cutting are evaluated in terms of material removal rate, kerf shape, and surface roughness. In addition, the occurrence of damage of the machined surface is recorded also, as an index of cutting quality. The correlation between the quality of the ceramics in slotting and the major machining parameters of the abrasive waterjet is presented.

The surface integrity includes microstructural change, cracking, work-hardening and heat-affected zones. The effects of operation parameters on surface texture, such as waviness, kerf taper, burr height, and surface finish in thin sheet metals are discussed in Sect. 6.4. The purpose of this current investigation is to explore the feasibility and approach of milling of composite materials by abrasive waterjet. Experimental results for single-pass milling are considered in terms of volume removal rate, depth, width, and width-to-depth ratio of cut. In addition, dimensional analysis is introduced to synthesize the effects of machining parameters. The results of double-pass milling and six-pass milling are discussed to produce a reproducible surface pattern.

6.2 Drilling in Composite Laminates

6.2.1 Mechanisms of Water jet-Induced Delamination

In water jet machining, materials are removed by the impingement of a continuous stream of high-energy water beads. The machined chips are flushed away by the water. Similarly to solid tools, the water jet exerts machining force on the workpiece during cutting. This force is transmitted by the water beads clashing with the cut. The direction of the force is given predominantly by the attack angle of the water jet and is insignificantly affected by the tail flow beyond the cut. The magnitude of the force is determined by the jet impetus, i.e. jet velocity, which can be calculated from the water pressure supplied by machine.

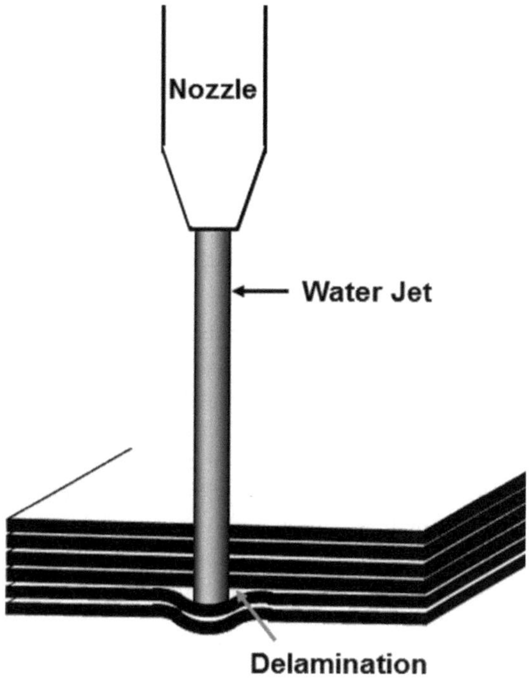

Fig. 6.1 Water jet-induced delamination at exit

In water jet drilling of composite laminate, there is a thrust jet force acting perpendicularly to the laminae that show flexural bending in response. The laminae under the water jet thus tend to be pushed away from the interlaminar bound around the hole. As the jet advances to the end, the uncut thickness decreases and the resistance to deformation weakens. At a certain point the thrust jet force causes such a deformation that the interlaminar bond strength can no longer hold and delamination around the hole exit occurs. This happens before the laminate is completely pierced by the water jet as shown in Fig. 6.1. Variable process parameters, such as water jet diameter and pressure, can affect the occurrence of delamination by changing the magnitude of the thrust jet force. The flexural rigidity and the opening-mode delamination fracture toughness of the composite laminate on the other hand determine the material response, i.e. deformation and separation of laminae during jet drilling.

As soon as an interlaminar crack is created, the water presses into the opening, which is called hydrodynamic pressurization. This pressurization is, as suggested by former researchers [5, 12], partly liable to cause damage around holes during water jet drilling.

Based on the discussions above, one can consider the delamination as a two-stage fracture process, i.e. the onset and the propagation. At the onset of delamination, the thrust jet force produces bending deformation which leads to laminae separation. An interlaminar crack around the bottom of hole is thus created.

At the second stage, the water pressurization together with the jet force widens this crack to a final, delamination damage. In the following section, an equation relating the onset of delamination and the water jet pressure as well as material properties will be developed.

6.2.2 Delamination Equation

Having discussed the mechanisms of delamination, one can construct a model based on fracture mechanics. Since composite laminates show extremely high moduli of elasticity and a lower degree of plasticity, it is assumed that the use of linear elastic fracture mechanics (LEFM) is proper in the modeling. As far as the applicability of LEFM to the composite materials is concerned, the reader is referred to review papers [26, 27], which confirm the approach provided that crack growth is collinear and the crack lies in a plane of material symmetry. Both these conditions are met in the current case. Interlaminar cracking is considered to be the explanation for the delamination failure produced during water jet drilling in the approach attained below.

As shown in Fig. 6.2, the cylinder in the middle represents the water jet in Fig. 6.1, of diameter D_w, applying thrust force F downwards. D_w is in the real case a very small value. x is the displacement, H is the thickness of the laminate, h is the uncut depth under water jet, and a is the radial size of an assumed existing circular crack. As the jet cuts downwards, the uncut laminate under it are pushed and deformed elastically by the central thrust force. If the resulting strain at the tip of the crack goes beyond the critical value, delamination occurs. The equation of energy balance, from linear elastic fracture mechanics, can be written as

$$G\,dA = F\,dx - dU \tag{6.1}$$

where G is the energy release rate per unit area, A is the area of delamination, U is the stored strain energy. dA is further expressed as

$$dA = \pi(D_w + 2a)da \tag{6.2}$$

To find the correlations linking F, x, and U together, classic plate bending theory for circular plate with concentrated load and clamped ends is used for this model [28]. (The condition of concentrated load is met, since in water jet cutting the jet diameter has a very small value compared to the workpiece dimensions). The strain energy

$$U = \frac{8\pi M x^2}{(a + D/2)^2} \tag{6.3}$$

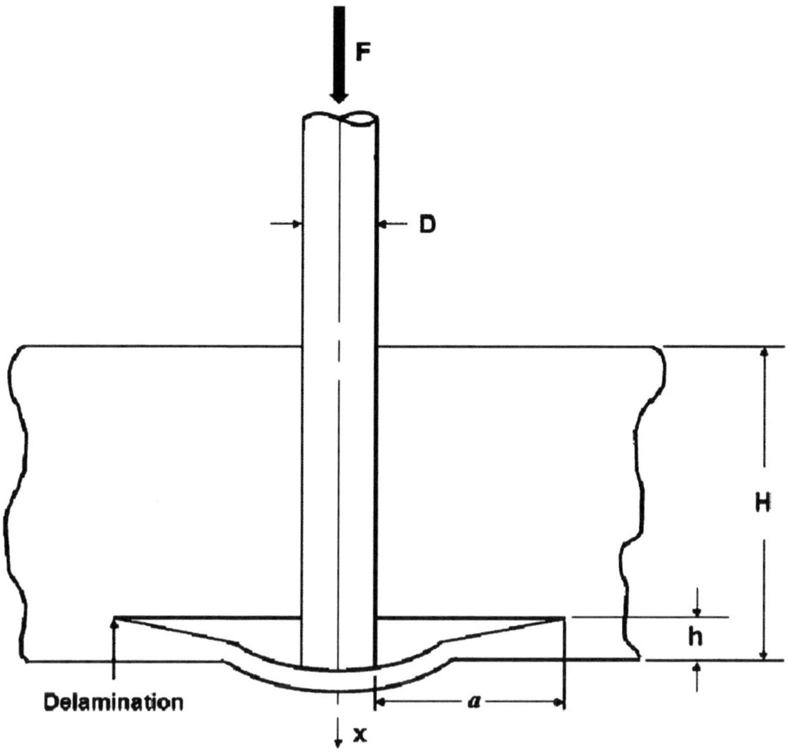

Fig. 6.2 Circular plate model for delamination analysis

where E is the young's modulus, v is the poisson's ratio

$$M = \frac{Eh^3}{12(l - v^2)} \tag{6.4}$$

The displacement x resulted from F is expressed as

$$x = \frac{F(a + D_w/2)^2}{16\pi M} \tag{6.5}$$

The thrust jet force F is assumed to be the product of water jet pressure P and cross-section area of the water jet

$$F = \pi P R^2 \tag{6.6}$$

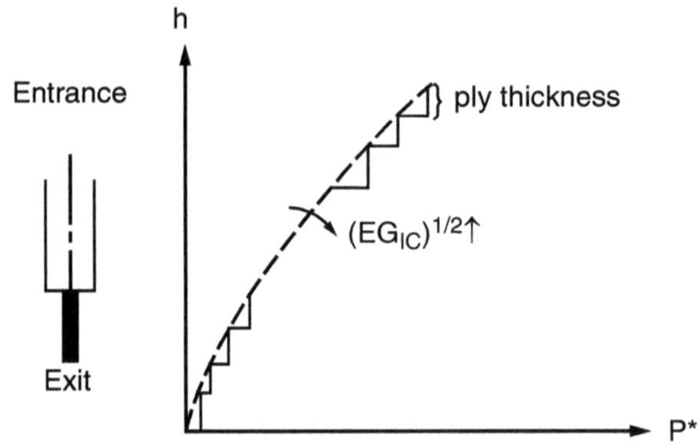

Fig. 6.3 Correlation between optical jet pressure and hole depth [30], reprinted with permission

where $R = D_w/2$. By substituting (6.2) to (6.6) into (6.1), one obtains the critical water jet pressure at the onset of delamination

$$P^* = \sqrt{\frac{8EG_{IC}}{3R^4(1-\nu^2)}} \times h^{\frac{3}{2}} \tag{6.7}$$

The applied water pressure should not exceed this value, which is a function of material properties and the uncut thickness, to avoid delamination.

It is noted that several simplifications are made in this model. Firstly, the isotropic calculation is adopted instead of applying a complicated algorithm for the anisotropic case, and the modulus of elasticity in the principal direction (also called i-direction), E, is used to represent the material property. This provides simplicity in the first stage of modeling and sacrifices only several percents of accuracy. Secondly, the value G_{IC} is used since it is easily measured, although plane conditions may not always be fulfilled. However, this value, which is lower than that for the plane stress case, gives the critical water pressure which is on the conservative side. Thirdly, the assumption of the constancy of the G_{IC} is based on earlier work which shows that G_{IC} is only a mild function of strain rate [29]. Fourthly, within a short standoff distance, the variation of water jet diameter is considered very small. The water pressure may drop across the water jet orifice, but this again gives conservative predictions for the critical load in real cases.

The results of the predicted thrust jet force are plotted in Fig. 6.3 [30] as a function at the hole depth. As can be seen, the jet pressure decreases toward exit to avoid water jets-induced delamination. The dotted line represents the continuous analytical predictions while the solid lines in steps describe the actual "digitalized" material response due to its laminate structure with finite thickness for each layer. This master curve can be used as the guide value to control the water pressure for maximal cutting rate and no delamination damage.

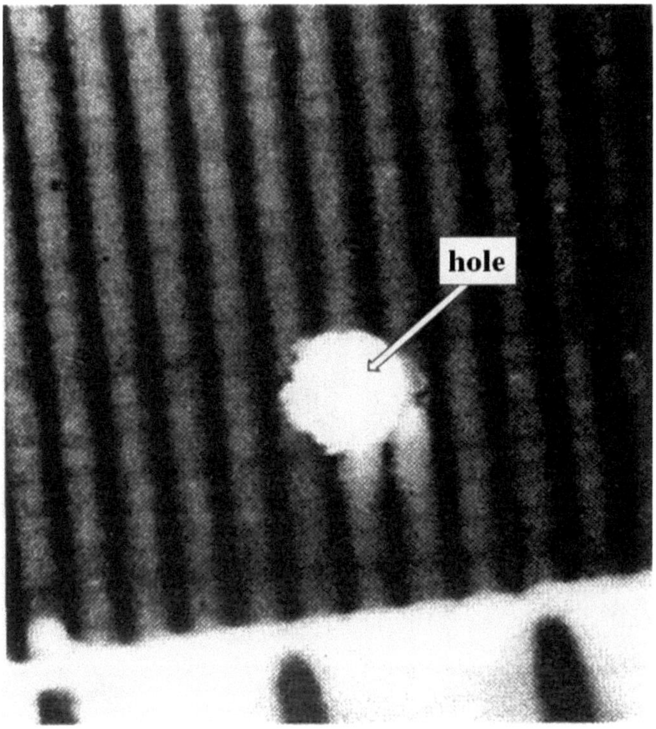

Fig. 6.4 Exit of water jet-drilled hole at $P = 241$ MPa (after Hashih [12]) [30], reprinted with permission

6.2.3 *Discussion*

To examine the validity of the current model, experimental studies should be compared with analytical predictions. Results from two independent investigators are discussed below. Koenig and coworkers ran a series of tests for water jet cutting of fiber-reinforced plastics [5]. They conclude that the jet force acting in the direction of fluid flow, thus perpendicular to the laminate surface, is mainly responsible for material damage at the tool exit. This approves the approach of central load-induced plate bending in the present model for delamination analysis.

Hashish published a rather complete set of experimental data of water jet machining of composites, including linear cutting, turning, milling, and drilling [12]. In the hole piercing section, he presented the clear consequence of the use of high-pressure and low-pressure respectively. Delamination damage of a 6.3 mm thick graphite epoxy composite is generated at pressure of 345 MPa while the exit shows a clean cut at pressure of 241 MPa as shown in Fig. 6.4 [30] and Fig. 6.5 [30]. This is the first quantitative report of the effect of water pressure on material delamination. Hashish also noted that, by reducing the jet diameter or by supporting

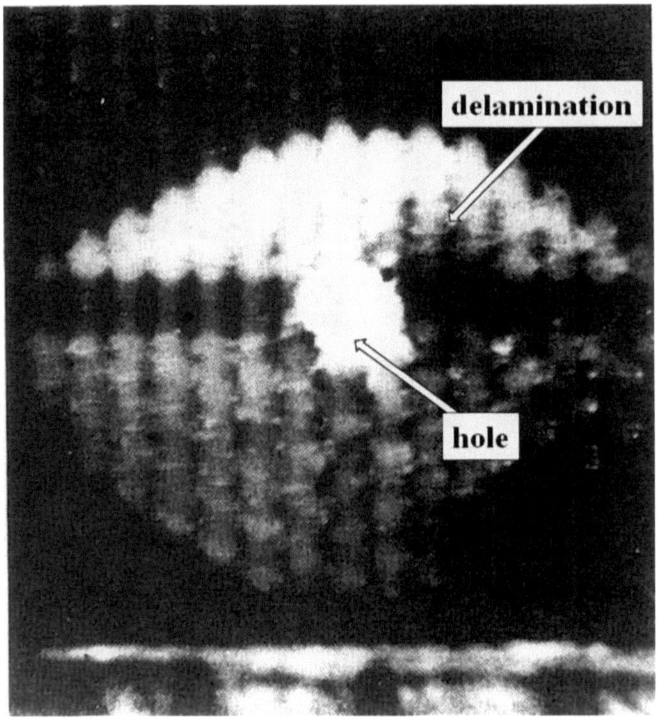

Fig. 6.5 Exit of water jet-drilled hole at $P = 345$ MPa (after Hashish [12]) [30], reprinted with permission

Table 6.1 Comparison of analysis and experiments

Theoretical prediction of critical jet pressure	295 MPa
Experimental results	241 MPa (not delaminated)
	345 MPa (delaminated)

the material at the bottom, the delamination can be eliminated. These conclusions are the logic deductions from the current analysis which clearly forsees the roles of the level of load and the resistance to plate bending in causing delamination (as shown in (6.7) and Fig. 6.3). For high-strength graphite/epoxy composite with fiber volume fraction of 0.6 and ply thickness of 0.1 mm, E is 124 GPa arid the energy release rate, G_{IC} is taken to be 200 J/m^2 from the reports of earlier works [29]. The water jet diameter is taken to be the orifice diameter from Hashish's report, which is 0.299 mm. The critical water pressure at the onset of delamination at exit is calculated from (6.7) with h = 0.1 mm (ply thickness) to be $P = 295$ MPa.

The comparison of theoretical and experimental results is shown in Table 6.1. As can be seen, the prediction falls ill the experimentally justified range with the maximum deviation, which occurs in extreme cases, of 17%. By considering the fact that the energy release rate of composite materials often shows great variation [31], one is convinced that the analysis is essentially correct. Several future works are expected. First, an anisotropic material model instead of simple isotropic calculation is to be constructed. Secondly, experiments giving more detailed information on the development of delamination should be carried out.

6.3 Kerf Analysis in Cutting Ceramics

6.3.1 Theoretical Background for Material Removal

Ceramic material impacted by projectiles displays several damage modes. Elastic response involves little detectable plastic deformation in the target, and occurs when the projectile is relatively soft or deformable. The high elastic stresses that are developed fracture the target locally, the extension of the cracks producing a small-scale chipping in the target. Plastic response without detectable fracture is prevalent for metals, but occurs also in ceramics at elevated temperature or with small projectiles. In this case, a depression develops at the impact center, and lips form around: these lips can be removed by subsequent impacts. An intermediate (elastic/plastic) mode of damage involves a localized plastic region around the impact center and fracture in both the elastic and plastic zones. The damage consists of a central plastic indentation and arrays of radial, conical, and lateral cracks. This type of damage occurs when the projectile is relatively hard, and it is a dominant mode in ceramics over wide impact conditions.

In 1978, Evans [32] and Hockey et al. [33] presented a theoretical treatment in terms of an elastic–plastic phenomenon to describe the erosion of brittle material by angular particle impact. They estimate the erosion rate from the depth of penetration of the impacting particle and the size of the lateral cracks. This analysis incorporates both the hardness, H, and the critical stress intensity factor, K_c. The amount of material removal per impact is

$$W \sim V_0{}^{2.5} R_P{}^4 \rho_P{}^{0.3} H^{-0.3} K_c{}^{-1.5} \tag{6.8}$$

where V_0 is the impact velocity, R_p is the particle radius, and ρ_P is the particle density. The exponent of impact velocity is 2.5. The exponents in this equation are obtained in part through mathematical analysis, and in part through experiment of the extent of damage. This expression can be applied to multiple impacts, provided that each impact event is similar. An equation is derived to describe the material removal rate, $\dot{W}$:

$$\dot{W} \sim \dot{m} V_0{}^{2.5} R_P^{4/3} \rho_P^{-0.7} H^{-0.3} K_c{}^{-1.5} \tag{6.9}$$

where $\dot{m}$ is the abrasive flow rate.

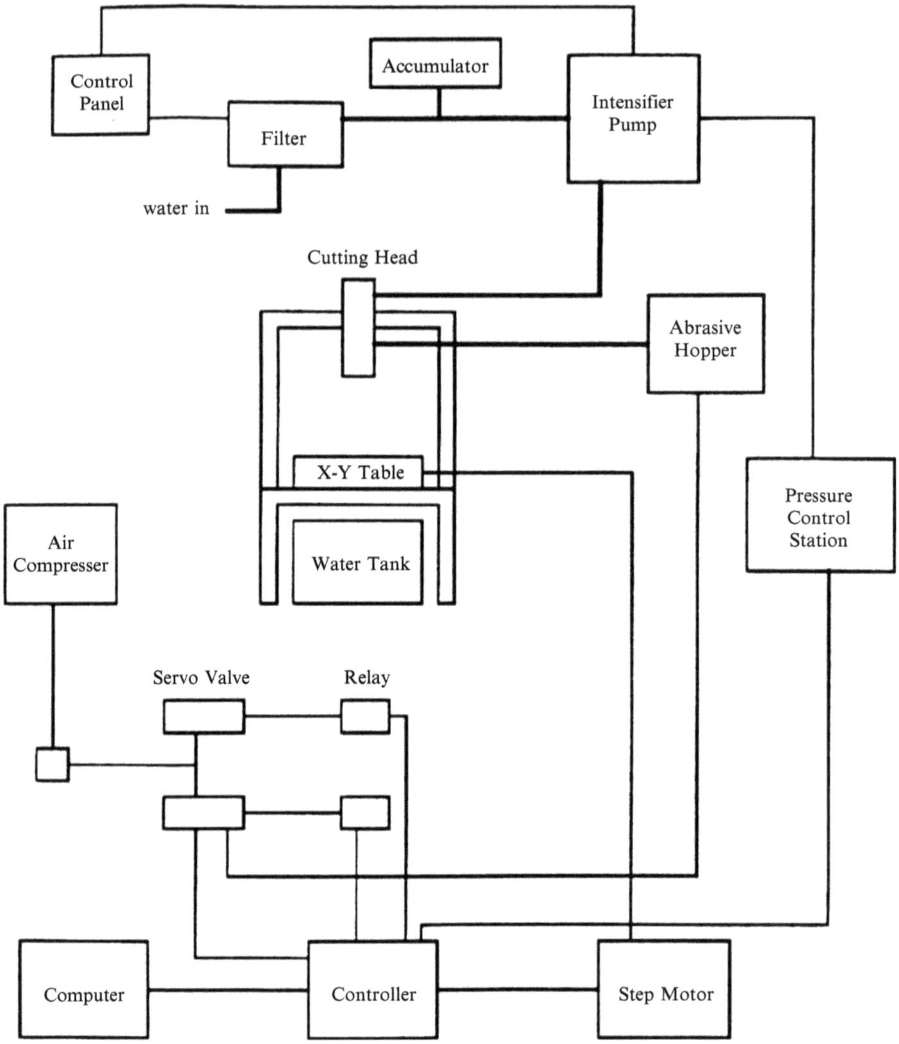

Fig. 6.6 Schematic of the experimental set-up [34], reprinted with permission

6.3.2 Experimental Setup and Procedure

The schematic experimental setup is shown in Fig. 6.6 [34]. The system includes the following basic components: a high-pressure pump, an abrasive waterjet cutting head, an abrasive delivery system, a water catcher, a pressure-control station, a translational table, and a numerical controller. A waterjet intensifier pump (Flow 6XS-55) is connected to the abrasive waterjet cutting head, this pump being capable of delivering water at up to 380 MPa and 1.93 l/min, the water pressure being set

Table 6.2 Properties of the ceramics used in this study

	Al_2O_3	Si_3N_4
Density	3.8 g/cc	3.26 g/cc
Strength	294 MPa	750 MPa
Young's modulus	3.2×10^6 kg/cm^2 (314 GPa)	3.1×10^6 kg/cm^2 (304 GPa)
Hardness	1,430 (Vickers)	1,500 (Vickers)
Fracture toughness	3.5 MN/m$^{3/2}$	7.0 MN/m$^{3/2}$

in advance at the pressure-control station. The translational table provides the necessary motion in cutting. The height of the abrasive cutting head is set manually by a screw and optical linear scale (Mitutoyo-AT2 100 mm). The abrasive delivery system consists of a hopper, an abrasive metering valve, and a delivery line. The abrasive metering valve is attached to the exit port of abrasive hopper, the function of the valve being to turn the flow of abrasive on and off, and to measure the amount of abrasive. A manually adjustable digital micrometer is used to set the abrasive flow rate.

The abrasive delivery line connects the outlet of the abrasive metering valve with the mixing chamber, where the primary waterjet and the abrasive are mixed to form a coherent high-energy stream.

Aluminum oxide and silicon nitride ceramics are cut in this study. The alumina specimens are of 96% purity, supplied in plates with thicknesses of 0.65 and 1.35 mm from Industrial Materials Laboratory, ITRI, in Taiwan. The silicon nitride specimens are also supplied in plates, with thicknesses of 5 and 10 mm, by the same laboratory. The mechanical properties of the ceramic specimens chosen for the present investigation are shown in Table 6.2.

The cutting characteristics are highly dependent on several system parameters, in the current slot-cutting experiments these parameters being: orifice diameter of 0.229 mm; mixing-tube diameter of 0.762 mm; mixing-tube length of 76.2 mm; standoff distance of 2 mm; garnet abrasive with two mesh sizes of #80 and #100; water pressure of 207, 259, and 310 MPa; traverse speed of 20, 60, and 150 mm/min; and abrasive flow rate of 100, 200, and 300 g/min.

After slot cutting, the upper and bottom kerf width were measured using a Mitutoyo PJ311 profile projector, and the surface roughness of the machined surface was measured parallel to the feed direction using a Mitutoyo 201 surface profilemeter. The material removal rate was measured by weight after non-through cutting. In addition, damage to the machined surface was observed by optical microscope.

6.3.3 Material Removal Rate

The material removal rate is defined as the volume removed per unit time. High hydraulic pressure and an abundant supply of abrasives produce a large material removal rate by increasing the number and the kinetic energy of cutting particles, as shown in Figs. 6.7 and 6.8, which figures demonstrate also the increasing capacity for material removal with a larger size of abrasives.

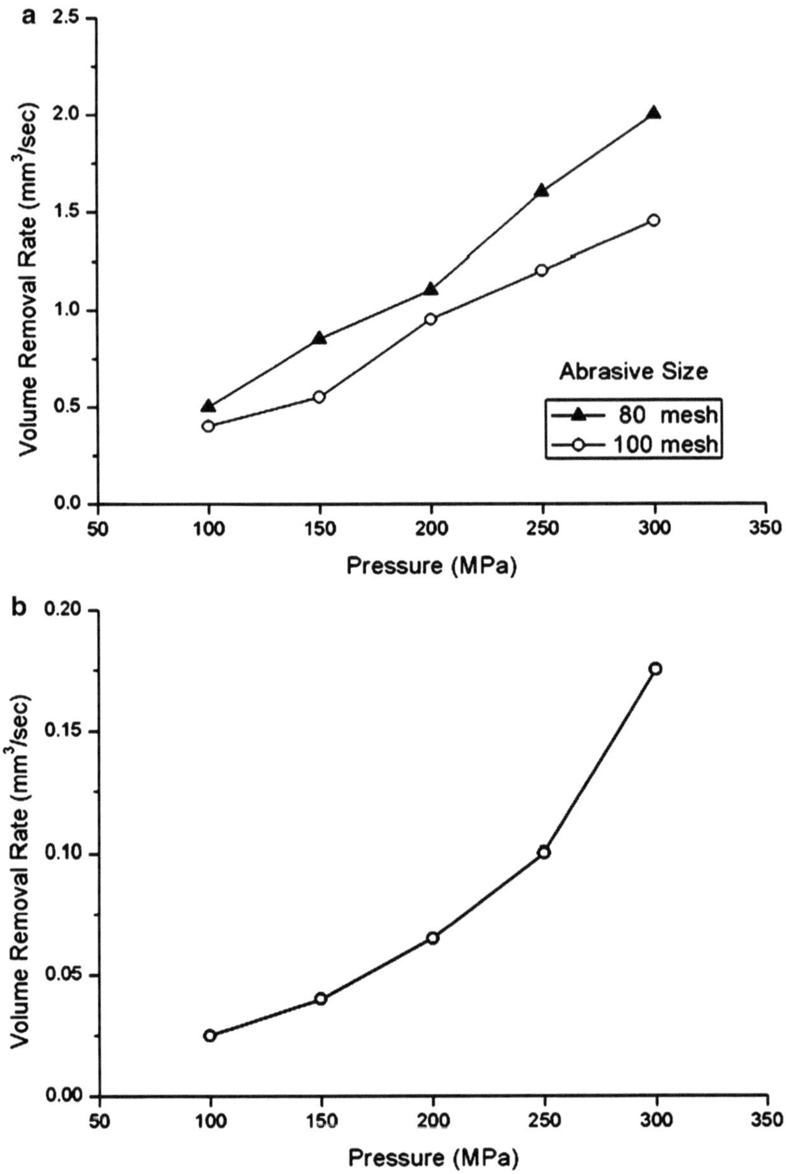

Fig. 6.7 Correlation between volume removal rate and hydraulic pressure: (**a**) Al_2O_3, abrasive flow rate 100 g/min traverse speed 150 mm/min; (**b**) Si_3N_4, abrasive flow rate 100 g/min, traverse speed 20 mm/min, abrasive size 80 mesh

An analysis of material removal by impact was presented by Evans, which predicts the removal rate to be proportional to the particle speed to the power of 2.5, another model based on cutting wear mechanism and derived by Hashish also proposing the same value. Figure 6.9 shows the experimental values of 2.42 and 2.47 for mesh

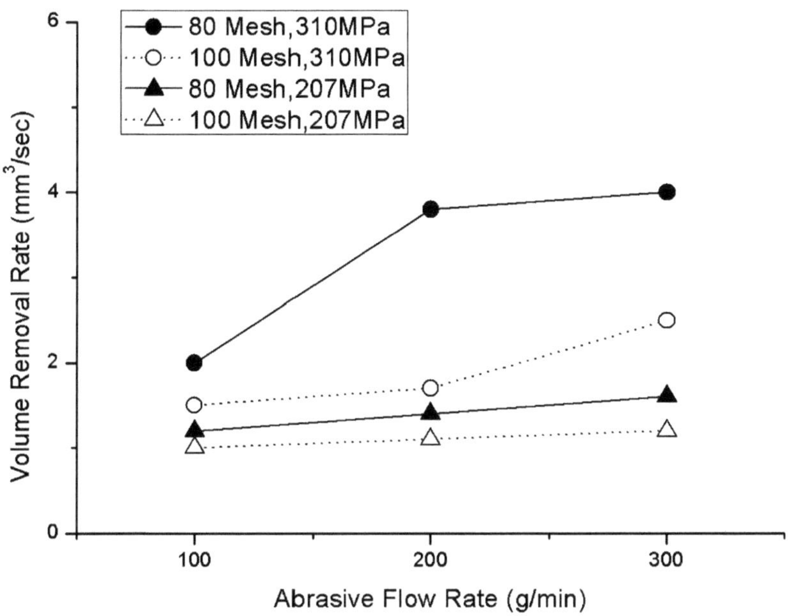

Fig. 6.8 Correlation between volume removal rate and abrasive flow rate, at a traverse speed of 150 mm/min

sizes of 80 and 100, respectively (the abrasives flow rate is 100 g/min), compared with the analytical value of 2.5: the agreement verifies that particle speed is the major factor of the waterjet system determining the material removal rate.

Basic fluid mechanics provide the correlation between the jet velocity and the applied hydraulic pressure with negligible energy loss:

$$V_j = \sqrt{\frac{P}{\rho_f}} \tag{6.10}$$

where V_j is the jet velocity, P is the water pressure, and ρ_f is the density of fluid. By using the law of conservation of momentum

$$V_a = \frac{\dot{m}_w}{\dot{m}_a + \dot{m}_w} V_j \tag{6.11}$$

and the expression of mass flow rate of fluid

$$\dot{m}_w \sim \frac{1}{4} \pi \rho_f d^2 V_j \tag{6.12}$$

where V_a is the velocity of the abrasives, $\dot{m}_w$ is flow rate of the fluid, $\dot{m}_a$ is the flow rate of the abrasives, and d is the jet diameter, it is possible to obtain the correlation between the velocity of the abrasive particles and the applied water pressure,

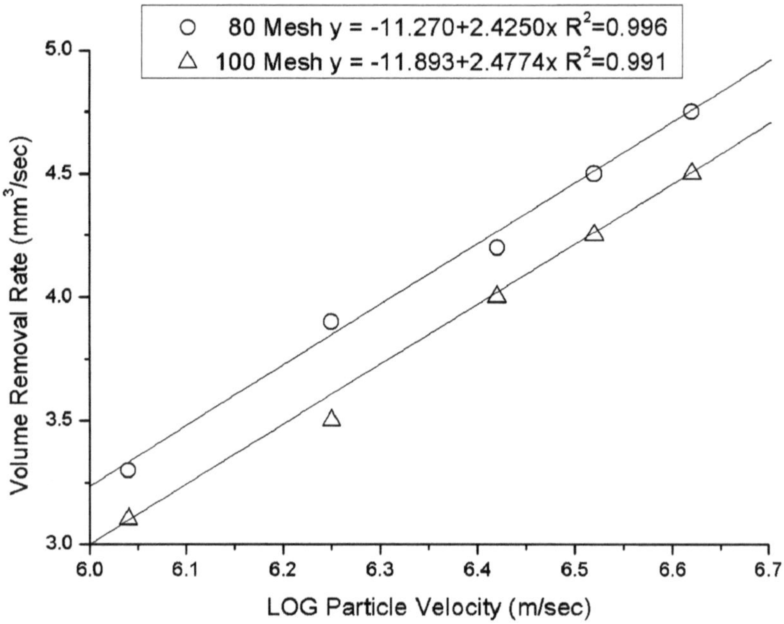

Fig. 6.9 Correlation between volume removal rate and particle velocity (*open circle* 80 mesh., $Y = -11.270 + 2.4250x$, $R^2 = 0.996$; *open triangle* 100 mesh, $Y = -11.893 + 2.4774x$, $R^2 = 0.991$)

illustrated in Fig. 6.10 for various amounts of abrasives. These curves are useful, when combined with the results for material removal rate versus velocity of abrasive particles, in operating the waterjet system for cutting at a desired removal rate.

The material removal rate with respect to abrasives consumption is a material property, as shown in Fig. 6.11. The alumina in the current tests is easily removed, partially due to its less degree of densification.

6.3.4 Through-Cut Domain

Table 6.3 contains the results from through-cut tests. The major factors are the level of the water pressure, the traverse speed and the amount of abrasive particles. Figure 6.12 shows the surface of the critical conditions for 1.35 mm thick alumina, below which through-cut cannot be achieved. The larger 80-mesh abrasives demonstrate a higher capability of material removal.

6.3.5 Kerf Form

The cuts produced are typically tapered with a wider entrance, the latter (upper edge) often being rounded by abrasives dispersed from the main cutting stream.

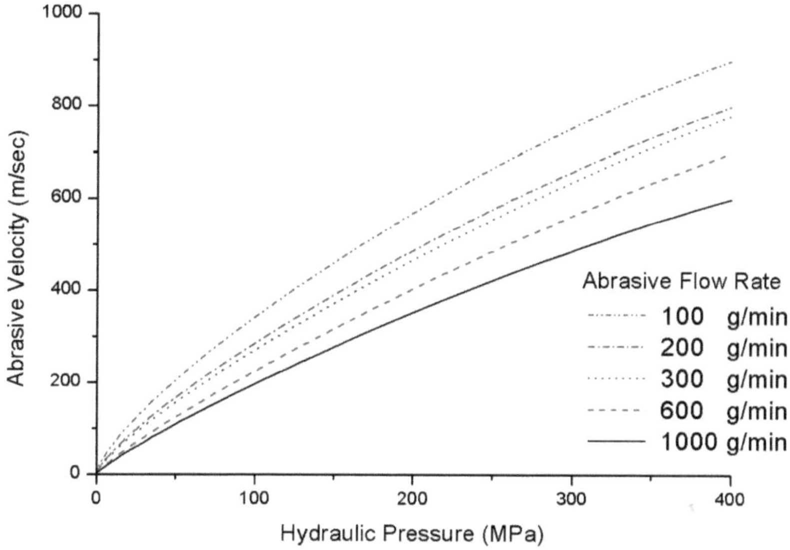

Fig. 6.10 Abrasive velocity at various hydraulic pressure and abrasive flow rate, with a nozzle diameter of 0.229 mm

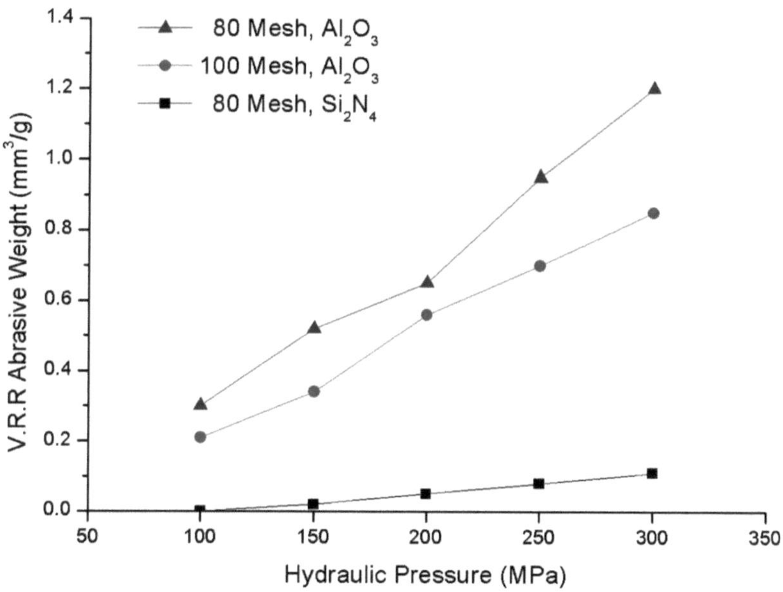

Fig. 6.11 Specific volume removal rate

Table 6.3 Results from through-cut tests

Pressure (MPa)	Abrasive flow rate (g/min)	Traverse speed (mm/min)	Abrasive size 100 mesh through-cut result	Abrasive size 80 mesh through-cut result
207	100	20	0	0
		60	x	0
		100	x	x
		150	x	x
259		20	0	0
		60	0	0
		100	x	x
		150	x	x
310		20	0	0
		60	0	0
		100	x	0
		150	x	x
207	200	20	0	0
		60	0	0
		100	x	x
		150	x	x
259		20	0	0
		60	0	0
		100	0	0
		150	x	x
310		20	0	0
		60	0	0
		100	0	0
		150	0	0
207	300	20	0	0
		60	0	0
		100	x	0
		150	x	x
259		20	0	0
		60	0	0
		100	0	0
		150	x	0
310		20	0	0
		60	0	0
		100	0	0
		150	0	0

(0) through-cut; (x) non-through-cut

The width of cut decreases toward the exit (bottom edge) forming a kerf, since the cutting capability of the jet is consumed continuously. The longer the cut is exposed to the abrasive jet (or the slower the traverse speed), the more evident is the rounding at the entrance, but the less is the kerf taper.

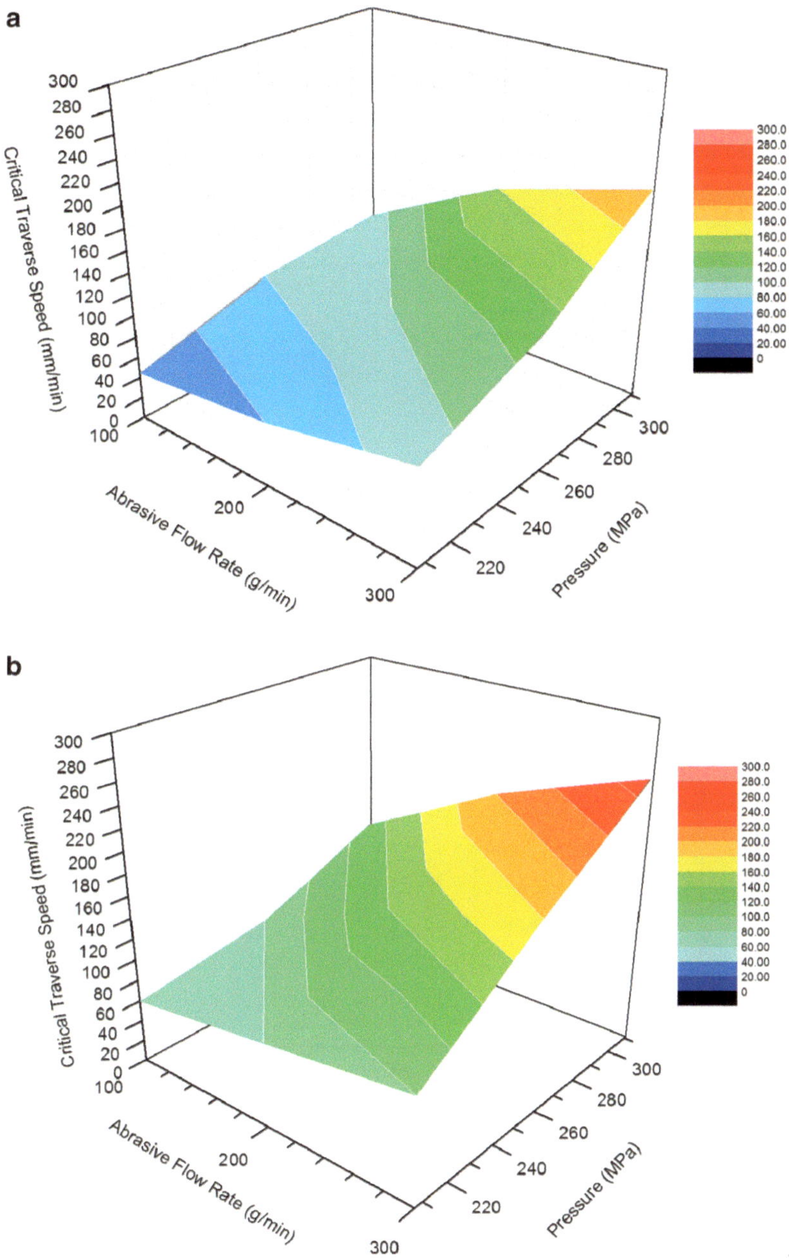

Fig. 6.12 Through-cut domain: (**a**) 100 mesh; (**b**) 80 mesh

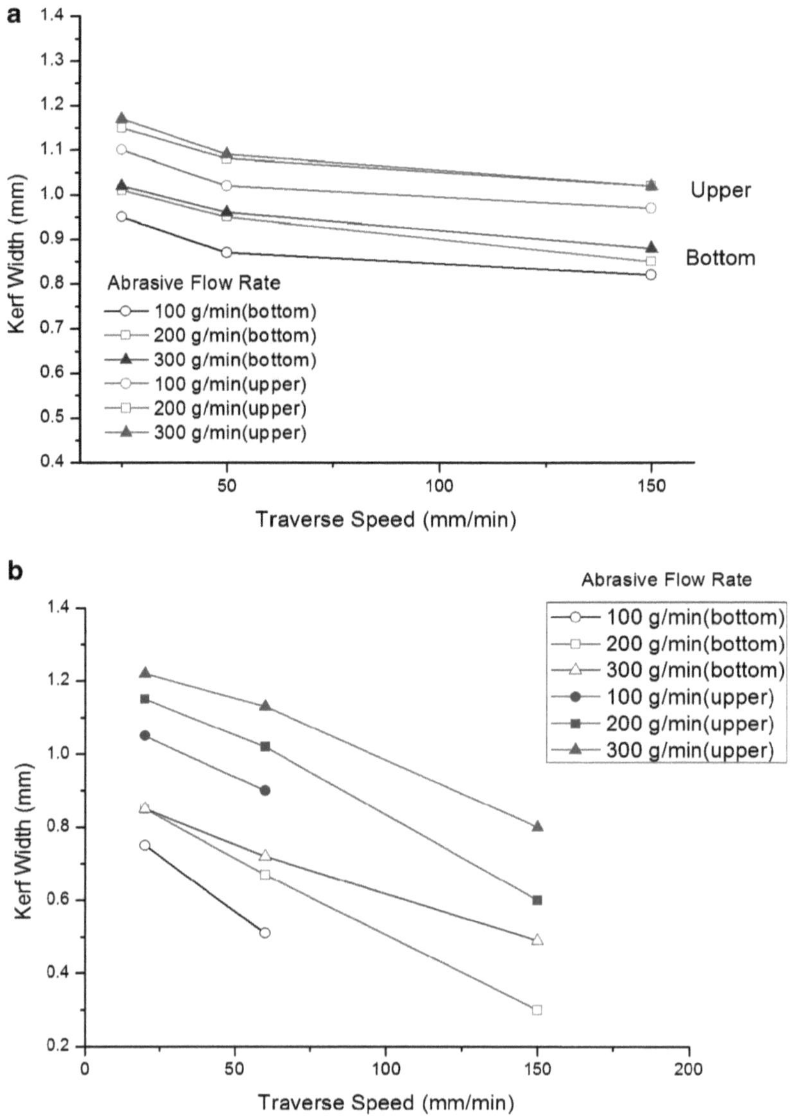

Fig. 6.13 Correlation between kerf width and traverse speed, with a pressure of 310 MPa and an abrasive size of 100 mesh, for alumina plate of thickness: (**a**) 0.65 mm; (**b**) 1.35 mm

6.3.5.1 Entrance Width

As seen in Fig. 6.13, the width of the entrance is inversely proportional to the traverse speed. Because a slower pass allows more abrasive particles to strike the jet target, a wider slot is produced. Higher hydraulic pressure opens a wider slot, since a greater jet kinetic energy impinges onto the ceramics. However, strong individual

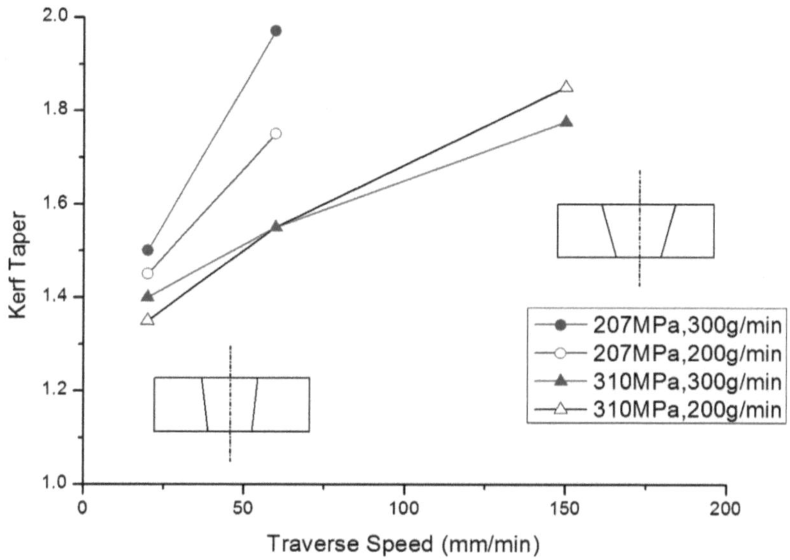

Fig. 6.14 Correlation between kerf taper and traverse speed, for alumina plate of 1.35 mm thickness and abrasive size 100 mesh

bombardment at high pressure is not as effective as more intensive consecutive bombardment, such as occurs with a lower traverse speed. A higher abrasive flow rate not only provides more abrasives, but the jet also becomes wider, therefore the slot produced is broader, as shown in the figure.

6.3.5.2 Exit Width

The width of the exit varied in the same manner as the width of the entrance. The opening is proportional to the abrasive flow rate and the hydraulic pressure and is narrowed by use of a large traverse speed. When the pressure is too low, or the speed of traverse is too high, through-cut of the material cannot be achieved, as indicated in Fig. 6.13 by the missing data points.

6.3.5.3 Kerf Taper

The kerf ratio is defined as entrance width divided by the exit width. At high traverse speed, through-cutting of the material is obtained with incomplete widening of the exit, thus the kerf ratio is larger. On the other hand, high water pressure provides the abrasives with abundant kinetic energy, therefore the bottom part of the kerf can be machined more effectively, resulting in a more uniform kerf width: these results are shown in Fig. 6.14. The abrasive flow rate causes no evident change to the taper ratio, since it increases or decreases the kerf width in a parallel manner.

6.3.6 Surface Roughness

The roughness of the cut increases with traverse speed, as shown in Fig. 6.15. Fast passage of the jet does not let the kerf walls be completely machined by the abrasive particles, thus striations marking the traverse of the jet are left. The measurement of the surface roughness along the direction of traverse reflects this observation.

The periodic appearance of striations also results in a rather low R_{max}/R_a ratio, since otherwise an incidental large roughness (R_{max}) located in a rather smooth region will result in a larger value of R_{max}/R_a. In metal cutting, materials are removed primarily by plastic deformation and the machined surface generally shows an R_{max}/R_a value of between 4 and 6. In the current tests, the alumina is removed by fracture, thus the roughness is less uniform with a larger R_{max}/R_a of from 6 to 8.

A sufficient supply of abrasives removes more material, hence the striations are lowered and the surface finish is improved, this effect being more evident at greater traverse speed. At small traverse speed, while the kerf walls can be completely machined, the supply of abrasives becomes less significant. As long as the hydraulic pressure provides the necessary machining force for an individual abrasive particle, the surface roughness remains unaffected by the level of pressure. Nevertheless, as the thickness of specimen increases, the requirement for a higher jet pressure becomes obvious. Three pieces of specimen were laminated in the current experiments. A low jet pressure, being unable to cut the final lamina successfully, produced a particularly coarse wall in that area. A greater abrasive flow rate helps to reduce roughness by providing the necessary cutting capability. These results are shown in Fig. 6.16.

Fine abrasives remove material in smaller amounts, therefore the surface finish obtained is better. If the traverse speed becomes so large that the striation is serious, the difference made by mesh size diminishes, as illustrated in Fig. 6.17.

6.3.7 Machinability

Table 6.4 integrates the results and discussions in the above sections comprehensively and enables an assessment of the machinability of advanced ceramics cut by an abrasive waterjet to be made.

6.4 Surface Machining

6.4.1 Existing Theory of Material Erosion by Solid Particle Impact

Abrasive waterjet cutting involves complex interaction between the individual abrasives, the flow development, and the material. The mechanism in which a

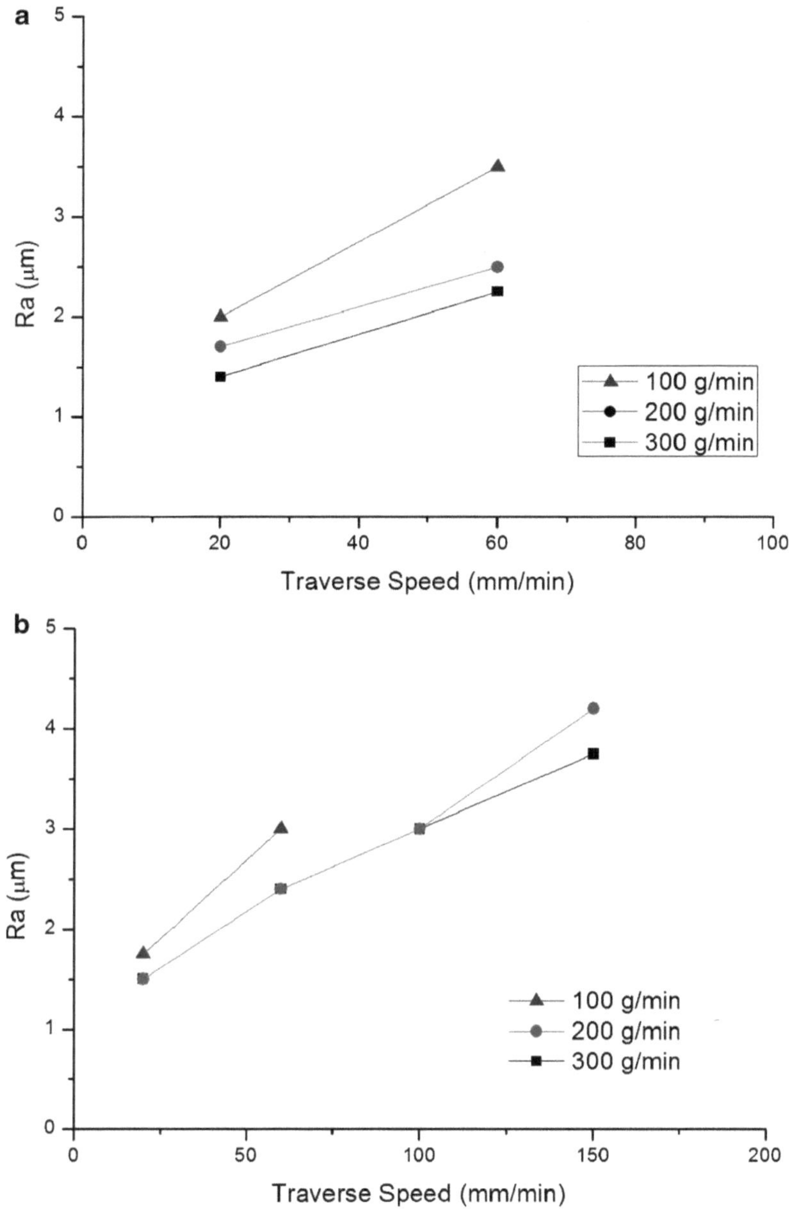

Fig. 6.15 Correlation between surface roughness and traverse speed, for alumina plate of 1.35 mm thickness and abrasive size 100 mesh, employing a pressure of (**a**) 207 MPa and (**b**) 310 MPa

high-velocity solid particle erodes ductile metal depends on its angle of impingement. At shallow angles, erosion proceeds as cutting wear; at large angles, the erosion is mainly due to deformation wear.

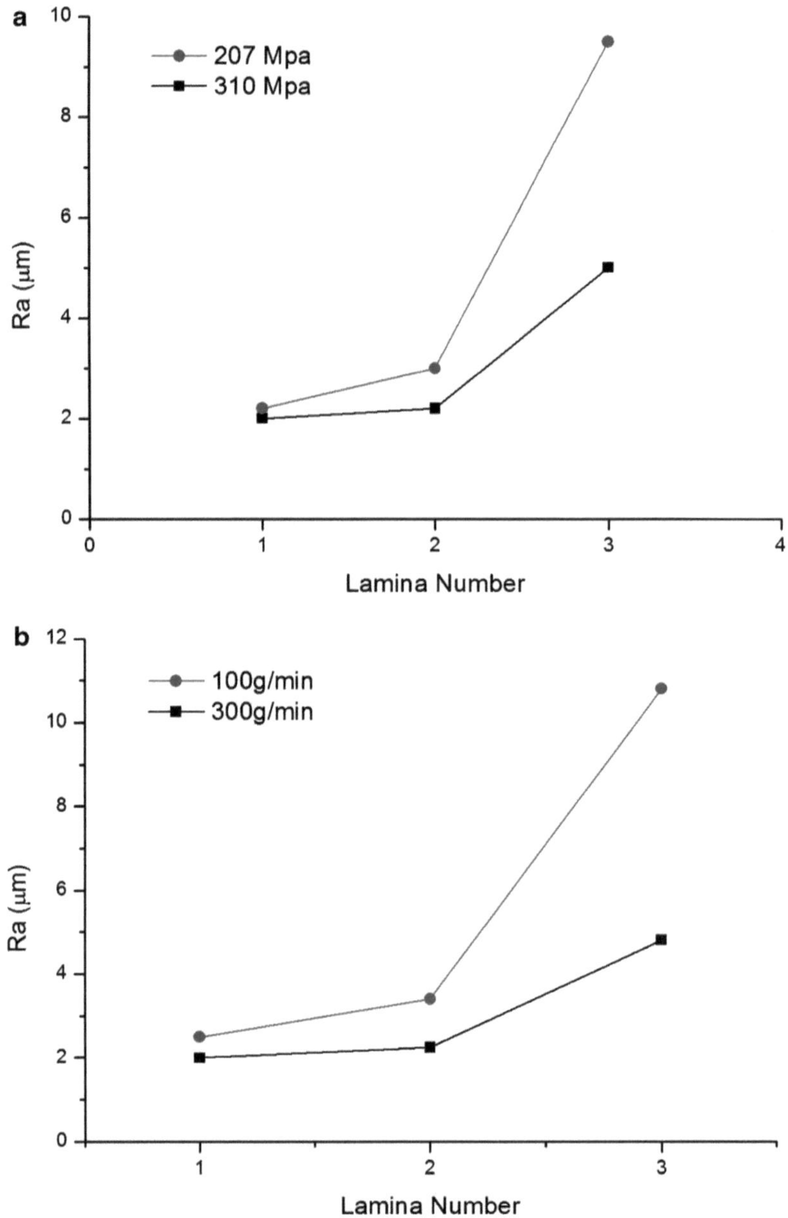

Fig. 6.16 Increasing surface roughness with kerf depth, for an abrasive flow rate of 300 g/min, a traverse speed of 20 mm/min and abrasive size 100 mesh: (**a**) at various hydraulic pressures and (**b**) at various abrasive flow rates

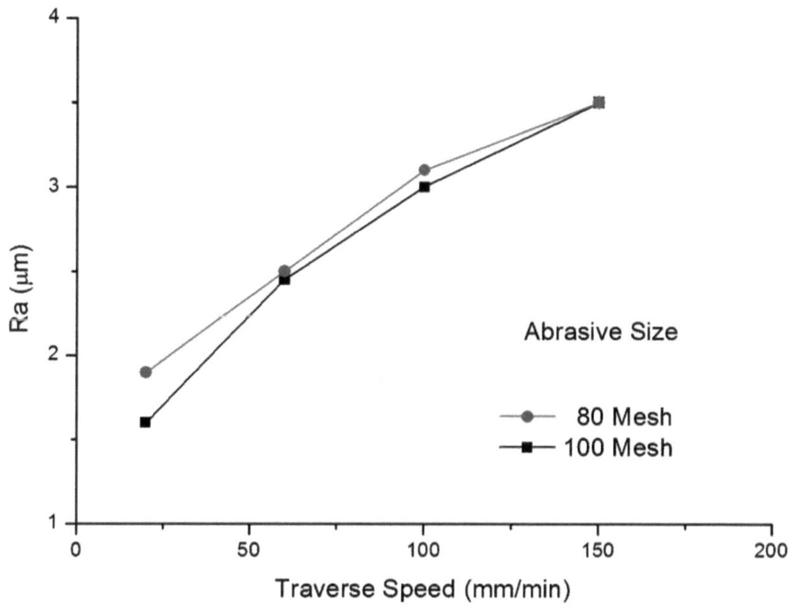

Fig. 6.17 Effect of abrasive size on surface roughness at different traverse speed, for a pressure of 310 MPa, an abrasive flow rate of 300 g/min and an alumina plate thickness of 1.35 mm

Table 6.4 Comprehensive results from the present study

	Cutting parameters			
Cutting results	Pressure	Traverse speed	Abrasive flow rate	Abrasive size
	↑	↑	↑	↑
Kerf width	↑	↓	↑	↑
Taper ratio	↓	↑	x	↓
Surface roughness	x	↑	↓	↑
Material removal rate	↑	↑	↑	↑
Through-cut capability	↑	↓	↑	↑

(↑) increase; (↑) decrease; (x) not obvious

6.4.1.1 Cutting Wear

Experimental study of the erosion of ductile metals by sharp particle impact shows that volume removal is related to the particle velocity to the power greater than 2 but less than 3 [35, 36]. The existing theoretical models, however, are not consistent with the experiment. Hashish includes the crater width variation as the depth of the trajectory varies [11]. The model that results shows a velocity exponent of 2.5 arid includes the particle shape expressed by its sphericity and roundness numbers. The improved model for the volume removal rate due to particle impact is best suited for shallow impact angles and is expressed as:

$$V = \frac{7m}{\pi \rho_p} \left(\frac{\upsilon_a}{C_k} \right)^{2.5} \sin 2\alpha \sqrt{\sin \alpha} \tag{6.13}$$

where V is the removed volume by one abrasive of mass m, υ_a is the impact velocity of the abrasive, ρ_p is the density of particle C_{K} is a characteristic velocity constant and α is the impact angle.

6.4.1.2 Deformation Wear

The volume removal mechanism in this mode of cutting is by impact at large angles, which causes excessive deformation and subsequent crack netting and material removal. For near-normal impacts, the equation of the volume removal rate (VRR) can be expressed as [37]

$$VRR = \frac{m(\upsilon_a - \upsilon_e)^2}{2\sigma_f} \tag{6.14}$$

where υ_e is the threshold velocity, and σ_f is the flow stress of material.

Hashish combines the theory of both cutting wear and deformation wear with a kinematic jet-solid penetration model to predict the depth of cut [11]. Kong et al. [38] used waterjet machining on NiTi shape memory alloys and developed an analysis of controlled-depth milling. Kong et al. [39] has also developed a model that can be generally applied to different machine system and to predict individual jet footprints that are one of the key steps for controlled-depth abrasive waterjet milling.

6.4.2 Experimental Setup and Procedure

The schematic experimental setup is shown in Fig. 6.6. The system includes the following major components: high-pressure pump, abrasive waterjet cutting head, water catcher, pressure-control station, translation table, and computer numerical controller. The waterjet intensifier pump (Flow 6XS-55) is connected with the abrasive waterjet cutting head. The intensifier pump is capable of delivering water up to 380 MPa and 1.93 l/mm. Water pressure can be set in advance on a pressure-control station. The translation table provides the necessary motion in cutting. The height of the abrasive cutting head is manually setup by screw and optical linear scale. The abrasive delivery system consists of hopper, abrasive metering valve, and delivery line. The abrasive metering valve is attached to the exit port of the abrasive hopper. The function of the metering valve measures the amount of abrasives. A manually adjustable digital micrometer is used to set

Table 6.5 Workpiece property and cutting parameters

Workpiece	Carbon/epoxy
Strength of material	1,315 MPa
Density of material	1,412.6 kg/m^3
Abrasive	Garmet mesh number 80#
Abrasive mass flow rate	100, 150, 200 g/min
Traverse rate	15, 25, 35 mm/s
Standoff distance	15, 20, 25, 30, 40, 50, 60, 65 mm
Pressure	105, 150, 175 MPa
Fiber direction	0°, 90°

the abrasive flow rate. The abrasive delivery line connects the outlet of abrasives metering valve with the mixing chamber, where the primary waterjet and abrasive are mixed to form a coherent high-energy stream.

Experimental material is the thermoset-based Carbon/Epoxy. Sixty-four layers of prepreg are laminated and cured with the following procedure:

1.	123°C at atmospheric pressure for 60 min, followed by
2.	183°C at 1.05 MPa for 120 min, followed by
3.	Air-cooling to room temperature

The laminate thickness is 8 mm and 230 mm by 210 mm in area. The fiber volume content is 60%.

The milling characteristics are highly dependent on several system parameters. In the current single-pass milling experiment, the fixed parameters are: orifice diameter of 0.229 mm, mixing rube diameter of 0.762 mm and mixing-tube length of 76.2 mm. the other parameters are shown in Table 6.5.

After single-pass milling, the depth and width of cut were measured by a Mitutoyo P3311 profile projector or by tracing paper, and the material removal rate was measured by weight.

Based on the single-pass milling, the lateral feed increments are designed 20, 30, 40, 50, and 60% of the width for double-pass milling. In some cases, 10 and 70% are also used. Upon examination of the cut profile, one can tell good results and the corresponding lateral feed by the naked eye. The good patterns from double-pass milling are extended to six-pass milling. In multiple-pass milling, a reproducible surface pattern and the affecting parameters are searched. The surface roughness is measured by a Hommel T1000. Based on the results of six-pass milling, a W10mm * D3mm milling is demonstrated.

6.4.3 Experimental Results and Discussions

6.4.3.1 Mechanism of Material Removal

In previous part, two cutting mechanisms are described. At shallow angles, material erosion proceeds as cutting wear: at large angles as in the current study, the erosion is due to deformation wear. In the single-pass milling process, the abrasives before and after cutting and the chips are observed, as shown in Fig. 6.18 [40]. Before cutting, the abrasives are uniform in size and sharp. After cutting, they remain sharp, but are often broken into several pieces. This reflects the impact between abrasives and material. The chips are fractured pieces with little plastic deformation. The plastic matrix covering fibers are removed first by the abrasive jet, then the almost bare fibers are cut by the abrasives since the strength of the matrix is much lower with respect to fibers. Some fibers are simultaneously broken away with the matrix. The size of the chips is largely between 0.1 and 0.4 mm. Because both the abrasives and the chips show the fractured debris as a result of impact, the deformation wear is presumed the major material removal mechanism in abrasive waterjet cutting of FRP, which is consistent with the large-angle material erosion theory.

6.4.3.2 The Material Removal Rate

The material removal rate (MRR) is defined as the volume removed per unit time. Larger standoff distance loses more kinetic energy of the jet in friction with air, the depth of single-pass milling is thus reduced, but the width of cut is significantly increased, thus a maximum of MRR correlated with the standoff distance will exist, as shown in Fig. 6.19. Hashish also found that this held true for mild steel and aluminum [11].

At faster traverse speed, the material is more effectively removed by exposing more fresh area of wear to the abrasives. This effect is expected to saturate at high traverse speed when the bombing density of the abrasives is fully developed. Compared with mild steel and aluminum, one finds that the specific removal rate of carbon/epoxy with respect to unit hydraulic power and abrasive flow rate by abrasive waterjet is larger than mild steel and aluminum, as shown in Table 6.6.

High hydraulic pressure and abundant supply of abrasives produce deeper cuts (i.e., large material removal rate) by increasing the number and the kinetic energy of cutting particles as shown in Fig. 6.19. The material removal rate is closely related to the velocity of abrasive particles at impact. Basic fluid mechanics provides the correlation between the velocity of abrasive and the applied hydraulic pressure for various amounts of abrasives with negligible energy loss, as illustrated in Fig. 6.20. Though more abrasives in the jet provide more impact accounts, they also reduce the impact velocity to conserve the momentum in total. The effects of abrasive flow rate on material removal rate are therefore less dominant than the applied pressure.

The fiber orientation is insignificant with respect to MRR, since the removal mechanism of impact fracture by the abrasives is independent of the lay-up of fibers.

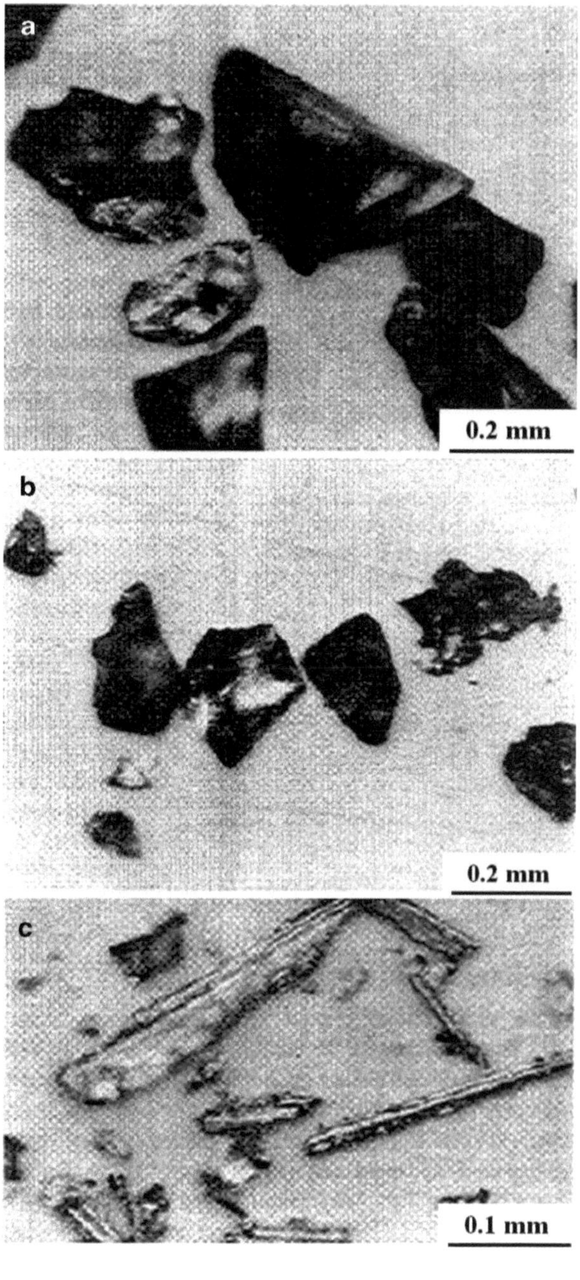

Fig. 6.18 The abrasives and chips [40], reprinted with permission. (**a**) Abrasives before cutting. (**b**) Abrasives after cutting. (**c**) Chips after cutting

6.4.3.3 Depth, Width, and Width-to-Depth Ratio of Single-Pass Milling

Every waterjet milling starts from a single-pass non-through cut, out of which multiple-pass is extended to generate a repeatable surface topography. Hence basic understanding of the depth and width, and especially the width-to-depth (WTD) ratio of single-pass cut is required.

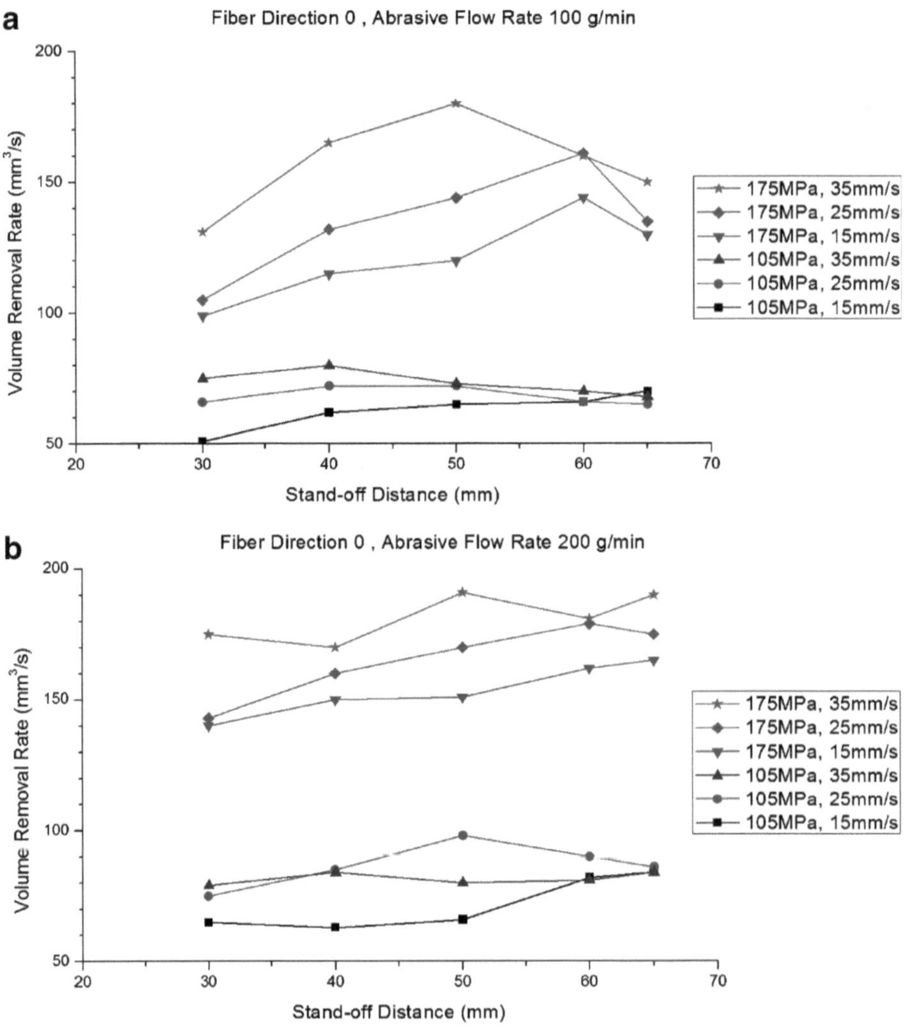

Fig. 6.19 Correlation between volume removal rate and standoff distance. (**a**) At fiber direction 0 and abrasive flow rate 100 g/min. (**b**) At fiber direction 0 and abrasive flow rate 200 g/min. (**c**) At fiber direction 90 and abrasive flow rate 100 g/min

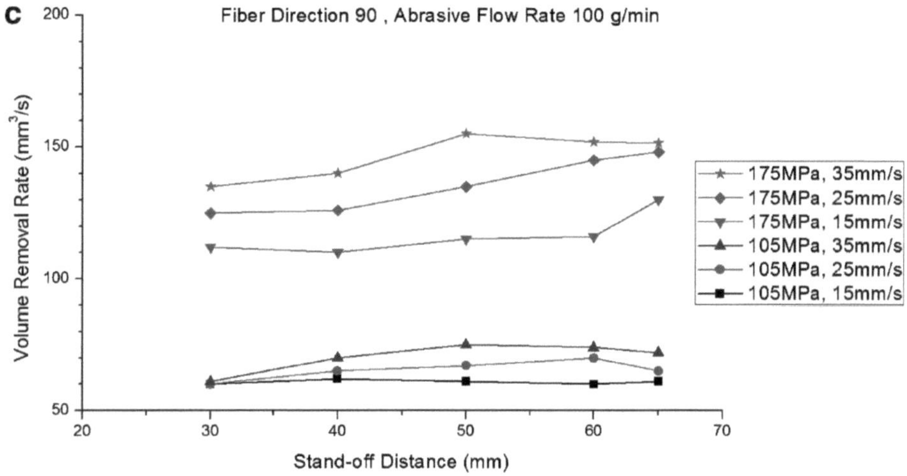

Fig. 6.19 (continued)

Table 6.6 Maximum volume removal rate of mild steel, aluminum, and carbon/epoxy

	Maximum volume removal rate (mm^3/s)	Standoff distance (mm)	Specific removal rate $mm^3/kw\ g$)
Mild steel	35	8	0.55
Aluminum	70	8	1.10
Pressure: 207 MPa Traverse rate: 5.1 mm/s Abrasive mass flow rate: 250 g/min Orifice diameter: 0.46 mm Mixing-tube diameter: 1.57 mm Garmet mesh No. 60 [11]			
Carbon/epoxy	177.7	50	36.0
Pressure: 175 MPa Traverse rate: 35 mm/s Abrasive mass flow rate: 100 g/min Orifice diameter: 0.23 mm Mixing-tube diameter: 1.0 mm Garmet mesh no. 80 [Current study]			

The depth of cut is shown in Fig. 6.21. It increases mainly with hydraulic pressure, and decreases with feed rate and standoff distance. Larger abrasive flow rates increase the frequency of impact, but the velocity of abrasive particles is reduced (as shown in Fig. 6.20). Hence its effect on the depth of cut is reduced. Nevertheless, more abrasives spread wider in air causing a wider cut.

Standoff distance mainly affects the width of cut, and the applied pressure is basically ineffective in changing the width, as illustrated in Figs. 6.21 and 6.22.

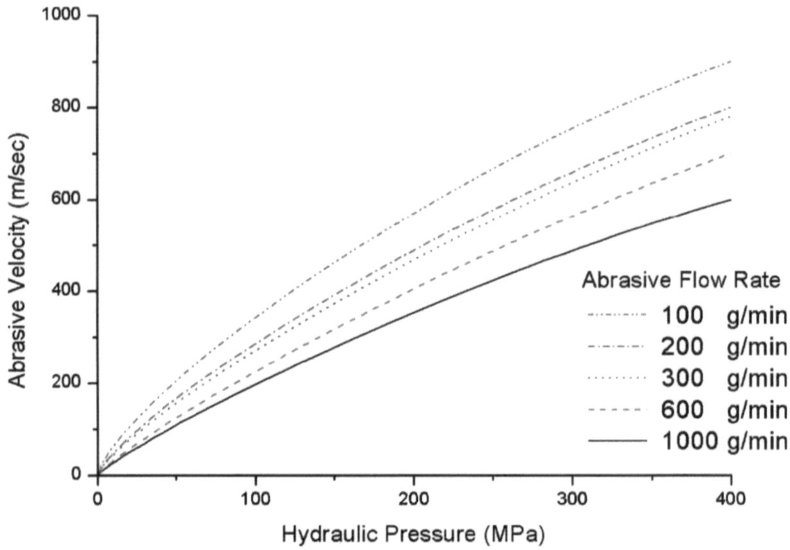

Fig. 6.20 Abrasive velocity at various hydraulic pressure and abrasive flow rate

At a higher traverse rate, the width is reduced since the accumulated impact probability of the abrasives at the outer skirt of the jet decreases. The fiber orientation does not affect the depth or width of cut as discussed in previous section.

From the results of single-pass miffing, one observes that the smaller the WTD ratio, the steeper the kerf. The correlation of WTD ratio and working parameters is shown in Fig. 6.22. The higher the pressure, the smaller the WTD ratio, since the capability of cutting is increased while the width is not apparently influenced. The larger the traverse rate and the abrasive flow rate are, the larger the WTD ratio. The former is due to little change in width with decrease in depth. When the abrasive flow rate increases, the cut becomes wider, while the depth is only slightly affected, as discussed in the last paragraph. As to the effects of the standoff distance, the larger the standoff distance, the larger the WTD ratio. The effects are very significant due to the widening and shallowing of the cut in parallel. Below is a standoff distance determined by the characteristic length of the supersonic waterjet in air—such a phenomenon is not fully developed. The cut profiles of various WTD ratios are shown in Fig. 6.23 [40].

6.4.3.4 The Effects of Width-to-Depth Ratio and Lateral Feed in Double-Pass Milling

To study the milling surface characteristics, double-pass cutting with various lateral feed at a certain WTD ratio of the single-pass cutting is tested. From Fig. 6.24, three categories of WTD ratio are discussed, i.e., less than 1.0, 1.0 to 2.0, and greater than 2.0. Values of lateral feed of 20–60% of the width of cut are used in the tests.

1. The WTD ratio is less than 1.0.

The bottom is extremely difficult to obtain plains. The second pass either cuts one-sided into the first pass at low lateral feed causing over plowing, or produces a separable second plow at large lateral feed, as shown in Fig. 6.25a [40]. Besides, the kerf of which the WTD ratio is less than 1.0 shows very steep slope, hence it is hard to make a flat bottom by the consecutive pass

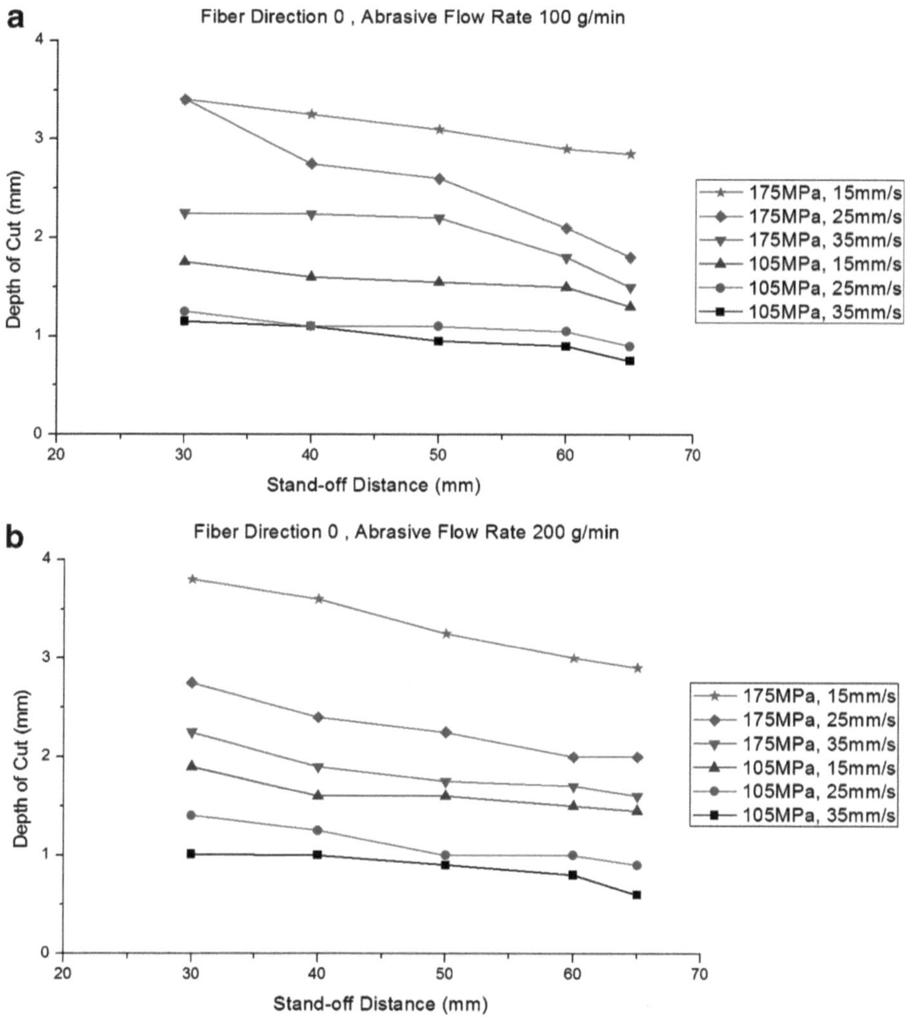

Fig. 6.21 Correlation between depth of cut and standoff distance. (**a**) At fiber direction 0 and abrasive flow rate 100 g/min. (**b**) At fiber direction 0 and abrasive flow rate 200 g/min. (**c**) At fiber direction 90 and abrasive flow rate 100 g/min

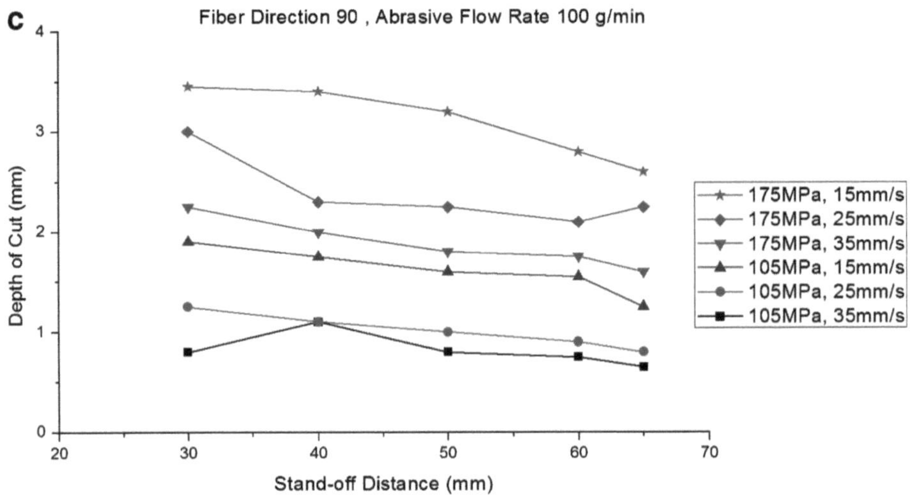

Fig. 6.21 (continued)

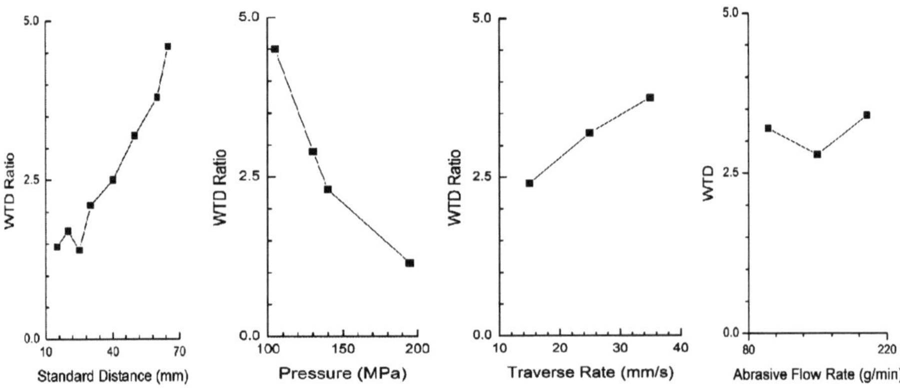

Fig. 6.22 Correlation between WTD ratio and pressure, abrasive flow rate, traverse rate, and stand-off distance

Fig. 6.23 The cut profile of various WTD ratio [40], reprinted with permission

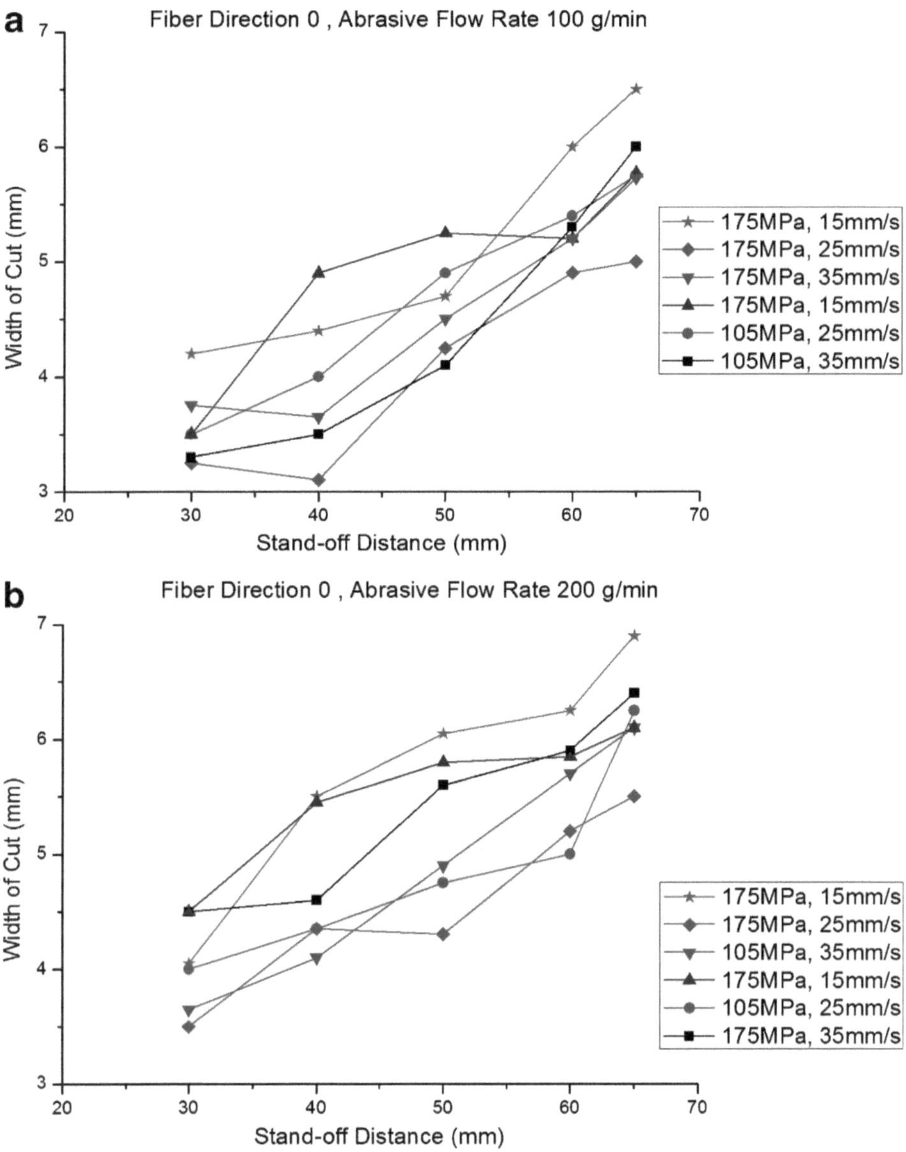

Fig. 6.24 Correlation between width of cut and standoff distance. (**a**) At fiber direction 0 and abrasive flow rate 100 g/min. (**b**) At fiber direction 0 and abrasive flow rate 200 g/min. (**c**) At fiber direction 90 and abrasive flow rate 100 g/min

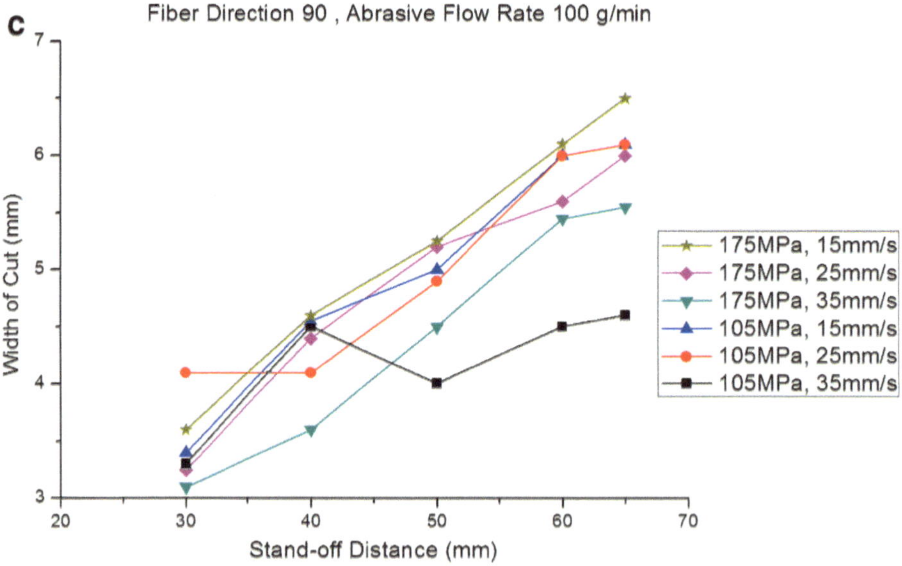

Fig. 6.24 (continued)

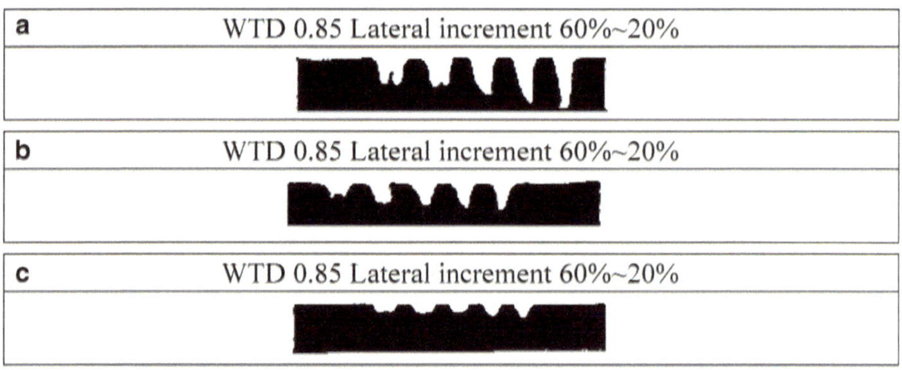

Fig. 6.25 The double-pass cut profile of various lateral feed increments [40], reprinted with permission. (**a**) WTD ratio less than 1.0. (**b**) WTD ratio between 1.0 and 2.0. (**c**) WTD ratio larger than 1.0

2. The WTD ratio is between 1.0 and 2.0.
   Most difficulties persist as in case 1, except that of some lateral feeds. Figure 6.25b [40] shows an example

3. The WTD ratio is larger than 2.0.
   The even surface is achievable at a large WTD ratio, as shown in Fig. 6.25c [40]. The better profiles occur at the lateral feed of about 40–50% of the cut width.

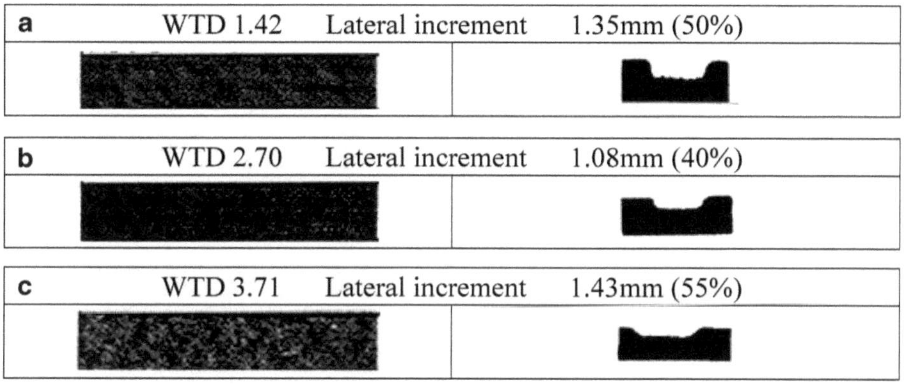

Fig. 6.26 Some examples of six-pass cut profile and surface [40], reprinted with permission. (**a**) WTD ratio 1.42 and lateral feed 50%. (**b**) WTD ratio 2.70 and lateral feed 40%. (**c**) WTD ratio 3.71 and lateral feed 55%

The results of double-pass cutting show large WTD ratio and optimal lateral feed of around 40–50% of cut width are desired for producing plain surface.

6.4.3.5 The Relationship Between Width-to-Depth Ratio, Lateral Feed, and Surface Roughness in Multiple-Pass Milling

The acquired knowledge of WTD ratio and optimal lateral feed are used to produce a plain surface by multiple passes. A reproducible section of plain surface is milled at the above-mentioned adequate conditions before six passes. Figure 6.26 [40] shows several examples. Figure 6.26a is the result of a WTD ratio 1.42 and a lateral feed 50%, Fig. 6.26b, of WTD ratio 2.70 and lateral feed 40%, Fig. 6.26c, of WTD ratio 3.71 and lateral feed 55%.

The relationship between the WTD ratio, lateral feed, and the obtained surface roughness is shown in Fig. 6.27. Above 2.0 of the width-to-depth ratio, the surface roughness decreases; between 0 and 70% of the lateral feed, the surface roughness shows the minimum of about 6 m at 40%. These results of a suitable range of working conditions are useful in the path planning of surface milling.

Based on the results of six-pass milling, a 10 mm wide and 3 mm deep square slot is machined with the optimal milling conditions. The roughness is 8.9 μm on average, ranging from 5.6–11.4 μm. The results are comparable to that of conventional milling. One should notice that depth of cut of the second layer may increase up to 25% more than the first layer. The increment comes from several sources. The kinetic energy is more effectively used as the waterjet being trapped in the deeper cut with less bouncing-up, the bumpy surface left by the first layer of cut is easier to be removed than the original smooth surface, and some surface microcracks may exist after the first cut and help the material removal in the consecutive cut. Thus the theory of control of depth of cut in multiple-layer machining is worth further investigation.

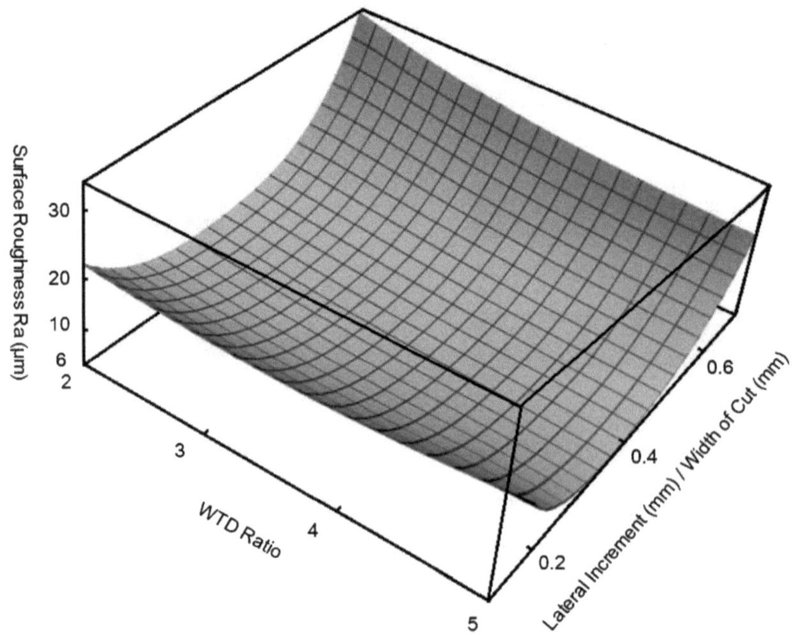

Fig. 6.27 The relationship between the WTD ratio, lateral feed, and the obtained surface roughness

6.4.4 Dimensional Analysis

Dimensional analysis provides a mathematical expression and a preliminary physical insight of the material removal process at minimum experimental expense. For the volume removal rate ($\dot{v}$, mm^3/s), depth of cut (h, mm) and width of cut (W, mm), the dominant parameters are shown in Table 6.7. The analysis will lead to correlation between the material removal rate and WTD ratio in milling of FRP (fiber-reinforced plastics) and the applied abrasive waterjet parameters. The knowledge will also be compared with the existing theory of cutting mechanism. The analysis is as following:

1. Volume removal rate

   The relationship is obtained by PI theorem [41] as:

$$\frac{\dot{v}}{\mu S^2} = k\left(\frac{\upsilon - \upsilon_e}{\mu}\right)^a \left(\frac{\mu S^2 \rho}{\dot{m}_a}\right)^b \left(\frac{D}{S}\right)^c \left(\frac{\sigma S^2}{\mu \dot{m}_a}\right)^d \tag{6.15}$$

   The value of threshold velocity υ_e is believed to be related to the modulus of elasticity [11]. With the examples of titanium and aluminum from the above reference, one estimates the υ_e for CFRP to be 70–90 mm/s. The current analysis

Table 6.7 Kinematic parameter design for copper wafers

Parameters	Dimension	
Abrasive diameter, D	L	
Abrasive density, ρ	ML^{-3}	
Abrasive flow rate, $\dot{m}_a$	MT^{-1}	
Traverse speed, μ	LT^{-1}	
Standoff distance, S	L	
Abrasive velocity, υ	LT^{-1}	
Material strength, σ	$ML^{-1}T^{-2}$	
	Theoretical	Experimental
Material removal rate	$\propto (\upsilon - \upsilon_e)^2$	$\propto (\upsilon - \upsilon_e)^{1.9}$
	$\propto \dot{m}_a$	$\propto \dot{m}_a^{0.6}$
Depth of cut	$\propto \upsilon^2$	$\propto (\upsilon - \upsilon_e)^{2.4}$
	$\propto \dot{m}_a$	$\propto \dot{m}_a^{0.1}$
Width of cut	Nonexisting	$\propto (\upsilon - \upsilon_e)^{0.06}$
		$\propto \dot{m}_a^{0.16}$
WTD ratio	Nonexisting	$\propto (\upsilon - \upsilon_e)^{-2.3}$
		$\propto \dot{m}_a^{0.5}$

assumed the value of 80 mm/s. Considering the maximum volume removal rate in Fig. 6.19, one obtains the relationship of the non-dimensionalized volume removal rate by regression as:

$$\frac{\dot{v}}{\mu S^2} = 0.018\left(\frac{\upsilon - \upsilon_e}{\mu}\right)^{1.9}\left(\frac{\mu S^2 \rho}{\dot{m}_a}\right)^{0.3}\left(\frac{D}{S}\right)^{0.5}\left(\frac{\sigma S^2}{\mu \dot{m}_a}\right)^{-0.9} \tag{6.16}$$

The dependence on the impact velocity agrees with the $(\upsilon - \upsilon_e)^2$ suggested by the theory of deformation wear. The coefficient of determination for data regression is 0.98.

2. Depth of cut
   The relationship can be obtained as:

$$\frac{h}{S} = k\left(\frac{\upsilon - \upsilon_e}{\mu}\right)^{a}\left(\frac{\mu S^2 \rho}{\dot{m}_a}\right)^{b}\left(\frac{D}{S}\right)^{c}\left(\frac{\sigma S^2}{\mu \dot{m}_a}\right)^{d} \tag{6.17}$$

The experimental results are then incorporated.

$$\frac{h}{S} = 0.12\left(\frac{\upsilon - \upsilon_e}{\mu}\right)^{2.4}\left(\frac{\mu S^2 \rho}{\dot{m}_a}\right)^{0.8}\left(\frac{D}{S}\right)^{0.9}\left(\frac{\sigma S^2}{\mu \dot{m}_a}\right)^{-0.9} \tag{6.18}$$

This result is consistent with the predicted proportionality of $(\upsilon - \upsilon_e)^2$ of the theoretical model. The coefficient of determination for data regression is 0.90.

3. Width of cut
The relationship can be obtained as:

$$\frac{w}{S} = k\left(\frac{\upsilon - \upsilon_e}{\mu}\right)^a \left(\frac{\mu S^2 \rho}{\dot{m}_a}\right)^b \left(\frac{D}{S}\right)^c \left(\frac{\sigma S^2}{\mu \dot{m}_a}\right)^d \tag{6.19}$$

The experimental results are then incorporated.

$$\frac{w}{S} = 0.17\left(\frac{\upsilon - \upsilon_e}{\mu}\right)^{0.06} \left(\frac{\mu S^2 \rho}{\dot{m}_a}\right)^{-0.14} \left(\frac{\sigma S^2}{\mu \dot{m}_a}\right)^{-0.02} \tag{6.20}$$

The results were not predicted in theoretical models. The coefficient of determination for data regression is 0.77.

(iv) Width-to-depth ratio
From (6.18) and (6.20), the WTD ratio is expressed as:

$$\frac{w}{h} = k\left(\frac{\upsilon - \upsilon_e}{\mu}\right)^a \left(\frac{\mu S^2 \rho}{\dot{m}_a}\right)^b \left(\frac{D}{S}\right)^c \left(\frac{\sigma S^2}{\mu \dot{m}_a}\right)^d \tag{6.21}$$

The experimental results are then incorporated.

$$\frac{w}{h} = 1.4\left(\frac{\upsilon - \upsilon_e}{\mu}\right)^{-2.3} \left(\frac{\mu S^2 \rho}{\dot{m}_a}\right)^{-0.95} \left(\frac{D}{S}\right)^{-0.9} \left(\frac{\sigma S^2}{\mu \dot{m}_a}\right)^{0.9} \tag{6.22}$$

The coefficient of determination for data regression is 0.89. The results show quantitatively the dominant parameters for WTD ratio and material removal rate. This fundamental relationship is important in the design of a milling procedure for FRP's by abrasive waterjet. Table 6.7 summarizes the results of dimensional analysis and the comparison with the theory of deformation wear.

6.5 Conclusion

Water jet machining of composite materials surpasses the conventional solid-tool machining with respect to tool wear, thermal damage, and environmental protection. Nevertheless, delamination during water jet drilling is a primary concern in applying this advanced technique. In the present paper, mechanisms of delamination are recognized, and the phenomenon is modeled by using classic plate bending theory and linear elastic fracture mechanics. The water jet force is identified as the main cause of delamination damage and the critical water pressure leading to the onset of delamination is predicted. The approach is justified by existing experimental data from independent researchers with good agreement. These results are expected to be valuable in applying the water jet drilling to composite materials. The production rate can be maximized without delamination.

The abrasive water jet is effective in the machining of modern ceramics. By careful control of the operating parameters, the response of the material to the sequential impact of the abrasives provides material removal with good kerf quality. The amount of material removal is determined primarily by the hydraulic pressure and the abrasive flow rate. Through-cut can only be achieved beyond a critical level of energy input rate per unit length. The traverse speed also affects the kerf width and the taper ratio by altering the specific cutting energy input. An optimum exists between narrow cuts and uniform kerf taper. Fast passage of the jet causes clear wall striations. A sufficient supply of hydraulic energy as well as fine-mesh abrasives at moderate traverse speed produces a smooth kerf surface. The knowledge acquired from linear cutting will be valuable for eventual two-dimensional machining applications involving modern ceramic materials.

The machinability of fiber-reinforced plastics in milling by abrasive waterjet is investigated. The material removal is found experimentally consistent with the existing theory of deformation wear. The machining experiment is divided into single-pass, two-pass, and six-pass stages; each reveals useful results of affecting cutting parameters, with respect to material removal rate, depth of cut, width of cut, and surface roughness. The hydraulic pressure and traverse speed effectively determines the MRR and depth of cut, while the standoff distance possesses significant influence on the width of cut. The fiber orientation does not affect the volume removal rate and surface roughness, since the mechanism of material removal is hardly anisotropic. The specific material removal rate with respect to unit hydraulic power and abrasive flow rate is several times higher than that of aluminum and steel. The WTD ratio and the lateral feed increment are found as two important parameters in the production of an acceptable plain surface. Large WTD ratio and around 50% of lateral feed increment with respect to the width of cut are desired. The acquired surface roughness of several microns is comparable to that of conventional milling. The dimensional analysis presents the experimental results and the dominant parameters in agreement with the theoretical prediction of material removal. The machinability of FRP in milling by abrasive waterjet is proven satisfactory regarding the material removal rate and surface roughness, and the process is worth further development for industrial applications.

References

1. Friend CA, Clyne RW, Valentine GG (1973) Machining graphite composite materials. In: Noton BR (ed) Composite materials in engineering design. Washington, DC, ASM, pp 217–224
2. Mackey BA (1980) How to drill precision holes in reinforced plastics in a hurry. Plastics Eng 36:22–24
3. Mcginty MJ, Preuss CW (1985) Machining ceramic fiber metal matrix composites. In: Sarin VK (ed) High productivity machining. Am. Soc. Metals, Washington, DC, pp 231–243

4. Miller JA (1987) Drilling graphite/epoxy at Lockheed. Am Mach Autom Manufact 131(10):70–71
5. Koenig W, Wulf CH, Grass P, Willerscheid H (1985) Machining of fibre reinforced plastics. Manuf Technol CIRP Ann 34:537–548
6. Ho-Cheng H, Dharan CKH (1988) Delamination during drilling in composite laminates. ASME winter annual meeting, Chicago, Illinois, Nov. 27–Dec. 2
7. Mohaupt UH, Burns DJ (1974) Machining unreinforced polymers with high-velocity water jets. Exp Mech 14:152–457
8. Snofys R, Staelens F, Dekeyser W (1986) Current trends in non-conventional material removal processes. Keynote Paper CIRP Ann 35:467–480
9. Ansorge A (1985) Fluid jet principles and applications. Nontraditional machining, Proc. Carbide and Tool Engineers, Cincinnati, Ohio. pp. 35–41
10. Jordan R (1985)Waterjets on the cutting edge of machining. Nontraditional machining, Proc. Carbide and Tool Engineers, Cincinnati, Ohio. pp. 13–21
11. Hashish M (1989) A model for abrasive-waterjet (AWJ) machining. Trans ASME J Eng Mater Technol 111(1):54–162
12. Hashish M (1988) Machining of advanced composites with abrasive-waterjets. ASME winter annual meeting, Chicago, Illinois, Nov. 27–Dec. 2
13. Koenig W, Trasser J, Schmelzier M (1987) Bearbeitung Faserverstaerkter Kunststoffe mit Wasserund Laserstrahl. VDI-Z 129:6–11
14. Saito S (1990) Fine ceramics. Elsevier, New York
15. Wilfrid K, Matthias P (1989) Precision machining of advanced ceramics. Am Ceram Soc Bull 68:550–555
16. Frederick RF, Vosely ET (1988) High power laser beam machining of structural ceramics. Symp. Machining of advanced ceramic materials and components, November. pp. 215–227
17. Petrofes NF, Gadalla AM (1988) Electrical discharge machining of advanced ceramics. Am Ceram Soc Bull 67:1048–1052
18. Moreland MA, Moore DO (1988) Versatile performance of ultrasonic machining. Am Ceram Soc Bull 67:1045–1047
19. Frederick M (1989) Water and sand. Machine Tool Technol 84–95
20. Kim TJ, Sylvia JG, Posner L (1985) Piercing and cutting of ceramics by abrasive waterjet. Int. Symp. machining of ceramic materials and components. 1985, ASTM Winter Annual Meeting, Proc. PED. vol 17. pp. 19–24
21. Hunt DC, Kim TJ (1986) A parametric study of waterjet processes by piercing. Proc. 8th Int. Symp., Jet Cutting Technology, Durham, England, September
22. Hunt DC, Burnhain CD, Kim TJ (1987) Surface finish characterization in machining advanced ceramics by abrasive waterjet. Proc. 4th U.S. Water Jet Conf., ASME. pp. 169–174
23. Hashish M (1991) Characteristics of surfaces machined with abrasive waterjets. ASME J Eng Mater Technol 113:354–362
24. Hashish M (1991) ASME J Eng Ind 113:29–37
25. Ho-Cheng H (1988) An analysis of drilling of composite materials. Ph.D. dissertation, UC Berkeley
26. Wilkins DJ, Eisenmann JR, Camin RA, Margolis WS, Benson RA (1982) Characterizing delamination growth in graphite-epoxy. In: Reifsnider KL (ed.) Damage in composite materials. ASTM STP 775. pp. 168–183
27. Dharan CKH (1978) ASME J Eng Mater Technol 100:233–247
28. Timoshenko S, Woinowski-Krieger S (1959) Theory of plates and shells, 2nd edn. McGraw-Hill, New York, pp 51–72
29. Saghisadeh H, Dharan CKH (1986) ASME J Eng Mater Technol 108:290–295
30. Hocheng H (1990) Int J Mach Tool Manuf 30(3):423
31. Garg A, Ishai O (1984) NASA technical memorandum No. 85935
32. Evans AG (1978) Impact damage in ceramics. In: Bradt BC, Hasselman DPH, Lange FF (eds) Fracture mechanics of ceramics, vol 3. Plenum, New York, pp 303–330

33. Hockey BJ, Wiederhorn SM, Tohnson H (1978) Erosion of brittle materials by solid particle impact. In: Bradt BC, Hasselman DPH, Lange F (eds) Fracture mechanics of ceramics, vol 3. Plenum, New York, pp 379–402
34. Hocheng H, Chang KR (1994) J Mater Process Technol 40(3):287
35. Engel P (1976) Impact wear of materials. Elsevier, New York, pp 104–158
36. Preece C (ed) (1979) Treatise on materials science and technology. Erosion, vol. 16. Academic, New York, pp 69–126
37. Bitter JGA (1963) Wear 6:169–190
38. Kong MC, Axinte D, Voice W (2011) J Mater Process Technol 211(6):959–971
39. Kong MC, Anwar S, Billingham J, Axinte DA (2012) Int J Mach Tool Manuf 53(1):58–68
40. Hocheng H, Tsai HY, Shiue JJ, Wang B (1997) J Manuf Sci Eng 119(2):133
41. Buckingham E (1914) Phys Rev 4(4):345–376

Chapter 7
Micromachining by Photonic Beams

H.Y. Tsai, H. Hocheng, K.Y. Wang, and S.W. Luo

Abstract We analyze a laser dragging process capable of ablating a groove pattern, and producing sophisticated 3D features, on a polycarbonate (PC) sheet through a shaped mask opening. To predict the machined profile during the dragging process, we developed a mathematical model that describes the relationship between laser machining parameters and the produced profile. In addition, we manufacture a miniature lamp lens by varying the mask shape, and dimensions based on the proposed model. The effects of the micro-size lamp lens on light efficiency are investigated. On the other hand, this chapter also introduces a study of sub-micron-structure machining on silicon substrates by a direct writing system using a femtosecond laser with the central wavelength of 800 nm, pulse duration of 120 fs and repetition rate of 1 kHz. Three types of experiment are studied: (1) The effect of photoresist on silicon, (2) The machinability of different orientations of silicon: spike morphologies were observed on all three orientations of silicon substrates without obvious directional difference of these spikes on the different silicon substrates, (3) Micro-structure size and cross-section shape: a numerical model of the machining parameters has been proposed to simulate the cross-section of the micro-structure resulting from a given ablation energy. The predicted shape, determined by simulation, fitted the profile of the cross-section shape well.

Keywords Excimer laser • Laser dragging • Micro lens • Mathematical model • Numerical simulation • Direct writing • Femtosecond laser • Sub-micron structure • Spike morphology • Silicon orientation • Ablation threshold • Predict shape

---

H.Y. Tsai (✉) • H. Hocheng • K.Y. Wang • S.W. Luo
Department of Power Mechanical Engineering, National Tsing Hua University, No. 101, Sec. 2, Kuang-Fu Rd., Hsinchu, Taiwan, ROC30013,
e-mail: hytsai@pme.nthu.edu.tw

H. Hocheng and H.-Y. Tsai (eds.), *Advanced Analysis of Nontraditional Machining*,
DOI 10.1007/978-1-4614-4054-3_7,

7.1 Introduction

About 50 years ago, the first LASER (Light Amplification by Stimulated Emission of Radiation) was invented by T. H. Mainman in 1960. Afterwards, lasers were widely applied in many industrial applications. Many different lasers, for example, CO_2 lasers, Nd-YAG lasers, Diode laser, and Excimer lasers, have been developed and their wavelength range are from the near ultraviolet (UV) to the far infrared (IR).

The first excimer laser was invented in 1970 by the group of N. G. Basov at the Lebedev Physical Institute in Moscow, using a xenon dimer (Xe_2) excited by an electron beam to give stimulated emission at 172 nm of the wavelength. Later, the availability of the excimer laser has drawn the attention and concerns of many scientists and researchers since Lambda Physik created the first commercial excimer laser in 1980.

The term excimer originated from a contraction of excited dimer. A molecule called dimer is created under appropriate conditions of electrical stimulation, which can only exist in an excited state. However, the excited molecules are very unstable in the excited state. They readily get back to the ground state from the excited state through laser light release. The released laser light is named excimer laser and its wavelength lies in the range of near UV to far UV. A schematic diagram of molecular transition for producing Krypton Fluoride (KrF) excimer laser is shown in Fig. 7.1 [1].

The excimer lasers are usually made up of the family of an inert gas (Argon, Krypton, or Xenon) and a halide gas (Fluorine or Chlorine), that is, Argon Fluoride (ArF), Krypton Chloride (KrCl), KrF, Xenon Fluoride (XeCl), and Xenon Chloride (XeF). Characteristics of these excimer lasers are listed in Table 7.1 [2]. From Table 7.1, one knows the excimer laser possesses not only the same advantages of conventional lasers, such as coherence, directionality, spectral purity (single wavelength), but also other advantages, such as shorter pulse duration, higher photon

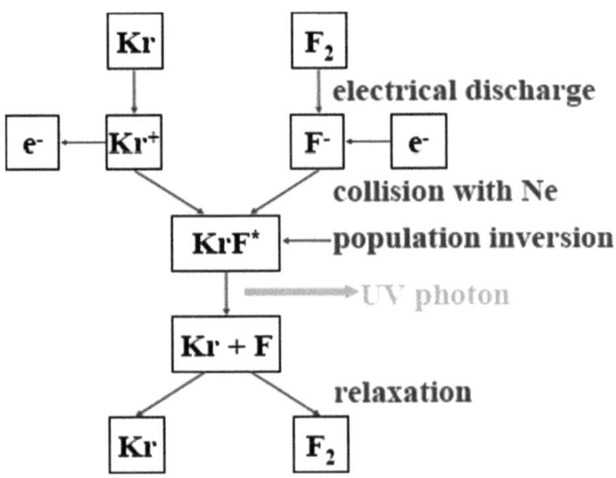

Fig. 7.1 Molecular transition in the KrF excimer laser [1], reprinted with permission

Table 7.1 Excimer lasers with different wavelengths

	ArF	KrCl	KrF	XeCl	XeF
Wavelength (μm)	193	223	248	308	351
Pulse energy (J)	0.8	0.2	1.5	2	0.7
Peak power (MW)	30	10	50	50	30
Average power(W)	60	10	150	200	70
Pulse duration (ns)	10–30 (up to 200 possible on XeCl)				
Beam divergence (mrad)	2–10				

Table 7.2 Dissociation energies of various bonds

Bond	Energy (KJ/mol)	Energy (eV)	Bond	Energy (KJ/mol)	Energy (eV)
C–C	344	3.58	C≡C	812	8.46
C–O	350	3.65	H–H	435	4.53
C–N	293	3.05	H–N	390	4.06
C=C	615	6.41	H–O	464	4.83
C=O	724	7.54	H–C	415	4.32
C=N	615	6.41	N≡N	942	9.81
C≡N	891	9.28	O=O	492	5.13

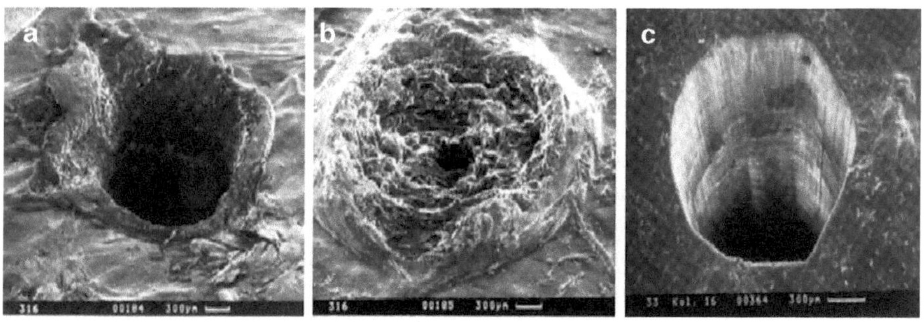

Fig. 7.2 Comparison of the machined results of polyimide using three different lasers: (**a**) Nd-YAG laser; (**b**) CO2 laser; (**c**) Excimer laser [1], reprinted with permission

energy generated due to short wavelength, and little thermal damage generated when machining organic materials.

Excimer laser is a useful nonconventional machining tool for material processing because of several advantages over conventional machining methods, including non-contact machining between the tool and the material, no chemical solvents used during machining process.

The range of energy output of the excimer lasers lies between 3.5 and 7.9 eV. The dissociation energies of various chemical bonds for polymers quite fit in the range, such as C–H bond, C–O bond and C–C bond etc. Table 7.2 shows examples of dissociation energies of chemical bonds [3].

Two examples of materials machined by various lasers are shown in Fig. 7.2 [4] and in Fig. 7.3 [5]. Figure 7.2 describes the polyimide machined by different lasers.

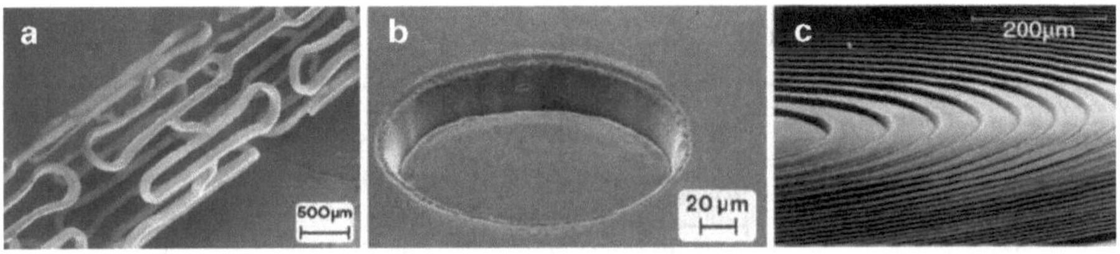

Fig. 7.3 Microstructures created with laser ablation: (**a**) A stand fabricated from a Ta tube by femtosecond machining; (**b**) Hole in a single crystalline LiNbO3 created by excimer laser irradiation; (**c**) 3D Fresnel lens in polyimide created with excimer laser irradiation applying a phase mask [1], reprinted with permission

The machining shows that polymer removal using excimer laser is better than the other two lasers and the thermal effect generated by excimer laser is nearly negligible. Figure 7.3 describes three materials structured with different lasers and machining techniques. The structure in Fig. 7.3a was fabricated by removing the material with a femtosecond laser. Figure 7.3b gives an example for the structuring of a $LiNbO_3$ manufactured using a 308 nm excimer laser with mask projection techniques. Finally, Fig. 7.3c shows a three-dimensional (3D) structure in polyimide. This structure was created by applying a phase mask which modulates the intensity of the laser beam and thereby controls the amount of ablation depth of polyimide. From the above-mentioned machining results, the excimer laser has great potential in the machining of a variety of materials, especially in polymer material, such as polyimide (PI), polycarbonate (PC), poly(ethylene terephthalate) (PET), and poly(methyl methacrylate) (PMMA).

Due to the versatile applications of optical components, the optical characteristics of micro-structure need to be improved and applied at the wide wavelength band. Regular array structures or periodical patterns can longer satisfy the requirement of the advanced optical applications. It is very important to develop the manufacturing technology for the optical component of certain pattern used at different wavelengths.

The machining mechanism of femtosecond laser was firstly proposed by T. B. Norris in 1990 [6]. He proposed that the laser ablation occurred after multiple-photon absorption to break the chemical bond in material. The mechanism is different from thermally melting or vaporization by conventional laser.

Femtosecond laser is characterized by (1) cold or heatless machining mechanism, (2) high resolution that beats diffraction limit, (3) clean machined surface and high efficiency, (4) any materials, and (5) machining/ablating on the focal spot. Compared with the traditional laser, femtosecond laser can fabricate nanostructures or complicated 3D micro-structures.

According to the above characteristics of femtoseond laser, this innovative machining method can be applied to many fields, including photonic crystal fabrication by two-photon absorption [7], array structures by one laser pulse lithography with micro array lenses [8], sub-30 nm structure machined by femtosecond laser with near field system [9, 10], and grating structures by laser interference [11, 12].

For the numerical modeling, L. Jiang and H.-L. Tsai proposed the Gaussian formula to express the laser intensity, which was combined with the heat-transfer process and the free-electron distribution in bulk material, from 1993 to 1996 [13–18]. With the numerical modeling, the equations can be used to calculate the ablation shape by a laser pulse correctly.

7.2 Excimer Laser

7.2.1 Model Construction of Laser Dragging

The desired and designed 3D microstructures on the surface of materials can be created by laser dragging technique with appropriate mask design, precise motion of the stage and control of the laser operating parameters.

In the laser dragging, a complex mask pattern is projected onto the surface of the workpiece. The laser dragging can be divided into three categories, for example, both the mask and workpiece are stationary, either the mask or workpiece is moving, and both the mask and workpiece are moving synchronously while the laser is firing.

The micro-patterning such as micro-pyramid array, aspheric microlenses, and microlens array have been manufactured by laser micromachining with various dragging methods [19–24]. Figure 7.4 depicts the microlens is produced on PC material by dragging the polymer surface along multiple circular paths with the excimer laser [21, 22]. Moreover, array of microstructures fabricated on PC material by linear dragging twice in different direction is observed in Figs. 7.5 and 7.6

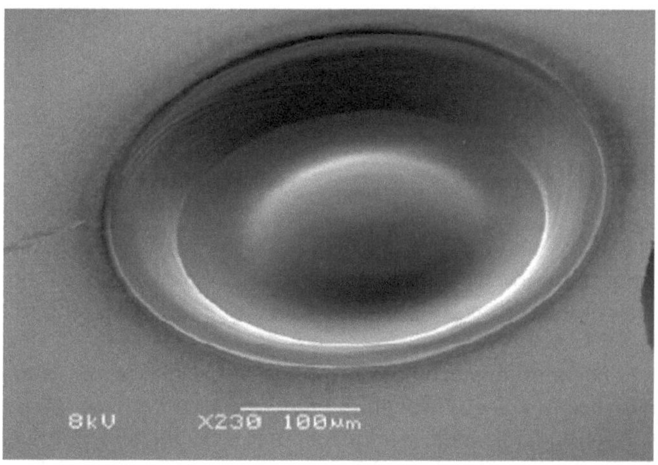

Fig. 7.4 SEM photograph of microlens made by dragging the polymer surface along multiple circular paths with excimer laser [1], reprinted with permission

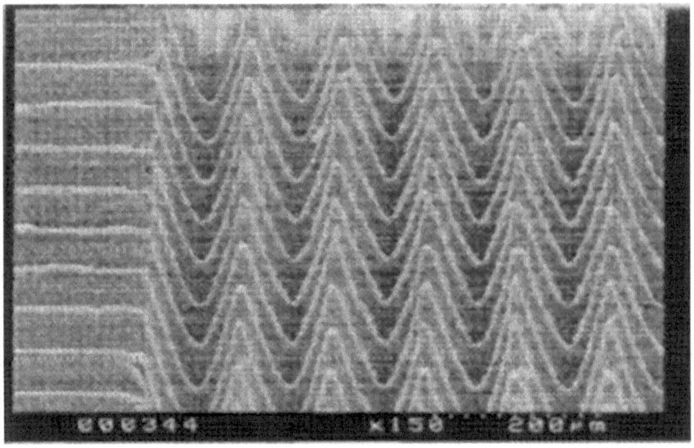

Fig. 7.5 SEM picture of array of micro pyramid [1], reprinted with permission

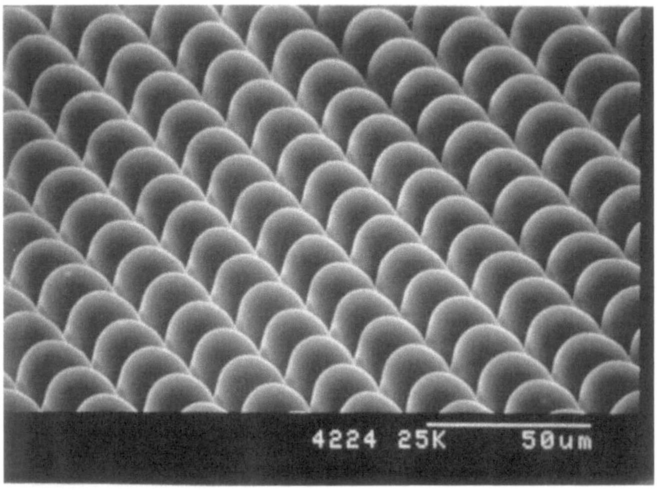

Fig. 7.6 SEM picture of array of microlenses [1], reprinted with permission

[19, 20, 23]. Furthermore, the convex and concave microlenses are fabricated on PC material by the circular dragging method with designed mask pattern as shown in Fig. 7.7 [24].

As the laser beam impinges on the elliptic, triangular or rectangular shape of the mask, the laser beam will go through the transparent portion of the mask onto the workpiece surface and ablate the workpiece. In this study, the workpiece moves at a constant velocity in one direction while the laser is firing. The KrF excimer laser dragging process with the triangular shape of the mask is illustrated in Fig. 7.8 [1].

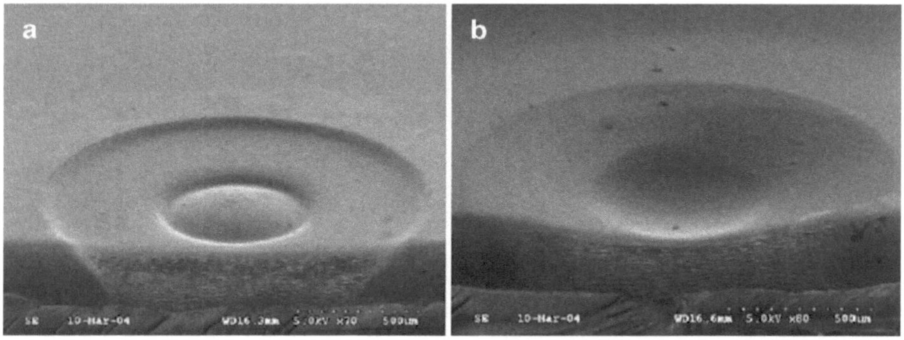

Fig. 7.7 SEM images for (**a**) convex microlens; (**b**) concave microlens [1], reprinted with permission

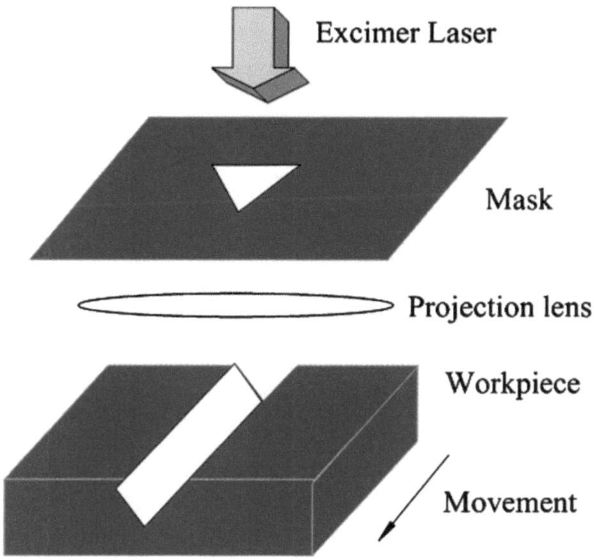

Fig. 7.8 Schematic view of laser dragging process [1], reprinted with permission

Since the ablation depth of the groove is related to the opening dimension of mask pattern during the dragging process, hence the profile of the groove pattern is dependent on the opening shape of the mask. In other words, the ablation depth increases as the dimension H of the mask pattern increases. Figure 7.9a [1] illustrates H is a function of x, and Fig. 7.9b [1] describes that the ablation depth D of the groove increases while the laser dragging velocity decreases in Y direction.

The operating parameters of the laser ablation process contain the opening of the mask pattern in the dragging direction, pulse number, workstation dragging

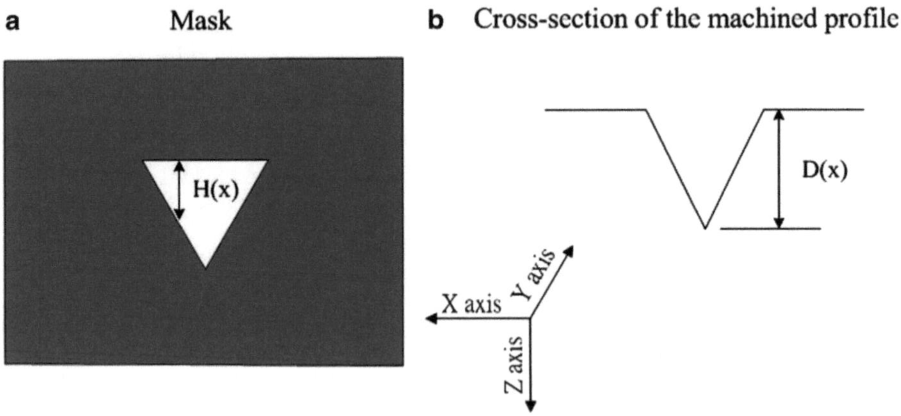

Fig. 7.9 Dragging mask and machined profile (**a**) opening dimension of the mask pattern $H(x)$; (**b**) ablation depth of machined profile $D(x)$ [1], reprinted with permission

velocity and pulse repetition rate. The relationship between these parameters can be expressed as [19]

$$f = \frac{S(x)}{H(x)/V} \tag{7.1}$$

where f is the pulse repetition rate (Hz), $S(x)$ is the laser pulse number of a function of x, $H(x)$ is the height of mask pattern of a function of x (μm), and V is the workstation dragging velocity (mm/min). Equation (7.1) can be rearranged as

$$H(x) = \frac{S(x) \times V}{f}. \tag{7.2}$$

Equation (7.2) reveals that $H(x)$ is proportional to $S(x)$ while V and f are kept constant. When larger H is exposed to the laser beam energy, deeper ablation depth of the workpiece will be produced. Based on the earlier findings, one finds that the ablation depth correlates closely with laser pulse number,

$$D(x) \propto S(x) \tag{7.3}$$

where $D(x)$ is the ablation depth of a function of x (μm). According to (7.2) and (7.3), the equation can be rearranged as

$$D(x) \propto \frac{f \times H(x)}{V}. \tag{7.4}$$

Equation (7.4) describes that $D(x)$ is proportional to $H(x)$ while V and f are kept constant. This means that the larger the opening of the mask pattern in the dragging direction, the deeper is the ablation depth.

From the previous literatures, one knows the main ablation mechanism for PC material through the excimer laser ablation at low fluence is regarded as a photochemical reaction [21, 25]. Besides, the characteristics namely reflectivity of PC material for the laser beam is also an influential factor in the ablation process. Consequently, the effects of the photochemical reaction and reflectivity of PC material upon the ablation depth are involved in the investigation. The equation is then developed to

$$D(x) \propto \frac{1}{\alpha} \ln\left[\frac{(1-R)\times F}{F_{TR}}\right] \times \frac{f \times H(x)}{V} \tag{7.5}$$

where α is an absorption coefficient determined from a curve fitting of the ablation rate vs. fluence; F is the fluence of the incident light on the surface of material (mJ/cm^2); F_{TR} is the threshold fluence; and R is the reflection coefficient.

Since the non-parallel laser beam passes through various sizes of the mask pattern, the energy density measured on the image plane by the energy detector differs. This causes the reduction of the ablation depth when decreasing the size of the mask pattern. Thus the effect of decreasing the size of the mask pattern on the ablation depth is also considered in the ablation process. The model can be rewritten as

$$D(x) \propto \frac{1}{\alpha} \ln\left[\frac{(1-R)\times F}{F_{TR}}\right] \times \frac{f \times H(x)}{V} \times \frac{d_{Af}}{d_{Ao}} \tag{7.6}$$

where d_{Ao} is the ablation depth with original size of the mask pattern; d_{Af} is the ablation depth with the decreasing size of the mask pattern.

Finally, the influences of various frequencies and various machining methods upon the ablation depth have to be taken into account. The proposed model for PC material removal in the current study is expressed as follows

$$D(x) = \frac{1}{\alpha} \ln\left[\frac{(1-R)\times F}{F_{TR}}\right] \times \frac{f \times H(x)}{V} \times \frac{d_{Af}}{d_{Ao}} \times \frac{f_s}{f_{20}} \times \frac{d_{scan}}{d_{fixed}} \tag{7.7}$$

where f_{s} stands for the frequency of 20, 40, 60, 80 Hz; d_{scan} is the ablation depth by the laser dragging process; d_{fixed} is the ablation depth by machining at fixed position.

The assumptions made in the proposed model are as follow.

1. Single-photon absorption process
2. No thermal damage
3. No plume shielding effect
4. Constant absorption coefficient
5. Uniform laser beam intensity

Under the assumption of uniform laser intensity through the opening of masks, the ablation depth on each position along the cross line indicates and determines the machined profile subject to the laser illumination. Hence, the current model, the simulation (Sect. 7.2.2) and experimental results (Sect. 7.2.3) of the ablation depth are adopted for the prediction and verification of machined profile in laser dragging.

The above-mentioned model of the ablation depth will be verified experimentally and discussed in Sect. 7.2.3.

The proposed model is also applied to the ArF excimer laser ablation. Due to the difference of optical delivery system and the wavelength between two excimer laser systems, the effects of various frequencies, various sizes of the mask pattern and the difference between laser dragging and machining at fixed location on the ablation depth have to be considered during the ArF laser dragging process. The details will be also described and discussed in Sect. 7.2.3.

7.2.2 Numerical Simulation of Dragged Profile

Numerical simulation is a tool which attempts to predict the behavior of the investigated system with a set of given operating parameters and initial conditions. The simulation would reduce the large quantity of the experiment. To effectively achieve the prediction of the laser-dragged groove profile of microstructure, a simulation method is applied in this study.

The procedure of the simulation is depicted briefly as follows.

Inputting the initial conditions, such as material parameters, laser operating parameters and mask parameters;

Constructing the adequate equation of the change of the dimension $H(x)$ of the mask in X direction based on various shapes of the mask;

Calculating the ablation depth of groove profile in one direction according to the constructed equation and initial conditions;

Sketching the profile of groove pattern produced by the single dragging with various shapes of the masks;

Calculating the machined depth after the overlap dragging in different direction;

Drawing the shape of microstructures based on the cross dragging at 0° and 45°, 0° and 60°, and 0° and 90°.

7.2.2.1 Single Dragging

As shown in Fig. 7.10 [1], there are three shapes of masks used for the simulation process, namely elliptic, triangular and rectangular shape. Their sizes are $0.2 \times 0.2\ \text{mm}^2$, $0.1 \times 1\ \text{mm}^2$ and $1 \times 20\ \text{mm}^2$ respectively.

The simulation of the dragging in one direction using three various shapes of the masks and various laser operating parameters are first executed. Figure 7.11 [1] displays the elliptic profile with two sets of various laser operating conditions; Fig. 7.12 [1] depicts the triangular profile with two sets of various laser operating conditions; and Fig. 7.13 [1] presents the rectangular profile with two sets of various

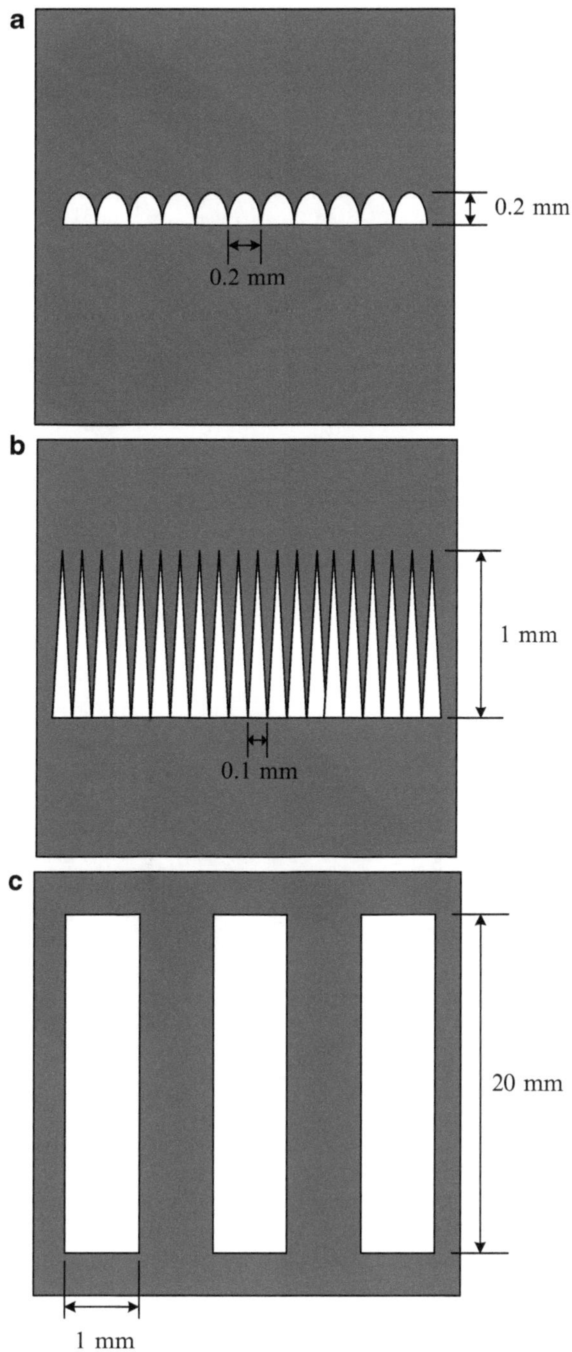

Fig. 7.10 Three geometric shapes of masks used for the simulation process (**a**) Elliptic mask; (**b**) Triangular mask; (**c**) Rectangular mask [1], reprinted with permission

a

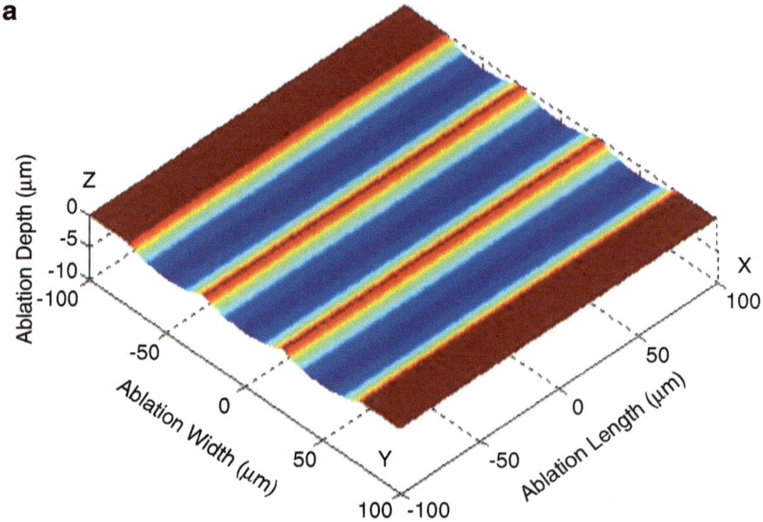

b

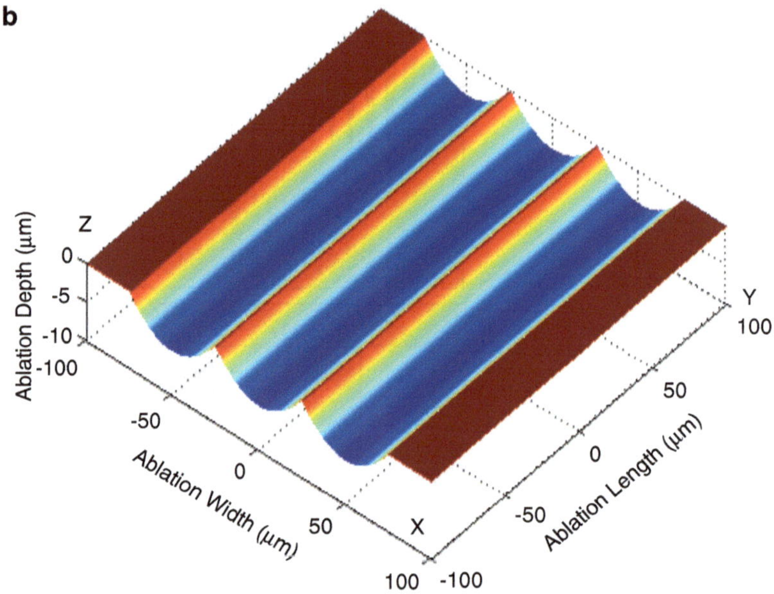

Fig. 7.11 Dragging with elliptic mask in one direction [1], reprinted with permission. (**a**) $F = 300$ mJ/cm^2, $V = 3$ mm/min, $f = 40$ Hz, $A = 1.96 \times 10-^3$ mm^2. (**b**) $F = 365$ mJ/cm^2, $V = 2$ mm/min, $f = 60$ Hz, $A = 1.96 \times 10^{-3}$ mm^2

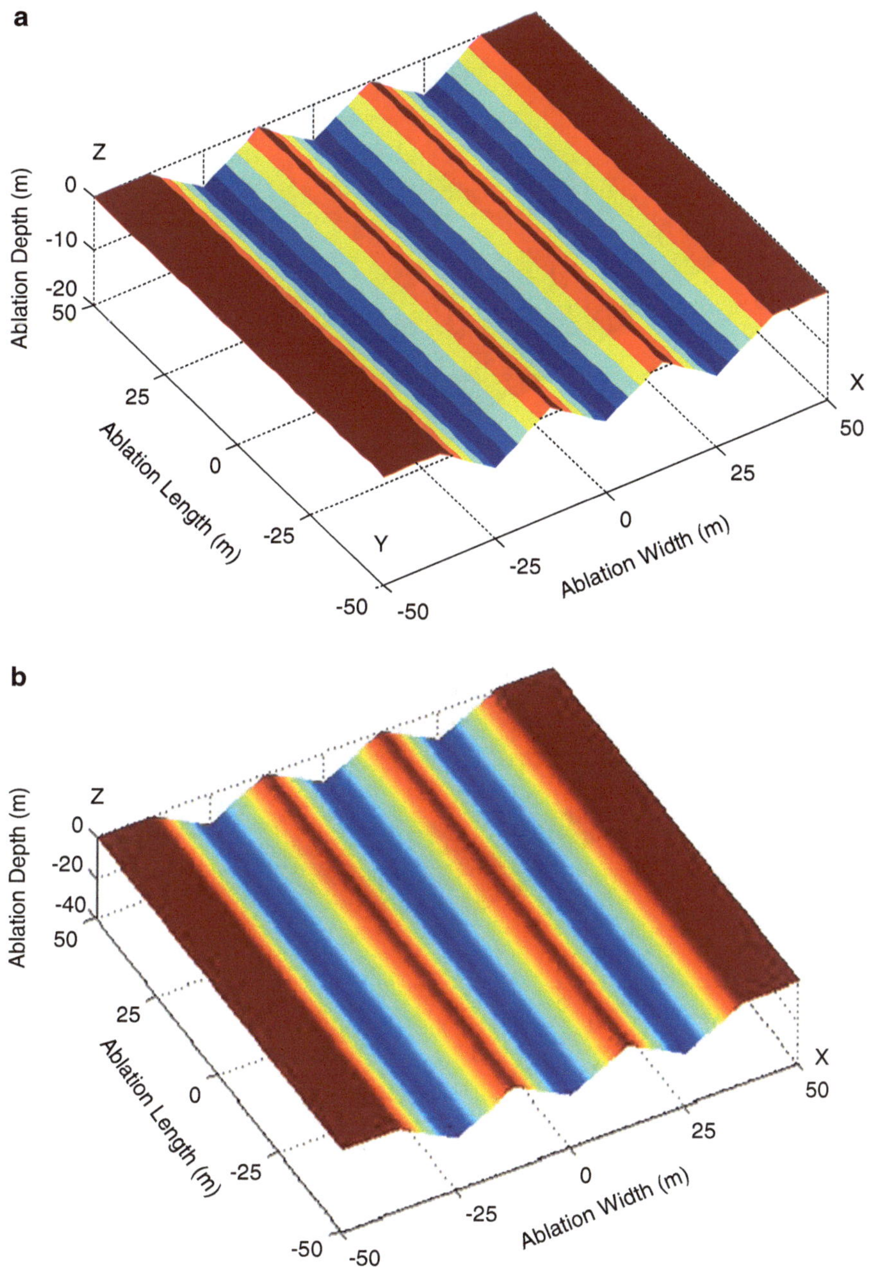

Fig. 7.12 Dragging with triangular mask in one direction [1], reprinted with permission. (**a**) $F = 300$ mJ/cm^2, $V = 4$ mm/min, $f = 60$ Hz, $A = 3.125 \times 10^{-3}$ mm^2. (**b**) $F = 330$ mJ/cm^2, $V = 2$ mm/min, $f = 60$ Hz, $A = 3.125 \times 10^{-3}$ mm^2

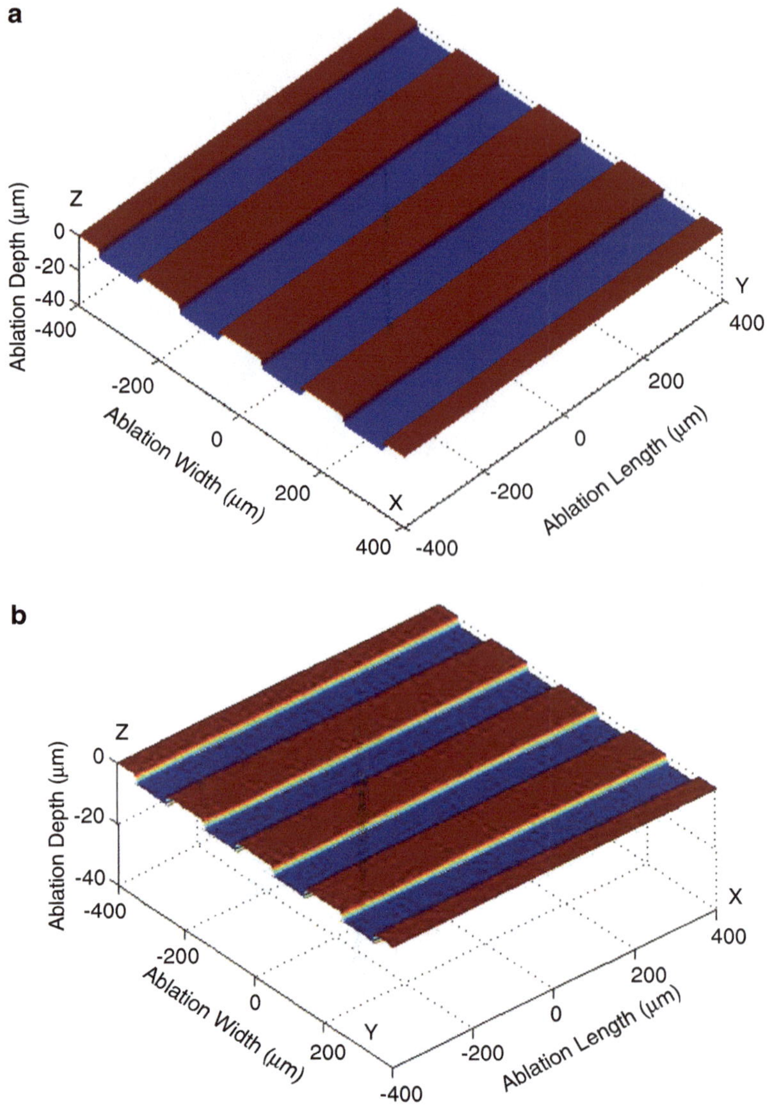

Fig. 7.13 Dragging with rectangular mask in one direction [1], reprinted with permission. (**a**) $F = 270$ mJ/cm^2, $V = 3$ mm/s, $f = 60$ Hz, $A = 0.1 \times 2$ mm^2. (**b**) $F = 300$ mJ/cm^2, $V = 4$ mm/s, $f = 40$ Hz, $A = 0.1 \times 2$ mm^2

laser operating conditions. The deepest groove depths in Figs. 7.11–7.13 are listed in Table 7.3. The results not only demonstrate that the groove profile can be well simulated by the proposed method, but also reveal that there is a strong correlation between the ablation depth and laser operating parameters.

Table 7.3 The groove depth obtained by single dragging with three various shapes of the masks and various laser operating conditions

Mask shape	Operating conditions	Depth
Elliptic mask	$F = 300$ mJ/cm^2, $V = 3$ mm/min, $f = 40$ Hz, $A = 1.96 \times 10^{-3}$ mm^2	1.49 μm
	$F = 365$ mJ/cm^2, $V = 2$ mm/min, $f = 60$ Hz, $A = 1.96 \times 10^{-3}$ mm^2	4.17 μm
Triangular mask	$F = 300$ mJ/cm^2, $V = 4$ mm/min, $f = 60$ Hz, $A = 3.125 \times 10^{-3}$ mm^2	6.7 μm
	$F = 330$ mJ/cm^2, $V = 2$ mm/min, $f = 60$ Hz, $A = 3.125 \times 10^{-3}$ mm^2	14.6 μm
Rectangular mask	$F = 270$ mJ/cm^2, $V = 3$ mm/s, $f = 60$ Hz, $A = 0.1 \times 2$ mm^2	2.78 μm
	$F = 300$ mJ/cm^2, $V = 4$ mm/s, $f = 40$ Hz, $A = 0.1 \times 2$ mm^2	5.2 μm

7.2.2.2 Cross Dragging

In cross dragging, the workpiece moves in one direction to form a machined profile and then repeats the dragging operation at another angle. When the two machined profiles cross each other, a certain 3D microstructure can be produced.

The simulation of dragging twice in different directions of 45°, 60° and 90° is performed. Examples of the cross dragging using three various shapes of mask and various angles are shown in Figs. 7.14–7.19 [1]. Figures 7.14 and 7.15 depict the simulated profile by cross dragging with elliptic mask and various laser operating conditions; Figs. 7.16 and 7.17 describe the simulated profile by cross dragging with triangular mask and various laser operating conditions; and Figs. 7.18 and 7.19 show the simulated profile by cross dragging with rectangular mask and various laser operating conditions The deepest cross depths in Figs. 7.14–7.19 are listed in Table 7.4.

From the simulation results, one finds that given the laser operating parameters and material parameters, as well as mask parameters, for example, fluence, threshold fluence, pulse repetition rate, workstation dragging velocity, the absorption coefficient, reflection of PC material and mask size, the desired microstructures can be effectively predicted with the help of the simulation.

7.2.3 *Experimental Verification*

This section introduces the experimental work to verify the proposed model and simulation discussed in previous chapters.

7.2.3.1 Experimental Setup

Two excimer laser systems are used for the laser experiments. One is the KrF excimer laser system, the other is the ArF excimer laser system. The experimental apparatus and related information will be described in detail in the following section.

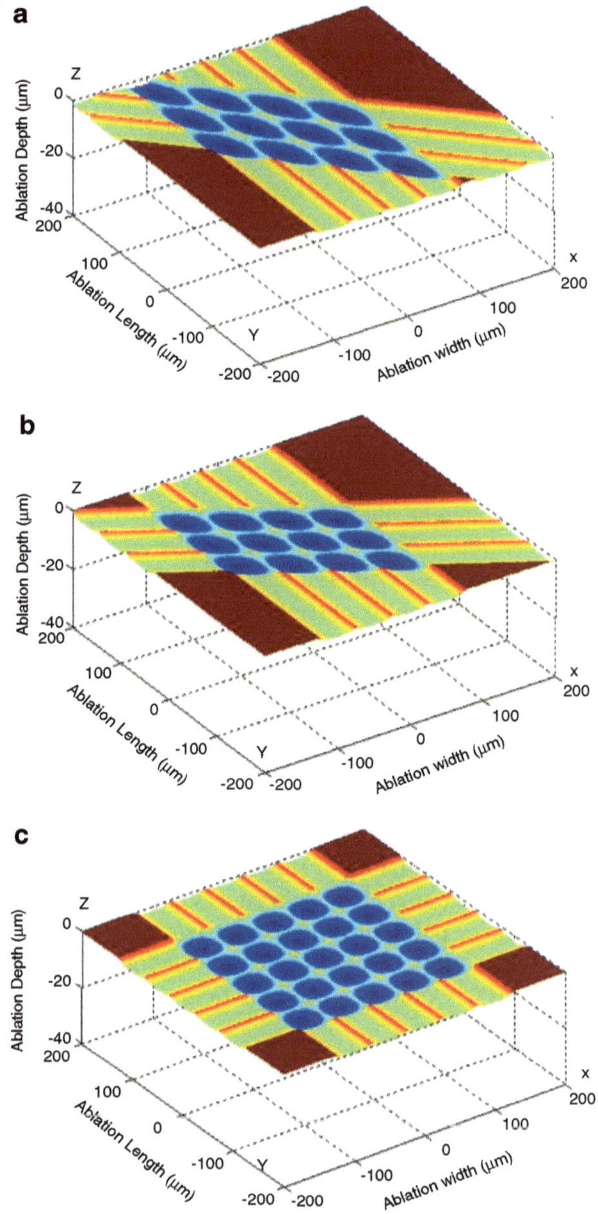

Fig. 7.14 Simulated cross dragging with elliptic mask ($F = 300$ mJ/cm^2, $V = 3$ mm/min, $f = 40$ Hz, $A = 1.96 \times 10^{-3}$ mm^2) [1], reprinted with permission. (**a**) 0° and 45°. (**b**) 0° and 60°. (**c**) 0° and 90°

Krypton Fluoride Excimer Laser System

The KrF excimer laser system can be divided into four modules, for example, a KrF excimer laser source, an optical delivery system, a moving stage, and a mask holder as shown in Fig. 7.20 [1]. The KrF excimer laser source is made by Lambda Physik.

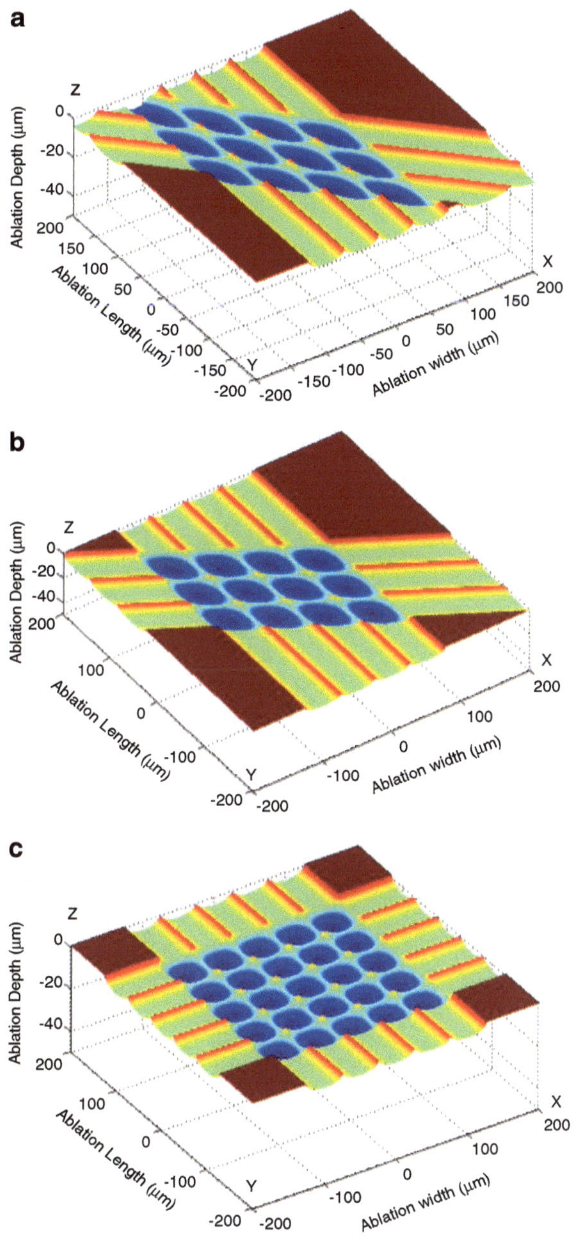

Fig. 7.15 Simulated cross dragging with elliptic mask ($F = 365$ mJ/cm^2, $V = 2$ mm/min, $f = 60$ Hz, $A = 1.96 \times 10^{-3}$ mm^2) [1], reprinted with permission. (**a**) 0° and 45°. (**b**) 0° and 60°. (**c**) 0° and 90°

Its wavelength, maximum pulse energy, pulse duration, and maximum repetition rate are 248 nm, 350 mJ, 25 ns, and 100 Hz respectively. The laser output mode contains constant discharge voltage and constant pulse energy. To maintain stable laser energy output during the ablation process, the experiments are operated at the constant pulse energy mode in the work.

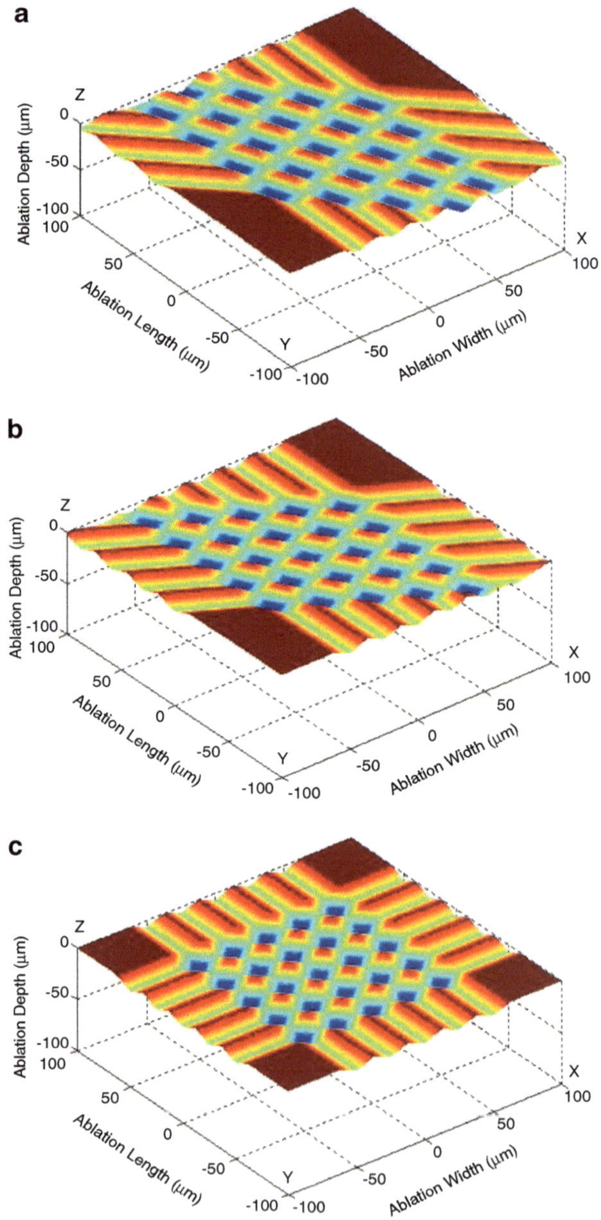

Fig. 7.16 Simulated cross dragging with triangular mask ($F = 300$ mJ/cm^2, $V = 4$ mm/min, $f = 60$ Hz, $A = 3.125 \times 10^{-3}$ mm^2) [1], reprinted with permission. (**a**) 0° and 45°. (**b**) 0° and 60°. (**c**) 0° and 90°

In general, there are two main methods used in the laser micromachining, namely direct writing and mask projection. The mask projection method in the study is applied to transfer patterns from a mask onto the workpiece. In the mask projection system, the laser beam is shaped and homogenized by use of some lenses

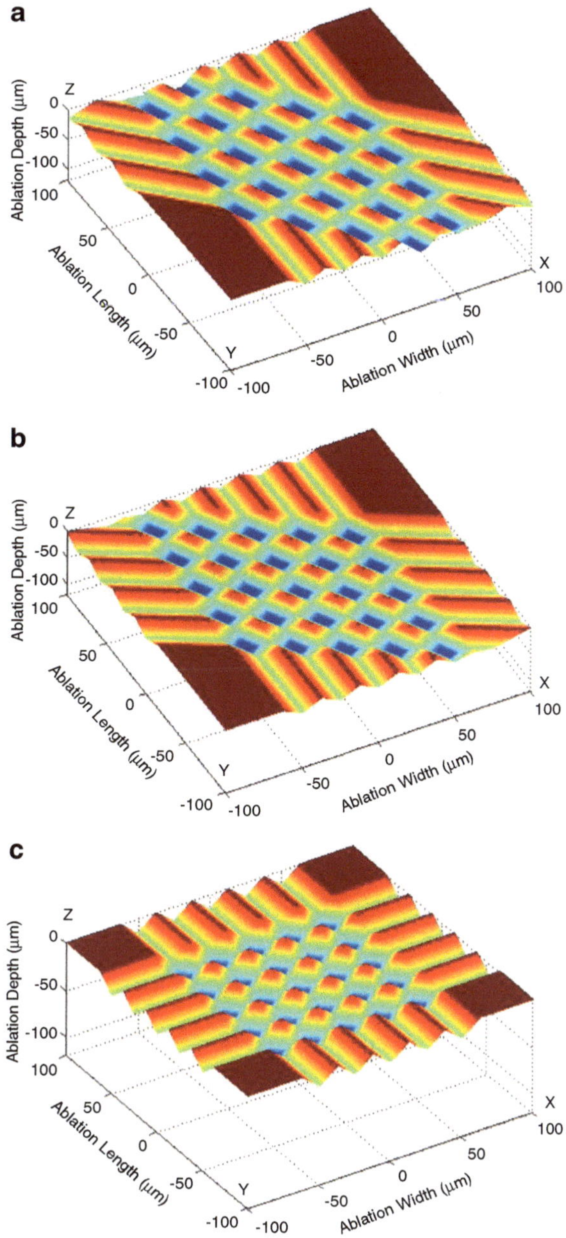

Fig. 7.17 Simulated cross dragging with triangular mask ($F = 330$ mJ/cm^2, $V = 2$ mm/min, $f = 60$ Hz, $A = 3.125 \times 10^{-3}$ mm^2) [1], reprinted with permission. (a) 0° and 45°. (b) 0° and 60°. (c) 0° and 90°

and a 6×6 fixed-array homogenizer to form a 10×10 mm^2 laser beam illuminating on the mask plane with the intensity variation less than ±5% root mean square (RMS). This ensures that the laser beam is capable of achieving a uniform energy distribution on the workpiece because the variations of energy lead

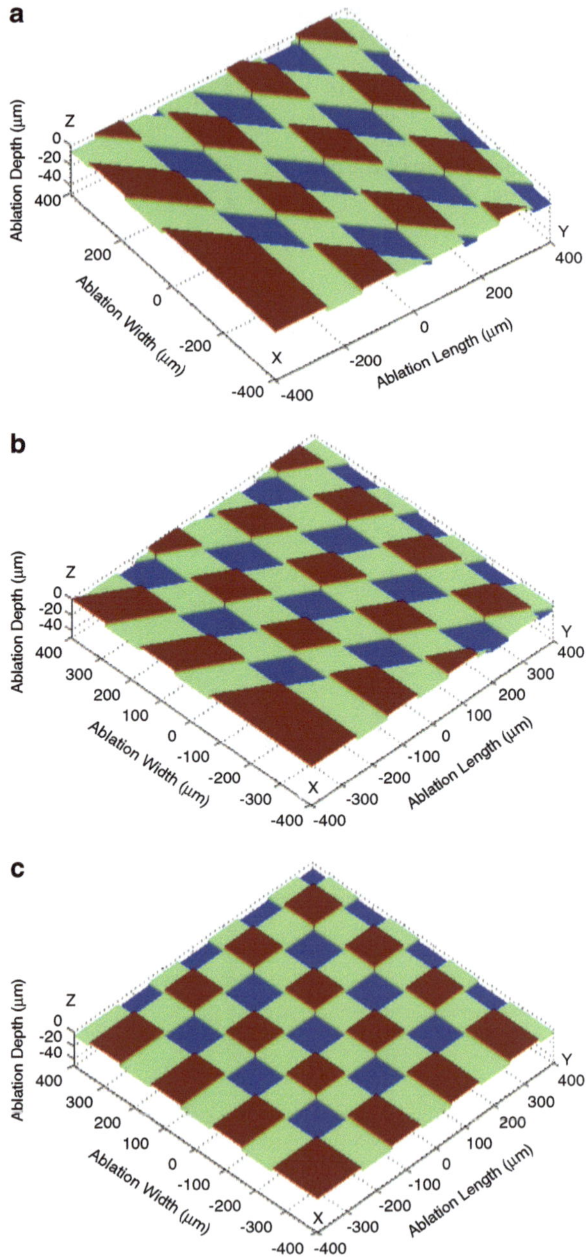

Fig. 7.18 Simulated cross dragging with rectangular mask ($F = 270$ mJ/cm^2, $V = 3$ mm/s, $f = 60$ Hz, $A = 0.1 \times 2$ mm^2) [1], reprinted with permission. (**a**) 0° and 45°. (**b**) 0° and 60°. (**c**) 0° and 90°

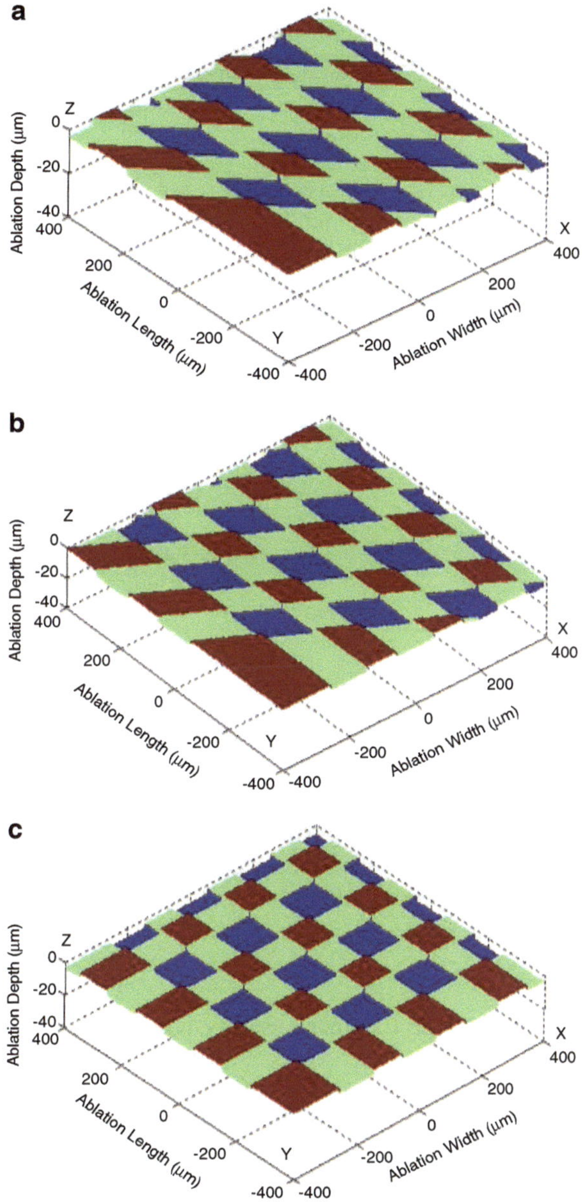

Fig. 7.19 Simulated cross dragging with rectangular mask ($F = 300$ mJ/cm^2, $V = 4$ mm/s, $f = 40$ Hz, $A = 0.1 \times 2$ mm^2) [1], reprinted with permission. (**a**) 0° and 45°. (**b**) 0° and 60°. (**c**) 0° and 90°

to the variations of the ablation depth on the machined workpiece surface during the laser ablation.

The mask pattern is projected onto the workpiece surface by utilizing a high-resolution projection lens with 4× demagnification factor in the optical delivery system. The beam size after going through projection lens is 0.25×0.25 cm^2 on

Table 7.4 The depth obtained by cross dragging with three various shapes of the masks and various laser operating conditions

Mask shape	Angle	Operating conditions	Depth
Elliptic mask	0°/45° 0°/60° 0°/90°	$F = 300$ mJ/cm^2, $V = 3$ mm/min, $f = 40$ Hz, $A = 1.96 \times 10{-}^3$ mm^2	2.98 μm
	0°/45° 0°/60° 0°/90°	$F = 365$ mJ/cm^2, $V = 2$ mm/min, $f = 60$ Hz, $A = 1.96 \times 10{-}^3$ mm^2	8.32 μm
Triangular mask	0°/45° 0°/60° 0°/90°	$F = 300$ mJ/cm^2, $V = 4$ mm/min, $f = 60$ Hz, $A = 3.125 \times 10{-}^3$ mm^2	13.4 μm
	0°/45° 0°/60° 0°/90°	$F = 330$ mJ/cm^2, $V = 2$ mm/min, $f = 60$ Hz, $A = 3.125 \times 10{-}^3$ mm^2	29.2 μm
Rectangular mask	0°/45° 0°/60° 0°/90°	$F = 270$ mJ/cm^2, $V = 3$ mm/s, $f = 60$ Hz, $A = 0.1 \times 2$ mm^2	5.56 μm
	0°/45° 0°/60° 0°/90°	$F = 300$ mJ/cm^2, $V = 4$ mm/s, $f = 40$ Hz, $A = 0.1 \times 2$ mm^2	10.4 μm

the workpiece. The workpiece is placed on a precision air-bearing stage. This stage can move in X, Y and Z directions and rotate about the θ-axis. To obtain higher operating quality, the position in Z axis should be adjusted slightly according to the thickness of each workpiece. During operation, all experiments are conducted in clean room of class 1000.

Argon Fluoride Excimer Laser System

The ArF excimer laser system consists of an ArF excimer laser source, an optical delivery system, a moving stage, and a mask holder as well. The schematic apparatus of the ArF excimer laser system is illustrated in Fig. 7.21 [1], and the properties of the system are listed in Table 7.5. In the system, the laser beam is shaped by lenses to form the size of 6×6 mm^2 and then uniformly delivered to the mask plane with the intensity variation less than ±5% RMS. After the laser beam passes through the projection lens with 10× demagnification factor, the size of the produced pattern on the workpiece surface contracts ten times compared to the size of the mask pattern. The workpiece is mounted on a precision XY stage. The irradiated surface is perpendicular to the direction of incident laser beam. The Z-axis is readjusted when varying the thickness of the workpiece.

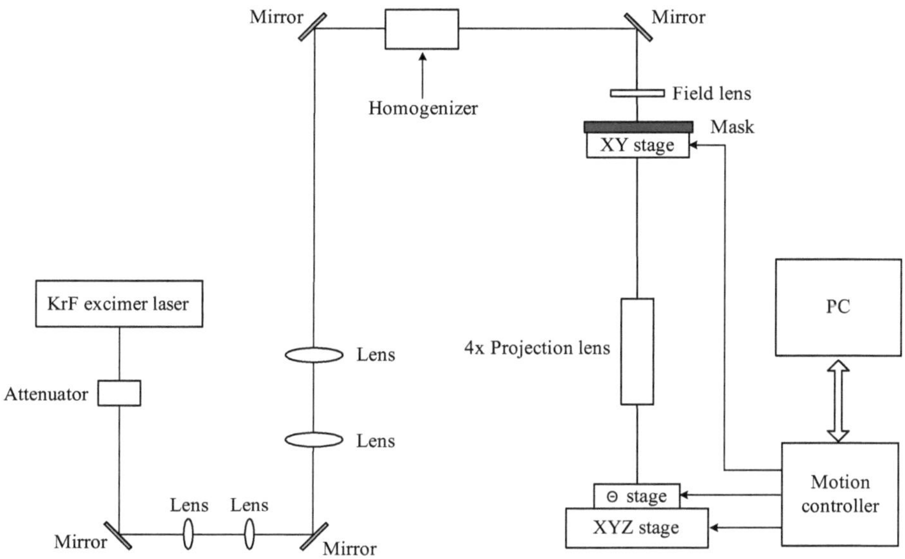

Fig. 7.20 Scheme of the KrF excimer laser system [1], reprinted with permission

7.2.3.2 Workpiece Preparation

The polycarbonate (PC) sheet with 1 mm thickness made by Goodfellow is used as the experimental material in this work because of its excellent ablation characteristics at 193 and 248 nm of the wavelength, and the good optical properties, as well as suitable for being utilized in the lamp.

The workpiece surface is cleaned by nitrogen guns prior to the experiments because the high-pressure nitrogen is an inert gas and does not easily cause other contaminations on the workpiece. After the experiments, the workpiece placed in isopropanol is oscillated by the ultrasonic oscillator for about 15 min to remove produced carbon residues or debris on the workpiece surface.

7.2.3.3 Mask Shape

To verify the validity of the proposed model of (7.7), elliptic and triangular shapes of mask are applied to the experiment of the KrF excimer laser machining, while rectangular shape of mask is applied to the experiment of the ArF excimer laser machining. These three shapes of masks are shown in Fig. 7.10 [1].

The mask manufactured by Taiwan Mask Corporation is made up of chromium (Cr) coated on quartz, which possesses higher transmission efficiency than that is made of Cr coated on glass. The Cr thickness is about 0.2 mm, while the total thickness of the mask is approximately 2 mm. The mask pattern is sketched by the

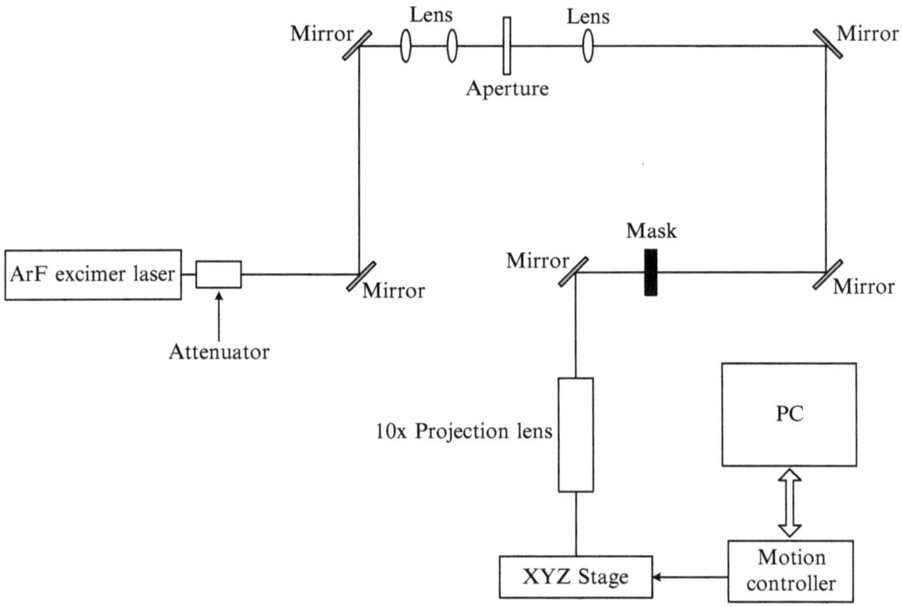

Fig. 7.21 Scheme of the ArF excimer laser system [1], reprinted with permission

Table 7.5 Characteristics of two excimer laser systems

	KrF	ArF
Laser model	Compex 110 (Lambda Physik)	Ex 50 (Gam Laser Inc.)
Max. pulse energy (mJ)	350	35.4
Pulse duration (ns)	25	16–20
Max. repetition rate (Hz)	100	250
Wavelength (nm)	248	193
Photon energy (eV)	5	6.4

Autocad software. The mask sizes are respectively $12.7 \times 12.7\ \mathrm{cm}^2$ and $2.54 \times 2.54\ \mathrm{cm}^2$ based on the sizes of the mask holder in the two KrF and ArF excimer laser systems.

7.2.3.4 Experimental Parameters

In the laser ablation process, the laser operating parameters include the dragging velocity (V), pulse repetition rate (f), pulse number (S), fluence (F), pulse duration, width of the mask pattern (W), and height of the mask pattern in the dragging direction (H). They are considered as experimental variables to generate various ablation profile and desired shape. The desired or designed shape of microstructure can be obtained effectively by adjusting the laser operating parameters and mask parameters.

7.2.3.5 Measurement

The ablation depth is measured by a surface profiler or WYKO optical profiler. The WYKO optical profiler provides a fast and accurate non-contact 3D surface measurement that measures the heights from 0.1 nm to several millimeters with a resolution down to 0.1 nm. The WYKO takes advantage of the principle of optical interference to measure the surface profile of workpiece. Noted that the acquired data are incomplete and inaccurate when the reflection light is insufficient or the curvature of the material surface is large. Moreover, the 3D images of the machined profile are captured by dragging electron microscope (SEM), which was operated at voltages in the range of 10–15 kV.

7.2.3.6 Experimental Results and Discussion

Ablation Depth

KrF Excimer Laser Experiments

A large number of experiments have been conducted to verify the proposed model of the ablation profile. The first experiment is to analyze the influence of varying fluence on the ablation rate for PC material. As illustrated in Fig. 7.22 [1], one learns that the ablation rate increases with increasing fluence. The threshold fluence and absorption coefficient of the material are obtained based on the formula of photochemical reaction presented in 1984 [26]. These values are 125 mJ/cm^2 and 8.5 μm^{-1} respectively.

The second experiment focuses on the correlation between the ablation depth and pulse number. Figure 7.23 [1] illustrates that greater laser pulse number and higher repetition rate result in deeper ablation depth when the fluence is kept constant. It is worth of noting that the relationship between the ablation depth and pulse number shows an excellent linearity.

The third experiment observes the relation of the ablation depth vs. reduction of size of the mask opening. The result shown in Fig. 7.24 reveals that the ablation depth decreases with decreasing the size of the mask. The reason is due to the difference of the energy absorbed on the workpiece surface. Less energy is absorbed on the workpiece surface when the laser beam goes through smaller opening of the mask, whereas more energy is absorbed on the workpiece surface when the laser beam goes through larger opening of the mask.

Figure 7.25 [1] compares the difference of the ablation depth produced by both the dragging method and machining at fixed position. The solid line stands for the ablation depth produced by the laser dragging method, while the dotted line

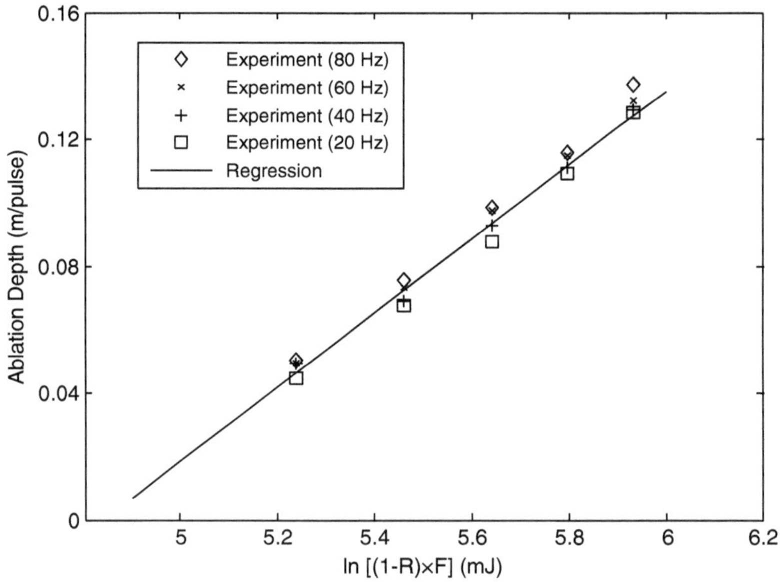

Fig. 7.22 KrF experimental ablation rate varying with fluence [1], reprinted with permission

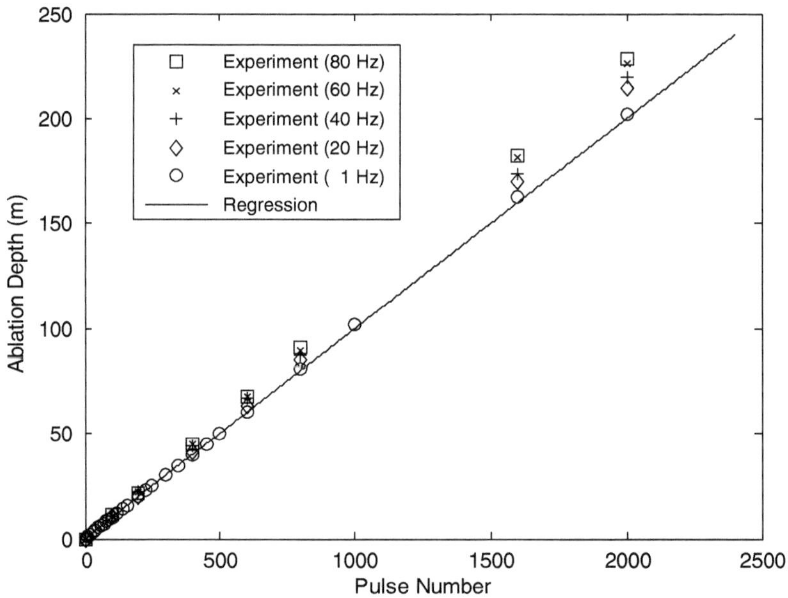

Fig. 7.23 KrF experimental ablation depth varying with pulse number [1], reprinted with permission

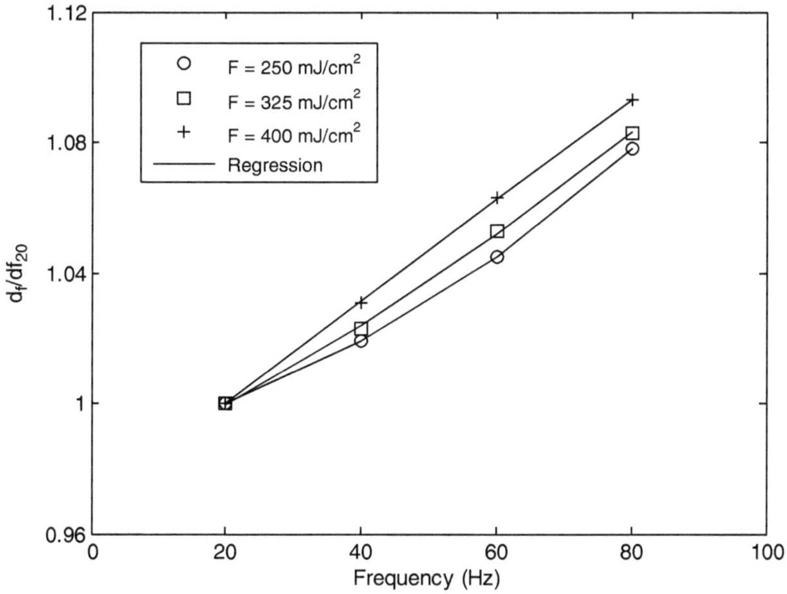

Fig. 7.24 Ratio of KrF experimental ablation depth varying with frequency [1]

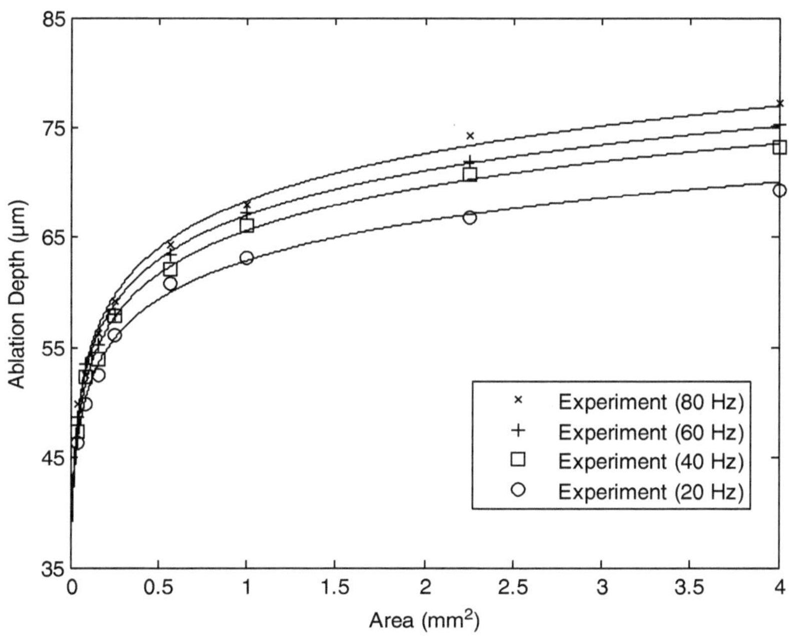

Fig. 7.25 KrF experimental ablation depth varying with area of mask opening [1], reprinted with permission

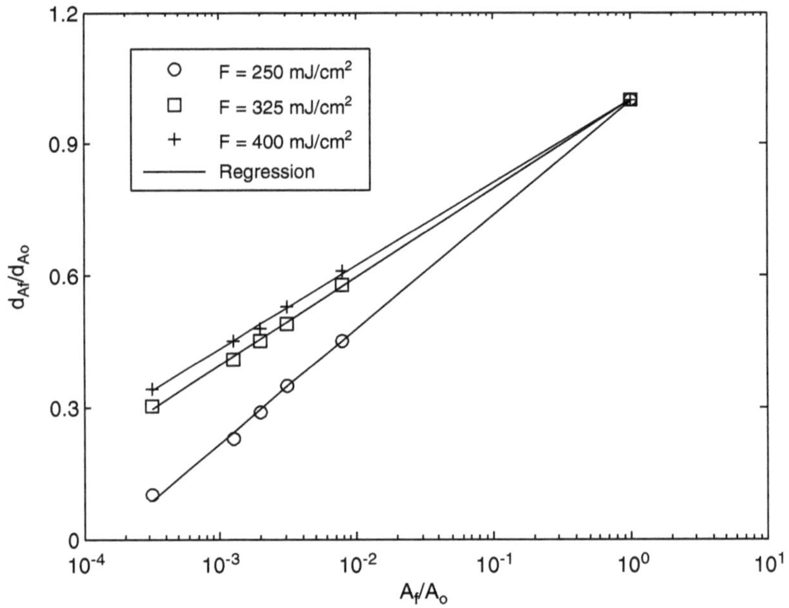

Fig. 7.26 Ratio of KrF experimental ablation depth varying with the variation of the ratio of area ($Ao = 1$ mm^2) [1], reprinted with permission

stands for the ablation depth produced by machining at fixed position. The reason causing the difference between the two methods is attributed to the effect of non-uniform beam distribution. The beam energy distribution is approximately a trapezoid impinging onto the workpiece surface. Figure 7.26a [1] describes that the energy absorbed at point A is a multiple of E when the workpiece is stationary, while Fig. 7.26b [1] characterizes the energy absorbed at point A as an $E(x)$ when the workpiece moves with a regular velocity in X axis. For the reason, the energy absorbed by the laser dragging is less than that by machining at fixed position. Correspondingly, the ablation depth produced by dragging method is less than that by machining at fixed position. The above-mentioned factor which is correlated with the ablation depth is involved in this study to correct the proposed model during the laser machining, namely the ratio of $d_{\text{drag.}}/d_{\text{fixed}}$. A good agreement of the proposed model with the experiment is showed in Fig. 7.27 [1].

ArF Excimer Laser Experiments

A few experiments are carried out to establish the effective model applied to the machining by an ArF excimer laser. The experimental parameters and results are shown in Table 7.6 and in Figs. 7.28–7.32 [1] respectively. From Fig. 7.32 [1], the

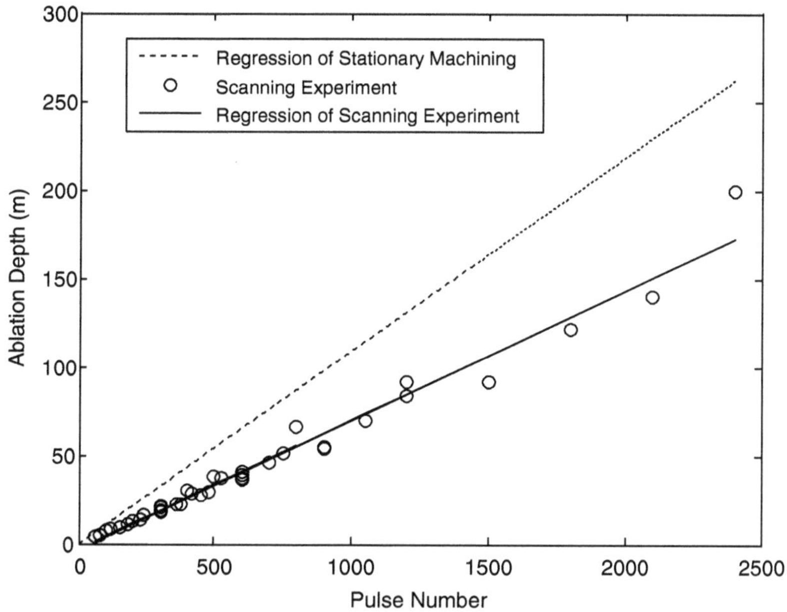

Fig. 7.27 KrF experimental ablation depth produced by dragging method and machining at fixed position [1], reprinted with permission

Table 7.6 ArF experimental parameters of laser ablation with rectangular shape of mask

	First set[a]	Second set[a]	Third set[a]	Fourth set[b]
F (mJ/cm^2)	116, 172, 228, 302, 344, 385, 448, 498	378	378	300, 422, 548, 600
f (Hz)	40	40, 60, 80, 100	40	40, 60, 80, 100
S	200	100, 200, 300, 400, 500, 600, 700, 800, 900, 1000	100, 200, 500 800, 1000	Obtained by the calculated of (3.1)
A (mm^2)	5 × 5	5 × 5	1 × 1, 2 × 2, 3 × 3, 4 × 4, 5 × 5	1 × 20
V (mm/min)	–	–	–	2.4, 4.8, 7.2, 9.6, 12

[a]Stationary machining
[b]Dragging machining

values of the threshold fluence and absorption coefficient of the material are obtained according to the formula of photolysis. They are 79 mJ/cm^2 and 10 μm^{-1} respectively.

Figure 7.29 [1] describes that greater laser pulse number leads to deeper ablation depth at constant fluence, while the influence of various frequencies on the ablation

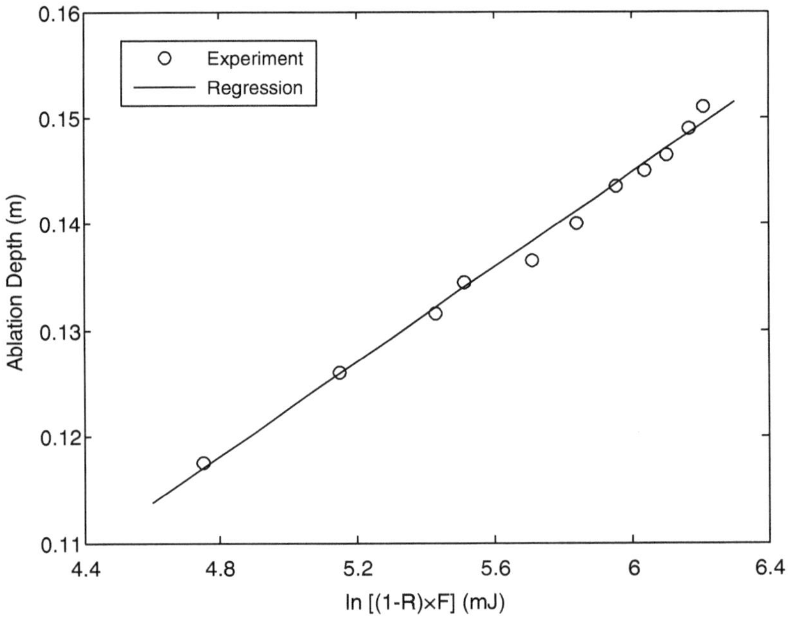

Fig. 7.28 ArF experimental ablation rate varying with fluence [1], reprinted with permission

depth is insignificant. As the fluences measured on the workpiece surface are quite approximate after laser beam goes through various sizes of the mask opening, the ablation depth varies slight as illustrated in Fig. 7.30 [1].

Figure 7.31 [1] gives the correlation between the ablation depth produced by both dragging method and machining at fixed position. The reason causing slight error between the two methods is also due to the difference of beam energy distribution, and it is the same as previous KrF experiments. Figure 7.32 [1] demonstrates the agreement of the proposed model with the experiment. Based on the results, the model used in the ArF excimer laser system can be expressed as

$$D(x) = \frac{1}{\alpha} \ln \left[\frac{(1-R) \times F}{F_{TR}} \right] \times \frac{f \times H(x)}{V} \times \frac{d_{drag}}{d_{fixed}}. \tag{7.8}$$

Based on the above-mentioned experiments, one finds that the proposed model used for the ArF excimer laser system can efficiently predict the ablation depth of PC material.

Machined Profile by Single Dragging

In the study, elliptic and triangular shapes of masks are used for the KrF experiment, while rectangular shape of mask is used for the ArF experiment. The groove

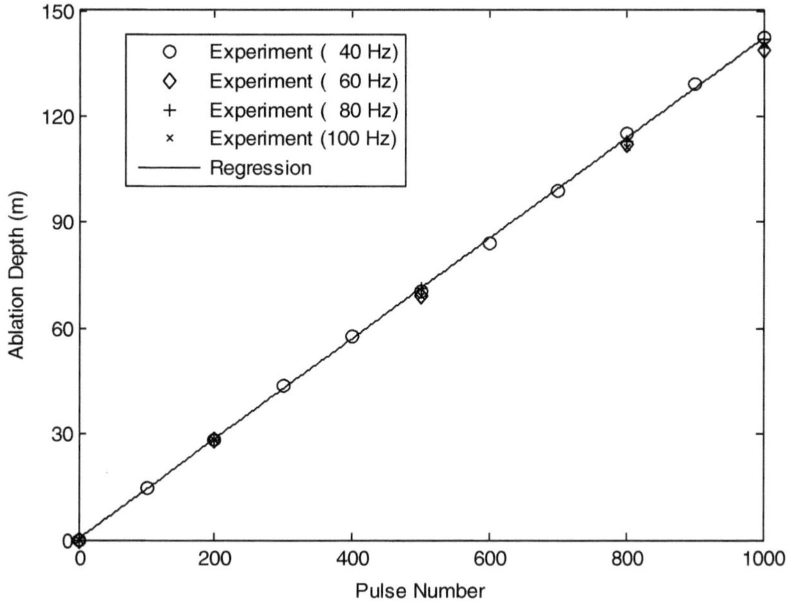

Fig. 7.29 ArF experimental ablation depth varying with pulse number [1], reprinted with permission

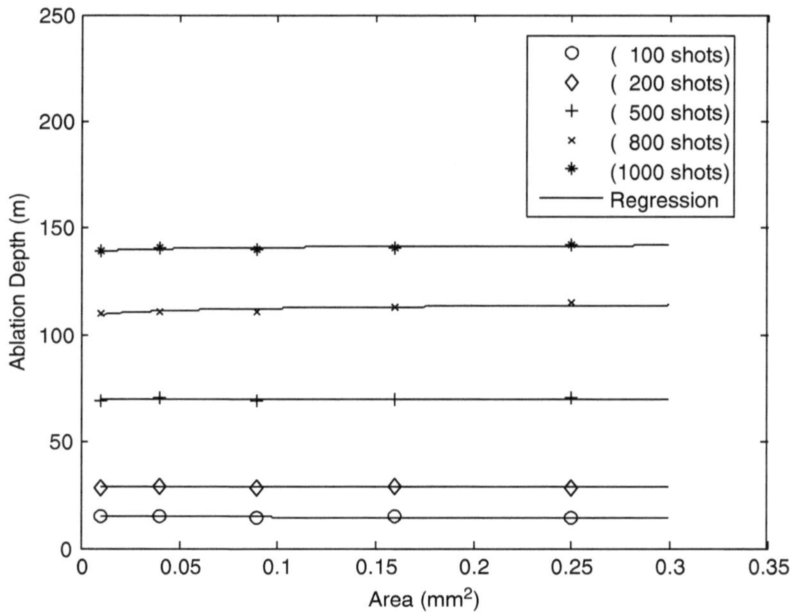

Fig. 7.30 ArF experimental ablation depth vs. various area of mask opening [1], reprinted with permission

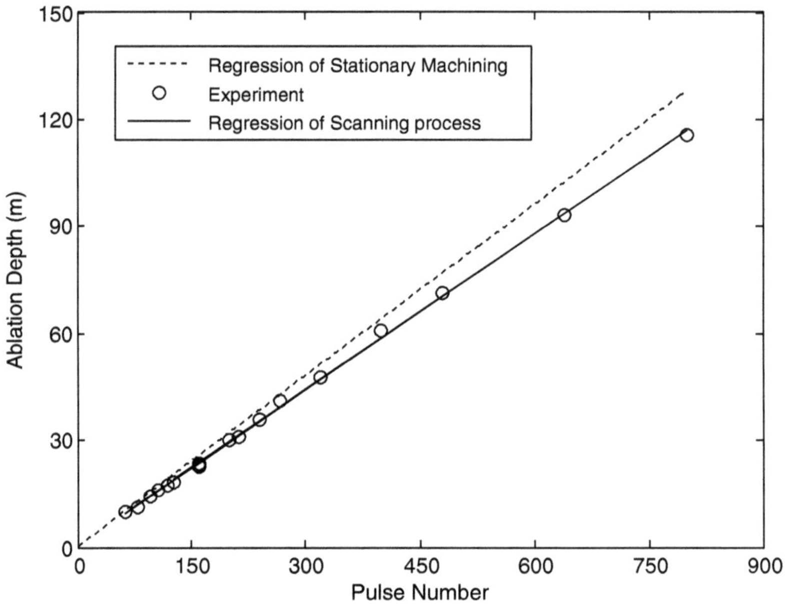

Fig. 7.31 ArF experimental ablation depth produced by dragging or machining at fixed position [1], reprinted with permission

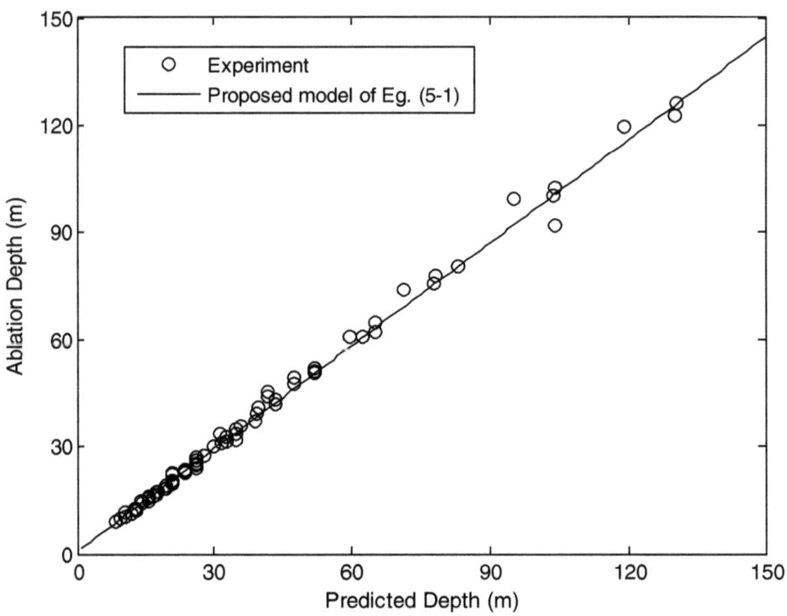

Fig. 7.32 ArF experimental verification of the proposed model [1], reprinted with permission

profiles by single dragging with the three shapes of masks are shown in Figs. 7.33–7.35 [1]. The groove depths in Figs. 7.37–7.35 measured by the WYKO are 4.4 μm, 13.9 μm and 5 μm respectively.

Table 7.7 lists the experimental and simulated depths of single dragging by utilizing various shapes of masks and various laser operating conditions. After comparing the experimental value with the predicted value, a good agreement with deviation under 5% can be verified. The deviation is defined as follows

$$Deviation = \frac{|Experimental\,value - Predicted\,value|}{Experimental\,value} \times 100\%. \tag{7.9}$$

Machined Profile by Cross Dragging

Figures 7.36–7.38 [1] display the machined results of cross dragging with various angles. The average depths in Figs. 7.36–7.38 are 8.1 μm, 26 μm and 9.6 μm respectively. Table 7.8 depicts the experimental and simulated depths of cross dragging by utilizing various shapes of masks and various laser operating conditions. The largest deviation of the cross depth between the experiment and prediction is around 13.8%. The reason causing larger deviation in the cross dragging than in the single dragging is attributed to the defocus effect occurred on the workpiece surface during the second dragging.

7.3 Sub-Micron-Structure Machining on Silicon by Femtosecond Laser

7.3.1 Experiment Setup

Direct writing system of femtosecond laser with the central wavelength of 800 nm, pulse duration of 120 fs, repetition rate of 1 kHz, and 10× object lens with NA 0.3 is used in the current study. The schematic diagram is shown in Fig. 7.39 [27]. Photoresist of SU8 is coated onto (110) substrates in thickness of 100 μm. Silicon substrates with (100), (110), and (111) orientation were machined. The passing of laser beam was switched by a mechanical shutter and different laser powers were adjusted to ablate the samples on a motion stage. The laser writing route is shown in Fig. 7.40 [27]. The machined surface is observed by scanning electron microscopy (SEM).

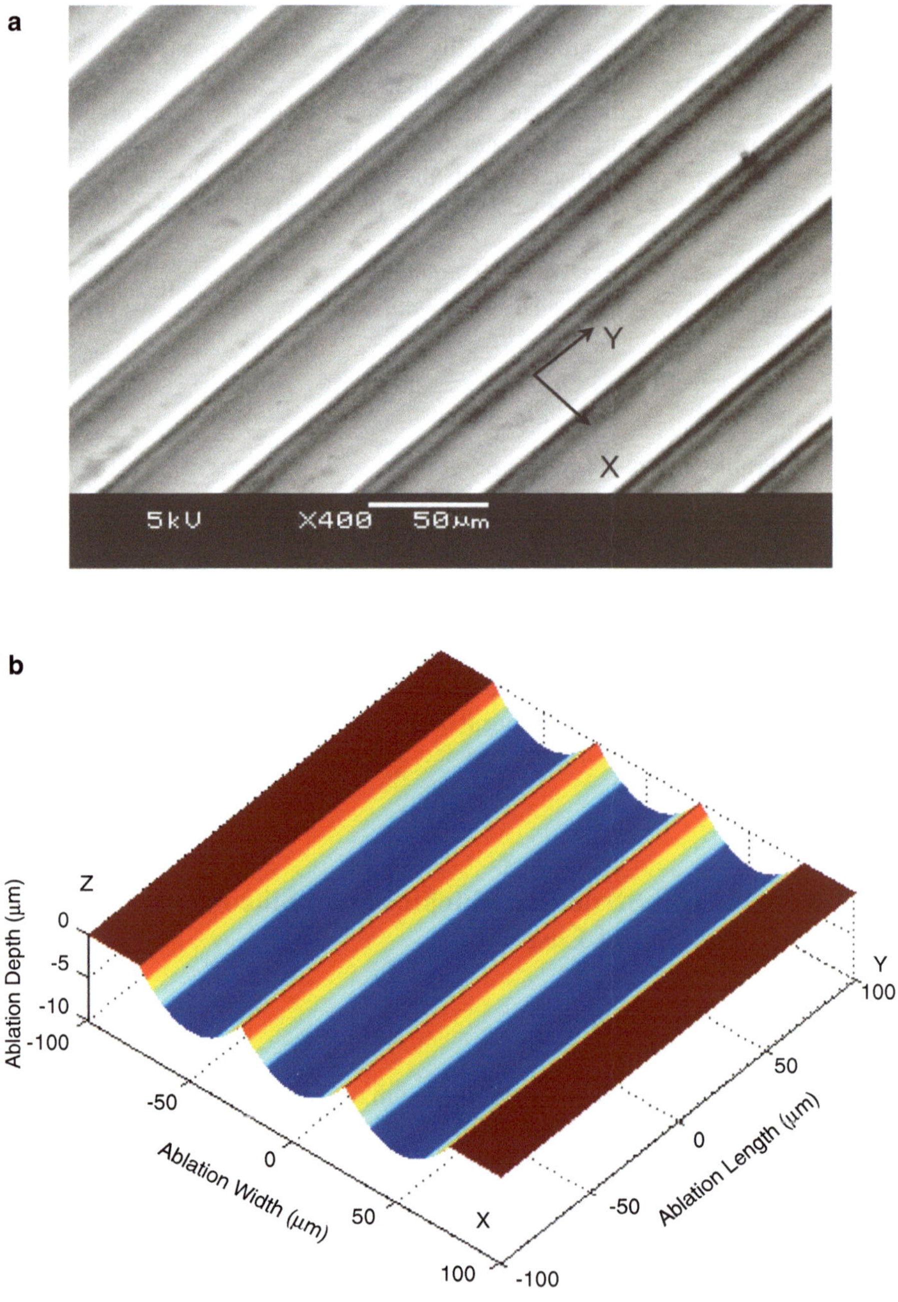

Fig. 7.33 Dragging with elliptic mask in one direction ($F = 365$ mJ/cm^2, $V = 2$ mm/min, $f = 60$ Hz, $A = 1.96 \times 10^{-3}$ mm^2) [1], reprinted with permission. (**a**) SEM image of machined profile. (**b**) Predicted profile

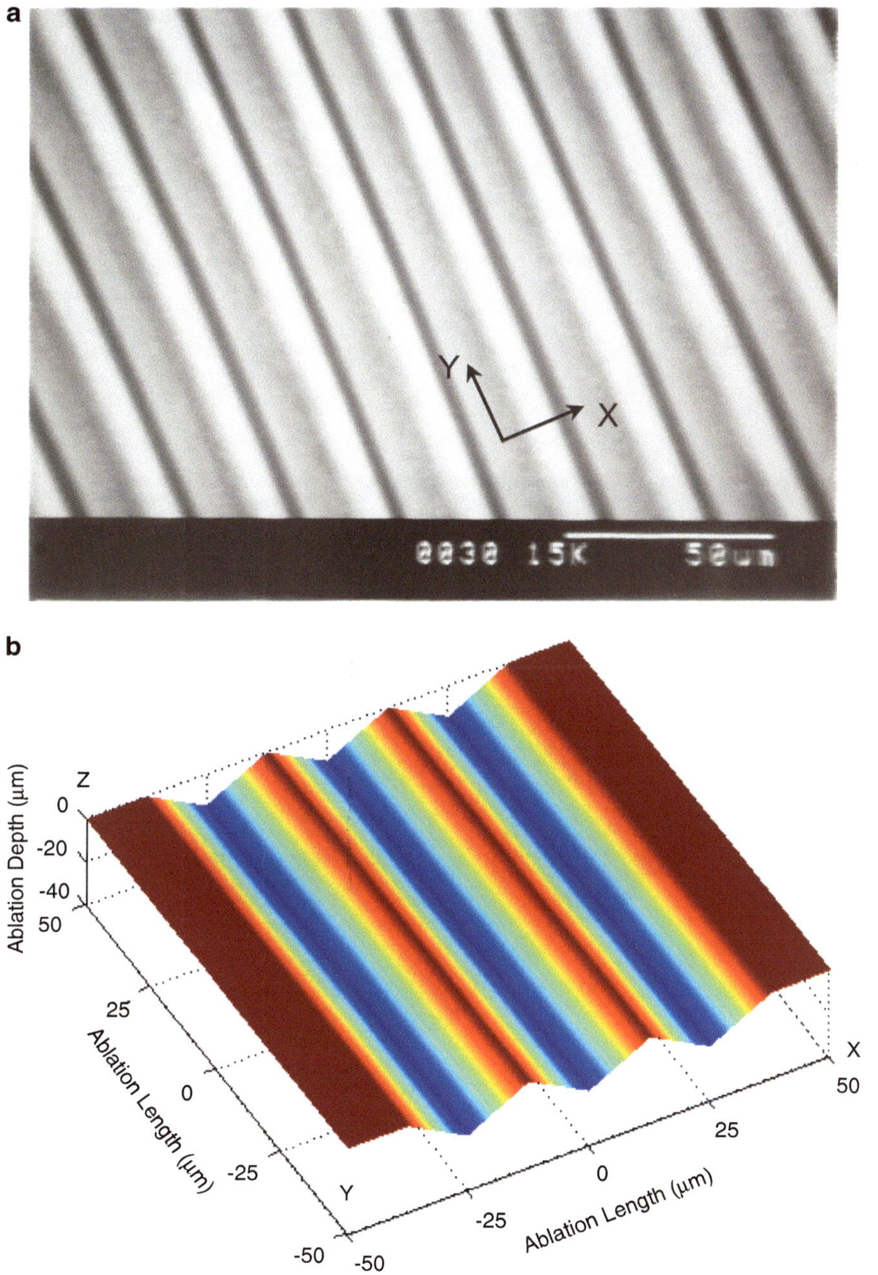

Fig. 7.34 Dragging with triangular mask in one direction ($F = 330$ mJ/cm^2, $V = 2$ mm/min, $f = 60$ Hz, $A = 3.125 \times 10^{-3}$ mm^2) [1], reprinted with permission. (**a**) SEM image of machined profile. (**b**) Predicted profile

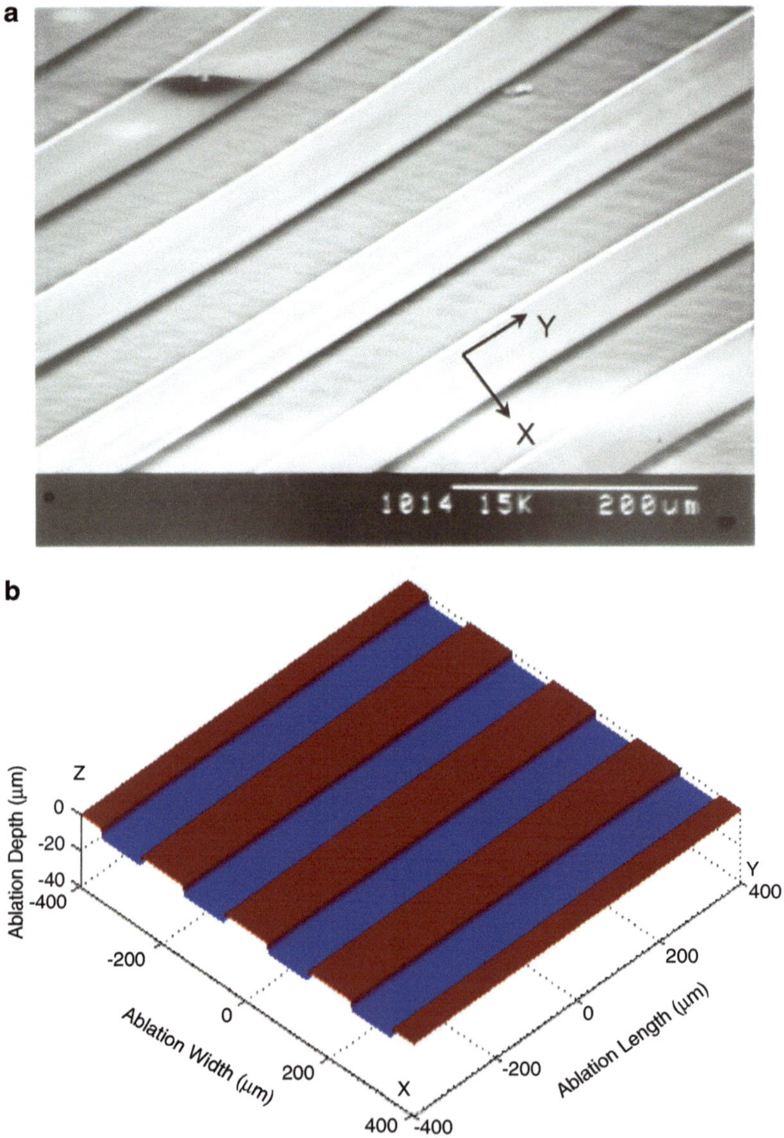

Fig. 7.35 Dragging with rectangular mask in one direction ($F = 270$ mJ/cm^2, $V = 3$ mm/s, $f = 60$ Hz, $A = 0.1 \times 2$ mm^2) [1], reprinted with permission. (**a**) SEM image of machined profile. (**b**) Predicted profile

Table 7.7 The comparison of the ablation depth between the experiment and simulation by single dragging

Mask shape	Experimental depth (μm)	Predicted depth (μm)	Deviation (%)
Elliptic mask	4.3	4.17	3
Triangular mask	13.9	14.6	5.0
Rectangular mask	5	5.2	4.0

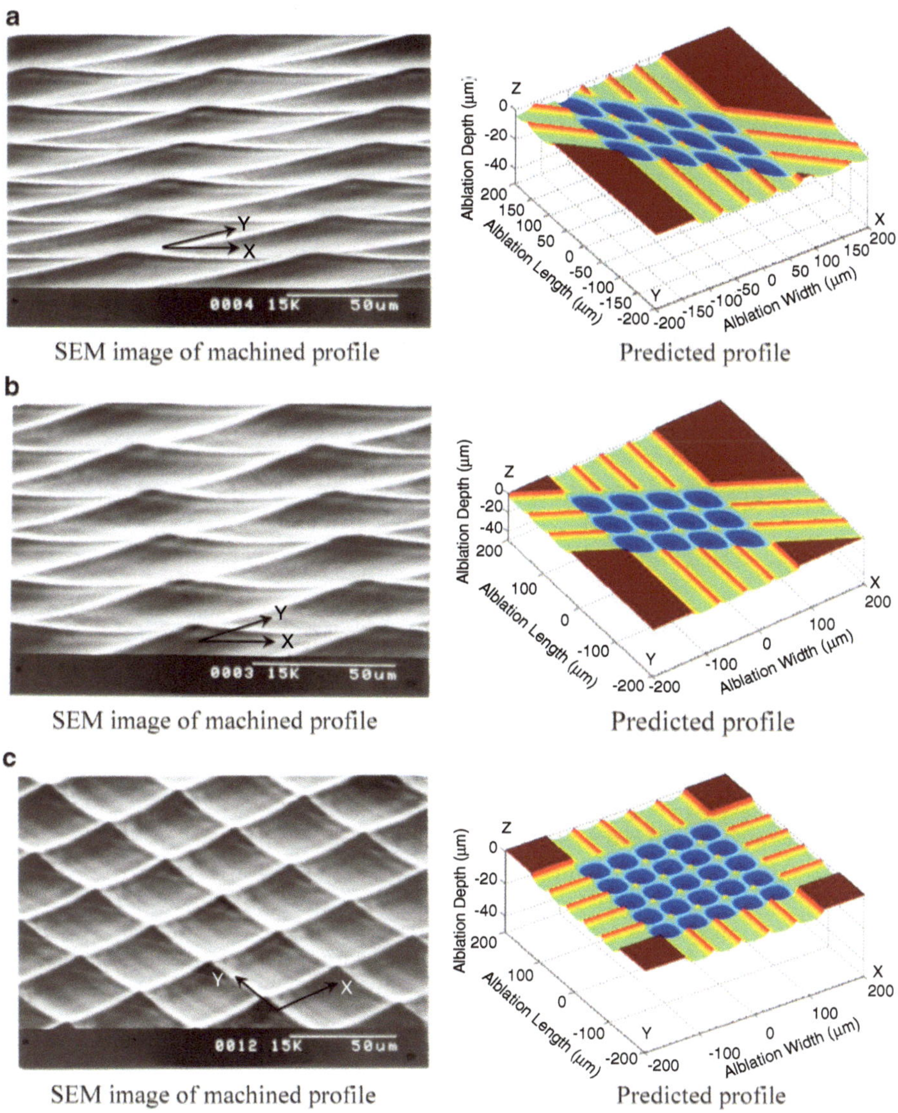

Fig. 7.36 KrF experimental cross dragging with elliptic mask ($F = 365$ mJ/cm^2, $V = 2$ mm/min, $f = 60$ Hz, $A = 1.96 \times 10^{-3}$ mm^2) [1], reprinted with permission. (**a**) 0° and 45°. (**b**) 0° and 60°. (**c**) 0° and 90°

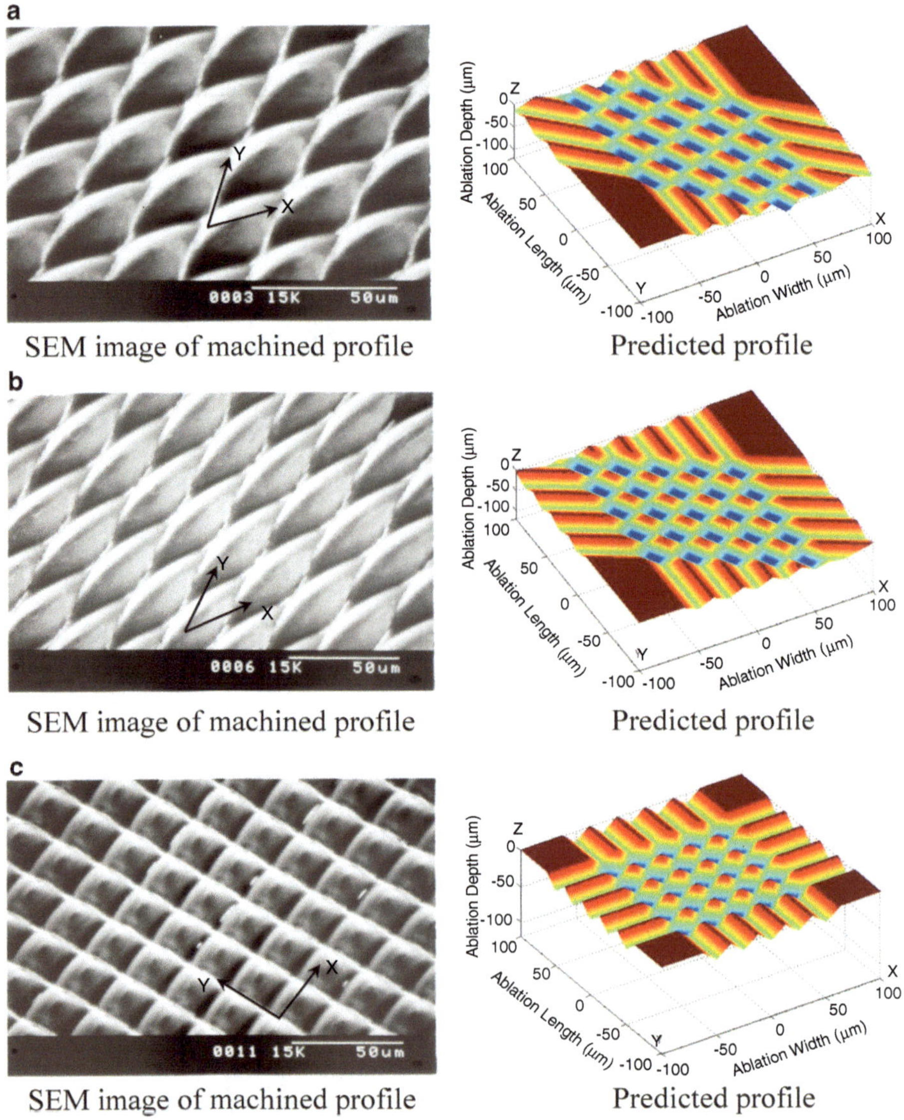

Fig. 7.37 KrF experimental cross dragging with triangular mask (F = 330 mJ/cm^2, V = 2 mm/min, f = 60 Hz, $A = 3.125 \times 10^{-3}$ mm^2) [1], reprinted with permission. (**a**) 0° and 45°. (**b**) 0° and 60°. (**c**) 0° and 90°

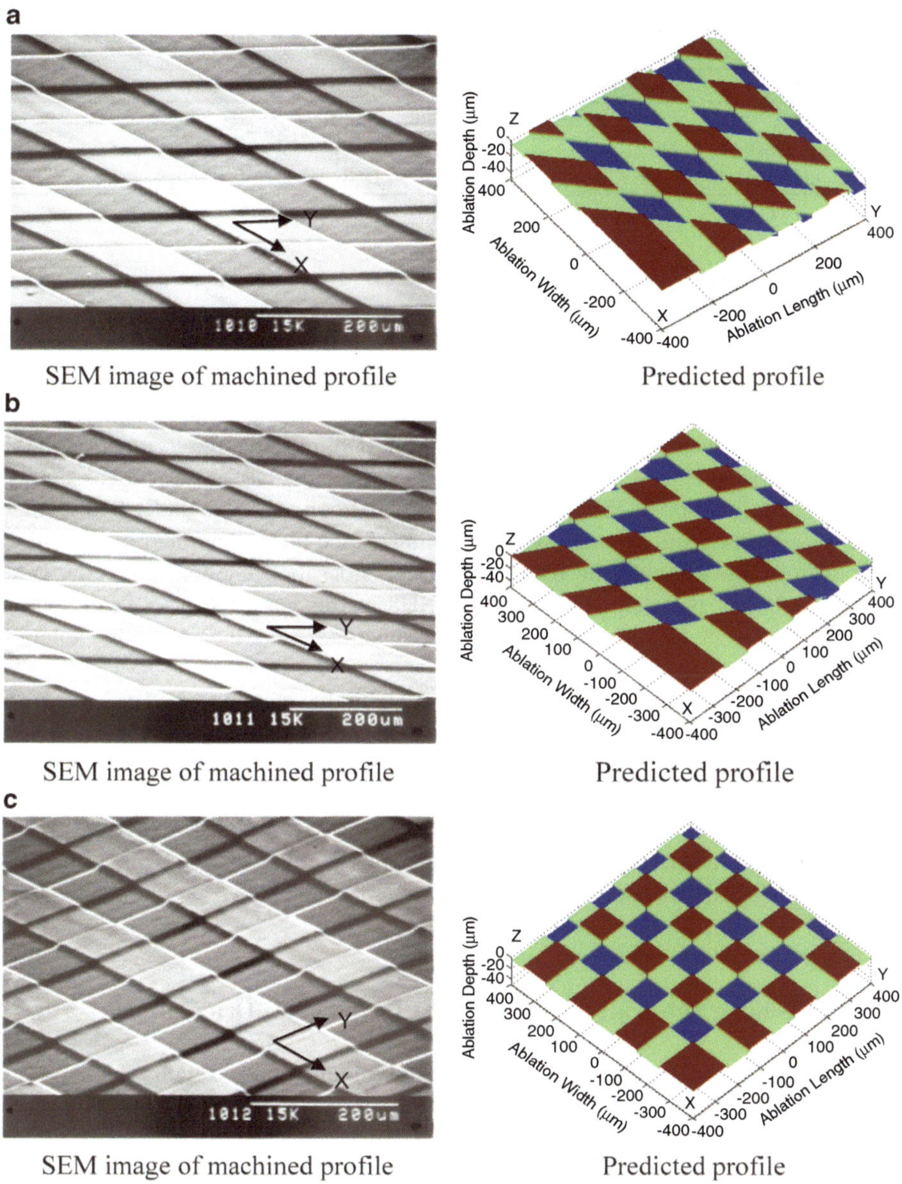

Fig. 7.38 ArF experimental cross dragging with rectangular mask ($F = 270$ mJ/cm^2, $V = 3$ mm/s, $f = 60$ Hz, $A = 0.1 \times 2$ mm^2) [1], reprinted with permission. (**a**) 0° and 45°. (**b**) 0° and 60°. (**c**) 0° and 90°

Table 7.8 The comparison of the ablation depth between the experiment and simulation by cross dragging

Mask shape	Experimental depth (μm)	Predicted depth (μm)	Deviation (%)
Elliptic mask	8.1	8.32	2.7
Triangular mask	26	29.2	13.8
Rectangular mask	9.6	10.4	8.5

7.3.2 Experimental Results

7.3.2.1 SU8

Figure 7.41a [27] shows the SEM image of SU8 machined by using the laser power of 60 μW and the scanning speed of 3 mm/min. It shows no obvious ablation in the central part of machined area except the edges where the laser beam moves slowly. Figure 7.41b [27] shows the SEM image of the silicon substrate under SU8, which is removed after femtosecond laser machining. Since the NA number of the object lens is not large enough, the focal spot size is not small enough. Once the laser fluence is below the ablated threshold of SU8 as this case, the laser beam passes through the SU8 to machine the silicon substrate, and therefore the ablation occurred.

Even the laser power is increased to 100 μW with the scanning speed of 3 mm/min, the similar machined results are shown in Fig. 7.42 [27]. The SU8 photoresist on the silicon substrate could be ablated as a rectangle structure when the laser power was increased to over 200 μW. Figure 7.43 [27] shows the ablated photoresist structures on Si (110) substrates with laser power of 200 μW, 600 μW, and 1 mW respectively. All SU8 coated substrates were etched by acetone to remove the SU8 photoresist off before taking SEM image.

7.3.2.2 Silicon Substrates with (100), (110), and (111) Orientations

These experiments were proceeded at different scanning speeds with identical laser power of 60 μW. Figure 7.44a [27] shows the result of the line-machining on Si (111) with the scanning speed of 1 mm/min. The depth of an area between two continuous holes by laser pulse is close to the depth. Figure 7.44b [27] is the result on Si (111) with the scanning speed of 3 mm/min. The two continuous holes are slightly connected by a shallow trench resulted from affected zone of single laser pulse. Figure 7.44c [27] shows the result on Si (111) with the scanning speed of 5 mm/min. The separated holes can be easily observed and there is nearly no interacted zone between two ablated holes.

From the previous experimental results, the scanning speed of 3 mm/min is chosen for laser machining on different orientations of silicon wafers. Figure 7.45a–c [27] shows the machining results of silicon substrates with (100),

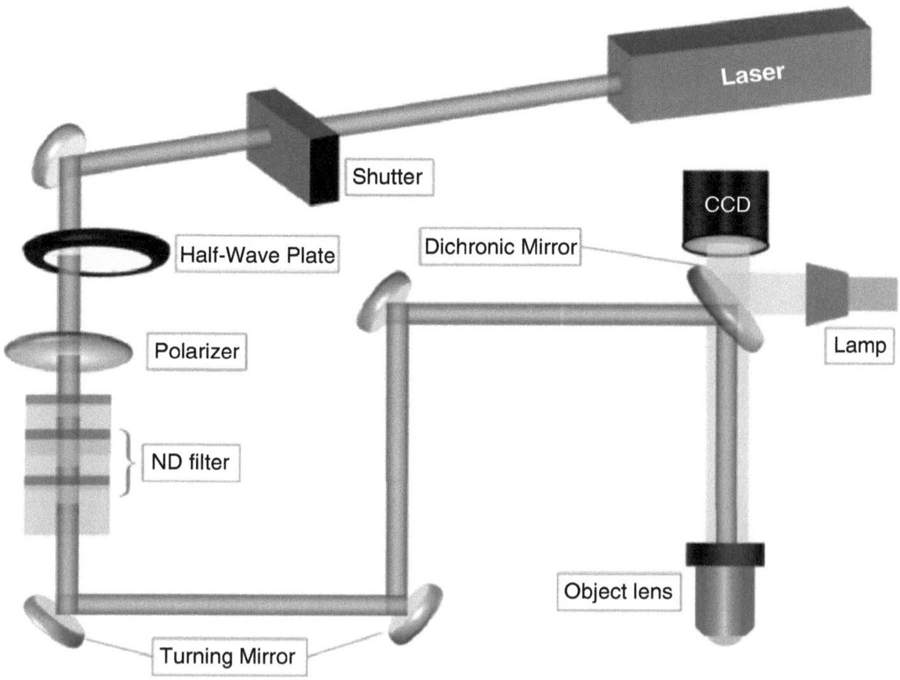

Fig. 7.39 Schematic diagram of experimental setup [27], reprinted with permission

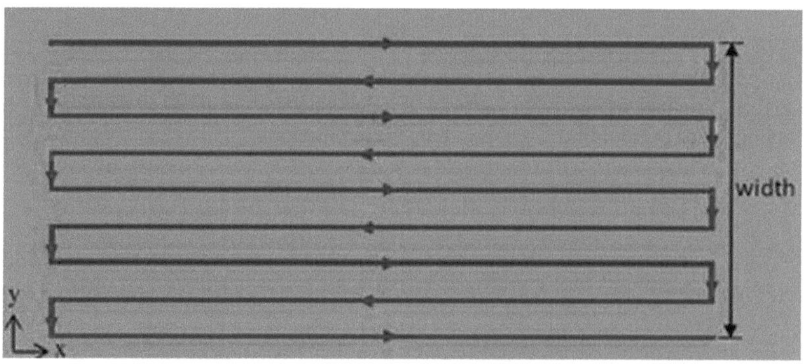

Fig. 7.40 The sketch image of scanning route [27], reprinted with permission

(110), (111) orientations by 60 μW laser power and 3 mm/min scanning speed. There is no obvious difference of ablated threshold energy between these three kinds of silicons; though the morphologies at the bottoms seem not similar.

Compared with the machined silicon substrate with (110) orientation in Fig. 7.43a–c, the spike morphology at the bottom can be obtained using the higher laser power than the threshold energy, and the whole rectangular shape shows clearer edge.

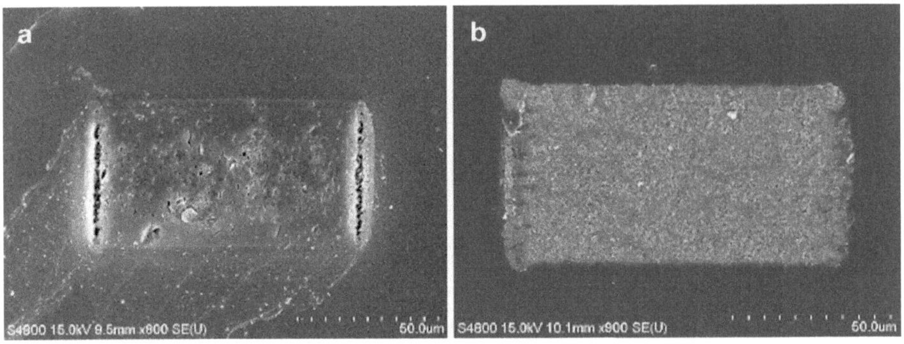

Fig. 7.41 SEM images of the SU8 result (**a**), and the substrate result (**b**) at laser power 60 μW and scanning speed 3 mm/min [27], reprinted with permission

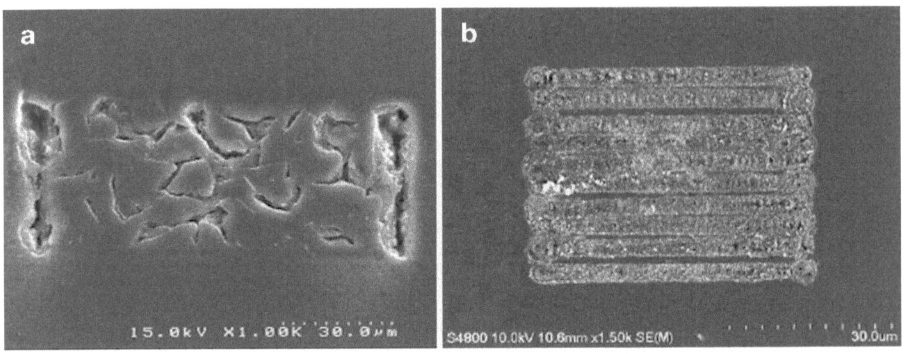

Fig. 7.42 SEM images of the SU8 result (**a**), and the substrate result (**b**) at laser power 100 μW and scanning speed 3 mm/min [27], reprinted with permission

Fig. 7.43 SEM images of the substrates under SU8 at scanning speed 3 mm/min and laser power (**a**) 200 μW, (**b**) 600 μW, and (**c**) 1 mW [27], reprinted with permission

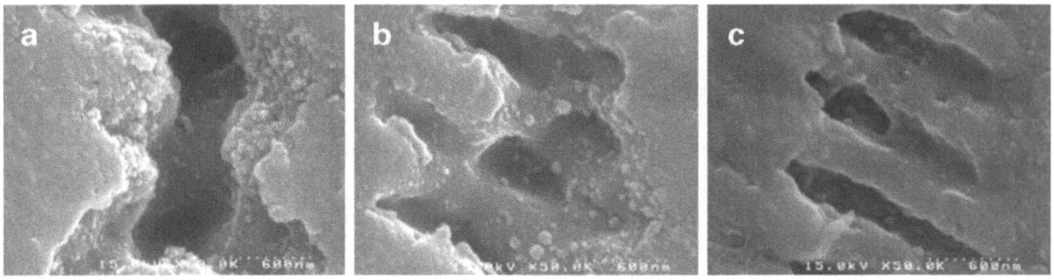

Fig. 7.44 Line machining on (111)Si by laser power 60 μW and scanning speed (**a**) 1 mm/min; (**b**) 3 mm/min; (**c**) 5 mm/min [27], reprinted with permission

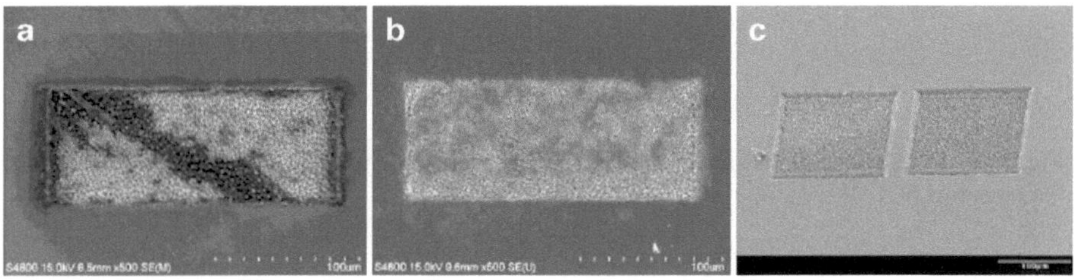

Fig. 7.45 SEM images of the rectangle structure machining on (**a**) (111); (**b**) (100); (**c**) (110)Si at laser power 60 μW and scanning speed 3 mm/min [27], reprinted with permission

7.3.2.3 Single Hole Structure

Figure 7.46 [27] shows the smallest hole structure with diameter of 400 nm obtained in the current study by a laser pulse of 60 μW, which is close to the ablated threshold.

7.3.3 Numerical Modeling

The process of photo-ionization by femtosecond laser can be divided into two parts: one is direct photoionization after electron absorbing enough photon energy; the other is impact ionization. When the kinetic energy of a free electron becomes sufficiently high by absorbing photons, part of the energy may transfer to a bound electron by collisions to overcome the ionization potential and produce two free electrons. The function of free electron distribution can be calculated by [15]:

$$\frac{\partial n_e(t,r,z)}{\partial t} = \beta(I) \cdot n_e(t,r,z) + P(I) \tag{7.10}$$

where $n_e(t,r,z)$ is the free electron density; $\beta(I)$ is the impact ionization term; $P(I)$ is the photoionization term; t is the time; r is the distance to the Gaussian beam axis; z

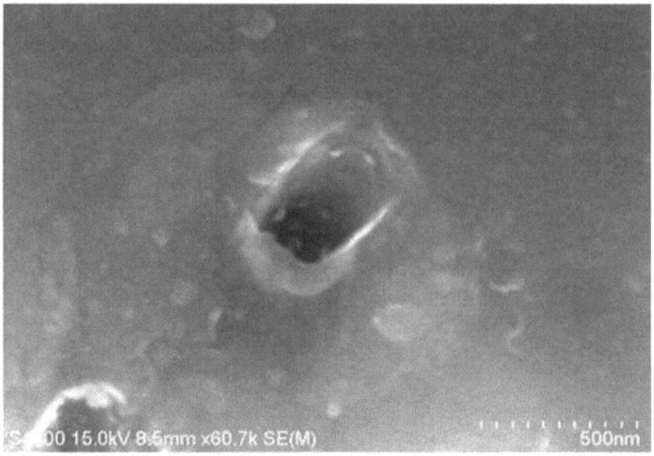

Fig. 7.46 SEM image of the smallest hole-structure with diameter of 400 nm by a laser pulse with the power of 60 μW [27], reprinted with permission

is the depth from the surface of the bulk material; I is the laser intensity inside the bulk material which is a function of t, r, and z.

Assuming that the impact ionization is linearly proportional to the laser intensity as follows [28]:

$$\beta(I) = a_i I(t, r, z) \tag{7.11}$$

where α_i is a constant, which is 5.88 cm^2/J for silicon. The photoionization rate can be expressed as a function of laser intensity by [15]:

$$P(I) = \delta_N (I(t, r, z))^N \tag{7.12}$$

where δ_N is the cross-section of N-photon absorption. For fused silica (bandgap = 9 eV) and the laser wavelength of 800 nm used in the present study, one photon is required to free a bound electron and, hence, $\delta_N = 1.6 \times 10^{18}\ cm^{-3} ps^{-1} [cm^2]$ [29, 30]; absorption number $N = 1$.

The original laser beam before it interacts with the material is assumed to be a Gaussian distribution in time and space. The laser intensity can be described by [15]:

$$I(t, r, z) = I_0(1 - R(t, r)) \times \exp\left(-\frac{r^2}{r_0^2} - (4 \ln 2)\left(\frac{t}{t_p}\right)^2 - \int_0^z \alpha(t, r, z)\, dz \right) \tag{7.13}$$

where $R(t, r)$ is the reflectivity of material, for silicon is 0.328 at wavelength of 800 nm; r_0 is the radius of the laser beam, 1.45 μm; t_p is the pulse duration, 120 fs; $\alpha(t,r,z)$ is the absorption coefficient of the material, for silicon is 7.8×10^2 cm^{-1} [31]; and I_0 is the peak intensity given by [13]:

$$I_0 = \frac{2F}{\sqrt{\pi/\ln 2} \times t_p} \tag{7.14}$$

where F is the fluence, the parameter is 1.8 J cm^{-2} for this study.

As for ablation depth calculation, it is assumed that the actual fluence at the ablation depth, d, equals to absorbed threshold fluence, $F(z = d) = (1 - r)F_{\text{th}}$. Based on this assumption, the ablation depth can be estimated by [13]:

$$d = \frac{1}{\alpha} \ln\left(\frac{F}{F_{th}}\right) \tag{7.15}$$

where F_{th} is the threshold fluence. It is widely assumed that ablation starts when the free electron density reaches the critical density. According to this assumption, threshold fluence can be considered as the minimal fluence to create the critical density. Based on this assumption, the theoretical threshold logarithmically depends on the value of critical density [28]:

$$F_{th} = \frac{2}{\alpha} \ln\left(\frac{n_{cr}}{n_o}\right) \tag{7.16}$$

where n_{cr} is the critical density, which relate to laser central wavelength and the permittivity of bulk material, given by [32]:

$$n_{cr} = \frac{4\pi^2 c^2 m_e \varepsilon_0}{\lambda^2 e^2} \tag{7.17}$$

where m_e is the electron mass, $9.10938188 \times 10^{-31}$ kg; c is the speed of light, 3×10^8 m/s; e is the electron charge, and λ is the laser wavelength of 800 nm. n_0 is the seed electrons generated by multiphoton ionization. The free electron distribution and ablation curve can be determined from the relationship between n_0 and threshold.

Figure 7.47 [27] shows the distribution of free electrons by laser pulse in bulk material. The ablated depth can be decided from the position of red broken line as shown in Fig. 7.48 [27]. The red broken line is determined from the intersection point of fluence (red line) and n_0 curve. According to the depth, the ablation shape can be drawn from the distribution of free electrons as shown in Fig. 7.47.

Figure 7.49a [27] is the result from the numerical modeling and Fig. 7.49b [27] is the AFM image of the cross section of the practical ablation profile in Fig. 7.44b. The red broken lines in Fig. 7.49a and b are the same profile of the approximate curve to the cross section. The simulation results agree well with the experimental ones.

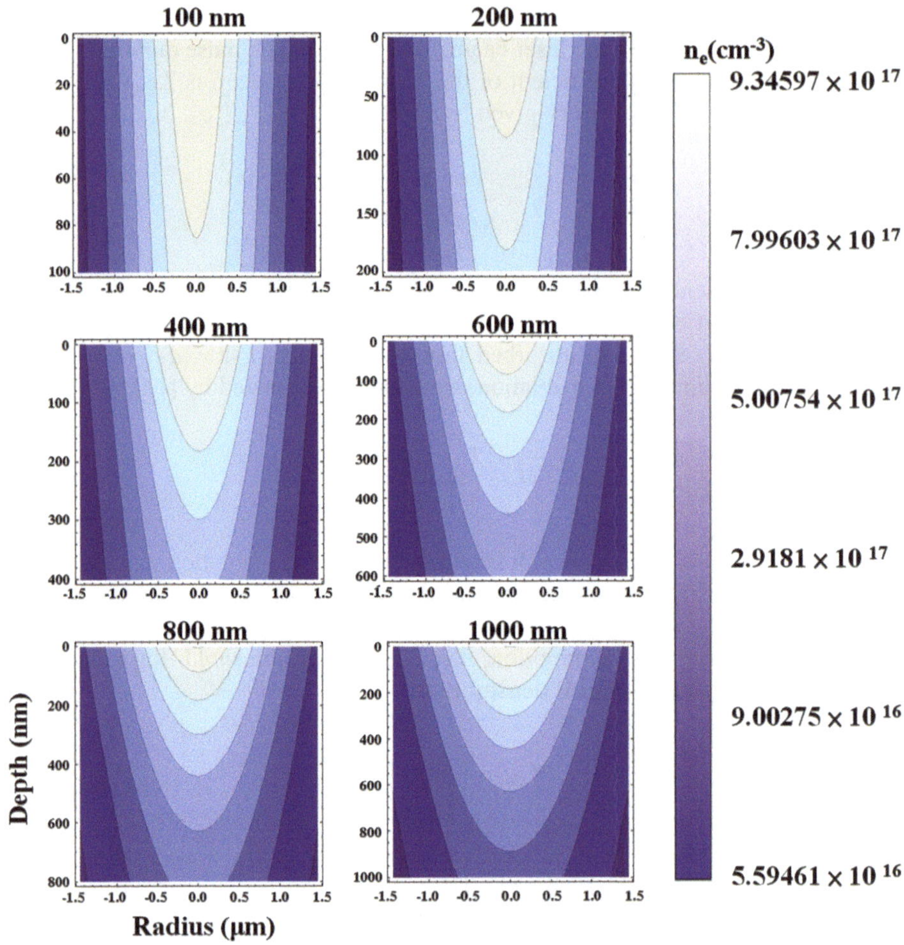

Fig. 7.47 The free electrons distribution in silicon by one pulse at the power of 60 μW [27], reprinted with permission

7.4 Conclusions

We established a mathematical model that takes account the ablation mechanism, material properties, laser operating parameters, and mask parameters to predict the ablation depth on a PC sheet using the laser dragging technique. A large number of experiments were carried out to verify the accuracy of the constructed model. The machined profile formed by laser dragging in one direction was simulated prior to the laser experiments, based on the constructed model. The paper discusses the profile formed by laser cross dragging using various shapes and sizes of mask.

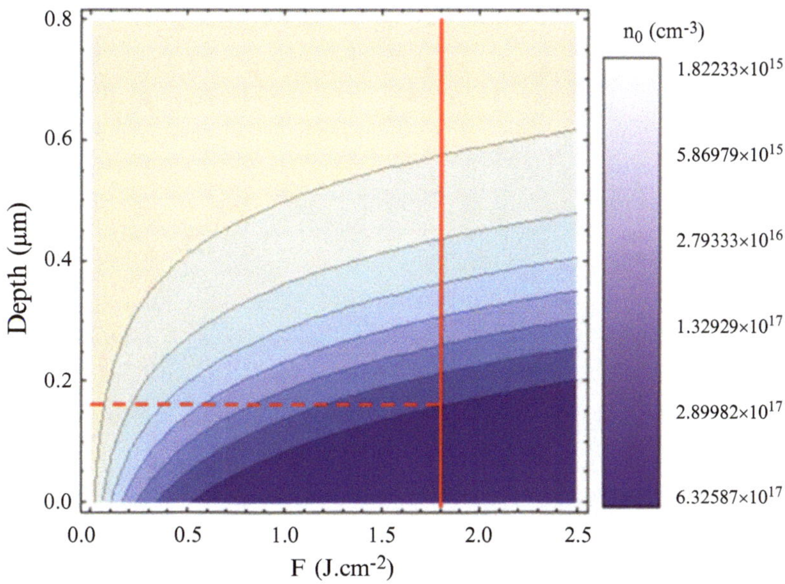

Fig. 7.48 The depth of the free electrons distribution as a function of laser fluence (The red line shows the fluence from the experimental data and the red broken line shows the machined depth.) [27], reprinted with permission

The experiments of both single dragging and cross dragging in different directions were carried out using the same operating parameters in the simulation process. Comparison of the experiment with simulation shows a satisfactory agreement with deviation of less than 15%.

A novel application of the proposed method to the fabrication of miniature lamp lens is presented. We conduct an analysis of the light efficiency of optical systems with either a macro-feature or micro-feature lamp lens. From the optical simulation results, one find outs that the optical system using micro-feature lamp lenses offers advantages of smaller divergence angle, greater illuminance, and more uniform distribution of light intensity in the illuminated region than the macro-feature lens systems do.

The desired lamp lens profile is predictable using the proposed model alongside adequate mask design, and control of the laser operating parameters. It is possible to fabricate spherical and aspheric lamp lenses with microsize features. We demonstrated the feasibility of making a micro-feature lamp lens for industrial applications.

In Sect. 7.3, we discussed the effect of different characteristics on the machinability of silicon substrates, and we proposed a numerical model to predict the cross-section shape using a femtosecond laser. In the case of a photoresist SU8 coating on silicon substrates, silicon beneath the SU8 photoresist was removed during laser machining, because the applied laser focal spot was too large. Silicon substrates with (100), (110), and (111) orientations were machined with at a laser power of 60 μW and scanning speed of 3 mm/min. Spike morphologies were

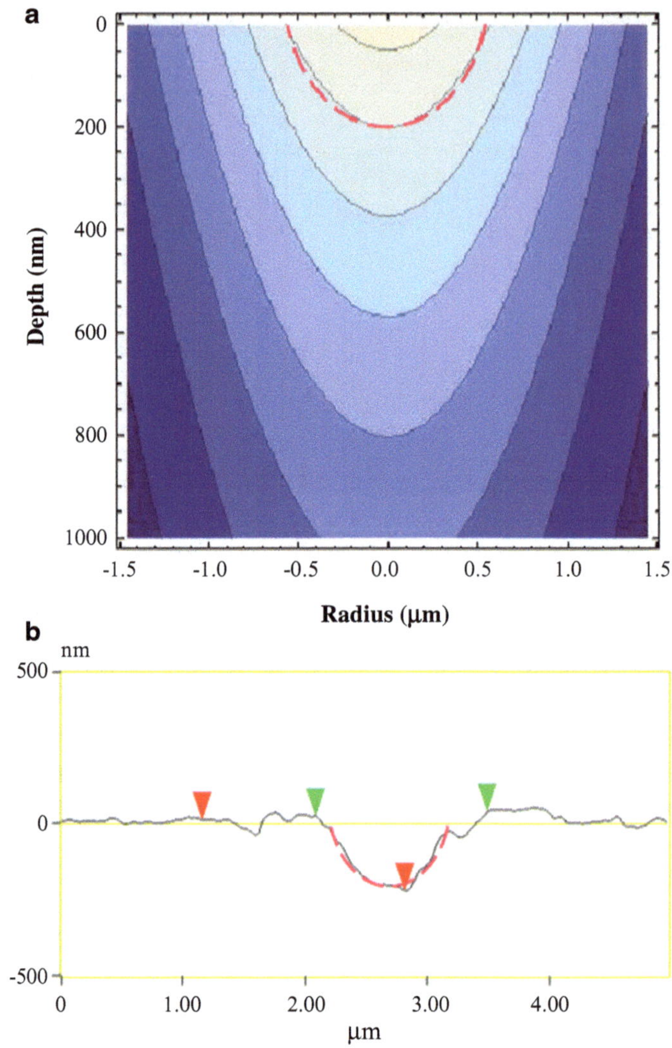

Fig. 7.49 The electrons distribution result in Fig. 7.47 at the depth of 200 nm (**a**) can be fitted to the AFM profile image of Fig. 7.44 (**b**) with the same shape of red dotted lines [27], reprinted with permission

observed on all three orientations of silicon substrates without obvious directional differences between the different silicon substrates. In this study, the smallest hole obtained by femtosecond laser had a diameter of 400 nm. In the case of ablation energy, we proposed a numerical model of the machining parameters to simulate the micro-structure cross-section. The shape predicted by simulation fitted the cross-section shape well. A femtosecond laser has the potential for both micro-scale and nano-scale fabrication to pattern different structures for optical applications such as CNC machining in the mechanical industries. Femtosecond laser machining will play an important role in micro- and nano-structure applications in the future.

References

1. Hocheng H, Wang KY (2011) Analysis of excimer laser machining of micropatterns. Lambert Academic Publishing, Germany
2. Crafer RC, Oakley PJ (1993) Laser processing in manufacturing. Chapman & Hall, London
3. Noggle JH (1989) Physical chemistry. Harper Collins College Publishers, New York
4. Znotins TA, Poulin D, Reid J (1988) Las Focus-Electro-Op 23:54
5. Hauer MR (2004) Laser ablation of polymers studied by time resolved methods, Department of Chemistry, Heinrich-Heine University Duesseldorf, Dissertation
6. Myaing MT, Urayama J, Norris TB, Braun A (2000) Opt Express 7(5):210
7. Seet KK, Mizeikis V, Juodkazis S, Misawa H (2006) J Non-Cryst Solids 352(23):2390
8. Matsuo S, Miyamoto T, Tomita T (2005) Proc CLEO/QELS 30:1731
9. Lin Y, Hong MH, Wang WJ, Law YZ, Chong TC (2005) Appl Phys A 80(3):461
10. Lin Y, Hong MH, Wang WJ, Wang ZB, Chen GX, Xie Q, Tan LS, Chong TC (2007) Sens Actuators A 133(2):311
11. Venkatakrishnan K, Sivakumar NR, Hee CW, Tan B, Liang WL, Gan GK (2003) Appl Phys A 77(7):959
12. Tan B, Sivakumar NR, Venkatakrishnan K (2005) J Opt A: Pure Appl Opt 7(4):169
13. Tsai HL, Jiang L (2003) Proceedings of the NSF Workshop on Research Needs in Thermal Aspects of Material Removal, Stillwater, OK, 163
14. Tsai H-L, Jiang L (2004) J Phys D: Appl Phys 37(10):1492
15. Tsai H-L, Jiang L (2005) Int J Heat Mass Transfer 48(3):487
16. Tsai H-L, Jiang L (2005) Appl Phys Lett 87(15):151104
17. Tsai H-L, Jiang L (2005) J Heat Transfer 127(10):1167
18. Tsai H-L, Jiang L (2006) J Appl Phys 100(2):023116
19. Wang SY, Huang CS, Chou HY, Lee TY, Chang RS (2001) Proc SPIE 4561:131
20. Cheng WL, Wang GJ, Yang JJ, Pan CT (2002) Aspheric micro-lens-array fabricated by excimer laser dragging, ASME, Micro-Electromech. Sys. Div. Publication, 463–467
21. Naessens K, Van Daele P, Baets R (2002) Proc SPIE 4426:124
22. Naessens K, Ottevaere H, Van Daele P, Baets R (2003) Appl Surf Sci 208:159
23. Pan CT, Shen SC, Hsieh CC, Chien CH, Chen YC (2004) J Microlith Microfab Microsyst 3(4):555
24. Lee YC, Chen CM, Wu CY (2005) Sensor Actuat A-Phys 117(2):349
25. Wei MK, Yang H (2003) Int J Adv Manuf Technol 21(12):1029
26. Srinivasan R (1983) J Vac Sci Technol B 1(4):923
27. Tsai HY, Luo SW, Wu CW, Wang SH (2009) Trans Nonferrous Metals Soc China 19(1):S171
28. Stuart BC, Feit MD, Herman S, Rubenchik AM, Shore BW, Perry MD (1996) Phys Rev B 53(4):1749
29. Ma W, Wua Y, Han J, Gu D, Gan F (2005) Chem Phys Lett 403(4):405
30. Bristow AD, Rotenberg N, Driel HMV (2007) Appl Phys Lett 90(19):191104
31. Aspnes DE, Studna AA (1983) Phys Rev B 27(2):985
32. Stuart BC, Feit MD, Rubenchik AM, Shore BW, Perry MD (1995) Phys Rev Lett 74(12):2248

Chapter 8
Material Shaping by Ion and Electron Nanobeams

J. Melngailis

Abstract Beams of electrons and ions of energies ranging from a few keV to over 100 keV and diameters in the single nanometers have become standard nanofabrication tools. Electron beams are routinely used to expose resist down to dimensions of 10 nm. In principle, ion beams can be and have been used similarly to expose resist. However, what motivates this chapter is that both ion beams and electron beams can be used to directly produce structures without the usual multi-step lithography process. Ion beams can simply locally sputter a surface, i.e., carve predesigned structures. In addition both electron beams and ion beams can be used to deposit material if a suitable precursor gas is adsorbed on the surface. The beam causes the adsorbed molecules to dissociate leaving some constituent behind. Similarly, if the adsorbate is a reactive gas, such as Cl_2 or XeF_2, a chemical reaction is induced where the beam is incident, and the material is locally etched. These material shaping techniques have found many applications and there are many types of structures have been built. However, the beam solid-interaction is complicated and factors other than the diameter of the beam limit the size of the structure that can be fabricated. In addition, these point-by-point fabrication techniques are slow. Typically to add or remove 1 cm^3 may require several tens of seconds. The finer the resolution needed the longer the fabrication time. Nevertheless these electron and ion nanobeam tools are widely used in research and in industry.

Keywords Beam system • Material shaping • Precursor gas • Surface coverage • Ion sputtering • Material removal • Model • Beam patterning

J. Melngailis (✉)
Institute for Research in Electronics and Applied Physics, University of Maryland, College Park, MD, USA
e-mail: melng@umd.edu

H. Hocheng and H.-Y. Tsai (eds.), *Advanced Analysis of Nontraditional Machining*, 453
DOI 10.1007/978-1-4614-4054-3_8, © Springer Science+Business Media New York 2013

8.1 Introduction

The shaping of materials has been a fundamental human activity from the Stone Age to the present. The structures made have depended on the availability of both the materials and the shaping techniques. The shaping techniques can be either additive, such as casting, or subtractive, such as machining. Our world today is characterized by a bewildering array of materials and shaping techniques. One of the most sophisticated structures fabricated today is the ubiquitous computer chip or integrated circuit (IC). Although its function is electrical, it is still fabricated as a physically shaped, mechanical structure using both additive and subtractive techniques. What has made our information age possible is the unprecedented ability to fabricate at ever smaller dimensions. The worldwide race to miniaturization has led to the development of sophisticated tools and techniques which have opened other micro/nanofabrication fields outside conventional integrated circuits. We will discuss one set of such techniques, namely those based on ion and electron nanobeams.

The fundamental step in the nanofabrication of IC's is photolithography [1]. The silicon wafer surface is covered by a thin light-sensitive polymer film, called photoresist, which is then selectively exposed to UV radiation, and the exposed parts are dissolved in an appropriate developer. The area laid bare is then treated in some way by having material added, subtracted, or altered compared to the neighboring covered surface. This is illustrated in Fig. 8.1 on the left. The early interest in focused ion and electron beams (focused to dimensions below, say, 100 nm, nowadays below 1 nm!) was that these material alteration steps could be carried out without the use of this multistep resist-based process. As illustrated on the right side of Fig. 8.1 a focused beam of ions [2] (or electron) can remove material, add material using the appropriate precursor gas [3], or, in the case of ions, implant and thereby alter the substrate. These functions are carried out in one step, albeit slowly and locally, rather than in the multistep, whole-wafer, global lithography-based processes (left side of Fig. 8.1). The interaction of an energetic ion (say, 30 keV) incident on the substrate surface and its effect on the surface is complex. The main attributes are summarized in Fig. 8.2. The ion can sputter off surface atoms, imbed itself in the substrate, induce a chemical reaction with an adsorbate on the surface, and cause electrons to be emitted. All of these effects can play a role, both enabling and limiting in the mechanical shaping of materials by focused ion beams (FIBs).

Electron beams can also add or remove material by locally inducing a surface chemical reaction with the adsorbed gas in the area where material is to be removed or added [3]. In some cases, electron beams can also alter the interior of the material chemically which can be exploited for shaping by preferential etching. (Another electron beam material fabrication technique is electron beam welding. It uses orders of magnitude more intense beams at orders of magnitude larger dimensions and will not be discussed here.) The main mechanism for inducing the surface chemical reactions is secondary electron emission [3]. When an electron strikes the surface, a number of low energy electrons are emitted from the surface, which interact with the adsorbed gas molecules producing a chemical reaction.

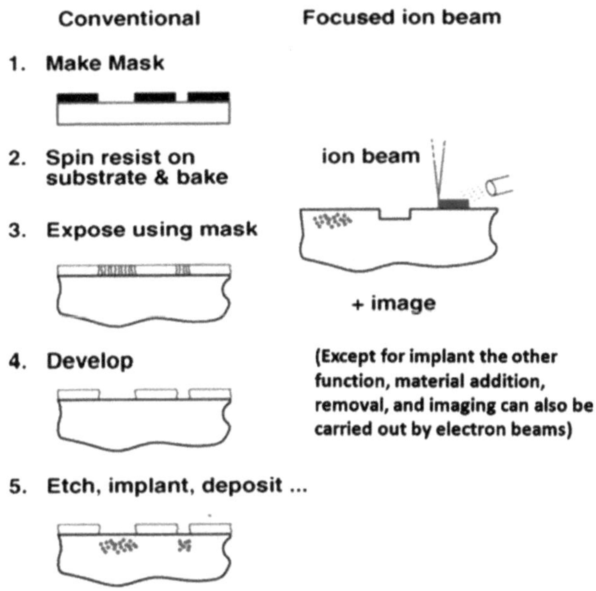

Fig. 8.1 The elementary patterning steps that characterize most micro/nanofabrication, as in the making of integrated circuits, are illustrated on the *left*. The process relies on resist which is exposed in a pattern, usually by UV radiation, and then the exposed part is chemically removed. The uncovered part of the sample is then treated in some way to create a permanent structure in the sample surface. The techniques we are discussing in this paper can in principle simplify the process and do the patterning directly on the material in one step. However, this is done in a point-by-point process which is of course slow

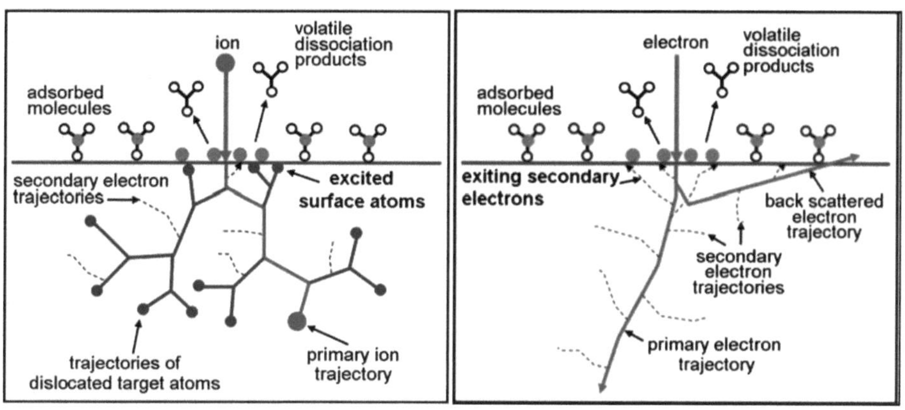

Fig. 8.2 The fundamental processes and ion-surface interaction are shown on the *left*. The incident ion embeds itself at some depth in the substrate depending on energy and on the nature of the random scattering. It also displaces substrate atoms creating both damage and sputtering from the surface. The electron–surface interaction is shown on the *right*. An electron in the tens of kilovolts energy range penetrates much deeper, several micrometers, than an ion of comparable energy. For both electrons and ions a local chemical reaction on the surface can be produced leading to either material removal or material addition. Secondary electrons are emitted from the surface in both cases and are used for imaging [3]

Both electron beams and ion beams can be used to image the sample. A scanning electron microscope (SEM) is usually used in this way. If no external gases or adsorbates are present, most materials are unaffected by electron beam imaging. Ion beams can also be used in the same way, namely, scanning the beam over the surface and collecting secondary electrons that are emitted to form an image. The only difference is that during this scanning the ion beam sputters the surface of the sample. However, with care a good image can be formed by sacrificing less than a monolayer of the substrate. Imaging is an important complement to the shaping of material. Thus the location where material is to be added or removed can be located and defined with very high precision.

8.2 Ion and Electron Beam Systems

8.2.1 How Are Focused Ion Beams Produced?

By far the most widely used FIB systems today are based on the Ga liquid metal ion source (LMIS) [2]. Briefly liquid gallium slides down a sharp needle from a reservoir, the needle is biased positively, for example at 10 kV, with respect to a facing, concentric, circular electrode. The beam from this, in effect, point source proceeds through this electrode and other biased electrodes which act as lenses and is refocused on the sample where it can be deflected (i.e., scanned over the surface) as well as turned on and off. A schematic of a system is shown in Fig. 8.3. The system can be thought of as having three parts:

1. The ion column which produces the beam on the substrate, typically below 10 nm in diameter with the current density in the beam of about 10 A/cm^2 at 30 kV. The beam profile on the surface is approximately Gaussian and when we speak of diameter we are considering full width at half maximum and the current density is simply the total beam current divided by the area of the circle defined by the beam diameter.
2. The vacuum chamber and stage which have to be vibration free with respect to the column otherwise the beam diameter will effectively appear to be enlarged. The vacuum chamber also contains an electron detector which collects the electrons which are emitted from the surface where the ions strike. This permits an object to be imaged in exactly the same way that it is imaged in the SEM. Of course, the surface is eroded during this imaging. However, if the length of the scan is appropriately controlled, a good image can be formed by sacrificing less than a monolayer of the surface. Many of the systems in use today include a SEM working in conjunction with the FIB column. These are called dual beam systems. As shown in Fig. 8.4, the two beams are arranged to focus to the same point on the sample with angles of incidence about 50–60° apart. Typically the sample stage can be translated, tilted, and rotated.

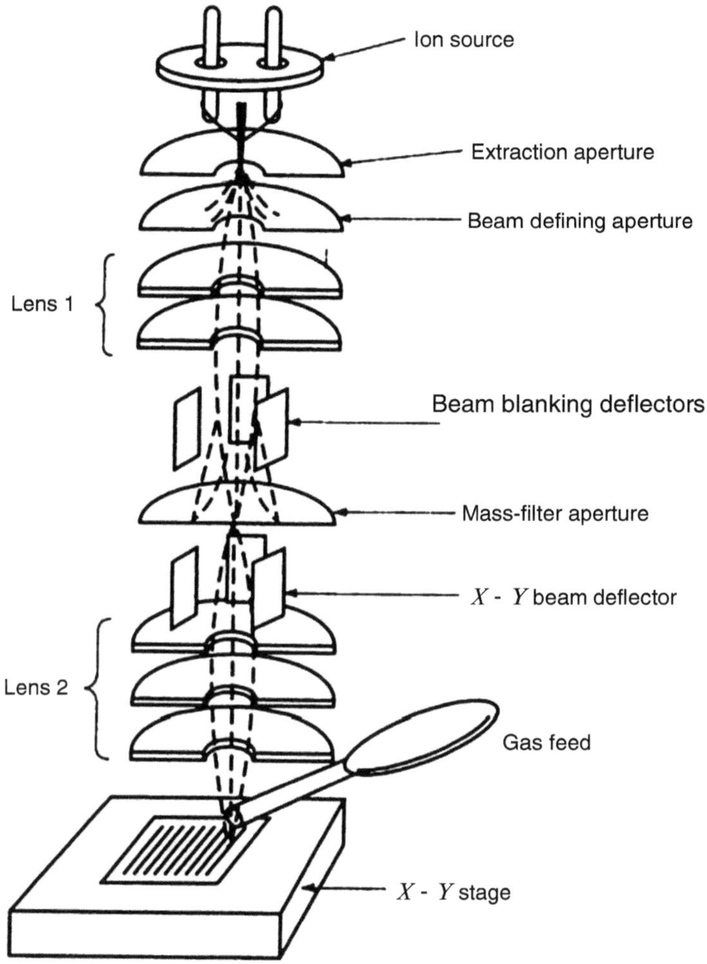

Fig. 8.3 A schematic of the focused ion beam (FIB) system consisting of a liquid metal (usually gallium) ion source, extraction aperture, beam blanking deflection electrodes, beam steering electrodes, and lenses. The lenses are concentric circular electrodes which both accelerate and focus the beam. The gas feed which provides an adsorbate for induced chemical reactions on the surface usually consists of a hypodermic needle-like stainless steel tubing aimed at the point of ion incidence on the sample

3. Beam-controlled electronics. The deflection and the turning on and off of the beam is typically under computer-control so that selected areas can be scanned and desired geometries can be generated. The programming of the beam scanning and stage motion is often an essential feature since the rate of material removal with the ion beam is slow, typically a few hundred seconds to remove 1 μm^3. Thus some operations can take many hours and can proceed without the operator. An example of the sophisticated structure is shown in Fig. 8.5.

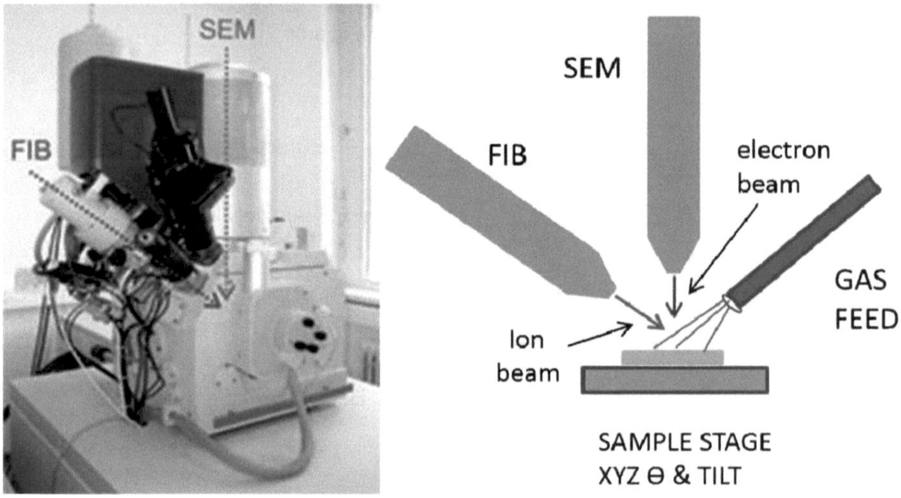

Fig. 8.4 On the *right*, a schematic of a dual beam system which permits both electron beams and ion beams to be incident on the sample. A gas feed is provided for the precursor gas. The sample stage can be displaced and tilted. On the *left*, a photo of a system is present (Nova 200 Nanolab, FEI) [4]. A number of companies supply these systems: FEI, Zeiss, Hitachi, Seiko, and Tescan

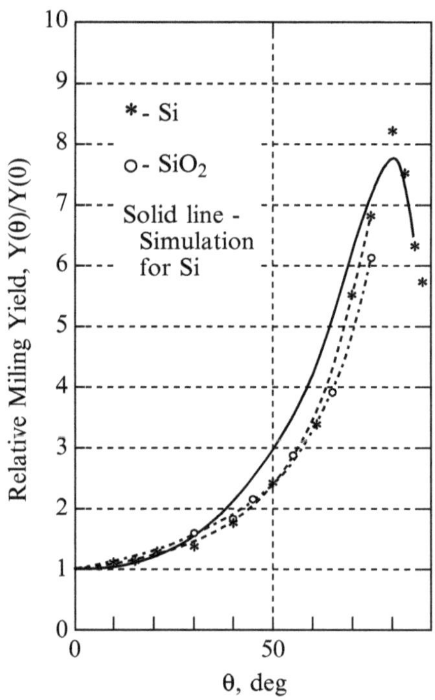

Fig. 8.5 Relative milling yield as a function of angle of incidence for silicon and silicon dioxide [5]. The *solid line* was calculated using TRIM

8.2.2 How Are Electron Beams Produced?

Two types of electron systems capable of producing nanometer diameter beams are commonly found: SEM's modified for materials shaping, and electron beam lithography systems. E-beam lithography systems are widely used in nanofabrication for resist exposure [1]. The fabrication then proceeds as illustrated in Fig. 8.1. Features in the 10 nm range have been defined in resist. These systems are specialized and very expensive and are not generally used for beam-induced processes. However, they would be ideal since excellent beam writing capabilities are built in.

SEMs are much less expensive and simply need to be modified by introducing a gas feed and by introducing more sophisticated beam control than is needed for microscopy. In the dual beam systems (FIB/SEM, Fig. 8.4) these capabilities are built in. Electron columns for producing the beam are common. The challenge is often in the peripherals: vibration free stage whose motion is coordinated with the beam, the gas feed, and sample neutralization. For highly insulating substrates an arriving electron beam (and also ion beam) will charge the surface, deflecting the subsequent incoming beam. Schemes for avoiding charging include, generating a plasma, and introducing a gas as in an environmental SEM.

Sophisticated electron beam tools have been developed adapted uniquely for photomask and extreme ultraviolet (EUV) mask repair [6].

8.3 Material Shaping with Ion Beams

For the time being we will restrict our discussion to material shaping with the gallium ion beam. There have been recent exciting developments (to be discussed below) such as the gas field ion source which can produce helium or neon ion beam diameters below 1 nm, and laser cooled ion sources promising ultrasmall beam diameters of various elements. A plasma ion source can be used to produce a higher current beam but with less resolution. In addition, the plasma source is a key component in the development of a multibeam system, which could extend beam processing to production. Incidentally the multibeam system has also been demonstrated with electrons. Our discussion of the gallium ion beam applications will help us understand how these other ion beams can be used.

Over much of the range of interest the beam current density in a Ga ion beam system remains constant (typically at about 10 A/cm^2) thus one can increase the beam current and increase the removal rate but at the cost of enlarging the beam diameter and sacrificing the minimum dimensions of features that can be fabricated [2].

8.3.1 Ion Milling

Typically an ion incident on the sample surface will cause several sample atoms to be sputtered off. You can think of this as sandblasting with atoms. The detailed mechanism is described as collision cascades. The incoming ion collides with an atom imparting substantial energy to it. The ion then further collides with other atoms, and each displaced atom of high energy again collides with other substrate atoms [7]. Some of these collision cascades will reach the surface. If as a result surface atoms have energies above a certain material-dependent threshold usually in the range of a few electron volts, then the surface atom is sputtered off. The number of atoms removed per incident ion is called the yield. It can also be expressed in terms of $\mu m^3/nC$. The yield, of course, is a critical parameter in any shaping of material with the ion beam. The measurement of the yield turns out to be more complex than one would think at first glance. The same complexity will be reflected in any practical application of the gallium ion beam to material shaping.

The simplest, unambiguous measurement of yield can be done by scanning the FIB repeatedly over a rectangle to produce a relatively shallow pit whose depth (i.e., volume removed) can then be measured and from the total number of ions (or coulombs) incident during the milling time, the yield can be calculated. What effects need to be taken into account in practical milling or in measurement of milling yield?

1. Angle of incidence. As one might intuitively expect the milling yield will increase as the angle of incidence departs from normal. This is indeed the case for most materials, and the increase in yield can be as much as a factor of 6 or 7. See Fig. 8.5. Perhaps somewhat surprisingly, the yield goes down as one approaches glancing incidence. This is predicted by the modeling using TRIM. This is due to the fact that the sputtering process is in fact due to a collision cascade produced by the incident ion. As the ion enters the substrate it scatters multiple times imparting significant momentum to atoms of the substrate which in turn scatter other atoms of the substrate. These scattered atoms have a probability of reaching the surface thus leading to sputtering. At glancing incidence the first collision of the incident ion with the substrate atom can lead to a 50% probability of the ion leaving the substrate thus producing no further scattering events. See Fig. 8.5.
2. Crystal structure of the substrate. Ions entering a single crystal along asymmetry axis will encounter open spaces between rows of substrate atoms. Thus they will penetrate much deeper before scattering. This is a well-known phenomenon and is called channeling. Obviously the yield will be reduced in such circumstances. The picture is further complicated by the fact that during the ion bombardment an initially single crystal sample surface is disordered, producing an amorphous surface layer. Milling of polycrystalline sample will result in the roughening of the surface since grains of different orientation will mill at different rates this as shown for gold in Fig. 8.6.

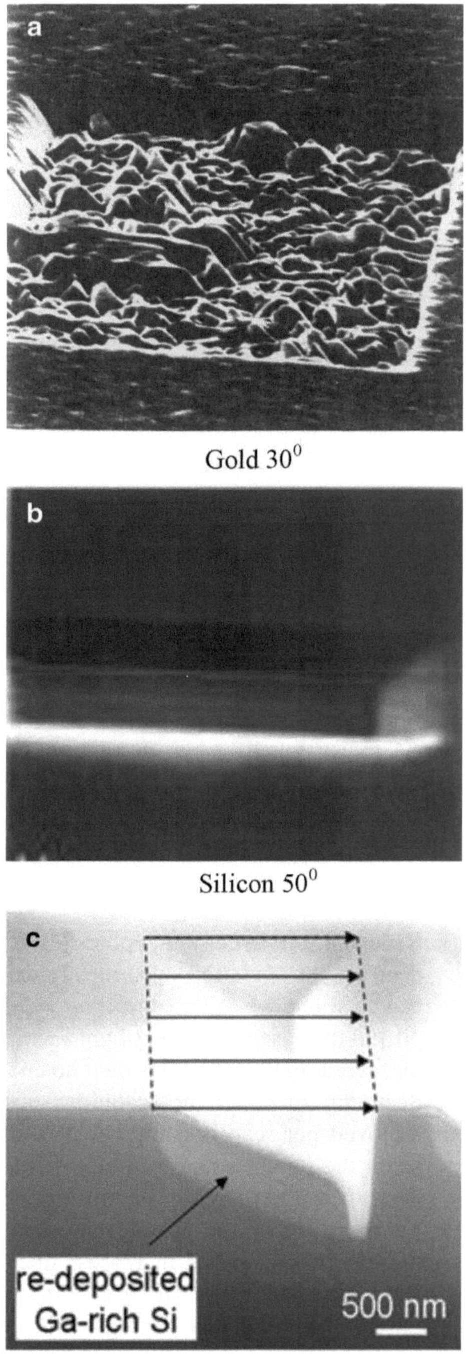

Fig. 8.6 On the left of top (**a**), rectangle sputtered in a polycrystalline gold film at 30° angle of incidence showing the roughening of the sputtered surface relative to the unsputtered surface. On the right of top (**b**) a similar rectangle sputtered in a single crystal silicon surface showing that no roughening occurs [5]. On the bottom (**c**), redeposition during milling. Ga FIB scanning is carried out only from *left* to *right*. Thus redeposition occurs "behind" the beam. From ref. [3] (original from P. Nellen and V. Callegari, Chimia60, A735 (2006)). If the beam is scanned back and forth in both directions repeatedly, redeposition can be minimized. For a fairly deep pit as shown it will still occur and reduce the overall material removal rate by the beam, but the pit will be symmetric

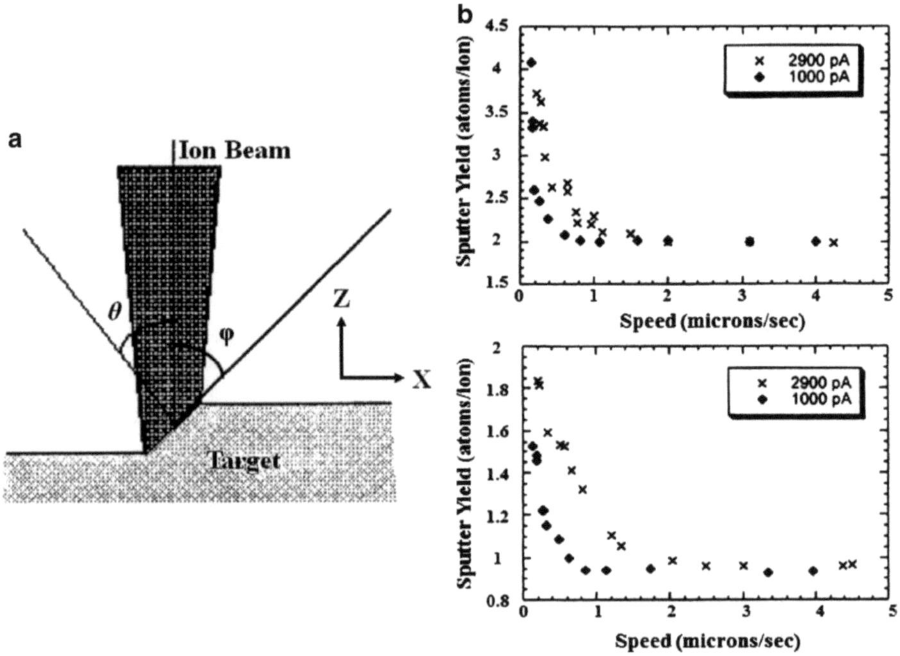

Fig. 8.7 (**a**) A schematic of the ion beam/target geometry illustrating why slow scan speed can lead to increased material removal rate. To simplify the analysis the beam was scanned rapidly in the *Y* direction, into the paper and advanced slowly in the *X* direction. If the beam is scanned slowly in both directions, the same reasoning applies but the analysis is more complex. (**b**) The sputter yield as a function of scan speed. *Top*, for Si, *bottom* SiO_2. If the beam is scanned so slowly that each pass removes a thickness comparable to the beam diameter (as illustrated in (**a**)), the yield will increase. Microscopically the ions in that case are not impinging at normal incidence [9]

3. Redeposition. In certain geometries, for example when attempting to mill a very deep pit, the sputtered substrate atoms will redeposit on the sides of the pit. (This is why above we said that milling yield should be measured by milling a shallow pit.) In the case of milling of gold with gallium, the angular distribution has been measured and has the expected cosine squared dependence [8].
4. Scanning speed. If the scanning speed is slow and the thickness of material removed per scan is comparable to the beam diameter, then locally under the beam the ions are not normally incident on the surface. See Fig. 8.7. This has been shown to increase the milling yield by about a factor of 2 [9].
5. Temperature of the sample can also affect the milling rate. As the thermal motion of the atoms increases, one would expect they would more readily leave the surface. But the reverse is sometimes true. At high enough temperature the sample will anneal during milling, reducing the amorphization of the surface and thus the milling yield [10]. One may also have to consider that under some circumstances the energy deposited by the ion beam may locally heat the

Table 8.1 Sputter yields at normal incidence, different ions, incident energies, and materials

Substrate	Ion	Energy (keV)	Yield (atoms/ion)	Yield (μm^3/nC)	References
Si	Ga^+	30	3.1 ± 0.8	0.388	[12]
Si	Ga^+	30	1.92 ± 0.16	0.24	[13]
Si	Kr^+	25	3.1	0.388	[14]
Si	Ga^+	30	1.8	0.22	[15]
Diamond (100)	Ga^+	20	2	0.072	[16]
Diamond	Ga^+	50	2.3	0.083	[17]
hard amorphous carbon	Ga^+	50	2.6	0.094	[17]
GaN	Ga^+	15	5.2		[18]
		30	6.2		
		50	6.8		
		70	6.7		
GaAs	Ga^+	30	4.9	6.9	[16]
Al_2O_3	Ga^+	30	1.9	0.1	[18]
Au	Ga^+	100	32	3.4	[13]
Au	Ga^+	40	15.7 ± 1.3	1.66	[19]
Au (plated)	Ga^+	25	18 ± 3	1.91	[5]
Au (evap.)	Ga^+	25	23 ± 5	2.44	[5]
Au	Kr^+	25	20	2.12	[14]
W (RF-sputt.)	Ga^+	25	5 ± 0.7	0.49	[5]
W	Kr^+	22	4.1	0.406	[14]
Al	Ga^+	30	2.7	0.28	[20]
SiO_2	Ga^+	30	0.8 molecules/ion	0.23	[15]
SiO_2	Ga^+	25	0.84	0.24	[5]
SiO_2	Ga^+	30	0.85	0.24	[21]

sample, thus affecting the milling rate. Heating under the beam can be estimated by a simple calculation [11]. The conclusion is that even for an insulating substrate such as silicon dioxide the heating is minimal unless the beam diameter and beam current are large, i.e., beam heating could play a role in some circumstances.

6. At low doses some swelling is observed for single crystal substrates [10]. This is due to the fact that the amorphized surface layer is less dense than the single crystal. This effect is negligible except when one is milling depths below, say, 10 nm.
7. Surface contamination may also affect the rate of milling. In a moderate vacuum, like 10^{-6} mbar, a monolayer is adsorbed in about a second. Thus if one is scanning a relatively large area such that the average ion impingement rate per unit area is comparable to the residual gas impingement rate (5×10^{14} molecules/cm^2 s), then the contamination can interfere with the removal process.

The milling yields of some typical materials are shown in Table 8.1. The numbers come from various published papers. Presumably the measurements were made for shallow pits rapidly milled and in ways that avoided the above-mentioned effects.

But probably this was not the case, and there is often considerable variation in the reported numbers.

A number of workers as well as the manufacturers of the FIB/SEM systems have developed strategies and computer programs for obtaining desired shapes. For brevity we include only two references [22, 23].

8.3.2 *Material Shaping Using a Precursor Gas*

The precursor gas used for both adding material and enhancing the material removal is beamed onto the surface near the point of ion impact through a small needle somewhat like a hypodermic needle [3]. See Fig. 8.4. In the case of ions the chemical reaction induced by the beam can be produced by three mechanisms:

(a) Collision cascades, as in the case of milling this results in energetic surface atoms which can induce chemical reactions.
(b) Secondary electron emission, these low-energy electrons can also induce chemical reactions.
(c) Direct ion collision with a precursor gas molecule on the surface.

Although all three mechanisms clearly can play a role, by far the dominant one in the case of gallium ions is collision cascades. (For electrons and possibly also light ion such as helium the secondary electron emission mechanism is likely dominant.)

8.3.2.1 Material Removal

The induced chemical reaction will produce material removal if the substrate is turned into a gas. For example, a commonly used gas is xenon difluoride, XeF_2. It turns silicon into SiF_4 which is a gas and is simply pumped away. Likewise one can think of silicon dioxide as being turned into oxygen and SiF_4. Chlorine gas will act similarly. This chemical removal process proceeds in parallel with the milling. The material removal rate with gas is typically about a factor of 10 higher than without gas and with ion impact alone. Clearly for materials like gold which have no stable and volatile compounds, the removal rate cannot be enhanced with a reactive precursor gas. Table 8.2 shows some of the gases that have been used and the enhancement rates.

In addition to the increased removal rate, an important advantage of material removal with a precursor gas is that it enables one to avoid, or at least minimize, redeposition. This permits one to, for example, fabricate deeper pits with more vertical sidewalls. A second important advantage is selectivity. One can take advantage of chemistry to, for example, remove part of a film that is on top of a substrate of a different material with minimal erosion of the substrate. For example, an organic film on top of the silicon dioxide can be etched very effectively with

Table 8.2 Gas-assisted FIB etching. The rate enhancement is the ratio of milling yield with gas to milling yield without gas

Substrate	Ion (energy, keV)	Gas (pressure on substrate or flux)	Rate enhancement	References
Si	Ga (30)	Cl_2 (4 mTorr)	20	[24]
Si	Ga (25)	H_2O (70 mTorr)	0.3	[25]
Si (111)	Ga (30)	Cl_2 (4×10^{17} molecules/cm^2 s)	11.8–15.8	[26]
GaAs (100)	Ga (30)	Cl_2 (4×10^{17} molecules/cm^2 s)	10 (14.8 for optimized conditions)	[26]
SiO_2	Ga (20)	XeF_2 (2×10^{19} molecules/cm^2 s)	10	[27]
W	Ga (20)	XeF_2 (2 Torr)	15–75 (depending on dwell time)	[28]
Al	Ga (30)	Cl_2 (4×10^{17} molecules/cm^2 s)	~3 up to 27 for optimized conditions	[26]
Cu	Ga (45)	$Cl_2 + NH_3$ 1:1, 0.1 Torr	6–12	[29]
Ti	Ga (30)	XeF_2	7	[30]
PMMA	Ga (25)	H_2O (70 mTorr)	20	[25]
C (diamond)	Ga (25)	H_2O (70 mTorr)	10	[25]
Si, SiO_2, Si_3N_4	Ga (25)	H_2O (70 mTorr)	~0.3	[25]

water vapor [25]. The difference between the removal rate of the organic film and the silicon dioxide is about 20, whereas with milling alone there may be little or no difference in the removal rate.

8.3.2.2 Material Addition

The inverse process occurs when a precursor gas absorbs on the surface which has the property that the incident ion causes a volatile compound to be removed from the precursor gas molecule leaving a solid behind. For example tungsten carbonyl, $W(CO)_6$, may dissociate into, CO, CO_2, O_2, C, and W. Unfortunately it turns out that both carbon and tungsten are left on the surface. Thus the "metal" left behind which is a mixture of tungsten, carbon, and gallium has a resistivity one or two orders of magnitude higher than pure tungsten. Nevertheless, this process is used for example in locally rewiring integrated circuits and in making various electrical connections in research device structures. Most of the common gases used for material deposition are organometallic, such as (Methylcyclopentadienyl) trimethylplatinum and likewise yield metal-carbon mixtures. Much purer deposits of copper and gold have been obtained by heating the substrate to temperatures of 80–100°C either by heating the entire stage [31, 32] or by locally heating the sample by focusing a laser beam near the point of ion impact [33]. In this way resistivity only about a factor of two higher than pure metal (rather than a factor

Table 8.3 Gallium ion beam-induced deposition of metals and silicon dioxide from various precursor gases

Precursor gas (desired material)	Ga+ ion energy (keV)	Yield	Deposit composition	Resistivity (μΩ cm)	References
$W(CO)_6$ tungsten carbonyl (tungsten)	25	2 atoms/ion	W:C:Ga:O 75:10:10:5	150–225 (pure W = 5.3)	[34]
$W(CO)_6$ tungsten carbonyl (tungsten)	30		W:C:Ga 51:37:12	200	[35]
(Methylcyclopentadienyl) trimethylplatinum	35	1–2 atoms/ion	Pt:C:Ga:O 45:22:25:4 25:55:19:2	70–700 (pure Pt = 10.4)	[36]
(Methylcyclopentadienyl) trimethylplatinum (platinum)	25	0.1–0.5 μm^3/nC		400–2,000	[37]
(TMVS)-Cu-hfa (copper)	25–35	10–30 atoms/ion	Cu:C 55:45 (25°C) 95:5 (100°C)	100 (25°C) 5 (100°C) (pure Cu = 1.7)	[31]
(TMVS)-Cu-hfa (copper)	50	0.1–0.35 μm^3/nC	Cu:C:Ga 60:30:10 (25°C)	18.8 (100°C)	[38]
Me2Au(hfa) (gold)	40	3–8 atoms/ion	Au:C:Ga 50:35:15 (25°C) 80:10:10 (120°C)	500–1,500 (25°C) 3–10 (120°C) (pure Au = 2.2)	[32]
Pentamethyl-cyclopentasiloxane $((CH_3)HSiO)_5 = Si_5O_5C_5H_{20}$ (silicon dioxide)	50			8×10^{11} Ω cm (pure SiO_2 = ~10^{15} Ω cm) (resistivity is nonlinear vs. voltage)	[39]

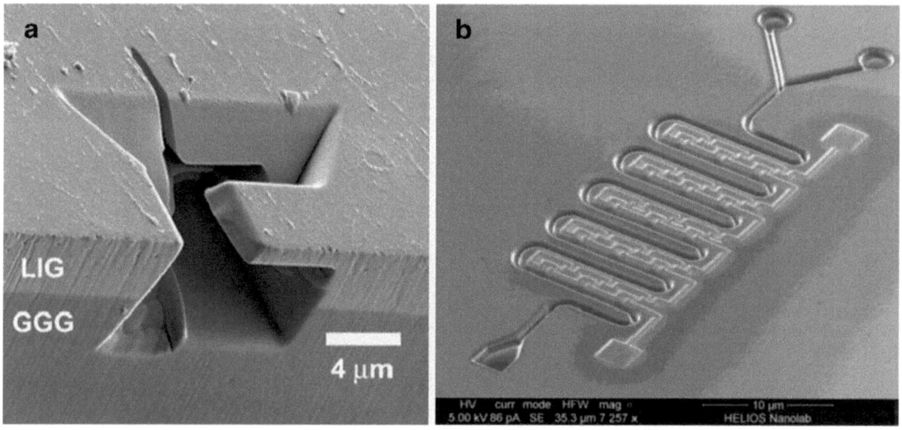

Fig. 8.8 (**a**) Example of FIB milled structure (30 keV Ga+ ion beam) Luthenium iron garnet (LIG) ferrimagnetic film on Gadolinium gallium garnet (GGG) [23]. (**b**) Microfluidics building block on silicon, FIB milled, serpentine channel with a conductor pattern on the surface fabricated by FIB-induced deposition (FEI Helios 600 FIB) photo courtesy of Jon Orloff, FEI

of 100) can be obtained. Some examples of precursor gases and deposited materials are shown in Table 8.3. Note that also insulators such as SiO_2 can be deposited. Again, the deposit is not pure. Probably largely due to the Ga ion implantation the resistivity of the oxide is lower than that of pure SiO_2. Still in many cases it is a useful insulator.

An important feature of both focused ion and electron beam-induced deposition is that structures can be made on non-planar surfaces. Conventional, lithography-based fabrication, see Fig. 8.1, can only be carried out on the planar surfaces. Beam-induced deposition does not have this limitation. Within reason material can be deposited at the bottom of the pit or on the end of the pillar. A frequent application utilizing both beam milling and deposition is that a pit is cut through several layers on an integrated circuit and then filled to make electrical contact to a buried conductor. This technique is used to locally rewiring a circuit during development phase to diagnose and/or correct malfunctions.

Other examples of fabricated structures include conductors deposited on surface, Fig. 8.8b, nanocantilevers [40] Fig. 8.9, and springs [41] Fig. 8.10. The fabrication of the spring structure may seem surprising. With the beam-induced deposition, one can fabricate a cantilever extending from the edge of a cliff at some particular angle of inclination. The beam is programmed to slowly advance and material is deposited on the end of the cantilever. If the beam is advanced more slowly, the cantilever is more vertical. (If the beam is not moved at all of vertical pillar can be grown. See for example Fig. 8.17. To fabricate the spring the same principle applies except that the ion beam is programmed to slowly move in a circle and since the beam diameter is small the windings below the point on the spring being fabricated do not have any Ga+ ions incident on them, i.e., the range of the Ga ions in the wire is less than the diameter of the wire. Thus one can maintain spacing between the windings. By the way, this kind of scheme would not

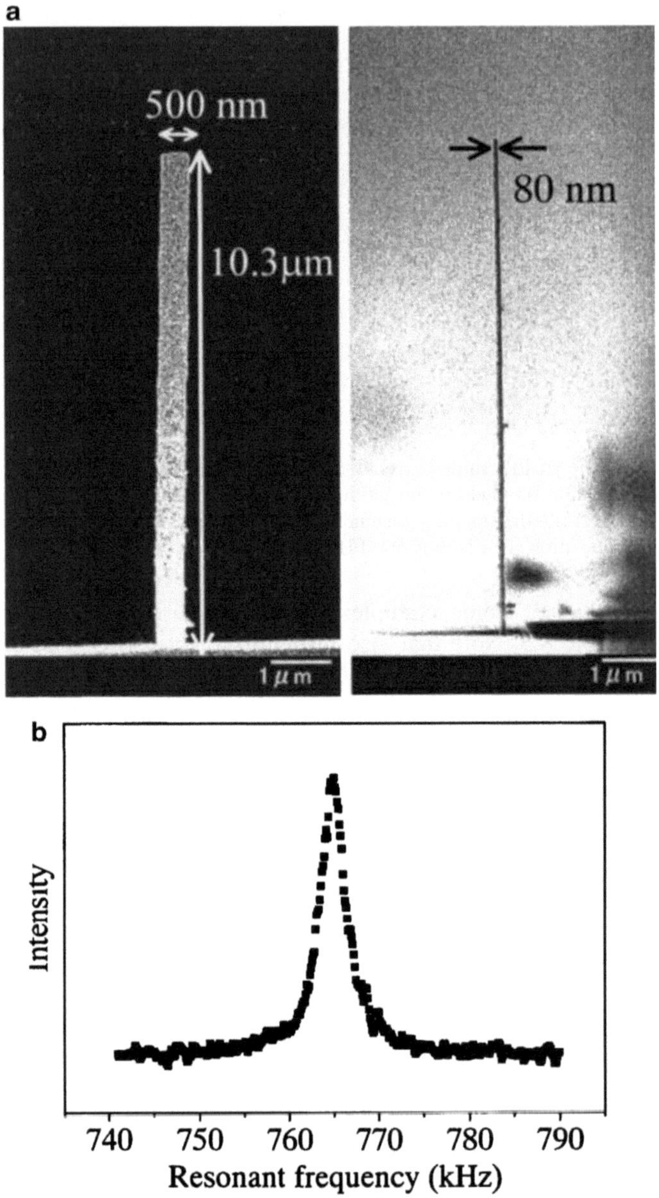

Fig. 8.9 (**a**) Diamond-like carbon cantilever fabricated by FIB-induced deposition from phenanthrene $C_{14}N_{10}$ precursor gas. A 30 keV Ga+ fixed, normal incidence ion beam at 0.5 pA was used [40]. (**b**) Resonant spectrum of the cantilever. A small deposit made on the end of the cantilever shifts the resonant frequency. In this way a mass of $12 \pm 3 \times 10^{-15}$ g was measured

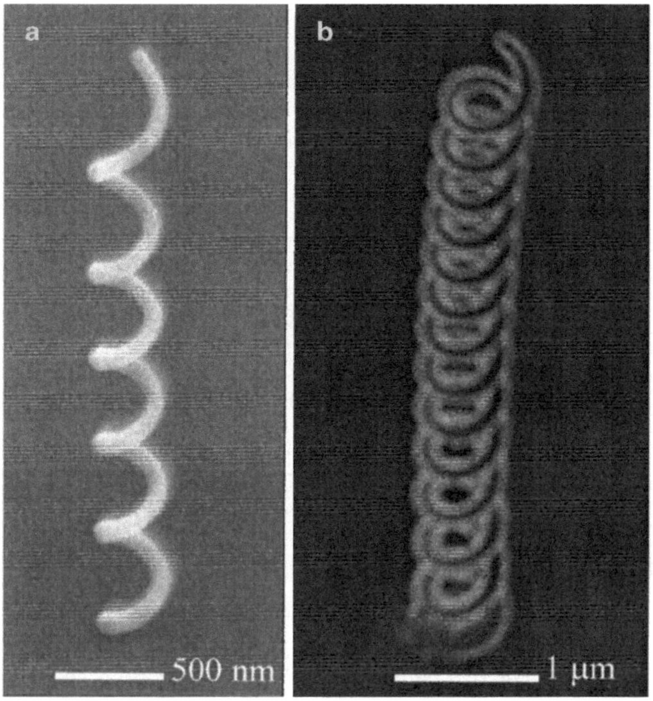

Fig. 8.10 (**a**) A scanning electron micrograph of a spring of diamond-like carbon with 130 nm "wire" diameter [41]. It was fabricated by FIB-induced deposition from a $C_{14}N_{10}$ precursor. (**b**) A scanning ion micrograph of a spring 20 μm long and 130 nm wire diameter. Both were fabricated using 30 keV Ga ions at 1 pA and 5 nm beam diameter

work with electron beams since the electron beam would penetrate through the spring winding and cause deposition on the levels below unless very low energy electrons are used.

8.4 Material Shaping with Electron Beams

In most cases, the electron beam systems used for direct material nanofabrication are simply a modified SEM. Beam diameters below 1 nm are available and a gas feed can be readily introduced. In fact, in the dual beam systems the same chamber, stage, and gas feed can be used for either electron beam or ion beam-induced processing. Electron beam-induced deposition actually predates ion beam-induced deposition. Recently electron beam-induced deposition and etching have gained technical importance because of the need to repair photomasks and EUV masks without damage due to gallium implantation. In addition electron beam-induced deposition may be favored for depositing electrical contacts on semiconductors where again damage and implantation may be undesirable.

Table 8.4 Electron beam-induced deposition with various precursor gases

Electron energy (keV)	Precursor gas	Composition of deposit	Resistivity (Ω cm)	References
2–20	$W(CO)_6$		2×10^{-3} to 0.1	[43]
25	$W(CO)_6$	W:C:O- 55:30:15	0.2	[44]
25	Dimethyl-gold-trifluoroacetylacetonate	Au:C:O 40:55:1.5		[44]
25	Dimethyl-gold-trifluoroacetylacetonate	25–40% Au	10^{-3} to 10^{-1}	[45]
15	(Methylcyclopentadienyl) trimethylplatinum (platinum)	Pt:C:O 21.5:73:5.5	3.0×10^{-5}	[46]
10	$Pt(PF_3)_4$	Pt:P:F- 81:17:1	3×10^{-5} to 6.5×10^{-4}	[47]
25	$AuClPF_3$	>95% Au	4.3×10^{-5}	[48]

8.4.1 Material Addition

With some exceptions the same precursor gases used with ion beams also work with electron beams. The mechanism, however, in this case is secondary electron-based. Because of their small mass and momentum, electron beams do not in general displace atoms and therefore collision cascades do not occur. Moreover, the process is less efficient. For example in many cases one molecule is dissociated per 100 incident electrons. While for ions, 10 molecules may be dissociated per incident ion. In addition, if organometallic precursors are used the large concentrations of carbon remain in the deposit and the electrical resistivity may be 4–5 orders of magnitude higher than that of pure metal. However, relatively pure deposits of gold and platinum, with the resistivities only 2–3 times higher than pure metal, have been obtained using inorganic precursor gases. However, the gold precursor $AuPF_3Cl$ is very unstable and difficult to use or obtain, and the platinum precursor $Pt(PF_3)_4$ has a low sticking coefficient and the deposition is inefficient. Limited samples of deposited materials are shown in Table 8.4. The dimensions of the deposits are larger than the beam diameter because the secondary electrons leave the surface at distances of many tens of nanometers, depending on the incident electron energy, from the point of electron impact. Note that the resistivity in Table 8.4 is given in Ω cm not μΩ cm. A far more complete listing of precursor gases and deposited materials can be found in the review article referenced earlier [3].

A nanothermometer built using electron beam-induced deposition from the organometallic platinum precursor is shown in Fig. 8.11. The resistance of this loop is a relatively sensitive to temperature so the structure can be used as a scanning probe to map the temperature of the surface [49].

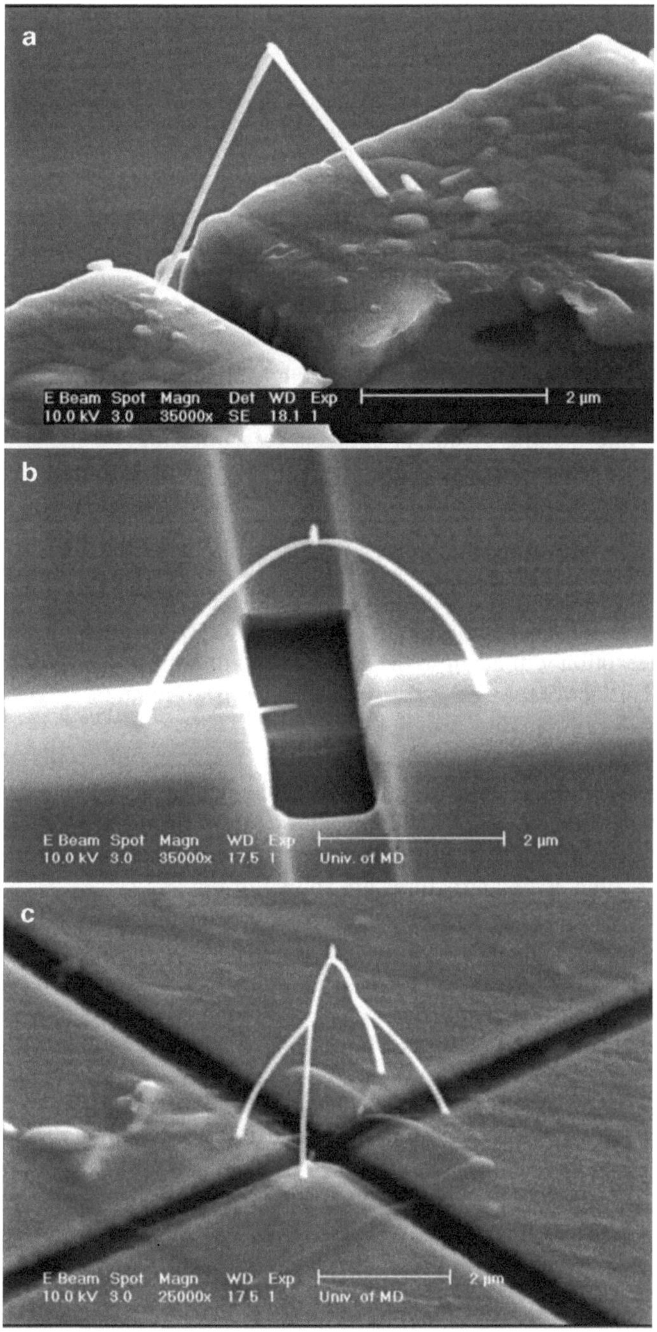

Fig. 8.11 Structures fabricated using electron beam-induced deposition using methylcyclopentadienyl-trimethylplatinum-platinum as a precursor gas. The structure in the top figure (**a**) was fabricated by tilting the stage at a given angle and keeping the beam stationary. The deposition then simply grows "up the beam." The stages then tilted again at an appropriate angle and the new beam is grown to connect to the previous one. The curved "wires" (**b**, **c**) are grown by programming the beam to scan in an appropriate slow rate. The structure in (**c**) allows for four-point measurement of resistance which is more sensitive and more accurate [49]

8.4.2 Material Removal

Electron beams can also be used to remove material using precursor gas. The process is entirely gas-dependent since negligible removal occurs with electron "milling." The secondary electrons produced by the incident beam induce a chemical reaction between adsorbed molecules and modules of the substrate leading to volatile compounds that are pumped away in the vacuum chamber. In some instances electron beams may be preferred to ion beams. Ion beams will damage whatever surface they are incident on. At the doses used here the top layer of any single crystal material will be amorphized by ions to a depth of a few hundred nanometers. Moreover, it will be heavily implanted with gallium ions. Thus in some applications like making electrical contacts to semiconductor materials or in the repair of quartz photomasks, i.e., the removal of an opaque chrome layer from the quartz, the ion damage and implantation are detrimental.

Most of the same precursor gases that are used with ion beams for material removal (Table 8.2) will also work with electron beams. It is of course necessary that the gas does not spontaneously react with the surface but rather reacts only at the point where the ion beam or in this case electron beam is incident. To be more precise, the reaction will occur in the area surrounding the point of incidence where the secondary electrons are emitted. An extensive list of precursor gases and of their applications is in the review article [3].

A technologically important challenge that has been addressed with electron beam-induced etching is the repair of lithography masks. Etchants are needed to be found for materials such as chromium and tantalum nitride [50]. (Because of the potential commercial sensitivity the etching gases have not been disclosed.) Examples of some etched structures are shown in Fig. 8.12.

An important advantage of electron beam-induced etching for mask repair is that the process can be made selective. In the case of ion beams complete selectivity is not possible because the ion beam by itself will also remove material. This selectivity is crucially important in the repair of EUV lithography masks [51]. These masks consist of delicate, thin multilayers which would be damaged by both the ion impact and the gallium implantation. An example of this kind of repair is the removal of an aluminum oxide particle using NOCl (Nitrosyl chlorine) gas. The electron beam does not induce etching of the substrate material with this precursor.

8.5 Models of Material Shaping by Ion and Electron Beams

Except for ion beam sputtering, the material removal and addition techniques depend on a precursor gas. The process can be viewed in two parts: Dynamics of the gas adsorption on the surface, and the details of the beam/solid interaction. Sputtering (ion milling) depends only on the beam solid interaction and will be discussed first.

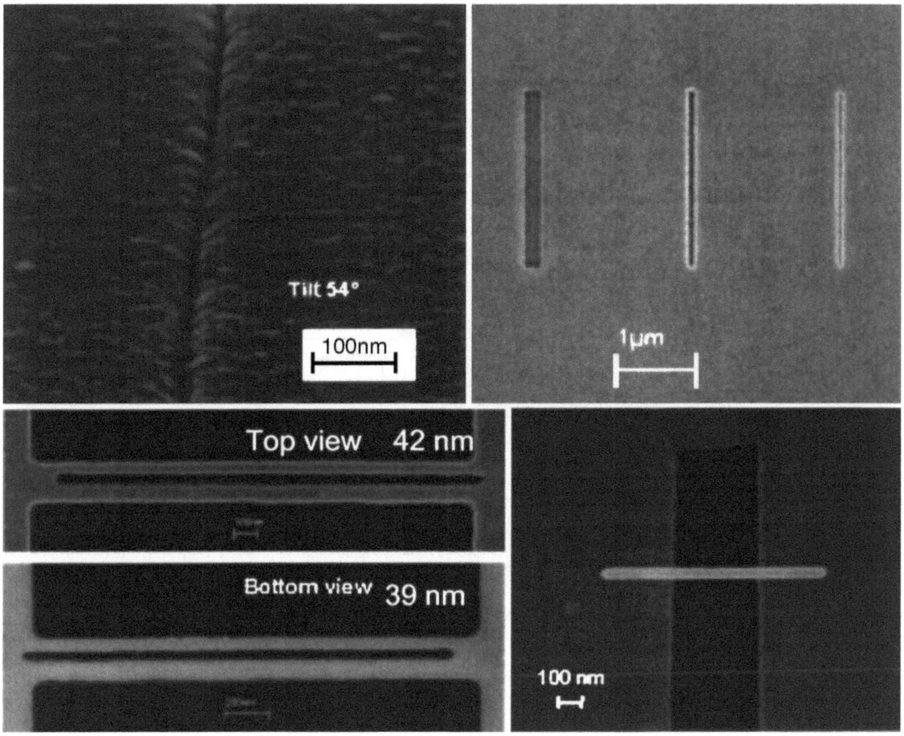

Fig. 8.12 *Top left*. Electron beam-induced etching of a chromium layer. *Top right* etching of a tantalum nitride layer bottom left slot *X* through a silicon carbide stencil mask viewed from top and bottom. *Bottom right* example of a freestanding bridge structure deposited across a gap [50]

8.5.1 Ion Sputtering

Material removal by ion bombardment is a long-standing process used in the research and industry, usually using energies between 100 and 1,000 eV. In this instance, the apparatus, usually called an ion miller, is designed to illuminate a large area, for example, an entire wafer, with the uniform current density per unit area. Patterning is achieved using photolithography, namely, covering part of the surface and leaving part of the surface uncovered to the incident ions.

There is extensive literature dating back many decades on the measurement of sputter yields of many substrates and many incident ions. See for example the three volume set, from Springer [52] and the earlier (1969) collection of data and theory [53]. In most cases, the milling yield increases monotonically with ion energy and peaks in the 50–100 keV range and then decreases.

Energetic ions entering a solid lose their energy in two ways: to the cloud of electrons, and by collisions with the nuclei in the solid. Since electrons are much, much lighter than ions. They do not perturb the path of the ions, but nevertheless,

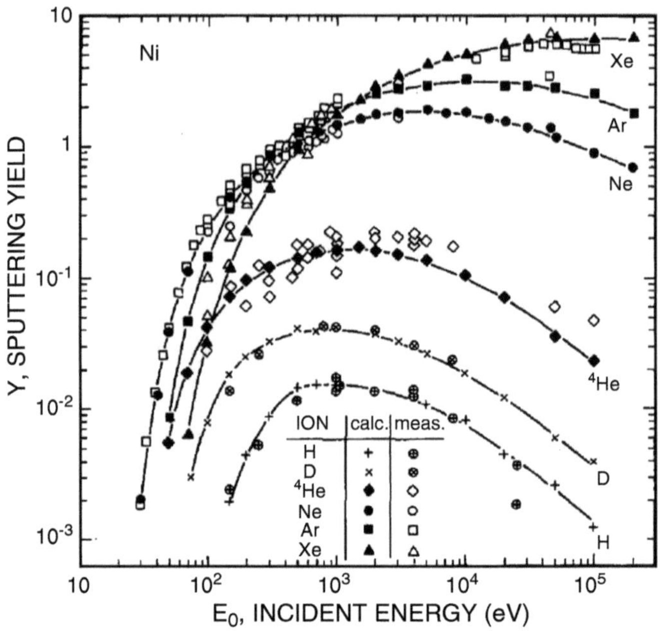

Fig. 8.13 Comparison of measured and calculated sputter yield of nickel bombarded by a variety of ions (from [55]). The *lines* are drawn through the calculated points as a guide to the eye. These are broad beam measurements not FIB measurements. Still there is a fair amount of scattered in some of the measured results shown by the open symbols. But the overall trends agree fairly well

extract energy from the incident beam. The collision with the nuclei is a more random process and results in the jagged path of the incoming ion. In addition, the substrate atoms participating in the collision absorb enough energy to be launched on a path in the solid where they produce more collisions. Thus, the incoming ion produces what is called a collision cascade. This is illustrated in Fig. 8.2 on the left side.

The energy, E, loss per unit length of path, z, is written as

$$dE/dz = dE/dz]_{electron} + dE/dz]_{nuclear} \tag{8.1}$$

stopping power is defined as $S(E) = -\mathrm{d}E/\mathrm{d}z$ and has, as in equation above both an electronic and nuclear complement $S(E) = \mathrm{Sn}(E) + \mathrm{Se}(E)$. Both of these terms strongly depend on the mass and energy of the incident ion and the mass of the atoms in the solid. For example, a 30 keV krypton ion (close in mass to gallium) loses 10 times as much energy to nuclei (1.5 MeV/μm) than to electrons (0.15 MeV/μm). At 1,000 keV the energy loss to both ions and electrons is about the same (0.9 MeV/μm). The energy lost to the electrons generally does not alter the material while sputtering and radiation damage, i.e., amorphization of the material if it is initially single crystal, is due to the nuclear stopping power. The sputter yield is very well predicted by the Monte Carlo models developed SRIM and TRIM [54]. This is illustrated in Fig. 8.13 for nickel bombarded by a variety of ions from

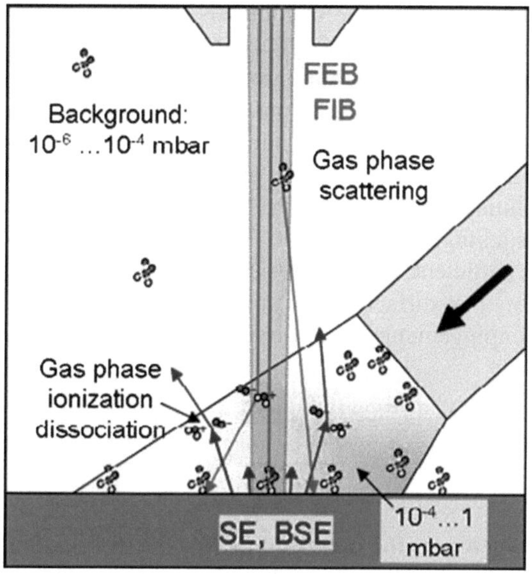

Fig. 8.14 From Utke et al. [3]. The gas is delivered by capillary tube from the *right* as shown, and impinges on the surface over a larger area, surrounding the point of electron or ion beam incidence

hydrogen, deuterium, and the noble gases. Also see Fig. 8.5 where both measured and calculated results for the relative milling yield of silicon and silicon dioxide are compared. Again, the agreement is quite good.

8.5.2 Surface Coverage by Precursor Gas

The precursor gas is generally delivered to the surface by capillary tube of internal diameter less than a millimeter in close proximity to the point where the ion or electron beam is incident. The geometry is shown in Fig. 8.14.

In general, the ion or electron beam traverses less than a few millimeters through the precursor gas in the gas phase. As analyzed in detail [3] the probability of scattering of the incident beam by the gas molecules is negligible. The relevant interaction of the electron or ion beam with the gas occurs on (or in) the surface.

The impingement rate of molecules on the surface or the flux, J (molecules/cm^2 s) at a pressure P is given by

$$J = P(2\pi kTm)^{-1/2} \tag{8.2}$$

where k is the Boltzmann constant, T the absolute temperature, and m the mass of the molecule.

For the gas used in the past for gold deposition DMG(hfac) (C7 H7F6Au) at. wt. = 434 g. At the pressures generally used, say 1 mbar, the impingement rate (or flux, J) of molecules on the surface is 7.5×10^{19} molecules/cm^2 s. The impingement rate of ions (or ion flux, *f*) for a beam current density of 10 A/cm^2 (typical for a Ga+ FIB) is 6.25 × 10^{19}/cm^2 s. Thus with a stationary beam any deposition or etching process would begin to be precursor gas limited. Moreover, one ion very likely consumes several gas molecules. (For incident electrons the situation is reversed, namely, a number of electrons are incident per dissociated molecule. However, the electron beam current density is much higher.). In fact, the situation is likely to be more complicated. An incident molecule has a certain sticking probability, *s*. The gas can diffuse on the surface with the diffusion coefficient, *D*, and a residence time, τ. Thus the density of adsorbed gas molecules on the surface as a function of radial distance from the center of the beam impingement and of time, $n(r, t)$, is given by the following equation [3]

$$\frac{\partial n}{\partial t} = \underbrace{sJ\left(1 - \frac{n}{n_0}\right)}_{Adsorption} + \underbrace{D\left(\frac{\partial^2 n}{\partial r^2} + \frac{1}{r}\frac{\partial n}{\partial r}\right)}_{Diffusion} - \underbrace{\frac{n}{\tau}}_{Desorption} - \underbrace{\sigma f n}_{Decomposition} \tag{8.3}$$

where σ is the cross-section for the dissociation of an adsorbed molecule, whether etching or deposition. We are assuming that at most one monolayer, characterized by a density n_0, can be adsorbed on the surface. Thus in the first term on the right hand side of the equation no more adsorption can occur if $n = n_0$. That only one monolayer is adsorbed on the surface at the impingement rates of interest was confirmed by experiment for the gold bearing precursor gas (DMGhfac) [56]. In steady-state the rate, $R(r)$, as a function of distance from the center of the beam at which material is added or removed (in units of height per second) is given by

$$R(r) = n(r) \cdot Vf(r) \tag{8.4}$$

where *V* is the volume of material added or removed per adsorbed and dissociated molecule. The quantity $f(r)$ is the radial distribution of incident ions measured from the center of the beam. $n(r)$ is the steady-state solution of the differential equation above, including the impingement and adsorption of gas molecules on the surface, and their diffusion toward the point of ion impact which in effect acts as a sink for the gas molecules. The influence of diffusion of the adsorbed precursor has been experimentally observed for Cl_2 on Si. A diffusion coefficient $D = 8.0 \times 10^{-6}$ cm^2/s was measured [57].

8.5.3 Ion Beam Precursor-Gas/Solid Interaction

We have discussed the ion-solid interaction above, illustrated in Fig. 8.2. The incident ion produces a number of effects: secondary electrons are emitted, atoms are sputtered off the surface, and substrate atoms are displaced from their normal crystal sites. The dissipation of the energy of the ion in the solid can also be thought

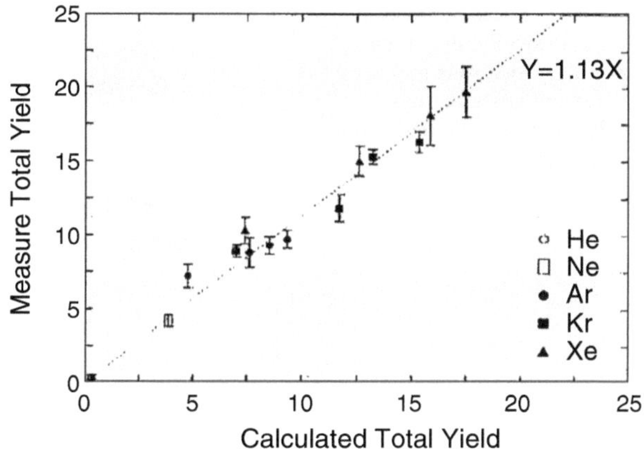

Fig. 8.15 A plot of the total decomposition yield vs. the total calculated yield using the binary collision model [58]. For each ion species, measurements were made at several energies. Thus there are several data points, i.e., the yield increased with increasing ion energy

of as very locally raising its temperature. How do these effects cause gas molecules on the surface to decompose and deposit material or chemically react with the substrate leading to etching? This has been studied to some degree for deposition using noble gas ions between 2 and 10 keV [58] and up to 100 keV [59]. The results show a strong correlation between the collision cascade energy reaching the surface, (calculated by TRIM) and the dissociation rate of the adsorbed precursor gas. This is the same mechanism that causes sputtering. If a collision cascade reaches the surface and imparts sufficient energy to the surface atom, of order 5 eV, that atom will be sputtered off. However, if an adsorbed gas is present, a lower energy, say, 1 eV is all that is necessary to cause the precursor molecule to dissociate. There is, of course, competition between the addition of material due to the dissociation of the precursor gas and the removal of material due to sputtering. Since dissociation occurs at a lower energy than sputtering, deposition dominates sputtering, provided there is sufficient surface coverage by the precursor gas.

The total dissociation yield is the sum of the measured net yield for material addition, i.e., how many atoms are deposited from the precursor gas per incident ion, and the sputter yield also experimentally measured simply by turning off the precursor gas and measuring how many atoms are removed per incident ion. This total dissociation yield correlates well with the calculated yield as shown in Fig. 8.15 [58]. For each ion species there are data points at various energies between 2 and 10 keV. The slope of the line drawn through the points is adjusted by 13% ($Y = 1.13X$), i.e., if the correlation were perfect, we would have $Y = X$. This is remarkably good agreement since there were no adjustable parameters and strongly suggests that the model based on the collision cascades reaching the surface and causing the decomposition of the precursor gas is valid. In the same study the thermal spike model was also analyzed and the correlation with the yield calculated in that way was not good. It is also possible that secondary electrons emitted when

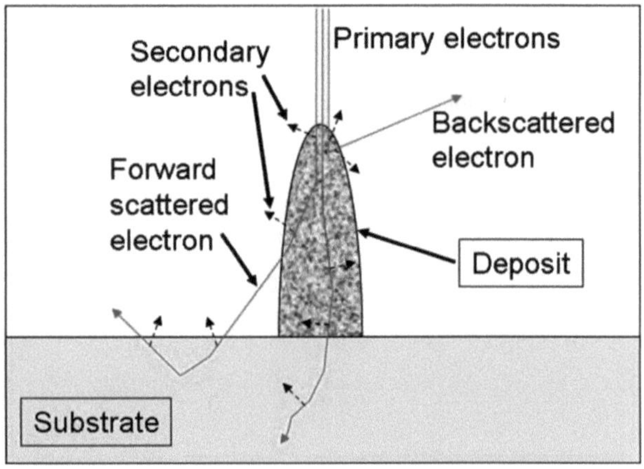

Fig. 8.16 Focused electron beam deposition due to the secondary electrons produced by the primary electrons which scatter to the side of the pillar [3]

the ion hits the surface can cause dissociation of the adsorbed precursor gas. However, in that instance correlation between the total dissociation yield and the secondary electron yield is completely absent.

8.5.4 Electron Beam Precursor-Gas/Solid Interaction

When an energetic electron strikes a solid surface, secondary electrons of lower energy (typically below 20 eV) are emitted. Other effects such as emission of photons or heating may also occur. But the secondary electrons play the major role in electron beam-induced deposition and etching. Because of the large difference in mass between an atom, or more precisely its nucleus, and the electron, virtually no disturbances to the lattice, such as sputtering, occur. As described in the review article [3] Monte Carlo simulations can effectively be used to model the deposition and etching. In particular the trajectories of the primary electrons and the locations where the secondary electrons are emitted can be used to model the growth of pillars. The diameters of pillars grown with a stationary electron beam in the presence of a precursor gas are always much larger than the diameter of the beam. This is due to the fact that the primary electron beam scatters in the pillar and produces secondary electrons that exit at the side of the pillar, dissociate the adsorbate, and increase the diameter. This is illustrated in Fig. 8.16. The electron trajectories can be simulated by the Monte Carlo models and the growth of surface film, as well as pillars can be predicted. As can be seen from the figure, the diameter of the pillar will be limited by the range of the primary electrons, i.e., far enough down the pillar no primary electrons will exit from the side and no secondaries will be produced.

The deleterious effects of the long electron trajectories in the solid can be avoided if deposition is carried out on an ultrathin membrane. This was in fact done and in that instance deposits, albeit very thin, with sub nanometer dimensions were produced [60].

8.6 Recent Developments

The first gallium ion beam nanofabrication was reported in the 1970s, in the decades since many practical applications have been explored and developed so that gallium ion beam nanofabrication can be thought of as a fairly mature technology.

In the past decade, four novel ion beam technologies have appeared based largely on ion source development.

8.6.1 The Gas Field Ion Source

The gas field ion source can be used to produce helium and neon ion nanobeams. The ions are produced by a cryogenically cooled, single crystal tungsten tip and are essentially emitted from a single tungsten atom. This is an exotic ion source with a long history of development. The first practical system was conceived and developed by the ALIS Corp [61]. The main application envisaged was imaging with the helium ion beam which has advantages over a conventional SEM. ALIS was acquired by Zeiss which now offers systems for sale. Nanofabrication is being explored. Typically the ion energy is about 35 keV with the beam diameter of about 1 nm and a beam current in the 1–5 pA range. Moreover, the milling yield of helium ions is very low (~0.1 atoms/ion), neon is somewhat higher. Gallium ion beams typically are in the 10 pA to 1 nA range, depending on the beam diameter desired, and have a much higher milling yield (~2–10 atoms/ion). Applications are now being explored which take advantage of the ultrasmall beam diameter and of the fact that the milling ions are inert and, unlike gallium, will not contaminate the substrate surface. Moreover, the area from which secondary electrons are emitted is smaller than for an incident electron beam, and likewise the area where collision Cascades reach the surface is likewise smaller than for gallium ions. The light helium ions impart less energy to atoms of the substrate so the straggler is more confined. Beam-induced deposition and milling have been demonstrated [42]. Some examples of fabricated structures are shown in Figs. 8.17, and 8.18. A special concern with helium ions is that they penetrate deep in the substrate resulting in trapped gas. In single crystal silicon for example macroscopic, micrometer-sized bubbles have been observed when irradiated with doses of $1.3 \times 10^{18}/\text{cm}^2$ at both 12 and 25 keV [62].

8.6.2 Plasma Ion Beam Source

Both the liquid metal gallium ion source and the cryogenic helium/neon source depend on emission from a small dot with a high current density. These sources have a high brightness. Sources based on an ionized gas are commonly used in for

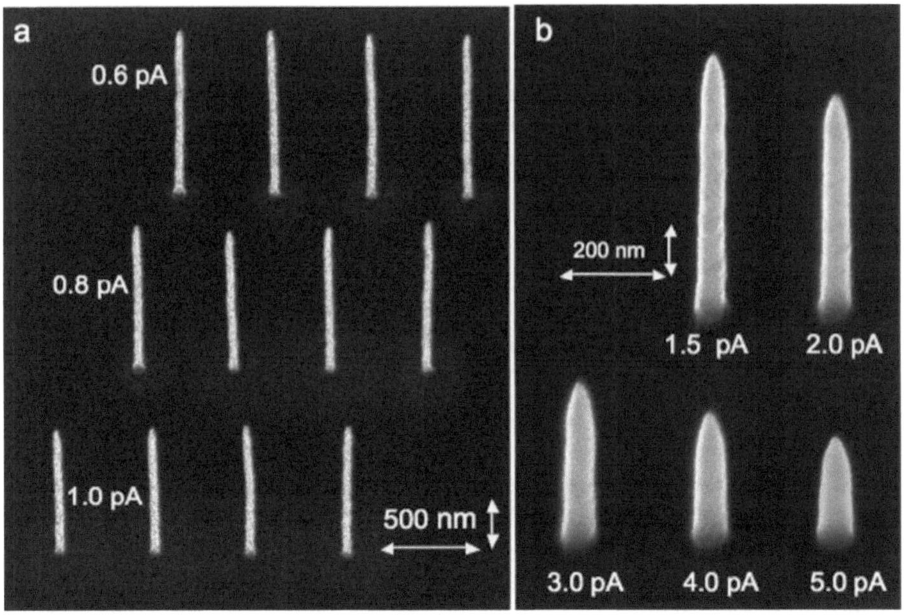

Fig. 8.17 Pillars grown using 25 keV Helium ion beam at various currents using methylcyclopentadienyl-trimethylplatinum precursor gas [42]. Minimum pillar diameter is 36 nm

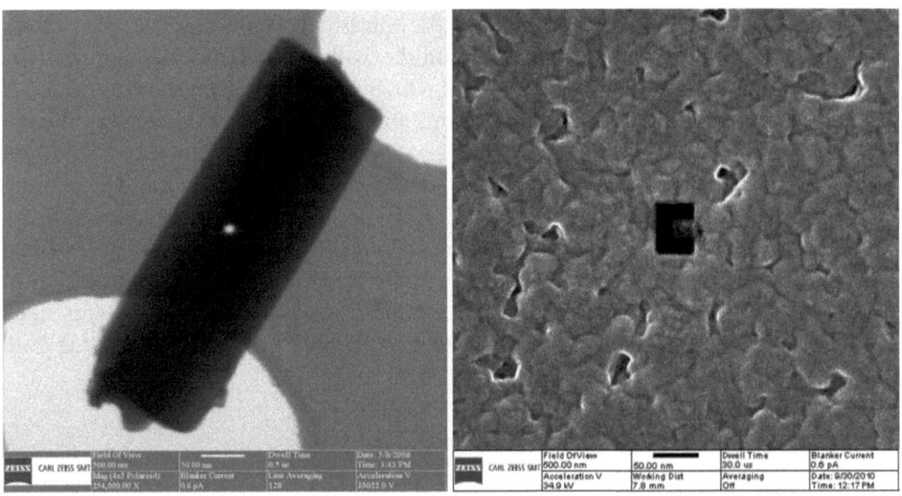

Fig. 8.18 (Supplied by L. Scipioni, Carl Zeiss NTS Peabody, Ma.) Examples of milling with He ion beam. *Left*: A 10 nm hole drilled through an asbestos fiber. *Right*: A 40 nm C-shaped aperture, done by Zeiss with Stanford U. Note the sharpness of the corners

Fig. 8.19 Cross-section of a 750 μm diameter solder ball used in flip chip packaging. With a 1.7 μA xenon beam this milling took 6 h. With the gallium ion beam at comparable resolution it would have taken 600 h. The image was also taken with a xenon beam but at the 100 pA where the resolution (beam diameter) is about 80 nm. The *dark* and *light* areas show crystal grains. This kind of sectioning is useful in diagnosing solder ball/substrate contact problems (image courtesy of Noel Smith, Oregon Physics, Hillsboro Oregon)

example, ion implanters, but are much less bright. However, they offer a very large total current, and almost any ion in the periodic table can in principle be used. If high-resolution milling is not needed and a beam diameter in the >100 nm range is acceptable, then the plasma ion source can outperform a gallium source particularly if heavy ions such as xenon are used. For example, 10 nA current can be obtained in a 300 nm diameter beam. According to simulations at a 20 μm beam diameter the xenon plasma source will deliver about two orders of magnitude more current than a gallium ion liquid metal source [63]. Plasma-based ion beam systems have been developed and are used for specific applications that require high removal rates but do not require high-resolution, for example in sectioning a solder ball in a flip chip integrated circuit as shown in Fig. 8.19 [63].

8.6.3 Laser Cooled Source

Laser radiation is used to cool atoms trapped in a magnetic field. The ion temperature can be reduced to the millikelvin range. Thus when they are ionized and extracted the energy spread is small permitting finer focus by minimizing chromatic

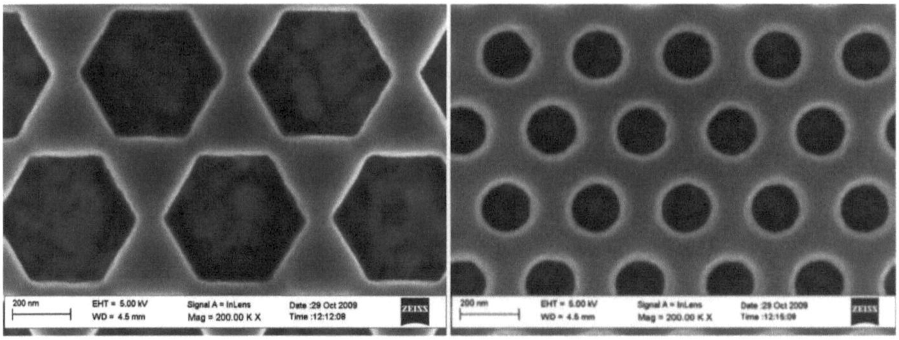

Fig. 8.20 Hexagons with 470 nm diagonal diameter and the minimum spacing of 100 nm milled in nickel to a depth of 80 nm and an array of circular holes with a diameter of 150 nm milled into a nickel surface to a depth of 80 nm [66]

aberration. Atomic spectral lines have to be matched to the laser wavelength. Most recently Li ion beams have been reported [64]. This is an exciting development but is still in the research stage.

8.6.4 Multibeam Ion Sources

One of the main shortcomings of FIBs (and electron beams) is the slow rate of material removal or addition. They are mainly used for local, single structure fabrication. Multiple beam systems for both ions and electrons have been developed in the past years. The systems are largely aimed at lithography, namely, exposure of resist. The systems use a plasma ion source to produce a relatively broad beam of collimated ions which are incident on a stencil mask that can have a static pattern of openings or programmable windows which allow complex patterns to be written [65]. These multibeam systems of either ions or electrons would be useful in instances where the intended fabrication is over a large area so that many sites can be addressed simultaneously. Obviously the systems are not suitable if one wants to make, for example, a single pit in the sample.

Figure 8.20 shows examples of structures milled in a nickel film with a 10 keV argon beams in a multibeam machine [66]. A silicon stencil mask with the pattern over a 5 mm $\times$ 5 mm area is back illuminated with a broad, collimated argon ion beam. The pattern passes through ion optics where it is demagnified 200 times resulting in a field size of 25 $\times$ 25 μm projected on the sample. The current density on the sample was 110 $\mu A/cm^2$ and the features shown in both parts of the figure are milled to a depth of 80 nm in 19 s. These patterns are used in a conventional injection molding machine to produce the reverse patterns in polymer materials. The patterns were produced with a static stencil mask. Programmable masks have also been produced which can direct and control 43,000 individual beams on the sample for writing more intricate, customized patterns. However, for this kind of patterning with a programmable mask rather than a fixed stencil mask the milling time would be much longer because the total window area in the programmable mask is much smaller than in the static mask. Considerable area is required to accommodate the switching circuitry, so that only about 1% of the ion current incident on the mask passes down to the sample. On the other hand, the static mask is exactly as shown in Fig. 8.20 enlarged 200 times, and has a large open area.

8.7 Beam Patterning in the Volume

The fabrication with ion and electron beams that we have discussed so far could be characterized as surface fabrication either material is removed from the surface or deposited on the surface. The beams can also produce chemical or physical changes in the volume of the material. This can also be exploited in nanofabrication. The standard exposure of resist, as in e-beam lithography, is an example of a chemical change produced by the beam. In addition, ion beams can produce physical change in the volume in the form of damage or local amortization. This can then be exploited in preferential etching, or resistance to etching. In addition, the presence of the implanted ions can also physically alter the material. For example, silicon implanted with the helium nanobeam discussed above can form bubbles [62]. See Fig. 8.21.

If a resist has good mechanical and chemical stability, it can itself present a useful structure. For example, an epoxy-based resist called SU8-2000 exposed with a 5 keV electron beam has been used to produce photonic waveguides down to 100 nm [67].

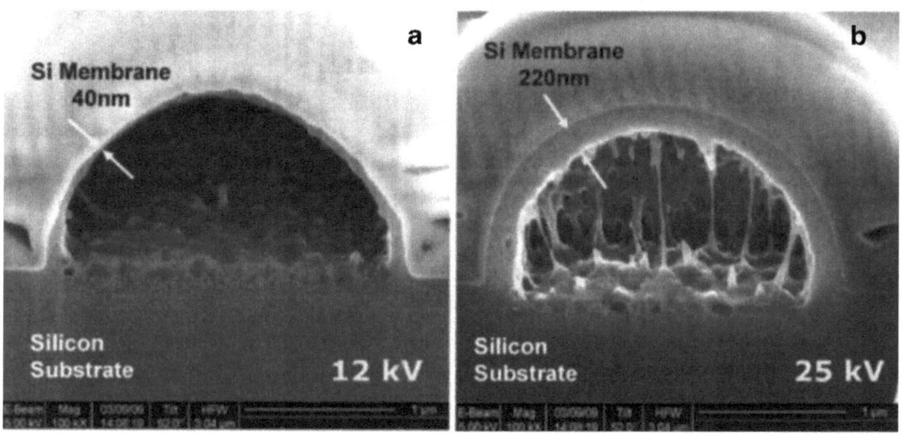

Fig. 8.21 From ref. [62]. Cross-section of silicon sample implanted with a helium ion dose of 1.3×10^{18} ions/ cm^2 at beam energies of (**a**) 12 keV and (**b**) 25 keV. A 2 μm × 2 μm area was exposed. Evidently the helium is not soluble in silicon and forms a bubble. Note that the bubble skin thickness is related to the ion penetration depth. At higher energy the bubble starts the form deeper in the substrate

High-energy, MeV proton beams have also been used to produce high aspect ratio structures in materials such as SU8. The advantage of the proton beam is that it can penetrate many tens of micrometers in material and alter the material along a straight line trajectory. This allows one to produce high aspect ratio structures, for example, 60 nm structures, 10 μm deep in SU8. For the review of this work, see ref. [68].

8.8 Conclusions

Focused ion and electron beams can be used to fabricate structures down into the 10 nm regime. Both material removal and material addition are possible. With ion beams, material can be removed by sputtering and by sputtering in the presence of a reactive gas which can greatly increase the removal rate. Material can be added by flooding the sample surface with an appropriate precursor gas which the ion beam dissociates to leave a deposit behind. With electron beams material can only be removed with the aid of a reactive gas. And material can be added with the aid of a precursor gas. However, since these are point beams, fabrication is slow. So in general they can be used only for limited volumes. For example, to mill a volume of 1 μm^3 with a 10 pA beam would take 5 min. By sacrificing resolution one can increase the current and decrease the milling time by as much as two orders of magnitude the minimum dimensions at which structures can be fabricated is larger than the beam diameter, often hundred nanometers, rather than the 1–10 nm ion or electron beam diameter. This is because the ion-surface interaction produces collision cascades some distance from the point of ion impact. Similarly electron-surface interaction causes secondary electrons to be emitted from an area surrounding the point of impact.

Much of the impetus for the development of the systems has come from the semiconductor industry where they are used for mask repair, for sample sectioning for failure analysis, and for circuit rewiring. However, many other nanofabrication applications have been demonstrated in diverse fields. Recent developments promise even finer resolution fabrication with helium or neon ions, rapid fabrication with xenon ion plasma sources, and multibeam fabrication with potential manufacturing applications.

References

1. Thompson LF, Willson CG, Bowden MJ (1994) Introduction to microlithography, 2nd edn. ACS, Washington, DC
2. Orloff J, Utlaut M, Swanson L (2003) High resolution focused ion beams: FIB and its applications. Springer, New York
3. Utke I, Hoffmann P, Melngailis J (2008) J Vac Sci Technol B 26(4):1197
4. http://portal.tugraz.at/portal/page/portal/felmi/research/FIB/Principles%20of%20FIB

5. Xu X, DellaRatta AD, Sosonkina J, Melngailis J (1992) J Vac Sci Technol B 10(6):2675
6. By Carl Zeiss: http://www.zeiss.de/c1256c1500431210/Contents-Frame/6531cac4d5c10402c1256df7003f3cbchttp://www.ticgroup.com.tw/menu/products/sem/mask/MeRit_MG_45/MeRiT_MG45.asp
7. TRIM (or SRIM) simulation of sputtering: http://www.srim.org/
8. Muller KP, Petzold HC (1990) Proc SPIE 1263:12
9. Santamore D, Edinger K, Orloff J, Melngailis J (1997) J Vac Sci Technol B 15(6):2346
10. Lugstein A, Steiger-Thirsfeld A, Basnar B, Hyun YJ, Pongratz P, Bertagnolli E (2009) J Appl Phys 105(4):044912
11. Melngailis J (1987) J Vac Sci Technol B 5(2):469
12. Yamaguchi H (1987) J Phys Colloques 48(C6):165
13. Muller KP, Petzold HC (1990) Proc SPIE 1263(12) (repeat with [8])
14. Anderson HH, Bay HL (1981) Topics in applied physics: sputtering by particle bombardment. In: Behrisch R (ed) Physical sputtering of single element solids, vol 47. Springer, New York, p 145
15. Young RJ, Cleaver JRA, Ahmed H (1993) J Vac Sci Technol B 11(6):234
16. Adams DP, Vasile MJ, Mayer TM, Hodges VC (2003) J Vac Sci Technol B 21(6):2334
17. Stanishevsky A (2001) Thin Solid Films 398:560
18. Steckl AJ, Chyr I (1999) J Vac Sci Technol B 17(2):362
19. Blauner PG, Ro JS, Butt Y, Melngailis J (1989) J Vac Sci Technol B 7(4):609
20. Ostadi H, Jiang K, Prewett PD (2009) Microelectron Eng 86(4):1021
21. Mulders JJL, de Winter DAM, Duinkerken WJHCP (2007) Microelectron Eng 84(5):1540
22. Vasile MJ, Xie J, Nasser R (1999) J Vac Sci Technol B 17(6):3085
23. Vick D, Sauer V, Fraser AE, Freeman MR, Hiebert WK (2010) J Micromech Microeng 20(10):105005
24. Ochiai Y, Gamo K, Namba S, Shihoyama K, Masuyama A, Shiokawa T, Toyoda K (1987) J Vac Sci Technol B 5(1):423
25. Stark TJ, Shedd GM, Vitarelli J, Griffis DP, Russell PE (1995) J Vac Sci Technol B 13(6):2565
26. Young RJ, Cleaver JRA, Ahmed H (1993) J Vac Sci Technol B 11(2):234 (repeat with [17])
27. Harriott LR (1993) J Vac Sci Technol B 11(6):2012
28. Kola RR, Celler GK, Harriott LR (1993) Mater Res Soc Symp Proc 279:593
29. Edinger K (1999) J Vac Sci Technol B 17(6):3058
30. Santschi C, Jenke M, Hoffmann P, Brugger J (2006) Nanotechnology 17(11):2722
31. DellaRatta AD, Melngailis J, Thompson CV (1993) J Vac Sci Technol B 11(6):2195
32. Blauner PG, Butt Y, Ro JS, Thompson CV, Melngailis J (1989) J Vac Sci Technol B 7(4):1816
33. Funatsu J, Thompson CV, Melngailis J, Walpole JN (1996) J Vac Sci Technol B 14(1):179
34. Stewart DK, Stern LA, Morgan JC (1989) Proc SPIE 1089:18
35. Luxmoore IJ, Ross IM, Cullis AG, Fry PW, Orr J, Buckle PD, Jefferson JH (2007) Thin Solid Films 515(17):6791
36. Tao T, Ro JS, Melngailis J, Xue Z, Kaesz HD (1990) J Vac Sci Technol B 8(6):1826
37. Puretz J, Swanson LW (1992) J Vac Sci Technol B 10(6):2695
38. Gannon TJ, Gu G, Casey JD, Huynh C, Bassom N, Antoniou N (2004) J Vac Sci Technol B 22(6):3000
39. Edinger K, Melngailis J, Orloff J (1998) J Vac Sci Technol B 16(6):3311
40. Igaki J, Nakamatsu K, Kometani R, Kanda K, Haruyama Y, Kaito T, Matsui S (2006) J Vac Sci Technol B 24(6):2911
41. Namatsu K, Nagase M, Igaki J, Namatsu H, Matsui S (2005) J Vac Sci Technol B 23(6):2801
42. Alkemade PFA, Chen P, Veldhoven EV, Maas D (2010) J Vac Sci Technol B 28(6):C6F22
43. Hoyle PC, Cleaver JRA, Ahmed H (1996) J Vac Sci Technol B 14(2):662
44. Koops HWP, Weiel R, Kern DP, Baum TH (1988) J Vac Sci Technol B 6(1):477
45. Utke I, Dwir B, Leifer K, Cicoira F, Doppelt P, Hoffmann P, Kapon E (2000) Microelectron Eng 53(1):261

46. Yavas O, Ochiai C, Takai M, Park YK, Lehrer C, Lipp S, Frey L, Ryssel H, Hosono A, Okuda S (2000) J Vac Sci Technol B 18(2):976
47. Barry JD, Ervin M, Molstad J, Wickenden A, Brintlinger T, Hoffman P, Meingailis J (2006) J Vac Sci Technol B 24(6):3165
48. Utke I, Hoffmann P, Dwir B, Leifer K, Kapon E, Doppelt P (2000) J Vac Sci Technol B 18 (6):3168
49. Edinger K, Gotsalk T, Rangelow I (2001) J Vac Sci Technol B 19(6):2856
50. Edinger K, Becht H, Bihr J, Boegli V, Budach M, Hofmann T, Koops HWP, Kuschnerus P, Oster J, Spies P, Weyrauch B (2004) J Vac Sci Technol B 22(6):2902
51. Bret T, Afra B, Becker R, Hofmann T, Edinger K, Liang T, Hoffmann P (2009) J Vac Sci Technol B 27(6):2727
52. Behrisch R (ed) (1981, 1983) Topics in applied physics "Sputtering by particle bombardment," vol 52, 47; Behrisch R, Wittmaak K (eds) (1991) Topics in applied physics "Sputtering by particle bombardment," vol 64. Springer Verlag, Berlin; also Behrisch R, Eckstein W (eds) ((2007)) Topics in applied physics, Sputtering by particle bombardment: experiments and computer calculations, vol 110. Springer, Berlin
53. Carter G, Colligon JS (1969) Ion bombardment of solids. Heinemann, London
54. Ziegler JF, Biersack JP, Littmark U (1984) The stopping and range of ions in solids, vol. 1 of series stopping and ranges of ions in matter. Pergamon Press, New York, see also http://www.srim.org/
55. Biersack JP, Eckstein W (1984) Appl Phys A 34(2):73
56. Dubner AJ, Wagner A (1989) J Appl Phys 66(2):870
57. Harashi S, Komoro M (1993) Jpn J Appl Phys 32:6168
58. Dubner AJ, Wagner A, Melngailis J, Thompson CV (1991) J Appl Phys 70(2):665
59. Ro JS, Thompson CV, Melngailis J (1994) J Vac Sci Technol B 12(1):73
60. van Dorp WF, Lazić I, Beyer A, Gölzhäuser A, Wagner JB, Hansen TW, Hagen CW (2011) Nanotechnology 22(11):115303
61. Ward BW, Notte JA, Economou NP (2006) J Vac Sci Technol B 24(6):2871
62. Livengood R, Tan S, Greenzweig Y, Notte J, McVey S (2009) J Vac Sci Technol B 27(6):3244
63. Tesch P, Smith N, Martin N, Kinion D (2008) Proceedings of the 34th ISTFA conference, vol 7
64. Knufman B, Steele AV, Orloff J, McClelland JJ (2011) New J Phys 13(10):103035
65. Platzgummer E, Loeschner H (2009) J Vac Sci Technol B 27(6):2707
66. Koeck A, Bruck R, Wellenzohn M, Hainberger R, Platzgummer E, Loeschner H, Joech P, Eder-Kapl S, Ebm C, Czepl P et al (2010) J Vac Sci Technol B 28(6):C6B2
67. De Vittorio M, Todaro MT, Stomeo T, Cingolani R, Cojoc D, Di Fabrizio E (2004) Microelectron Eng 73:388
68. Watt F, Breese MBH, Bettiol AA, van Kan JA (2007) Mater Today 10(6):20

ERRATUM

Chapter 5
Ultrasonic Machining

K.L. Kuo, H. Hocheng, and C.C. Hsu

H. Hocheng and H.-Y. Tsai (eds.) *Advanced Analysis of Nontraditional Machining*,
DOI 10.1007/978-1-4614-4054-3, pp. 325–357, © Springer Science+Business Media New York 2013

DOI 10.1007/978-1-4614-4054-3_5

The publisher regrets that the names of the fourth and fifth authors of Chapter 5, N.H. Tai and C.S. Liu, were not included in the chapter.

The online version of the original chapter can be found at
http://dx.doi.org/10.1007/978-1-4614-4054-3_5

Index

H. Hocheng and H.-Y. Tsai (eds.), *Advanced Analysis of Nontraditional Machining*,
DOI 10.1007/978-1-4614-4054-3, © Springer Science+Business Media New York 2013

D

MIX
Papier aus verantwortungsvollen Quellen
Paper from responsible sources
FSC® C105338

If you have any concerns about our products,
you can contact us on
ProductSafety@springernature.com

In case Publisher is established outside the EU,
the EU authorized representative is:
Springer Nature Customer Service Center GmbH
Europaplatz 3, 69115 Heidelberg, Germany

Printed by Libri Plureos GmbH
in Hamburg, Germany